Environmental Science and Engineering

Environmental Science and Engineering

Edited by **Chuck Lancaster**

SYRAWOOD
PUBLISHING HOUSE
New York

Published by Syrawood Publishing House,
750 Third Avenue, 9th Floor,
New York, NY 10017, USA
www.syrawoodpublishinghouse.com

Environmental Science and Engineering
Edited by Chuck Lancaster

International Standard Book Number: 978-1-68286-232-2 (Hardback)

Printed in China.

Contents

Preface

Every book is a source of knowledge and this one is no exception. The idea that led to the conceptualization of this book was the fact that the world is advancing rapidly; which makes it crucial to document the progress in every field. I am aware that a lot of data is already available, yet, there is a lot more to learn. Hence, I accepted the responsibility of editing this book and contributing my knowledge to the community.

Environmental engineering is a multi-disciplinary branch of engineering and is an essential component of sustainable development as well as resource management. It combines subjects from diverse branches of engineering and environmental science. The aim of this book is to educate the reader about various theories and practical applications of environmental engineering, such as environmental preservation, control and effective management of waste from human and animal activities, waste water management, etc. It strives to provide a better understanding of the interactions between human beings and their environment. This book is highly recommended for the students of various branches of engineering and those pursuing environmental sciences.

While editing this book, I had multiple visions for it. Then I finally narrowed down to make every chapter a sole standing text explaining a particular topic, so that they can be used independently. However, the umbrella subject sinews them into a common theme. This makes the book a unique platform of knowledge.

I would like to give the major credit of this book to the experts from every corner of the world, who took the time to share their expertise with us. Also, I owe the completion of this book to the never-ending support of my family, who supported me throughout the project.

Editor

1

Experimental investigations on the phase equilibrium of semiclathrate hydrates of carbon dioxide in TBAB with small amount of surfactant

Abhishek Joshi[1], Jitendra S Sangwai[1*], Kousik Das[1] and Nagham Amer Sami[2,3]

Abstract

Experimental studies are carried out on a semiclathrate hydrate system of carbon dioxide in tetra-*n*-butyl-ammonium bromide (TBAB) with a small amount of surfactant, sodium dodecyl sulfate (SDS), for 5, 10, and 20 wt.% TBAB to determine the phase equilibrium temperature and pressure conditions. It is observed that the presence of SDS did not influence the equilibrium conditions of the semiclathrate hydrate. Re-nucleation (memory) effect of semiclathrate hydrates of CO_2 is studied for few cases of TBAB concentration in an aqueous solution. The equilibrium pressure and temperature conditions obtained for memory effect and regular experimental run without memory effect were observed to be quite close. It is concluded that in the case of no memory effect, with increasing TBAB percentage in the system, the time required for nucleation is reduced. For the same TBAB concentration, the incipient pressure and temperature required for nucleation and re-nucleation of semiclathrate hydrates increase while the time required for re-nucleation decreases.

Keywords: Carbon dioxide, Clathrate hydrate, Phase equilibria, Semiclathrate hydrate, Surfactant

Background

Global warming of the Earth's atmosphere has increased concerns for the conservation of Mother Nature. The melting of glaciers, the encroaching sea line on the coastal areas, decreasing agricultural outputs, health effects on human beings are considered to be some of the major effects of this phenomenon. The burning of fossil fuels and even the use of renewable energies have drastically increased carbon dioxide (CO_2) emission in the atmosphere. It is, therefore, important to look at the proper disposal of carbon dioxide so as to reduce post-use implications of fossil fuels. Carbon capture and sequestration (CCS) is considered to be one of the prime remedy for global warming. However, it is more expensive and technically challenging to have CCS performed in the formation. Additionally, CO_2 leakage from reservoirs may pose a risk to overlying fresh groundwater and also may suddenly increase CO_2 percentage in subsurface environments affecting the marine life [1-3]. In order to mitigate CO_2 leakage and make it more stable in the reservoirs, several methods are being investigated. One of the proposed methods of CO_2 sequestration is using it in the form of gas hydrate and injecting them into the reservoir [4,5]. Additionally, CO_2 can be transported in the form of gas hydrate slurries. However, more efforts have to be made to make the process more economical and safe.

Gas hydrates, also referred to as clathrate hydrates, form under extreme conditions of low temperature (typically <10°C) and high pressure (typically >3 MPa), when gases (guest) like carbon dioxide, methane, nitrogen, come in contact with water (host) [6]. Precise conditions in terms of pressure and temperature depend on the composition of the guest gas. These hydrates get destroyed by destabilizing the phase equilibrium condition, typically by raising the temperature or decreasing the pressure or by employing chemical inhibitors. Semiclathrate hydrates are similar to gas hydrates but have a different lattice structure. This structural difference arises because semiclathrate hydrates are formed when the gas hydrate system contains some thermodynamic promoter, such as tetra-*n*-butyl-ammonium

* Correspondence: jitendrasangwai@iitm.ac.in
[1]Petroleum Engineering Program, Department of Ocean Engineering, Indian Institute of Technology Madras, Chennai 600 036, India
Full list of author information is available at the end of the article

bromide (TBAB), tetra-*n*-butyl-ammonium chloride, tetra-*n*-butyl-ammonium fluoride. TBAB in water forms a semiclathrate hydrate which shares similar physical and structural properties as true clathrate hydrates. The principal difference is that, unlike true clathrates where the guest molecules are not physically bonded within the water structure, in semiclathrate hydrates, the host/guest molecules of TBAB may form the part of the water lattice (host) as well as occupy cages (guest) along with the gas molecules [7-10]. Typically, semiclathrate hydrates are formed at lower conditions of pressure and temperature as compared with the gas hydrate system for the same gas as a guest molecule. Typically, in the semiclathrate structure of TBAB, one TBAB and 38 H_2O molecules form a hydrate structure. Semiclathrate hydrates have a wide range of engineering applications such as carbon dioxide sequestration, transportation and storage of natural gas, and flue gas separation [11-15]. Understating the phase behavior of these hydrate systems forms a precursor for their successful applications.

Arjmandi et al. [7] conducted experiments to determine the equilibrium temperature and pressure of semiclathrate hydrates of methane, carbon dioxide, and nitrogen with varying percentages of TBAB in the aqueous system. The TBAB weight percentage (wt.%) used in this study was in the range of 5 to 42.7 wt.%. A shift in the equilibrium curve of semiclathrate hydrates to the right as compared with that for pure clathrate hydrate systems is observed. This shift implies that semiclathrate hydrates are one of the potential 'thermodynamic' promoters of hydrate formation. The thermodynamic promoter changes the Gibbs energy of hydrate formation which shows implications for the stability condition of hydrates at complex reservoir conditions. Here, the stability of hydrate refers to the equality of fugacity of the hydrate phase and the gas phase. Kinetic promoters also refer to the catalyst which does not affect the phase equilibrium. However, the thermodynamic promoter does affect the phase equilibrium. Several researchers [11-20] conducted experiments to determine the equilibrium conditions of semiclathrate hydrate systems. Duc et al. [11] conducted phase stability experiments on carbon dioxide semiclathrate hydrates for varying TBAB weight percentages from 4.95 to 65 wt.%. Lin et al. [12] studied the phase equilibrium and dissociation enthalpy of carbon dioxide semiclathrate hydrates formed in the presence of TBAB. Sakamoto et al. [13] studied the thermodynamic behavior of hydrogen semiclathrate hydrates formed in an aqueous solution of TBAB. Li et al. [14] conducted experiments for semiclathrate hydrates of carbon dioxide for 5 and 10 wt.% TBAB. Sun et al. [15] determined equilibrium conditions of methane hydrates for 5 to 45 wt.% TBAB in the system. Li et al. [16] and Ding et al. [17] performed experiments to study the formation and dissociation of methane semiclathrate hydrates in an aqueous system containing TBAB. Mohammadi et al. [18,19] conducted experiments to determine the phase behavior of an aqueous system of methane, hydrogen sulfide, carbon dioxide, and nitrogen in the presence of TBAB. Ye and Zhang [20] studied the phase equilibrium of CO_2 in TBAB aqueous solution. The effect of the kinetic promoter (surfactant) such as sodium dodecyl sulfate (SDS) is observed to increase the formation of clathrate hydrates of methane while not affecting the phase equilibria [21]. The use of SDS (kinetic promoter) along with TBAB (thermodynamic promoter) may help to increase the efficiency of the system, making it more useful for CO_2 sequestration and gas storage and transportation. However, the phase behavior of CO_2 hydrates containing both agents is not reported and needs more investigation.

Memory effect has been studied in the crystallization of clathrate hydrates [22]. It is a phenomenon in which the re-nucleation of hydrate crystals occurs at a lower pressure of at least 1 MPa and a higher temperature of at least 2 K than the initial nucleation of hydrates. Parent and Bishnoi [23] conducted experiments to study the memory effect of methane + water system. Takeya et al. [24] performed experiments to analyze memory effect of carbon dioxide + water system. Ohmura et al. [25] observed that the re-nucleation probability and induction time depend on the temperature after the dissociation of initial gas hydrate. Oshima et al. [26] found that the re-nucleation phenomena of semiclathrate hydrates occurred at conditions of temperature 1 K higher than the nucleation temperature of the initial semiclathrate hydrate. It is observed that the effect of the concentration of TBAB on incipient pressure, temperature, and time required for re-nucleation of semiclathrate hydrates has not yet been examined and thus requires investigation.

In this work, an experimental study is performed on the phase behavior of semiclathrate hydrates of carbon dioxide in TBAB aqueous solution with a small amount of surfactant, SDS, for varying weight percentages of TBAB to understand the effect of SDS on the phase stability of the system. Re-nucleation (memory) effect is examined for semiclathrate hydrates of carbon dioxide, and the effect of TBAB concentration on incipient conditions (pressure, temperature, and time) required for re-nucleation is discussed. The results obtained on the phase stability of the combination of CO_2 + TBAB + SDS + H_2O are new and not reported in the literature.

Methods

A detailed description about the experimental setup is given, followed by discussion on the experimental procedures.

Experimental setup

The heart of the experimental setup is a high-pressure reactor shown in Figure 1. The volume of the high-pressure reactor is 1 L. The maximum operating pressure of the reactor is 10 MPa. The reactor has a magnetic stirrer which has a maximum rotation speed of 1,000 rpm. The high-pressure reactor is installed with pressure transducers and temperature sensor, Pt-100. The reactor also has a jacket within which ethylene glycol + water solution is circulated from the Julabo® (Julabo Gmbh, Seelbach, Germany) water bath at a desired temperature of about 263.15 K (±1 K). The reactor and the Julabo® water bath is connected to a personal computer (PC; Intel® Core i5, 2 GB RAM; Intel Corporation, Santa Clara, CA, USA) and operated online. The data on pressure and temperature as a function of time are acquired at an interval 30 s and stored in the PC.

TBAB (in powder form) and SDS used in this work are of ultrapure quality and supplied by Sisco Research Laboratory Private Limited (Mumbai, India). Deionized water obtained from the Millipore® (EMD Millipore, Billerica, MA, USA) deionized setup is used to make the TBAB solution of varying weight percentages as per the requirement. Carbon dioxide gas used in this work is 99.5% pure and is supplied by Bhuruka Gas Agency (Banglore, India). The carbon dioxide cylinder is equipped with a heater at the outlet which helps the smooth flowing of gas to the reactor at the desired pressure.

Experimental procedure

Typically, two methods, the 'isochoric' and the 'isobaric' methods, are used for the determination of phase equilibrium of semiclathrate hydrates [27-31]. In this work, the isochoric method is used. The temperature of the system is reduced, which results in the reduction of the system pressure. An abrupt fall in the pressure of the system is observed at the onset of hydrate nucleation. After attaining sufficient pressure drop, the system is heated slowly until it attains initial pressure condition. The point of equilibrium is decided when the intersection of tangent is drawn between the heating and cooling lines. The isochoric method is much simpler to conduct and is relatively more accurate as compared to other methods [27,28].

An aqueous solution of a desired TBAB weight percentage with 0.1 g (0.025 wt.%) of SDS is filled in to the reactor. Carbon dioxide is purged in to the reactor to a pressure of 0.1 to 0.2 MPa for a couple of times to remove dissolved air from the aqueous TBAB solution. The reactor is then pressurized using carbon dioxide and left overnight to allow the system to become stable [31]. The reactor system is then stirred at 480 to 500 rpm and cooled to a lower temperature of about 263.15 K using the Julabo® water bath to form semiclathrate hydrates. As the semiclathrate hydrate forms, the pressure of the reactor falls rapidly. The temperature of the system is then gradually increased at the rate of 2 K/h, followed by an increase at the rate of 0.2 K/h. It is to be noted here that, with the help of several exploratory run *a priori*, we observed a slow heating rate of 0.2 K/h of the reaction mass, which was suitable for our *well-stirred* reactor system to get a *reliable* equilibrium point. To study the re-nucleation phenomena, the system is heated slightly (roughly 0.5 K) above the equilibrium temperature. This was done to ensure that all the

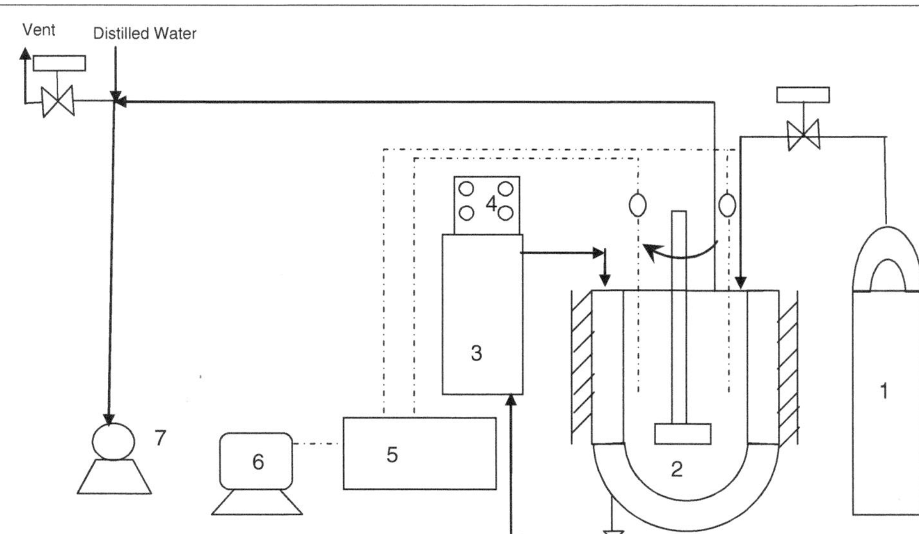

Figure 1 Details of the experimental setup used in this work. (1) CO_2 gas cylinder, (2) high-pressure reactor, (3) Julabo® water bath, (4) temperature control panel, (5) temperature and pressure data acquisition unit, (6) computer, and (7) vacuum pump.

gas from the liquid phase in the system has completely escaped to the gas phase. The reactor is then re-cooled to re-form (re-nucleate) the semiclathrate hydrates. Once a semiclathrate hydrate formation is confirmed by sudden drop in the system pressure, the temperature is raised at a rate of 2 K/h followed by 0.2 K/h until the equilibrium point is reached.

Results and discussion

In this work, an experimental study is performed on the phase behavior of semiclathrate hydrates of carbon dioxide in TBAB aqueous solution with a small amount of surfactant for varying weight percentages of TBAB. To verify the experimental procedures, experimental investigations were carried out on the pure CO_2 hydrate system for various pressure and temperature conditions to get the phase equilibrium points, followed by studies on semiclathrate hydrates of CO_2 in 5 wt.% TBAB in aqueous solution (without SDS). Figure 1 shows the results obtained for two cases along with literature data. The equilibrium data obtained for the pure CO_2 hydrate system is well matched with those in the literature. A similar behavior is observed for semiclathrate hydrates of CO_2 in 5 wt.% TBAB aqueous solution, confirming the experimental procedures. As the experimental procedure is verified, several experimental runs were performed to accumulate data sets on equilibrium pressure and temperature conditions of CO_2 semiclathrate hydrate in TBAB aqueous solution with a small amount of SDS for 5, 10, and 20 wt.% TBAB. Table 1 gives the details of experiments performed in this study.

Results on phase stability

An isochor is defined as the variation of the system pressure with temperature while keeping the volume constant. Each data point on equilibrium temperature and pressure for each TBAB weight percent (as in Table 1) generates separate isochors. A sample isochor is presented in Figure 2 for the experimental run of 20 wt.% TBAB in aqueous solution. The aqueous solution of 20 wt.% TBAB containing 0.025 wt.% SDS and the CO_2 gas in the reactor, at an initial pressure condition as shown by point A (see Figure 2), is cooled at a fast rate (−10 K/h) using cold water from the Julabo® water bath circulated through the jacket of the stirred reactor. Point

Table 1 Details of the experiments performed in this study

System	TBAB (wt.%)	Data points on equilibriums *P* and *T*
CO_2 + TBAB + H_2O	5	6
CO_2 + TBAB + H_2O	10	5
CO_2 + TBAB + H_2O	20	7

SDS weight percentage used in all experimental runs is 0.025 wt.%.

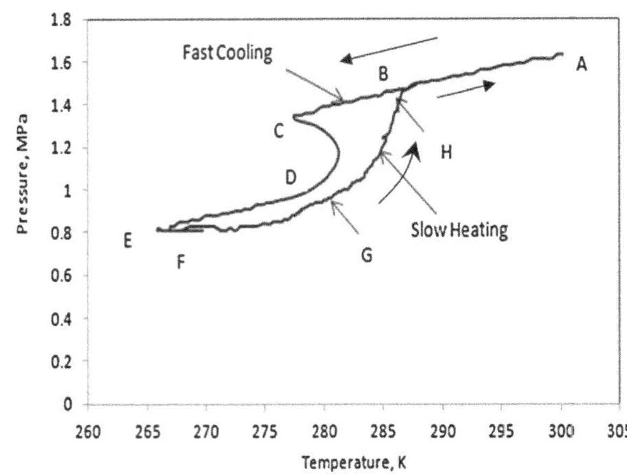

Figure 2 Typical isochor of CO_2 semiclathrate hydrate (for 20 wt.% TBAB in aqueous solution).

C represents the onset of nucleation where the system pressure is observed to decrease rapidly, deviating from the normal cooling behavior. Since semiclathrate hydrate formation is an exothermic process, an increase in temperature is observed from points C to D. The cooling is continued until point E, after which the system is heated at the rate of 2 K/h until point G, followed by a rate of 0.2 K/h until the end of the experimental run. At the end of the slow heating process, it is observed that the system retraces its path at point H. The system follows the initial path, B to A, beyond point H. The equilibrium point is decided by the intersection of tangents drawn on the heating

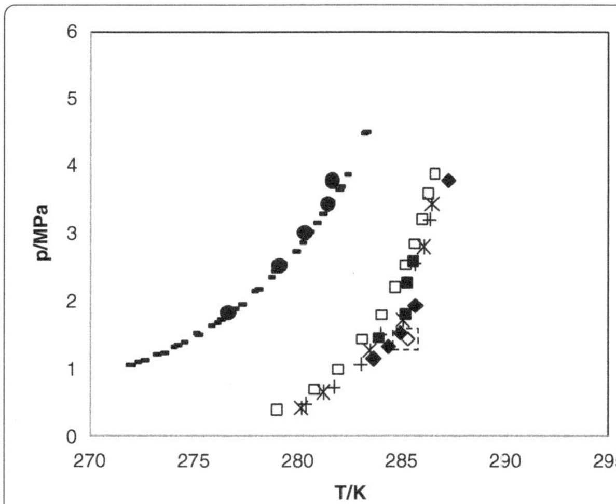

Figure 3 Experimental results for CO_2 hydrate and semiclathrate hydrates with 5 wt.% TBAB. Pure CO_2 hydrate: (black circle) this work and (small black square) Sloan and Koh [6]. Semiclathrate hydrate with 5 wt.% TBAB: (black square) this work; (asterisk) Li et al. [14]; (plus sign) Mohammadi et al. [19]; and (white square) Ye and Zhang [20]. Semiclathrate hydrate with 5 wt.% TBAB + SDS: (black diamond) this work and (white diamond) this work with memory effect.

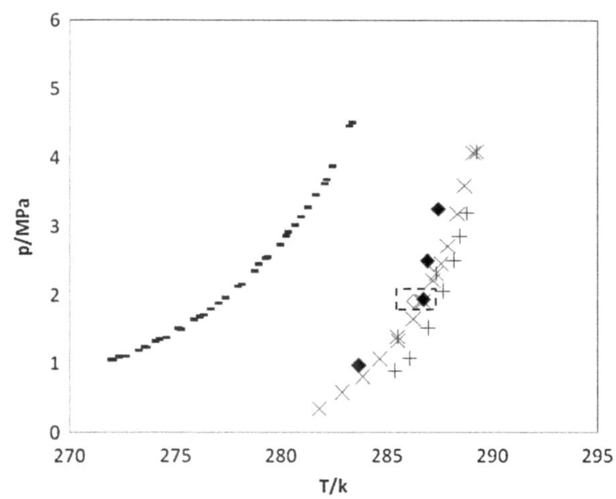

Figure 4 Experimental results for carbon dioxide hydrate and semiclathrate hydrates with 10 wt.% TBAB. Pure CO_2 hydrate: (small black square) Sloan and Koh [6]. Semiclathrate hydrate with 10 wt.% TBAB: (asterisk) Arjmandi et al. [7]; (plus sign) Mohammadi et al. [19]; and (cross sign) Ye and Zhang [20]. Semiclathrate hydrate with 10 wt.% TBAB + SDS: (black diamond) this work and (white diamond) this work with memory effect.

line and cooling line near point H as followed by other researchers [27-30]. It is to be noted here that the heating observed under semiclathrate formation can be used to determine heat of formation along with better analytical techniques, such as calorimetry or differential thermal analysis [12,32,33]; however, this measurement does not form the focus of this work.

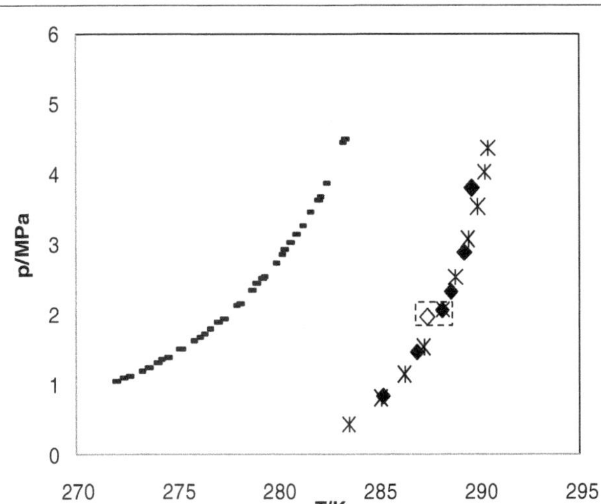

Figure 5 Experimental results for carbon dioxide hydrate and semiclathrate hydrates with 20 wt.% TBAB. Pure CO_2 hydrate: (small black square) Sloan and Koh [6]. Semiclathrate hydrate with 19 wt.% TBAB: (asterisk) Ye and Zhang [20]. Semiclathrate hydrate with 20 wt.% TBAB + SDS: (black diamond) this work and (white diamond) this work with memory effect.

Figures 3, 4 and 5 show the experimental data on phase equilibrium for the carbon dioxide semiclathrate hydrate system with 5, 10, and 20 wt.% TBAB with SDS (0.025 wt.%) in aqueous solution, respectively. The equilibrium data obtained in this work is compared with the equilibrium data available in an open literature [7,11-20]. It is found that the data obtained in this work match satisfactorily with the available equilibrium data from the literature for 5 and 10 wt.% TBAB in the system. As mentioned, we have performed experiments with an addition of 0.025 wt.% SDS in TBAB aqueous solution. The obtained results from this work show that the presence of SDS in the system does not influence the phase equilibrium of semiclathrate hydrates of carbon dioxide. The observed results are in accordance with the clathrate hydrate system of methane in the presence of a small amount of SDS [21]. SDS typically helps to increase the rate of formation of the clathrate hydrate system without affecting the equilibrium conditions, so it is termed as 'kinetic' promoters [21]. It is to be noted here that the focus of the present study is to check the effect of the presence of a surfactant on the phase equilibrium of semiclathrate hydrate system. The observed phase equilibrium data points in this study are tabulated in the Table 2.

Table 2 Equilibrium data for various TBAB percentages for semiclathrate hydrates of CO_2

TBAB (wt.%)	Equilibrium temperature (K)	Equilibrium pressure (MPa)
5	285.25	0.84
	286.90	1.47
	287.47	1.96
	288.18	2.07
	288.60	2.33
	289.29	2.89
10	283.68	0.97
	285.79	1.92
	286.73	1.94
	286.93	2.49
	287.46	3.27
20	285.25	0.87
	286.90	1.52
	287.47	2.03
	288.19	2.14
	288.60	2.41
	289.29	2.98
	289.66	3.91

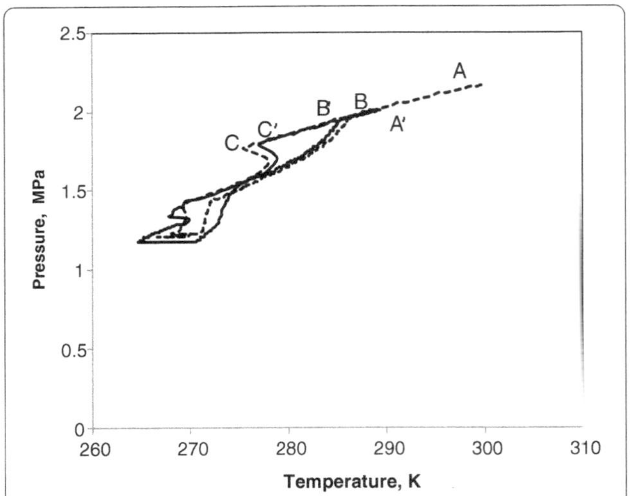

Figure 6 Re-nucleation (memory) effect for semiclathrate hydrates of CO_2 for sample case of 20 wt.% TBAB. Isochor for first run (dashed lines); isochors for memory effect (solid line). The notations on points A, B, and C are the same as those in Figure 2. A ', B', and C' represent experimental points for re-nucleation.

Re-nucleation effect in semiclathrate hydrate

The purpose of this study is to get an inference of the effect of the presence of TBAB concentration on the re-nucleation effect in the semiclathrate hydrate system, not to analyze the re-nucleation effect in detail. To understand the memory effect in detail, one may require more sophisticated analytical/visual techniques. A sample isochor for the re-nucleation effect for semiclathrate hydrate of CO_2 is shown in Figure 6 for a case of 20 wt.% TBAB in the system. It is to be noted that the isochor here (Figure 6) is separate from the isochor presented in Figure 2, which is representing another equilibrium point. The isochor with dashed lines in Figure 6 is for initial experimental runs (without memory effect) while the solid line represents the re-nucleation (memory effect) run. The notations for points A, B, and C are the same as those in Figure 2. A′, B′, and C′ represent the points for experimental run with re-nucleation effect. It is to be noted that the formation of semiclathrate hydrate is achieved first and then is dissociated to reach up to equilibrium point B as in Figure 6. After reaching point B, the system temperature is raised by about 0.5 K to ensure that all gas molecules from semiclathrate hydrate crystals have escaped to the bulk gas phase. This is validated by the fact that the P to T line traces back to its original cooling curve after point B (path B to A). This is then followed by re-cooling of the system to form the semiclathrate hydrates again. An analysis is done to see an effect of the TBAB concentration on nucleation condition from several cases of TBAB concentrations studied in this work. For this, out of several experimental runs (as in Table 1), three cases, one each from 5, 10, and 20 wt.% TBAB concentrations, are considered. In order to draw realistic and practical conclusions and inferences, the same initial conditions of system pressure and temperature are considered for the three cases. The three cases, thus, considered for this examination have initial conditions (point A in Figure 6) of system pressure and temperature that are nearly same. This is around 2.25 MPa of initial system pressure and nearly in the range 298.15 to 300.15 K of initial temperature. Table 3 gives a comparative study between the difference in insipient temperature and pressure conditions at which nucleation starts (point C in Figure 6) in the case of the first run and a run with memory effect (second experimental run) along with the corresponding time of re-nucleation of the semiclathrate hydrates in the system for 5, 10 and 20 wt.% TBAB. The equilibrium point obtained for memory effect and the regular experimental run are observed to be quite close (also shown in Figures 3, 4 and 5). It is observed from Table 3 that for the case of no memory effect with increase in TBAB percentage in the system, the time required for nucleation reduces and so is for the case of memory effect. Additionally, for the same TBAB concentration, the incipient pressure and temperature required for nucleation and re-nucleation of semiclathrate hydrates increase, while the time required for re-nucleation decreases. This study, in general, indicates that with the increase of TBAB concentration in the system, it does help to form semiclathrate hydrates at an early stage.

Conclusions

The experimental study on the phase behavior of semiclathrate hydrate of carbon dioxide in TBAB aqueous solution with a small amount of surfactant for 5, 10, and 20 wt.% TBAB are discussed. The data on phase equilibria of semiclathrate hydrates for 5 and 10 wt.%

Table 3 Temperature and pressure variation with time during the memory effect phenomenon

Number	TBAB (wt.%)	*P* and *T* condition at first run			*P* and *T* condition at re-nucleation (memory effect)		
		T_1^a (K)	P_1^a (MPa)	t_1^a (s)	T_2^b (K)	P_2^b (MPa)	t_2^b (s)
1	5	275.5	1.97	510	277.05	2.06	450
2	10	276.65	1.88	420	278.57	1.97	300
3	20	277.15	1.87	330	279.85	1.89	240

P, pressure; *T*, temperature. $^aT_1$, P_1, t_1 = incipient temperature, pressure, and time at nucleation of semiclathrate hydrates (at point C in Figure 6); $^bT_2$, P_2, t_2 = incipient temperature, pressure, and time at nucleation of semiclathrate hydrates (at point C′ in Figure 6).

TBAB are in good agreement with the published data from the open literature. The presence of the surfactant, SDS, did not influence the equilibrium conditions for the semiclathrate hydrate system. The memory effect phenomenon is studied for few cases of TBAB concentration. The increase in the concentration of TBAB in the system helps to reduce the time required for re-nucleation of semiclathrate hydrates. Additionally, for the same TBAB concentration, the incipient pressure and temperature required for nucleation and re-nucleation of semiclathrate hydrates increase, while the time required for re-nucleation decreases.

Competing interests
The authors declare that they have no competing interests.

Authors' contributions
AJ developed the experimental setup and carried out the experimental studies. JSS initiated, guided and supervised the work, and drafted the manuscript. NAS and KD carried out validation studies and helped in literature search. All authors read and approved the final manuscript.

Authors' information
AJ is currently working as a field engineer in Schlumberger. He finished his M.Tech. in Petroleum Engineering from Indian Institute of Technology Madras (IITM). JSS is working as an assistant professor at IIT Madras. His areas of research interest are in gas hydrates, flow assurance, and enhanced oil recovery. NAS is a student at Anna University and is also a lecturer at the University of Baghdad, Iraq. KD is a project officer working at IIT Madras and is under the M. Tech. program in Chemical Engineering in IIT Madras.

Acknowledgments
The authors would like to thank the director of the National Institute of Ocean Technology (NIOT) and NIOT-IITM cell for the encouragement towards setting up the laboratory facilities. Thanks to Dr. S Ramesh and Dr. Ramadass from NIOT and Professor SK Bhattacharyya of IIT Madras for their valuable comments and extended cooperation during the course of our work. Financial support from NIOT, Chennai, India, through grant OEC/10-11/105/NIOT/JITE is also gratefully acknowledged.

Author details
[1]Petroleum Engineering Program, Department of Ocean Engineering, Indian Institute of Technology Madras, Chennai 600 036, India. [2]Department of Petroleum Engineering, College of Engineering, University of Baghdad, Baghdad, Iraq. [3]Department of Chemical Engineering, AC Tech Campus, Anna University, Chennai 600 025, India.

References
1. Hawkes, C.: Geomechanical factors affecting geological storage of CO_2 in depleted oil and gas reservoirs. J Can Petro Technol **44**(10) (2005). doi:10.2118/05-10-05
2. Little, M.G.: Jackson, RB: Potential impacts of leakage from deep CO_2 geosequestration on overlying freshwater aquifers. Environ Sci Technol **44**(23), 9225–9232 (2010)
3. Zappa, P., Schreiber, A., Marx, J., Haines, M., Hake, J.F., Gale, J.: Overall environmental impacts of CCS technologies—a life cycle approach. Inter J Greenh Gas Contr **8**, 12–21 (2012)
4. Holder, G., Cugini, A., Warzinski, R.: Modeling clathrate hydrate formation during carbon dioxide injection into the ocean. Environ Sci Technol **29**, 276–278 (1995)
5. Qanbari, F., Pooladi-Darvish, M., Tabatabaie, S.: CO_2 disposal as hydrate in ocean sediments. J Natu Gas Sci Eng **8**, 139–149 (2012)
6. Sloan, E.D., Koh, C.A.: Clathrate Hydrates of Natural Gases, 3rd edn. CRC Press, New York (2008)
7. Arjmandi, M., Chapoy, A., Tohidi, B.: Equilibrium data of hydrogen, methane, nitrogen, carbon dioxide, and natural gas in semiclathrate hydrates of tetra-*n*-butyl ammonium bromide. J Chem Eng Data **52**, 2153–2158 (2007)
8. Shimada, W., Ebinuma, T., Oyama, H., Kamata, Y., Takeya, S., Uchida, T., Nagao, J., Narita, H.: Separation of gas molecule using tetra-*n*-butyl ammonium bromide semi-clathrate hydrates crystals. Jap J App Phys Part 2 Lett **42**, L129–L131 (2003)
9. Shimada, W., Ebinuma, T., Oyama, H., Kamata, Y., Narita, H.: Free-growth forms and growth kinetics of tetra-*n*-butyl ammonium bromide semi-clathrate hydrate crystals. J Crys Growth **274**, 246–250 (2005)
10. Shimada, W., Shiro, M., Kondo, H., Takeya, S., Oyama, H., Ebinuma, T., Narita, H.: Tetra-*n*-butylammonium bromide-water (1/38). Acta Crystall Sect C **61**, 65–66 (2005)
11. Duc, N.G., Chauvy, F., Herri, J.M.: CO_2 capture by hydrate crystallization - a potential solution for gas emission of steelmaking industry. Ener Conv Manag **48**, 1313–1322 (2007)
12. Lin, W., Delahaye, A., Fournaison, L.: Phase equilibrium and dissociation enthalpy for semiclathrate hydrate of CO_2 + TBAB. Fluid Phase Equili **264**, 220–227 (2008)
13. Sakamoto, J., Hashimoto, S., Tsuda, T., Sugahara, T., Inoue, Y., Ohgaki, K.: Thermodynamic and Raman spectroscopic studies on hydrogen + tetra-*n*-butyl ammonium fluoride semi-clathrate hydrates. Cheml Eng Sci **63**, 5789–5794 (2008)
14. Li, S., Fan, S., Wang, J., Lang, X., Wang, Y.: Semiclathrate hydrate phase equilibria for CO_2 in the presence of tetra-*n*-butyl ammonium halide (bromide, chloride, or fluoride). J Chem Eng Data **55**, 3212–3215 (2010)
15. Sun, Z.G., Sun, L.: Equilibrium conditions of semi-clathrate hydrate dissociation for methane + tetra-*n*-butyl ammonium bromide. J Chem Eng Data **55**, 3538–3541 (2010)
16. Li, D.L., Du, J.W., Fan, S.S., Liang, D.Q., Li, X.S., Huang, N.S.: Clathrate dissociation conditions for methane + tetra-*n*-butyl ammonium bromide (TBAB) + water. J Chem Eng Data **52**, 1916–1918 (2007)
17. Ding, Y., Gong, J., Peng, Y.: Formation-dissociation characteristics of TBAB hydrate. J China Univ Petrol (Edi of Natu Sci) **35**, 150–153 (2011)
18. Mohammadi, A.H., Richon, D.: Phase equilibria of semi-clathrate hydrates of tetra-*n*-butyl ammonium bromide + hydrogen sulfide and tetra-*n*-butyl ammonium bromide + methane. J Chem Eng Data **55**, 982–984 (2010)
19. Mohammadi, A.H., Eslamimanesh, A., Belandria, V., Richon, D.: Phase equilibrium of semiclathrate hydrates of CO_2, N_2, CH_4, or H_2 + tetra-*n*-butylammonium bromide aqueous solution. J Chem Eng Data **56**, 3855–3865 (2011)
20. Ye, N., Zhang, P.: Equilibrium data and morphology of tetra-*n*-butyl ammonium bromide semiclathrate hydrate with carbon dioxide. J Chem Eng Data **57**, 1557–1562 (2012)
21. Gayet, P., Dicharry, C., Marion, G., Graciaa, A., Lachaise, J., Nesterov, A.: Experimental determination of methane hydrate dissociation curve up to 55 MPa by using a small amount of surfactant as hydrate promoter. Chem Eng Sci **60**, 5751–5758 (2005)
22. Makogon, Y.K.: Hydrates of Natural Gas. PennWell Books, Tulsa, Oklahoma (1981)
23. Parent, J.S., Bishnoi, P.R.: Investigations into the nucleation behavior of natural gas hydrates. Chem Eng Commu **144**, 51–64 (1996)
24. Takeya, S., Hori, A., Hondoh, T., Uchida, T.: Freezing-memory effect of water on nucleation of CO_2 hydrate crystals. J Phy Chem B **104**, 4164–4168 (2000)
25. Ohmura, R., Ogawa, M., Yasuoka, K., Mori, Y.H.: Statistical study of clathrate-hydrate nucleation in a water/hydrochlorofluorocarbon system: search for the nature of the 'memory effect'. J Phy Chem B **107**, 5289–5293 (2003)
26. Oshima, M., Shimada, W., Hashimoto, T.A., Ohgaki, K.: Memory effect on semi-clathrate hydrate formation: a case study of tetragonal tetra-*n*-butyl ammonium bromide hydrate. Chem Eng Sci **65**, 5442–5446 (2010)
27. Bishnoi, R.J., Natarajan, V.: Formation and decomposition of gas hydrates. Fluid Phase Equili **117**, 168–177 (1996)
28. Nixdorf, J., Oellrich, L.R.: Experimental determination of hydrate equilibrium conditions for pure gases, binary and ternary mixtures and natural gases. Fluid Phase Equili **139**, 325–333 (1997)
29. Masoudi, R., Tohidi, B., Anderson, R., Burgass, R.W., Yang, J.: Experimental measurement and thermodynamic modeling of clathrate hydrate equilibria

and salt solubility in aqueous ethylene glycol and electrolyte solutions. Fluid Phase Equili **219**, 157–163 (2004)

30. Masoudi, R., Tohidi, B., Danesh, A., Todd, A.C.: A new approach in modeling phase equilibria and gas solubility in electrolyte solutions and its applications to gas hydrates. Fluid Phase Equili **215**, 163–174 (2004)
31. Sa, J.H., Lee, B.R., Park, D.Y., Chun, H.C., Lee, K.H.: Amino acids as natural inhibitors for hydrate formation in CO_2 sequestration. Environ Sci Technol **45**, 5885–5891 (2011)
32. Deschamps, J., Dalmazzone, D.: Dissociation enthalpies and phase equilibrium for TBAB semi-clathrate hydrates of N_2, CO_2, N_2 + CO_2 and CH_4 + CO_2. J Therm Anal Calorim **98**, 113–118 (2009)
33. Oyama, H., Shimada, W., Ebinuma, T., Kamata, Y., Takeya, S., Uchida, T. Nagao, J., Narita, H.: Phase diagram, latent heat and specific heat of TBAB semiclathrate hydrate crystals. Fluid Phase Equili **234**, 131–135 (2005)

2

Adsorption of catechol, resorcinol, hydroquinone, and their derivatives: a review

Sundaramurthy Suresh[1*], Vimal Chandra Srivastava[2] and Indra Mani Mishra[2]

Abstract

In recent years, there has been an increasing interest in finding innovative solutions for the efficient removal of contaminants from water, soil, and air. The present study reviews the adsorptive removal of catechol (C), resorcinol (R), hydroquinone (HQ), and their derivatives from various adsorbents. As an effective, efficient, and economic approach for water purification, adsorbents and adsorption processes have been widely studied and applied in different aspects for a long time. The role of various adsorbent materials like activated carbon, activated carbon cloth, carbon nanotubes, polymeric resins, organic clays, $Fe(OH)_2$, and TiO_2 was discussed together with that of other experimental parameters. In all the synthetic resins, particularly, aminated hypercrosslinked polymers have good adsorption capability for phenols. These polymeric adsorbents are suitable for industrial effluents containing C, R, HQ, and their derivatives. The adsorption capacities of the adsorbents reviewed here vary significantly depending on the characteristics of the individual adsorbent, the extent of chemical modifications, and the concentrations of solutes.

Keywords: Catechol, Resorcinol, Hydroquinone, Adsorbents, Wastewater

Review

Introduction

Phenol (P) and substituted P are important organic intermediates for the products of industry and agriculture [1]. For example, hydroxy aromatic compounds, such as catechol (C), resorcinol (R), and hydroquinone (HQ), were used widely as industrial solvents. C (1,2-dihydroxybenzene) is also widely used to produce food additive agents, hair dyes, and antioxidants [2]. R (1,3-dihydroxybenzene) is usually employed to produce dyes, plastics, and synthetic fibers [3-5]. Phenolic compounds, such as C, R, and HQ, were found in the effluents of industries such as textile, paper and pulp, steel, petrochemical, petroleum refinery, rubber, dye, plastic, pharmaceutical, cosmetic, etc. and in the wastewater of synthetic coal fuel conversion processes [6,7]. The effluents from synthetic coal fuel conversion processes may contain C and R concentrations ranging from 1 to 1,000 mg/l [8].

* Correspondence: sureshpecchem@gmail.com
[1]Department of Chemical Engineering, Maulana Azad National Institute of Technology Bhopal, Bhopal 462 051, India
Full list of author information is available at the end of the article

H. Reinsch first isolated C in 1839 by distilling catechin (from catechu, the juice of *Mimosa catechu* (*Acacia catechu* L.f)). The pyrocatechol is formed by the heating of catechin above its decomposition point. C is produced industrially by the hydroxylation of P using hydrogen peroxide [9]. C has also been produced by the hydrolysis of 2-substituted phenols, especially 2-chlorphenol, with hot aqueous solutions containing alkali metal hydroxides. Its methyl ether derivative, guaiacol, converts to C via hydrolysis of the CH_3-O bond as promoted by hydriodic acid. R is obtained by fusing many resins (galbanum, asafoetida, etc.) with potassium hydroxide or by the distillation of brazilwood extract. It may be prepared synthetically by fusing 3-iodophenol, phenol-3-sulfonic acid, or benzene-1,3-disulfonic acid with potassium carbonate; by the action of nitrous acid on 3-aminophenol; or by the action of 10% HCl on 1,3-diaminobenzene [10]. Many *ortho*- and *para*-compounds of the aromatic series (for example, the bromophenols, benzene-*para*-disulfonic acid) also yield R by fusion with potassium hydroxide. HQ is obtained from few methods: oxidation of aniline with manganese dioxide and sulfuric acid, followed by reduction with iron dust and water [11]. The second method consists of the alkylation of benzene with propylene to produce a mixture of diisopropylbenzene isomers from which, in a first step, the *para*-isomer is isolated. This is oxidized with oxygen to produce the corresponding dihydroperoxide, which is treated with an acid to produce HQ and acetone [12]. Finally, the oxidation of P with hydrogen peroxide can be used to produce a mixture of products from which both HQ and C (1,2-dihydroxybenzene) can be isolated [12].

C occurs freely in kino and in beechwood tar; its sulfonic acid has been detected in the urine of horses and humans [13]. HQ exists in a free state in pear leaves [14]. Arbutin, a glucoside of HQ, occurs widely in the leaves, barks, buds, and fruits of many plants [11], especially the Ericaceae [15]. C (*o*-benzenediol, 1,2-benzenediol, or 1,2-dihydroxybenzene) is a natural polyphenolic compound that widely exists in higher plants such as teas, vegetables, fruits, tobaccos, and some traditional Chinese medicines [16]. The smoke from 100 cigarettes contains about 10 mg of P and 50 mg of C besides other phenols [17]. HQ has been detected in cigarette smoke [18,19]. This compound occurs in the effluents resulting from different industrial activities such as photoprocessing [20], coal-tar production [21], and the paper industry [22].

The toxicity of C for microorganisms has been demonstrated in the past years [23-25] and has been suggested to be the reason for the difficulties in cultivating microorganisms on benzene, toluene, or chlorobenzene [23]. Several studies additionally indicated the toxicity of C for water flea, zebra fish, trout, rabbit, cat, rat, and mouse and for human cell lines [26]. C is strongly irritating to the eyes, skin, and respiratory tract, and it has been proven to cause DNA damage, vascular collapse, coma, and death. Between R and C, C is considered more toxic [11,27]. However, these compounds are considered as the primary pollutants in wastewater due to their high toxicity, high oxygen demand, and low biodegradability [8,28]. Consequently, their removal from wastewater has attracted significant environmental concerns. C, like other phenols, is of particular interest from a sanitary point of view due to its toxicity and deleterious effect on the quality of water supplies. In human medicine, HQ is used as a topical application in skin whitening to lighten the color of the skin as it does not have the same predisposition to cause dermatitis as metal does. There is inadequate evidence in humans for the carcinogenicity of HQ; however, it causes toxicity in several organs, notably the kidney and forestomach [12].

Additional file 1: Table S1 shows the hazard ranking of C, R, and HQ. The compound is readily absorbed from the gastrointestinal tract, causes hemolysis, degenerates the renal tubes, diminishes liver function, and accumulates in the bone marrow [29]. Its metabolites may initiate many cancers and neurodegenerative diseases [30]. C is even more toxic than P since it provokes changes in the function of erythrocytes at doses as low as 50 μg/l compared to 250 μg/l of P [31].

The United States Environmental Protection Agency (USEPA) has designated granular activated carbon (GAC) adsorption as the 'best available technology' for removing organic pollutants [32]. P and associated compounds have been listed as priority pollutants by the Ministry of Environment and Forests (MoEF), Government of India, and USEPA. MoEF has prescribed that the concentration of phenols should not exceed 1.0 mg/l for their discharge into surface waters and 5.0 mg/l for their discharge into public sewers, on land for irrigation, and on marine coastal areas. These limits have generally been on the basis of the total phenols present in the effluent.

Conventional methods used in the remediation/degradation of C, R, and HQ are biodegradation [33,34], anaerobic biodegradation [35-37], anodic oxidation [38,39], photocatalysis [40], oxidative catalysis [41], etc. Adsorption techniques have gained favor in recent years because they are considered efficient for the removal of trace organic pollutants from water that cannot be removed using other treatment processes.

In addition, adsorption and desorption kinetics are technologically important because the diffusion within solid particles is a phenomenon of great importance in catalysis, metallurgy, microelectronics, materials science, and other numerous scientific and technological applications. Chemical kinetics explains how fast the rate of chemical reaction occurs and also on the factors affecting the reaction rate. The nature of the sorption process will depend on physical or chemical characteristics of the adsorbent systems and also on the system conditions.

Adsorption using GAC has been found to be an attractive process for the removal of phenols [6,42-50]. However, the cost of GAC and the loss of adsorption efficiency after regeneration of the exhausted GAC have limited its use in effluent treatment. Therefore, alternative low-cost, non-conventional adsorbents such as activated carbon cloth [51,52], waste Fe(III)/Cr(III) hydroxide [53], hypercrosslinked resin [53-55], TiO_2 surface [56,57], organoclays [58], bagasse fly ash [59], activated cashew nut shell [60], etc. have been investigated for the treatment of effluents. So, in this context, the present review is made for the adsorption of C, R, and HQ, and their derivatives onto different adsorbents.

Properties of C, R, and HQ

C, R, and HQ are available in the form of colorless crystalline and white crystals. C and R have phenolic odor and unpleasant sweet taste and become brown on exposure to air and light and pink on contact with air and light [61]. Small amounts of C occur naturally in fruits and vegetables, along with the enzyme polyphenol oxidase. R crystallizes from benzene as colorless needles which are readily soluble in water, alcohol, and ether, but insoluble in chloroform and carbon disulfide. With concentrated nitric acid, in the presence of cold concentrated sulfuric acid, it yields trinitro-resorcin (styphnic acid), which forms yellow crystals, exploding violently on rapid heating. It reduces Fehling's solution and ammoniacal silver solutions. It does not form a precipitate with lead acetate solution, as the isomeric pyrocatechol does. Iron(III) chloride colors its aqueous solution a dark violet, and bromine water precipitates tribromoresorcin. Some of the physicochemical properties of adsorbents affect the adsorption process like water solubility, acid dissociation constant (pK_a) value, and octanol/water partition coefficient. R has more solubility (1,100 g/l) than C and HQ. The octanol/water partition value of R is 0.93, which is more than those of C and HQ. The pK_a values of C, R, and HQ were 9.25, 13; 9.4, 12.3; and 9.9, 11.6, respectively. Some more properties are given in Table 1. C can form stable complexes with various di- and trivalent metal ions, the complexes with trivalent ions being the most stable. C can also undergo redox reactions (Scheme 1 of [7]), cycling between C, semiquinone radicals, and *ortho*-benzoquinone.

Physicochemical characteristics of various adsorbents for removal of C, R, and HQ

Physicochemical characteristics of various adsorbents for the removal of C, R, HQ, and their derivatives from water were listed in Tables 2 and 3.

Usually, there are different processes for the preparation and pretreatment of adsorbents, namely either thermal procedures (physical) or chemical routes. In comparison with physical activation, there are two important advantages of chemical activation: One is the lower temperature in which the process is accomplished, and the other is that the global yield of the chemical activation tends to be greater since burn-off char is not required. For impregnation, the precursor with dehydrating agents was widely used as a chemical agent in the preparation of adsorbents. Knowledge of different variables during the activation process is very important in developing the porosity of materials which is sought for a given application. For example, chemical activation done by using H_2SO_4, HCl or HNO_3, $ZnCl_2$, $AgNO_3$, H_3PO_4, H_2O_2, etc. can improve the pore distribution and increase the surface area of adsorbents in the structure due to the use of different chemicals [42,44,51-53,58,70] and then heat treatment at moderate temperatures in a one-step process.

Table 1 Some of the physicochemical properties of C, R, and HQ

Parameters	C	R	HQ	References
Chemical structure	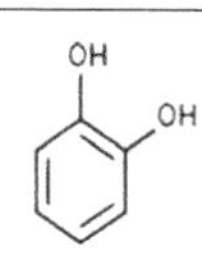 $C_6H_6O_2$	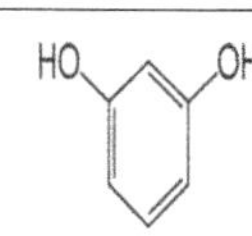 $C_6H_6O_2$	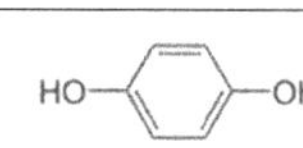 $C_6H_4(OH)_2$	-
Other names	Pyrocatechol; 1,2-benzenediol; 1,2-dihydroxybenzene	1,3-Dihydroxybenzene; 1,3-benzenediol; 3-hydroxyphenol; resorcin	*p*-Benzenediol; 1,4-benzenediol; dihydroxybenzene; 1,4-dihydroxybenzene; quinol	-
λ_{max} (nm)	275	273	289	[63]
Boiling point (°C) at 101.3 kPa	245.5	277	287	[57]
Octanol/water partition coefficient (log P_{ow})at 25°C	0.88	0.79 to 0.93	0.59	[64]
Density (g/cm^3)	1.344	1.28	1.3	-
Molecular weight (MW; g/mol)	110.11	110.11	110.1	-
Water solubility (g/l) at 25°C (C_s)	430	1,100	59	-
Molecular size (nm)	0.55 × 0.55	0.56 × 0.47	-	[57]
pK_a	9.25, 13	9.2, 10.9	9.9, 11.6	
Dipole moment (Debye; D)	2.620	2.071	0.0	[65]
Polarity/polarizability parameter (π; cm^3)	$11.89 \pm 0.5 \times 10^{-24}$	59.11×10^{-24}	0.21 ± 0.02	[63,66]
Hydrogen-bonding donor parameter (α_s)	0.85	1.10	-	[63]
Hydrogen-bonding acceptor parameter (β_m)	0.58	0.52	-	[63]

Table 2 Physicochemical characteristics of AC for removal of C, R, and HQ, and their derivatives from water

References	Rajkumar et al. [66]	Kumar et al. [6]	Rodriguez et al. [67]	Mondal and Balomajumder [68]	Mohamed et al. [42]	Blanco-Martinez et al. [44]	Richard et al. [43]	Richard et al. [45]	Suresh et al. [46-48]
Absorbent	AC	AC	AC	GAC-SAB	Treated AC	Treated AC	AC	AC	GAC
Absorbates	C, R, P, and others	C and R	R	R and P	C, R, P, and others	C, R, and HQ	C and R	C	C, R, and HQ
Pre-treatment	-	Oven at 105°C for 72 h	-	-	-	HNO_3solution for 9 h	-	-	-
Moisture (%)	3 to 6	-	-	-	-	-	-	-	9.04
Ash (%)	3 to 5	-	-	-	-	-	-	-	-
Average particle size (mm)	12/40	0.536	-	1.4 to 5	-	-	-	-	1.67
BET surface area of pores (m^2/g)	1,100	579.23 ± 17.02	864	778.12, 583.23, 393.61	43, 115, 152, 182, and 210	1,140	1,370	1,370	977.65
BET average pore diameter (Å)	-	-	-	-	-	-		-	18.79
Pore volume (cm^3/g)	-	-	0.485	0.2371, 0.2044, 0.1692	-	0.51, 0.12		-	0.072
Boehm titration (μmol/m^2)	-	-	-	-	0.6, 2.15, 3.65, 4.05, 4.13	0.30, 0.60	0, 0.1, 0.4, and 0.02 (COOH, COO, OH, and CO, respectively)	-	0.4, 1.8, and 3.3 (COOH, -COO, and -OH, respectively)
pH_{PZC}	-	-	9.25	-	-	9.8		-	10.3

Volatile matter (%), fixed carbon (%), heating value, and ultimate analysis are not reported; AC, activated carbon; GAC, granular activated carbon; SAB, simultaneous adsorption biodegradation; C, catechol; R, resorcinol; P, phenol; HQ, hydroquinone, BET, Brunauer-Emmett-Teller.

Table 3 Physicochemical characteristics of other adsorbents for removal of C, R, HQ, and their derivatives from water

References	Bayram et al. [69]	Huang et al. [55]	Liao et al. [70]	Shakir et al. [58]	Sun et al. [54]	Ayranci and Duman [52]	Yildiz et al. [51]	Namasivayam and Sumithra [53]	Vasudevan and Stone [56]
Absorbent	ACC	HJ-1	MWCNT	CTAB-B	AH and NDA-100 resins	ACC	ODTMA-B, HDTMA-B	Fe(III)/Cr(III) OH	TiO_2
Absorbates	C and R	C and R	C, P, and others	C	C and R	P, 4NP, HQ, and others	HQ and others	C	C and others
Pre-treatment	Warm (60°C) water	1% HCl and ethanol	HNO_3 (63%) and refluxed in oil bath at 140°C for 12 h	85°C for 3 h and 105°C for 1 h	-	Washed with water and then dried under vacuum at 120°C	-	Washed with distilled water for three times and 60°C for 15 h	-
Average particle size	-	0.4 to 0.6 mm	-	-	0.4 to 0.6 mm	-	TOC = 19.8% to 24.28%	250 to 500 μm	-
BET surface area of pores (m^2/g)	1,870	750; 727	72; 121	-	934; 819; 726; 483	1,464 to 2,500	28.92 to 55.37	156	39.5
BET average pore diameter (Å)	6.51, 5.78	-	-	-	2.4, 2.4, 2.6, 3.0			7.6	-
Pore volume (cm^3/g)	0.709	-	0.41; 0.49					0.297	-
pH_{PZC}	7.2		-	-	-	7.4		8.1	6.3

ACC, AC cloth; HJ-1,hypercrosslinked resin; MWCNT, multi-walled carbon nanotubes; CTAB-B, cetyltrimethylammonium bromide-modified bentonite; AH, aminated hypercrosslinked polymers; NDA-100, hypercrosslinked resin without amino groups; ODTMA-B, octadecyltrimethylammonium bromide; HDTMA-B, hexadecyltrimethylammonium bromide; TiO_2, TiO_2 Degussa P-25 surface; C, catechol; R, resorcinol; P, phenol; 4NP, 4-nitophenol; HQ, hydroquinone, TOC, total organic carbon; BET, Brunauer-Emmett-Teller.

The raw sawdust was carbonized at 873 K in a nitrogen atmosphere, and concentrated H_2SO_4 was impregnated with carbonized materials in different concentrated H_2SO_4/sawdust ratios from 3:1 to 6:1 (*w*/*w*) at 575 K for 4 h. The product was cooled to room temperature, washed with 5% H_2SO_4 and then with bi-distilled water until free from sulfate ions, and dried at 393 K for 6 h. GACs are placed in the reducing system with hydrogen under vacuum condition and then warmed at 573 K for 6 days. Finally, the AC is placed with HNO_3 solution in Soxhlet equipment for 9 h. The sample is then washed with distilled water until a constant pH value is obtained, and the AC is then dried at 383 K for 24 h. These GACs are used for the removal of C [42], HQ, and other phenols [44]. The chloromethylated PS beads were dried at 323 K in vacuum for 8 h and then swollen with nitrobenzene at 298 K overnight with anhydrous $ZnCl_2$ added into the reaction flask with the temperature at 323 K, and the beads were obtained after 12 h. The polymeric beads are washed with 1% HCl (*w*/*w*) aqueous solution and ethanol until the solution becomes transparent. Finally, they were extracted with ethanol for 8 h [55]. Shakir et al. [58] prepared the organobentonite by treating natural bentonites which are crushed and oven-heated at 85°C for 3 h and dried for 1 h at 105°C. The treated clay was washed with $AgNO_3$. Finally, the organobentonite was at last separated from water by vacuum filtration and dried for 1 h at 105°C, and a similar procedure was reported for the removal of HQ onto AC by drying under vacuum at 120°C [51,52].

Figures 1a,b,c and 2a,b,c show the yearly progress of publications and citations, respectively, in the late twentieth century and during the present decade. These figures clearly indicate the increasing interest in the removal of C, R, and HQ by adsorption.

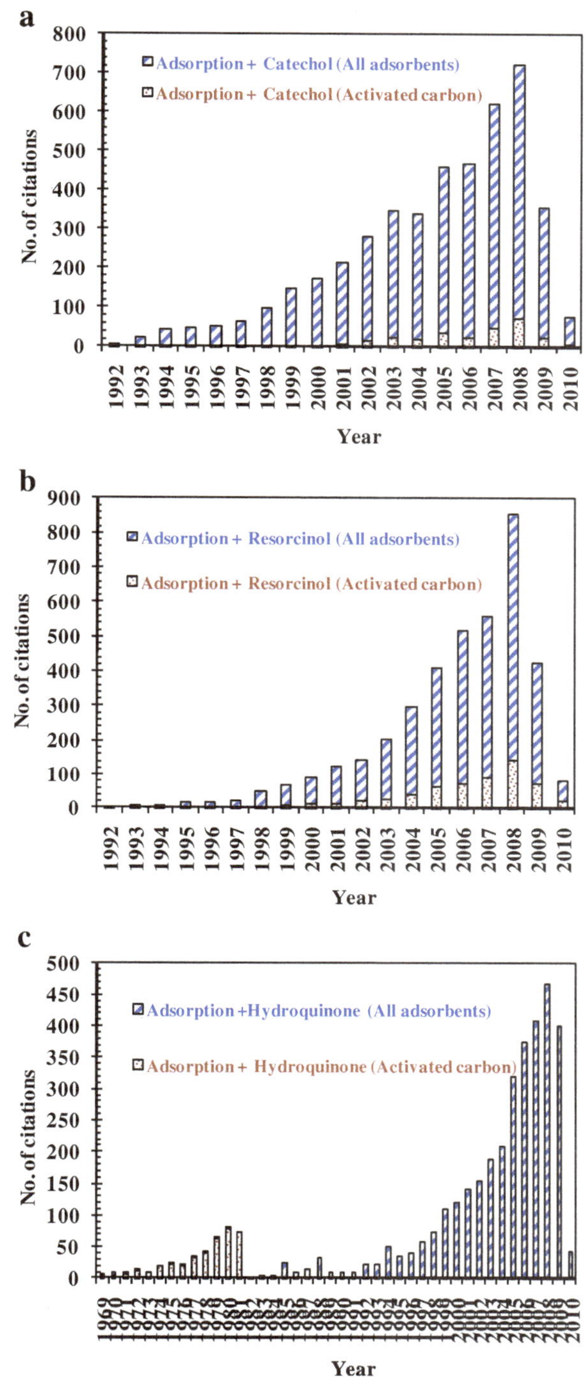

Figure 1 Number of research papers appearing with 'adsorption' and (a) 'catechol,' (b) 'resorcinol,' and (c) 'hydroquinone.' In the topic as listed in [71] for 'All Years' (1983 to 2010; out of a total of 267 and 14; 282 and 49; and 239 and 18, respectively, articles appearing). Note that early adsorption articles appeared in the late 1990 and 1983. The small number of papers in pre-1990 and 1982 means that they cannot be properly shown in the figure.

Removal by adsorption onto various adsorbents

Adsorption of organic solutes from the aqueous phase is a very important application of ACs, and it has been cited by the USEPA as one of the best available environmental control technologies [72]. The removal by adsorption onto various adsorbents of C, R, HQ, and their derivatives can be found in the literature [6,43-45, 52-57,68,70,73-75].

Activated carbon

Adsorption onto AC is widely used for wastewater treatment. Various adsorbents prepared from AC and low-cost adsorbents used in wastewater treatment were reported by Lin and Juang [76]. Thus, AC is used in the control of color and odors and in the removal of organic compounds or trihalomethane precursors, chlorine, and toxic compounds in general. A large amount of work has been devoted to the study of phenolic compound adsorption onto AC for water treatment purposes. The

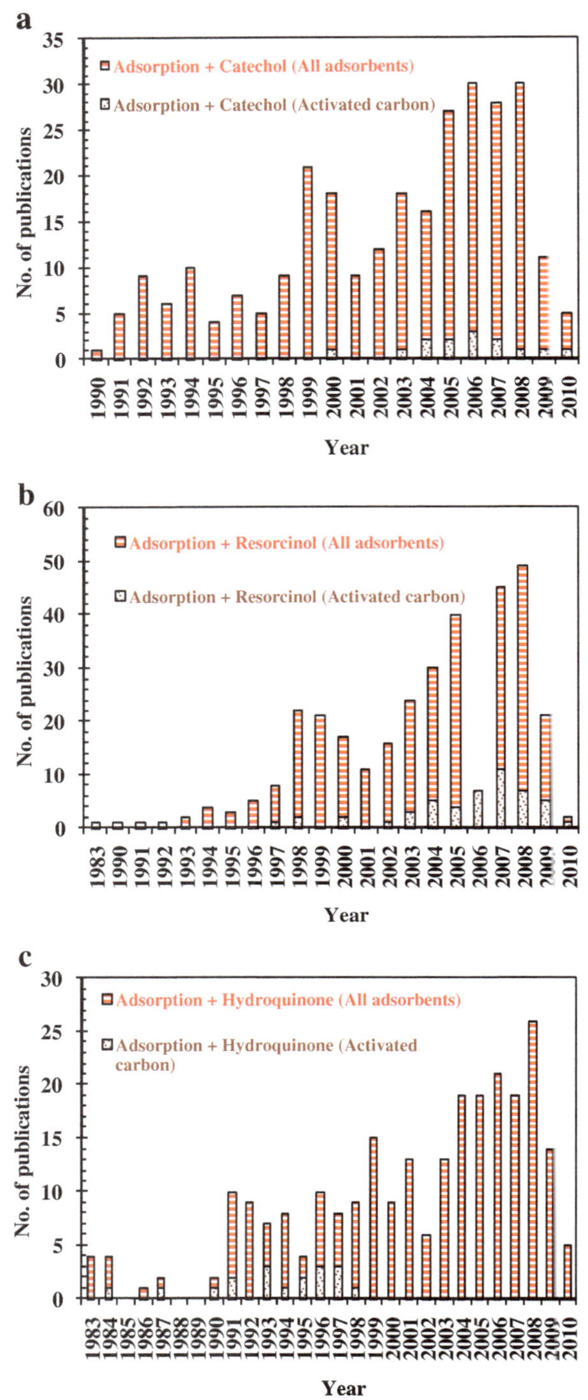

Figure 2 Number of citations appearing with 'adsorption' and (a) 'catechol,' (b) 'resorcinol,' and (c) 'hydroquinone.' In the topic as listed in [71] for 'All Years' (1983 to 2010; out of a total of 4,734 and 326; 3,812 and 686; and 3,358 and 394, respectively, articles appearing). Note that early adsorption articles appeared in the late 1991 and 1992. The small number of papers in pre-1991 and 1968 means that they cannot be properly shown in the figure.

development of the investigation on the subject has been recently reviewed by Dabrowski [77]. Dobbs and Cohen [78] reported an extensive list of isotherm data of several toxic compounds. Ozkaya [79] compared different isotherm models used to describe the adsorption of phenols on AC. Garcia-Araya et al. [80] studied the adsorption of some phenolic acid on AC.

ACs present an outstanding adsorption capacity that stems from their high surface area, pore structure, and surface chemical properties. These materials are effective adsorbents for priority pollutants, therefore being suitable for the decontamination of water and wastewater. The AC surface is usually divided in three zones: basal planes, heterogeneous superficial groups (mainly oxygen-containing surface groups), and inorganic ash. For aromatic compounds, most of the adsorption sites are found on the basal planes [81]. However, heterogeneous groups have a higher activity and define the chemical characteristics of the carbon surface. The nature of the surface groups can be modified through physical, chemical, and electrochemical treatments. The most common are liquid phase treatments using HNO_3 and H_2O_2, gas-phase oxidation with O_2 or N_2O, and heat treatment under inert gas to selectively remove some of the functional groups [42,81,82].

ACs possess excellent adsorption ability for relatively low molecular weight organic compounds such as phenols [82]. From a general point of view, an AC to be used in such processes must have adequate adsorptive capacity, mechanical strength, and chemical purity. Furthermore, all of these specifications should coexist with a low production cost. The most common materials used for the preparation of ACs are wood, coal, lignin, petroleum coke, and polymers [83]. Recently, many investigators have studied the feasibility of using many agricultural by-products and wastes, which are available at very little or no cost to prepare AC [84]. Both the texture and surface chemistry of ACs determine performance, and the final application of the carbon material will depend on its characteristics. The porous structure of ACs is a function of the precursor used in the preparation, the activation method followed, and the extent of activation. AC types produced by applying chemical activation processes contain mainly mesopores, while carbons obtained via gas activation are of the microporous type [85], although, depending on the nature of the parent material and by adjustment of the process conditions, different pore sizes that cover the micro-, meso-, and macropore ranges can be obtained [86]. Therefore, in recent years, it has prompted a growing research interest in the production of AC from renewable and cheaper precursors which are mainly agricultural by-products, such as corncob [87], rattan sawdust [88], rice straw [59,89], apricot shell [90], jute fiber [91], rubber wood sawdust [33,92], bamboo [93], oil palm fiber [94], bagasse fly ash, coconut shell [59], and activated cashew shell nut [60].

Table 4 compares various parameters like adsorbent dosage (*m*), time (*t*), pH, etc. for the adsorption of C, R, HQ, and their derivatives onto GAC. Kumar et al. [6] investigated in a batch mode the adsorption behavior of R and C on GAC from a basic salt medium at optimum pH ≈ 7.1, optimum temperature ≈ 30°C, and optimum *m* ≈ 10 g/l for both R and C with 48 h as equilibrium time. However, Mohamed et al. [42] found 5 g/l as the optimum dosage of modified AC with 48 h as equilibrium time for the removal of C and R. Mondal and Balomajumder [68] observed adsorption and simultaneous adsorption biodegradation (SAB) for the removal of R and P from two separate synthetic wastewater over a 2- to 4-mm particle size at 28°C with 10 h as equilibrium time. They conducted the experiments over various adsorbent doses (2 to 12 g/l). They found that the removal of R reaches its constant value at the dose of around 8 g/l. Richard et al. [43] have reported that the adsorption of polyfunctional phenols encountered in olive oil mill wastewater was carried out onto AC at 20°C. They studied the effect of *m* in the range of 2 to 15 g/l and the effect of pH in the range of 5 to 9. They found that 16 g/l and 6.25 are the optimum dosage and pH, respectively, for the removal of C from olive oil mill wastewater. Huang et al. [55] reported the adsorption of C and HQ on modified GAC at pH 7, 9, and 11 with 48 h as equilibrium time. They observed that as the pH of the solution increases from 7 to 11, the amount of the phenols adsorbed diminishes. This indicates that adsorption increases when more protonated species are present in the solution. The adsorption isotherms show a linear behavior at C_o< 200 mg/l, and later, the retained quantity increases slightly at pH 9. The adsorbed species are the protonated and monobasic anions, and due to the positive charge of the carbon surface, attractive and dispersive electrostatic forces begin to interact between the surface and the anion, favoring their accumulation. At pH 11, the surface is loaded negatively, the pH > pH_{PZC} and the species present are the anionic monohydroxylated phenols. Therefore, the anionic species present in the solution repulse the surface AC. The adsorption occurs when the repulsive electrostatic forces predominate, and the adsorption at low concentrations is less than that at high concentrations [55].

Activated carbon cloth

Generally, utilization of AC can be in the form of powder, granule, and fiber or cloth. Activated carbon cloth (ACC), having very high specific surface area and adsorption capacity, uniform pore size, and mechanical strength, has attracted attention in recent years. Not only the adsorption but also the electrosorption characteristics of ACC when acting as a working electrode material have been investigated on various adsorbates [94-97]. Table 5 compares various parameters like *m*, *t*, pH, etc. for the adsorption of C, R, HQ, and their derivatives onto ACC.

Bayram et al. [69] investigated the removal of C and R from aqueous solutions by adsorption and electrosorption onto high-area ACC with a dosage of 0.9 g/l, with 25°C and 24 h as operating temperature and contact time, respectively. They found that the extent of electrosorption of C was higher than R. It was attributed partly to a higher dipole moment of the former. It was demonstrated that ACC can partly be regenerated by an electrodesorption process. It was predicted that adsorption and electrosorption of C and R resulted mainly from dispersion interactions between surface charges and π electrons on adsorbate molecules. Ayranci and Duman [52] reported the adsorption of P, HQ, *m*-cresol, *p*-cresol, and *p*-nitrophenol from aqueous solutions onto ACC. The effect of ionization on the adsorption of these ionizable phenolic compounds was examined by studying the adsorption from acidic, basic, and natural pH (7.6) solutions at an operating temperature of 30°C.

Carbon nanotubes

Carbon nanotubes (CNTs), because of their chemical and physical properties, attract wide applications such as in the field of polymer composites [98], field emissions [93], energy storage [99], and sensors [100]. The large specific surface area of CNTs makes them suitable for the adsorption of gas [101], metal ions [102,103], and organic compounds [104]. Recent studies suggest that CNTs can serve as good adsorbent for a variety of gas and liquid molecules, like hydrogen, butane, dioxins, dichlorobenzene, and heavy metal ions [105-110]. The considerable attentions about CNTs depend on their unique structural and mechanical properties, high thermal stability, and large specific surface area [62,90,111].

Carbon nanomaterials include fullerenes, single-walled carbon nanotubes (SWCNTs), and multi-walled carbon nanotubes (MWCNTs). The monomer structure of fullerenes is a closed graphite ball, while those of CNTs are rolled-up graphite sheets forming a coaxial tube. A single rolled-up graphite sheet forms the SWCNT structure, while several rolled-up graphite sheets form the MWCNT structure. They consist of sheets of carbon atoms covalently bonded in hexagonal arrays that are seamlessly rolled into a hollow cylindrical shape [112]. The cylindrical surface geometry of CNTs has been well defined with outer diameters in the range of approximately 1 to 100 nm and length up to several tens of micrometers. Knowledge of environmental risk assessment of both toxic chemicals and CNTs once they are released to the environment [113-116] is also important.

Table 4 Pseudo-first- and second-order kinetic constants for adsorption of C, R, HQ, and their derivatives on various adsorbents

Adsorbent	Adsorbate	pH	T (°C)	C_o (mg/l)	m (g/l)	t (h)	Pseudo-first-order constants			Pseudo-second-order constants		References
							k_f (min^{-1})	q_e (mg/g)	s^2 (%)	q_e (mg/g)	k_s (g/mg min)	
Waste Fe(III) OH	C	6.08	32	10	10	2.7	0.0198	0.42 to 2.11	0.993	0.015 to 0.931	0.16 to 2.73	Namasivaya and Sumithra [54]
NDA-100	C	-	25 to 40	1,000	1	24	0.026 to 0.042	-	0.985	-	-	Sun et al. [55]
AH-1	C	-	25 to 40	1,000	1	24	0.0238 to 0.0395	-	0.988	-	-	Sun et al. [55]
GAC	C	2 to 12 (6)	30	50 to 1,000	1 to 30 (10)	24	0.036 to 0.898	5.17 to 100.32	0.944 to 0.982	5.06 to 99.66	0.024 to 2.60	Suresh et al. [48]
ACC	HQ	7.4	30	-	0.15	1.3	0.0323	-	0.988	-	-	Ayranci and Duman [53]
GAC	HQ	2 to 12 (6.1)	30	50 to 1,000	1 to 30 (10)	24	0.011 to 0.224	4.95 to 49.8	0.930 to 0.988	4.92 to 104.5	0.018 to 9.03	Suresh et al. [49]
HJ-1	C	-	20	100	6	7.5	0.002 to 0.007	-	-	-	0.01 to 0.027	Huang et al. [56]
CTAB-B	C	5 to 12 (≥9.9)	30	88 to 330	5	4.16	0.1858 to 0.7359	-	0.988	-	-	Shakir et al. [59]
ACC	C	6.44	25	-	0.9	24	0.0056	0.004	-	0.006	2.43	Bayram et al. [69]
ACC	C[a]	6.44	25	-	0.9	24	0.0094	0.007	-	**0.007**	2.26	Bayram et al. [69]
NDA-100	R	-	25 to 40	1,000	1	24	0.034 to 0.053	-	0.985 to 0.989	-	-	Sun et al. [55]
AH-1	R	-	25 to 40	1,000	1	24	0.030 to 0.049	-	0.991to 0.995	-	-	Sun et al. [55]
HJ-1	R	-	20	100	6	24	0.0078 to 0.0249				-	Huang et al. [56]
ACC	R	6.55	25	-	0.9	7.5	0.0023	0.0024	-	0.0057	2.81	Bayram et al. [69]
ACC	R[a]	6.55	25	-	0.9	24	0.0069	0.0042	-	0.0064	2.11	Bayram et al. [69]
GAC	R	2 to 12 (6)	30	50 to 1,000	1 to 30 (10)	24	0.036 to 0.773	4.84 to 113.6	0.933 to 0.996	4.73 to 112.2	0.285 to 4.79	Suresh et al. [48]

[a]Obtained from electrosorption data; intraparticle diffusion coefficients D_e (m^2/s) for C and R are 2.33 to 6.14 × 10^{-13} and 3.19 to 3.96 × 10^{-13}, respectively [6]; NDA-100, hypercrosslinked resin without amino groups; AH-1, aminated hypercrosslinked polymers; GAC, granular activated carbon; ACC, activated carbon cloth; HJ-1,hypercrosslinked resin; CTAB-B, cetyltrimethylammonium bromide-modified bentonite; C, catechol; R, resorcinol; HQ, hydroquinone.

Table 5 Freundlich, Langmuir, and other constants for adsorption of C, R, HQ, and their derivatives on various adsorbents

Adsorbent	Adsorbate	*t* (h)	K_F (mg/g)/(mg/l)$^{1/n}$	1/*n*	q_m (mg/g)	K_L (l/mg)	References
GAC	C	48	42.40	0.205	143.47	0.08	Kumar et al. [6]
Waste Fe(III) OH	C	2.33	0.39 to 0.62	0.681	4.04	0.12	Namasivaya and Sumithra [53]
NDA-100	C	24	0.51	0.493	111.43	-	Sun et al. [54]
ACC	HQ	24	350.9	0.371	204.6	0.0028	Ayranci and Duman [52]
CTAB-B	HQ	-	-	-	17 to 22	0.007 to 0.009	Yildiz et al. [51]
GAC	HQ	24	2.59 to 2.69	0.43 to 0.62	102.3 to 135.3	19.8 to 36.3	Suresh et al. [46]
AH-1	C	24	1.015	0.308	171.33	-	Sun et al. [54]
MWCNT	C	10	16.49	0.411	14.20	0.904	Liao et al. [70]
CTAB-B	C	4.16	0.30	0.305, 0.299, 0.277	56.83	0.002	Shakir et al. [58]
CTAB-B	C	4.16	33.02	-	53.41	0.003	Shakir et al. [58]
CTAB-B	C	4.16	0.29	-	49.78	0.031	Shakir et al. [58]
AC	C	48	-	0.134	320	0.095	Richard et al. [43]
ACC	C	24	0.015	0.23	0.0013	0.000007	Bayram et al. [69]
CZ	C	48	-	-	2.202	-	Mohamed et al. [42]
TiO_2	C	2	-	0.337	11.172	0.0589	Arana et al. [57]
TiO_2-H_2O_2	C	2	-	0.38	11.140	0.0150	Arana et al. [57]
HJ-1	C	7.5	1.40 to 2.55	0.574 to 0.630	125 to 133	0.003 to 0.004	Huang et al. [55]
GAC	C	48	1.49	0.81	238.1	0.004	Blanco-Martinez et al. [44]
GAC	C	24	0.59 to 1.12	52.8 to 62.7	1.09 to 1.41	492.8 to 749.1	Suresh et al. [46]
GAC	R	48	34.83	0.226	142.8	0.047	Kumar et al. [6]
CZ	R	48	-	-	0.880	-	Mohamed et al. [42]
GAC	R	10	34.52	0.236	140.72	0.046	Mondal and Balomajumder [68]
GAC-SAB	R	10	23.43	0.877	49.75	0.074	Mondal and Balomajumder [68]
TiO2	R	2	-	0.703	7.870	0.0089	Arana et al. [57]
TiO_2-H_2O_2	R	2	-	0.781	7.605	0.0083	Arana et al. [57]
ACC	R	24	0.005 to 0.014	0.23 to 0.25	0.0013 to 0.0021	0.046 to 0.143	Bayram et al. [69]
HJ-1	R	7.5	1.44 to 2.65	0.55 to 0.61	128.2 to 129.9	0.002 to 0.003	Huang et al. [55]
GAC	R	24	0.49 to 0.93	57.2 to 108.9	1.21 to 1.27	229.9 to 391.6	Suresh et al. [46]

GAC, granular activated carbon; NDA-100, hypercrosslinked resin without amino groups; ACC, activated carbon cloth; CTAB-B, cetyltrimethylammonium bromide-modified bentonite; AC, activated carbon; AH-1, aminated hypercrosslinked polymers; MWCNT, multi-walled carbon nanotube; HJ-1,hypercrosslinked resin; SAB, simultaneous adsorption biodegradation; C, catechol; R, resorcinol; HQ, hydroquinone.

Table 5 compares various parameters like *m*, *t*, pH, etc. for the adsorption of C, R, HQ, and their derivatives onto GAC by CNTs.

Liao et al. [70] studied the adsorption of P, C, R, HQ, and pyrogallol onto untreated MWCNTs and HNO_3-treated MWCNTs. The uptake of R was 19.7 mg/ml (60% of total amounts adsorbed) within 1 min, when the total contact time was 660 min. Liao et al. [70] have suggested that the adsorption mechanism of MWCNTs is not totally the same as that of AC. The adsorption onto AC is dependent on the porous structure, so it takes time for adsorbates to diffuse through pores [117]. It is confirmed by the adsorption of N_2 to CNTs that most available spaces of CNTs for adsorption are the cylindrical external surfaces, neither the inner cavities nor the inner-wall spacings [116].

Liao et al. [70] observed a pH < 6. There is a slight increase in the uptake of R with the decrease of pH because the solubility of R is dependent on pH; due to the weak acidity of R, the solubility of R decreases with the decrease of pH. The adsorption onto AC suggested that the uptake of phenolic derivatives was inversely proportional to solubility [118]; thus, the uptake of R on MWCNTs increased with the decrease of pH. Lin and

Xing [75] have conducted experiments for the adsorption of P, C, and pyrogallol onto CNTs over a range of pH. It was found that the final solution pH (4.0 to 6.5) was lower than the pK_a of the phenolics and close to the point of zero charge (pH_{PZC}) of CNTs, which suggested that phenolics used for adsorption should be primarily in neutral form. Therefore, electrostatic interaction might not greatly influence the sorption of phenolics to CNTs in the weakly acidic environment. Sorption was found to increase sharply from P to C and then to pyrogallol, while their hydrophobicity decreased. Thus, the hydrophobic interaction can be insignificant as a major influencing factor regulating the sorption of these phenolics to CNTs.

Liao et al. [70] found that the acid-treated MWCNTs showed decreased adsorption capacity for R compared to untreated MWCNTs because of the increased electrostatic repulsion between carboxylic groups weakening the π-π interaction and water adsorption. The uptake of the adsorbates increased with the increase in the number of hydroxyl groups and their location in *meta*-position on the aromatic ring. Four possible solute-sorbent interactions are possible for the adsorption of phenolics onto CNTs as follows: (a) hydrophobic interaction, (b) electrostatic attraction or repulsion, (c) hydrogen bonding between the -OH and the tube surface -OH or -COOH groups, and (d) -OH substitution-enhanced π-π interactions between the phenolics and the CNTs. The -OH substitution on the phenolics and the hydroxy/carboxylic groups on the CNT surface may form hydrogen bonds; the hydrogen bonds may also form between the surface-adsorbed and dissolved phenolics. However, very low hydrogen and oxygen contents were detected on the CNTs, indicating that hydrogen bonding (if any) might not be significant between the phenolics and the functional groups on the CNTs [70,119,120]. An insignificant effect of hydrogen bonding on the sorption of nitroaromatics to CNTs was also recently reported [121], and hydrogen bond formation may have favorable adsorption at high concentration.

Polymeric resins

Adsorption onto polymeric resins has been widely studied for the treatment of effluents containing phenolic compounds [54,122]. In comparison to classical adsorbents such as silica gel, alumina, and AC, polymeric adsorbents have high chemical stability, easy regeneration ability, excellent selectivity, and longevity. Among them, the commercial resin Amberlite XAD-4 has been considered as one of the best for the removal of phenolic compounds from wastewater [54,123]. Many researchers have made more efforts on the chemical modification of polymeric adsorbents with functional groups such as phenolic hydroxyl, acetyl, benzoyl, and hypercrosslinked polymers [124,125] to improve their adsorption properties by increasing the interactions between adsorbates and adsorbents [54,55,126,127]. Styrene-divinylbenzene matrix has also been used for the removal of hydrophobic organic pollutants [128-130]. Table 4 compares various parameters like *m*, *t*, pH, etc. for the adsorption of C, R, HQ, and their derivatives onto various polymeric resins.

Adsorption of C and R from an aqueous solution onto AH-1, AH-2, AH-3, NDA-100, and HJ-1 resins was reported by Sun et al. [54] and Huang et al. [55]. Sun et al. [54] found that the q_m of C and R onto aminated hypercrosslinked resins AH-1, AH-2, and AH-3 are higher than that onto hypercrosslinked resin NDA-100, and more time was required to reach the equilibrium for higher C_o and the required time for the adsorption of R is shorter than that of C [54,55]. Thus, effects of the solubility and position of the hydroxyl group at the *ortho*-position may probably account for higher adsorbability of C than R onto HJ-1. Specific surface area, micropore structure, and content of tertiary amino groups play a combined role during the adsorption of both compounds onto the aminated hypercrosslinked polymers [54,55]. Generally, the removal of these compounds mainly results from van der Waals interaction between the solute and the sorbent phase [131]. However, many hydrophilic or water-soluble organic compounds cannot be readily removed by those sorbents partly due to the strong solute-water interaction [130].

Organoclays

Over the last few decades, organoclays have gained much importance in the removal of organic pollutants from aqueous solutions. Among the various types of clays, montmorillonite, which is the main constituent of the low-cost and naturally abundant mineral bentonite, possesses several properties that make it very appropriate for organoclay preparation. Montmorillonite has a 2:1 type of layer consisting of one octahedral sheet of alumina inserted in between two silica tetrahedral sheets. In the tetrahedral sheets, Al^{3+} can substitute for Si^{4+}, and in the octahedral sheets, Mg^{2+} or Zn^{2+} can replace Al^{3+}. These isomorphous substitutions in the clay lattice result into a net negative charge on the clay surface [132]. Montmorillonite, with its lower surface charge density, has substitutions mostly in the octahedral sheet [133].

The adsorption properties of the organoclay surfaces may be significantly altered by exchange reactions, thus making the clay more organophilic in nature, and this increases its capability to remove organic pollutants from aqueous solutions [134-136]. Hydrophilic clays can be changed to organophilic clays through organic chemicals [137]. The modified organoclays have acted as a partition medium in the sorption of organic pollutants [138-141]. Table 4 compares various parameters like *m*, *t*, pH, etc. for the adsorption of C, R, HQ, and their derivatives onto organoclays.

The removal of C from aqueous solutions onto cetyltrimethylammonium bromide-modified bentonite (CTAB-B) surfaces was reported by Shakir et al. [58]. They carried out the experiment with a wide range of pH (5 to 12), contact time (1 to 250 min), and concentration (0.8 to 15.3 mmol/l) at 30 ± 1°C. The percentage adsorption is ≤ 11% at pH ≤ 7.5, whereas it increases sharply with pH, attaining about 100% at pH ≥ 9.9. However, the sorption capacities increase with decreasing pH. Yildiz et al. [51] reported the removal of HQ and benzoic acid on synthesized organobentonites (ODTMA-B, HDTMA-B). Zhu et al. [134] and Smith and Grum [135] indicated that the magnitude and mechanism of sorption are functions of the cation-exchange capacity of the clay, the molecular structure of the exchanged organic cation, the extent of cation exchange, and the molecular structure of the solute. Sorption may take place either by partition or by adsorption depending mainly on the characteristics of the exchanged organic cation and the molecular structure of the solute. Adsorption of the large cationic surfactant molecule, CTAB, greatly modifies the nature of the clay surface which may exhibit both hydrophilic and hydrophobic as well as electrostatic properties and van der Waals interaction between the -R group of the surfactant and adsorbate [136-142], and it is also observed that the percentage adsorption increases from 81.5% to approximately 100% as the initial adsorbate concentration is reduced from the initial concentration of 88 mg/l. This observed increase in the percentage adsorption is due to the availability of larger sorbent surface sites for a relatively smaller amount of C at lower C_o [58].

Metal surface

TiO_2

The adsorption of organic compounds onto oxides, hydroxides, and other minerals is known to retard their migration in soils and aquifers and to alter their susceptibility toward chemical and biological transformations [143]. Few literatures on the adsorption of organic compounds with hydroxyl or amino groups as the sole ligand donor groups are available [144]. It has been indicated that certain adsorbates can interact with active centers such as hydroxyl groups or bridging oxygen on TiO_2 surface, resulting in a different catalytic activity [145]. Also, some of them can act as poisons [146]. Table 4 compares various parameters like *m*, *t*, pH, etc. for the adsorption of C, R, HQ, and their derivatives onto TiO_2 surface.

The adsorption removal of phenols was 90% at 1 g/l onto TiO_2 surface. 4-Nitrocatechol adsorbs to a significantly greater extent than 4-nitro-2-aminophenol over the entire pH range at 1 g/l TiO_2 dosage [56]. When pH < pH_{PZC} of the surface, positive surface charge increases with decreasing pH, and when pH > pH_{PZC}, negative surface charge increases with increasing pH [56]. Generally, the adsorption behavior of a number of organic compounds possessing carboxylic acid groups and the effects of pH have been successfully modeled by choosing appropriate stoichiometries and equilibrium constants for adsorption [147,148].

Waste Fe(III)/Cr(III) OH. Adsorption has gained wide acceptance and popularity for the removal of phenolic compounds including industrial solid wastes [149]. Chromium(VI) compounds are used as corrosion inhibitors in cooling water systems in industries. Fe(II), which is generated electrolytically, reduces chromium(VI) in wastewater to Cr(III) under acidic conditions [150]. The Fe(III)/Cr(III) ions, produced in solution, are precipitated as Fe(III)/Cr(III) hydroxide by the use of lime. The resultant sludge is discarded as waste by the industries. Waste Fe(III)/Cr(III) hydroxide has been investigated in the laboratory for the removal of heavy metals [151], dyes, and pesticides [152]. The percentage C removal at equilibrium decreased from 80% to 65% as the C_o increased from 10 to 40 mg/l onto 'waste' Fe(III)/Cr(III) hydroxide [53].

Which (C, R, HQ, and their derivatives) gets more adsorbed by different adsorbents?

Kumar et al. [6] found that C was adsorbed more than R, suggesting that a lesser quantity of AC would be required to remove the same amount of C as compared to R in the individual compounds. The results may differ when both R and C are simultaneously present in solution [6]. Mohamed et al. [42] found that for various carbons, the amount of adsorbate adsorbed follows the order P > HQ > R > C. Mondal and Balomajumder [68] concluded that the efficiency of SAB is more than that of adsorption and the percentage removal of P is more than that of R for both adsorption and SAB because of the increase of specific surface area as well as micropore volume in the SAB system. Kumar et al. [6] concluded that C is adsorbed to a greater extent than R because of the difference in adsorbability which can be explained in terms of the compound's solubility, pH, density, and the presence and position of the hydroxyl group on the aromatic benzene ring [63]. Liao et al. [70] have suggested that the uptake of R and other phenols increased with the increasing number of hydroxyl and its location in *meta*-position on the aromatic ring, which resulted in the highest adsorptive capacity. The solubility of R in water is higher than that of C [153]; thus, it has more affinity towards water, i.e., hydrophilic. A similar result was shown by Sun et al. [54] on different polymer resins as adsorbents. C adsorption is much higher than those of the other phenolics, and its interaction occurs preferentially through the formation of a catecholate monodentate. R and the cresols interact by means of hydrogen bonds through the hydroxyl group, and their adsorption is much lower than that of C [57]. According to the Lewis acid–base theory, the benzene ring of the aromatic-based resin and the tertiary amino groups on it can be viewed as Lewis bases, while phenolic compounds can be viewed as Lewis acids [134]. Therefore, the Lewis acid–base interaction may occur between phenolic compounds and the benzene ring as well as the tertiary amino group of the resins, which leads to the formation of a hydrogen-bonded complex. In addition, the matching of polarity between adsorbent and adsorbate is also an important factor affecting adsorption of phenolic compounds. The tertiary amino nitrogen on the resins has a large dipole moment, and the dipole moment of C is larger than that of R (2.620 and 2.071 D, respectively [64,65,154]); therefore, the interaction between the resins and C is expected to be stronger than that between the resins and R. Furthermore, some researches revealed that the same functional group but at the *ortho*-position greatly enhances the adsorption energy of C [6]. These may be possible reasons for its lower adsorbability. Kumar et al. [6] have also observed that the same functional group but at the *ortho*-position greatly enhances the adsorption energies of these compounds, such as C. However, R was adsorbed more compared to C which may refer to the geometry and molecular size of the adsorbate molecule [42]. The changes induced in the texture of the prepared ACs as a result of H_2SO_4 activation do not seem to have a significant role in the adsorption of phenols (P, C, R, and HQ), although activation with sulfuric acid leads to a continuous increase in both micropore volume and surface area [42].

Kinetics study

The study of the adsorption equilibrium and kinetics is essential to supply the basic information required for the design and operation of adsorption equipment [59]. A mass transfer occurs during the adsorption process; the first step is the solute transfer through the adsorbent external surface film, and the others are the solute fluid diffusion into the pore holes and the adsorbed molecules' migration along the pore surfaces, if it takes place. The former is characterized by the external mass transfer coefficient, and the last ones, by the internal pore and surface diffusivities. Available bulk adsorbate concentration in the liquid phase and adsorbed solute concentration in the solid phase are considered time-dependent.

Various kinetic models, namely pseudo-first-order, pseudo-second-order, and intraparticle diffusion models, have been used to test their validity with the experimental adsorption data for C, R, and HQ onto GAC and other adsorbents. The most commonly used kinetic expressions to explain the solid/liquid adsorption processes are the pseudo-first-order and pseudo-second-order kinetic models [59,155]. The pseudo-second-order expression as proposed by Ho [156] and Srivastava et al. [59] was found to explain the kinetics of most of sorption systems very well for the entire range of sorption period. Weber and Morris had presented an intraparticle diffusion model in 1962 [47,59,157]. Recently, the intraparticle diffusion model was reviewed by Wu et al. [158]. Various models are reported for the removal of C, R, and HQ onto various adsorbents [6,43-45,52-57,68,70,73-75]. Table 5 shows a list of the first- and second-order kinetic constants for the adsorption of C, R, HQ, and their derivatives on various adsorbents.

The first-order kinetic model was fitted for the adsorption of C and R onto different polymers and ACC over a period of 80 and 90 min, respectively, at natural pH [52,54], and it was found by Sun et al. that the k_1 value of R was higher than that of C [54]. Similar results were found by Ayranci and Duman [52] in the rate constants which decreased in the order *p*-nitrophenol > *m*-cresol > *p*-cresol > HQ > P onto ACC. Various researchers have proposed and fitted the first-order kinetic model for the removal of C, R, and other phenols onto different adsorbents [54,58,94,95]. The adsorption and electrosorption of C and R onto ACC follows the pseudo-second-order model for adsorption periods of 1,042 and 1,508 min, and the electrosorption of C was higher than that of R which was attributed partly to the higher dipole moment of the former [69]. The pseudo-second-order model rather than the pseudo-first-order model was fitted by Bayram et al. [69] because of smaller error and the R^2 value was close to 1. Namasivayam and Sumithra [53] and Huang et al. [55] have reported an adsorption dynamic curve which followed the second-order rate kinetics onto metal surface and HJ-1 resin.

Isotherm and thermodynamic study

Adsorption equilibrium measurements are used to determine the maximum or ultimate adsorbed capacity. Adsorption equilibrium is established when the amount of solute being adsorbed onto the adsorbent surface is equal to that being desorbed from the surface to the bulk fluid [59,159]. At this point, the equilibrium solution concentration remains constant. Adsorption equilibrium data are formulated into an isotherm model. Six types of adsorption isotherms exist including types I to VI [59,160].

The various adsorption isotherms were reviewed [161]. Equilibrium isotherms are measured to determine the capacity of the adsorbent for the adsorbate. The equilibrium of adsorption processes is usually dependent on the amounts adsorbed on the fluid phase composition for a constant temperature [162]. To design and optimize separations using adsorption isotherms, knowledge on the powerful concept adsorption isotherms is mandatory [163-167].

An increase in temperature increases the chemical potential of the organic molecules to penetrate through the surface pores of GAC. Also, the mobility of the adsorbate increases with an increase in temperature. Together, these result in the enhancement in the adsorptive capacity of the GAC at higher temperature. An increase in the phenol adsorption capacity of the carbonaceous adsorbents with an increase in temperature has also been reported by other investigators [59,168,169]. The investigators have ascribed different reasons for the endothermic nature of the adsorption of phenolics onto ACs. A rise in adsorption temperature weakens hydrogen bonds formed among water molecules and between water molecules and the solute or the adsorbent [170] and enhances pore diffusion [59,80,171]. Therefore, an increase in temperature favors the dehydration of adsorbate molecules, which makes them more planar and gives them a larger dipolar moment. An increase in planarity provides solute molecules a greater access to the pores of the GAC, while an increase in dipolar moment results in enhanced adsorbent-adsorbate interactions. As a result, the adsorption is found to be endothermic because of the endothermicity of dehydration adsorbate molecules. Therefore, the adsorptive uptake increases with an increase in temperature.

Adsorption isotherm modeling

Various isotherm equations like those of Freundlich, Langmuir, Temkin, Dubinin-Radushkevich, and Redlich-Peterson have been used in the literature to describe the equilibrium characteristics of adsorption for the removal of C, R, HQ, and their derivatives onto different adsorbents. Some researchers already explained the theory, assumption, and equation or models for the removal of phenolic compounds [43,59].

The R^2 values alone are not sufficient in determining the best isotherm model to represent the experimental data because they are generally found to be >0.91 for all the four models. Table 5 lists the Freundlich, Langmuir, and other constants for the adsorption of C, R, HQ, and their derivatives on various adsorbents.

The most important parameter to compare in Table 5 is the q_{max} value of the Langmuir model because this measures the adsorption capacity of AC and others for the adsorbates. By comparing the results for the adsorption capacity of AC for C, R, or HQ, it is seen that the q_{max} value is clearly greater for C at higher temperature (45°C), whereas it is greater for R at lower temperatures (15°C and 30°C).

The adsorption capacity as measured by K_F of the Freundlich model was best fitted for R and C onto GAC as compared to other models [6,69]. The Freundlich model is more suitable for characterizing adsorption than the Langmuir model [55], indicating that the adsorption may be a multilayer process and the adsorbent possesses a heterogeneous nature [119]. However, the adsorption was not a pure monolayer type, and the mixed model was expected for the isotherms of R and HQ as reported by Liao et al. [70].

There were higher adsorption constant (K_L) and average adsorption energy for the adsorption of C than those of R and other phenolic compounds onto Degussa P-25 TiO_2. The Langmuir model was best fitted for the adsorption of C and other phenolic compounds having K_L which was much larger than those of P and R. However, the Langmuir equation is valid up to a concentration limit, which is a characteristic of each phenolic compound [53,57,172].

The adsorption isotherms of C, R, HQ, and pyrogallol have been well fitted by both the Freundlich and Langmuir models [58,70], which means that physical adsorption occurred [173] and the adsorbed amount of phenolic derivatives follows the order R >> C onto MWCNTs [70]. Lu et al. [102] indicated that the adsorption capacity of AC was closely related with the molecular weight and boiling point of the compound. The difference in boiling point of the three phenolic derivatives (HQ, 285.2°C > R, 281°C > C, 245°C) might be the reason for the lowest adsorbability of C. However, the uptake of R is the highest, which might be due to other complex factors and need further exploration. The adsorption capacity of MWCNTs was more than that of the AC treated by Mohamed et al. [42], even though the specific surface area of AC (SSA = 210 m^2/g) was much larger than that of MWCNTs (SSA = 72 m^2/g) in our experiments.

Adsorption isotherms of pyrogallol, C, and P show that the effect of -OH substitution on sorption was not proportional to the number of -OH groups. The higher K_D ratios of pyrogallol/C than C/P at high equilibrium concentrations were due to the possible greater hydrogen-bonding interaction between the surface-adsorbed pyrogallol and the ones dissolved in water; the hydrogen bonding interaction may not be significant between the surface-adsorbed and dissolved C or P molecules due to fewer -OH groups on the molecules. Multilayer sorption of pyrogallol, but not C and P, on the CNTs could be the evidence for the sorption mechanism difference between these phenolics at high concentrations [75].

Adsorption capacity of C was higher than that of R [52,55,69] due to its smaller molecular size. The value of ΔH for C was more negative than that for R which can be explained in terms of the solubility and the polarity of the two adsorbates [55]. However, the adsorption capacity decreases with increase in temperature [54]. The Langmuir isotherm was best fitted for C, and the Freundlich isotherm, for the other compounds [43]. The adsorption behavior of phenolic compounds on the same kind of AC was also studied [42], and it was also concluded that the adsorption capacities are of the same order of magnitude, but the other parameters appear to be strongly dependent upon the chemical groups around the aromatic ring. However, acidic phenols exhibit a more progressive isotherm than the non-acidic species [42,43].

Adsorption thermodynamics

The Gibbs free energy change (ΔG_0) of the adsorption process is related to the equilibrium constant (K_D) by the classical Van't Hoff equation and also related to the entropy change (ΔS_0) and heat of adsorption (ΔH_0) at constant temperature. The equilibrium adsorption constant (K_D) can be related to as follows:

$$\ln K_D = \frac{-\Delta G_0}{RT} = \frac{\Delta S_0}{RT} - \frac{\Delta H_0}{R}\frac{1}{T} \quad (1)$$

where T is the absolute temperature (K), R is the universal gas constant (8.314 × 10^{-3} kJ/mol K), and K_D (=q_e/C_e) is the single point or linear sorption distribution coefficient. Thus, ΔH_0, which is the enthalpy change (kJ/mol), can be determined from the slope of the linear Van't Hoff plot, i.e., ln K_D versus ($1/T$). This ΔH_0 corresponds to the isosteric heat of adsorption ($\Delta H_{st,0}$) with zero surface coverage (i.e., q_e = 0) [174]. K_D at q_e = 0 can be obtained from the intercept of the ln q_e/C_e versus q_e plot [175]. The adsorption ΔH_0, ΔS_0, and ΔG_0 corresponding to different percentages of the adsorbent surface, adsorption free energies, and adsorption entropies are reviewed in Table 6.

Generally, the ΔG_0 value is in the range of −20 to 0 kJ/mol for physisorption and in the range of −400 to −80 kJ/mol for chemisorption [176]. ΔG_0 values were negative, indicating that the adsorption process led to a decrease in ΔG_0 and that the adsorption process is feasible and spontaneous [59,177]. The ΔG_0 values achieved at four different temperatures are negative, indicating a spontaneous physisorption process [58]. The ΔH_0 of the adsorption of organic molecules from an aqueous solution onto AC is usually within the range of 8 to 65 kJ/mol [178]. The positive values of ΔH_0 indicate the endothermic nature of the adsorption process. The ΔH_0 value decreases with increase in the percentage of the adsorbent surface which resulted from the energetic heterogeneity of the adsorbent surface [119]. In physisorption, the bond between adsorbent and adsorbate is the van der Waals interaction, and ΔH_0 is typically in the range of 5 to 10 kJ/mol for liquid-phase adsorption. In the case of chemisorption, a chemical bond is formed between adsorbate molecules and the surface, and the chemisorption energy is, generally, in the range of 30 to 70 kJ/mol [179]. The value of ΔH_0 of C was little more negative than that of R at the same percentage of the adsorbent surface [55]. The initial adsorption enthalpy of C is greater than that of R, displaying that the interaction of the adsorbent with C is a little stronger [54]. The negative adsorption free energies imply a favorable and spontaneous process, and the adsorption free energy is independent on the occupancy percentage of the adsorbent surface. The ΔH_0 indicates exothermic adsorption, and the positive value of ΔS_0, indicates increased randomness at the solid/solution interface with some structural changes in the adsorbates and the adsorbents [57]. The positive ΔS_0 value also corresponds to an increase in the degree of freedom of the adsorbed species [180]. The negative value of ΔS_0 reveals that a more ordered arrangement of the adsorbates is shaped on the adsorbent surface and also a weaker activity of adsorbate molecules on the adsorbents than on the aqueous solution [54].

Conclusions

Chemical contamination of water from a wide range of toxic compounds, in particular aromatic molecules, is a serious environmental problem owing to their potential human toxicity especially C, R, HQ, and their derivatives which appear to be the major organic pollutants globally. They, derived from industrial effluent discharges, present an ongoing and serious threat to human health and to natural water. Most of the researchers concluded that there is higher adsorbability of C than R and HQ because of the effects of solubility and the position of the (−OH) functional group at the *ortho*-position.

The role of AC and other adsorbents in the removal of C, R, HQ, and their derivatives from water and wastewater was discussed. However, it is only able to remove few milligrams of phenols per gram of AC, and there are still some problems encountered in the regeneration process. This makes AC an expensive adsorbent for this purpose. The solid waste can be converted into low-cost adsorbents for the treatment of discharged wastewater; the cost of removal might decrease. Although the organoclays revealed good adsorption capability, they were still non-economic. The ability and efficiency of the adsorption technologies in water treatment depend on the characteristics and functions of adsorbents. Therefore, to transfer the pollutants in promoting the adsorption rate, we can design and prepare some special composite adsorbents with good adsorptive functions.

Table 6 Thermodynamic properties for adsorption of C, R, HQ, and their derivatives on various adsorbents

Adsorbent	Adsorbate	Temperature (°C)	ΔH_0 (kJ/mol)	ΔG_0 (kJ/mol)	ΔS_0 (J/mol K)	References
Waste Fe(III) Cr(III) OH	C	32 to 60	15.43		63.82	Namasivaya and Sumithra [53]
AH-1	C	10	−33.76	−7.64	−92.30	Sun et al. [54]
		25	−33.76	−7.71	−87.42	
		40	−33.76	−6.96	−85.62	
AH-2	C	10	−33.26	−7.90	−89.61	Sun et al. [54]
		25	−33.26	−7.93	−85.00	
		40	−33.26	−7.60	−81.98	
CTAB-B	C	30	−7.3 ± 0.148	−12.446 ± 0.094	16.985 ± 0.467	Shakir et al. [58]
		40	−7.3 ± 0.148	−12.616 ± 0.196	16.985 ± 0.467	
		50	−7.3 ± 0.148	−12.787 ± 0.199	16.985 ± 0.467	
		60	−7.3 ± 0.148	−12.956 ± 0.203	16.985 ± 0.467	
HJ-1	C[a]	20	−22.53	−4.2418	−62.42	Huang et al. [55]
		30	−22.53	−4.2639	−60.28	
		40	−22.53	−4.1272	−58.89	
	C[b]	20	−16.44	−4.2418	−41.63	
		30	−16.44	−4.2639	−40.19	
		40	−16.44	−4.1272	−39.34	
	C[c]	20	−15.19	−4.2418	−37.37	
		30	−15.19	−4.2639	−36.06	
		40	−15.19	−4.1272	−35.34	
GAC	C	15 to 45	17.99	−16.84 to −20.45	121.63	Suresh et al. [46]
NDA-100	R	10	−24.93	−5.38	−69.08	Sun et al. [54]
AH-1	R	10	−31.37	−5.99	−89.68	Sun et al. [54]
		25	−31.37	−5.74	−86.01	
		40	−31.37	−5.99	−81.09	
AH-2	R	10	−31.17	−5.88	−89.36	Sun et al. [54]
		25	−31.17	−5.75	−85.30	
		40	−31.17	−5.25	−82.81	
HJ-1	R[a]	20	−20.02	−4.4591	−53.11	Huang et al. [55]
		30	−20.02	−4.3919	−51.58	
		40	−20.02	−4.2396	−50.42	
	R[b]	20	−14.40	−4.4591	−33.93	
		30	−14.40	−4.3919	−33.03	
		40	−14.40	−4.2396	−32.46	
	R[c]	20	−13.06	−4.4591	−29.35	
		30	−13.06	−4.3919	−28.61	
		40	−13.06	−4.2396	−28.18	
GAC	R	15 to 45	28.32		152.04	Suresh et al. [46]
Modified bentonite	HQ	20	21.1974	−16.5869 to −17.0896	0.1268	Yildiz et al. [51]
		40	8.9448	−18.4797 to −18.3909	0.0873	
GAC	HQ	15 to 45	35.28	−19.69 to −25.39	191.25	Suresh et al. [48]

[a]20%; [b]30%; [c]40% [$(q_e/q_m) \times 100\%$]; AH-1 and AH-2, aminated hypercrosslinked polymers; CTAB-B, cetyltrimethylammonium bromide-modified bentonite; HJ-1, hypercrosslinked resin; GAC, granular activated carbon; NDA-100, hypercrosslinked resin without amino groups; C, catechol; R, resorcinol; HQ, hydroquinone.

Additional file

Additional file 1: Table S1. Hazard ranking of C, R, and HQ [181].

Competing interests
The authors declare that they have no competing interests.

Authors' contributions
SS, VCS, and IMM conceived the concept and procedures of the analysis and drafted the manuscript. All authors read and approved the final manuscript.

Author's information
Dr. SS obtained his Ph.D. in 2010 from the Indian Institute of Technology Roorkee. Currently, he is an assistant professor in the Department of Chemical Engineering, Maulana Azad National Institute of Technology Bhopal. He has more than 20 papers in peer-reviewed journals. His research interests include reactor design, biochemical engineering, and energy and environmental engineering. He received a Young Scientist Award from the Uttarakhand State Council for Science and Technology, Government of Uttarakhand, India, and a Visiting Scientist fellowship from the Jawaharlal Nehru Centre for Advanced Scientific Research, International Centre for Materials Science, Bangalore, India, and was associated with Prof. CNR Rao. Dr. VCS obtained his Ph.D. in 2006 from the Indian Institute of Technology Roorkee. Currently, he is an assistant professor in the Department of Chemical Engineering, Indian Institute of Technology Roorkee. He has published more than 45 papers in peer-reviewed journals and more than 40 papers in conferences/seminars. His major research interests are separation processes, physicochemical treatment of industrial wastes, multicomponent adsorption, electrochemical treatment, etc. He was awarded with the ProSPER.Net-Scopus Young Researcher Award 2010 - First Runner-up Prize for research work on 'Treatment of industrial wastes' and the 'INSA Young Scientist Medal 2012' by the Indian National Science Academy in the Engineering Science category. He has more than 900 citations of research papers with an *h*-index of 15. Prof. IMM obtained his Ph.D. (Chemical Engineering) in 1977 from BHU, India, and finished his postdoctoral research at the University of Hannover, Germany, from 1980 to 1982. Currently, he is the Dean of the Indian Institute of Technology Roorkee, Saharanpur Campus. He served as the Head of the Department of Chemical Engineering for two terms (1991 to 1994 and 1999 to 2002). He has published more than 110 papers in peer-reviewed journals and more than 150 papers in conferences/seminars. His research interests include transport phenomenon, biochemical engineering, environmental engineering, and biomass energy engineering. He has received several awards to his credit and more than 1,150 citations to his research papers with an *h*-index of 17.

Acknowledgements
This study was financially supported by MHRD, Government of India.

Author details
[1]Department of Chemical Engineering, Maulana Azad National Institute of Technology Bhopal, Bhopal 462 051, India. [2]Department of Chemical Engineering, Indian Institute of Technology Roorkee, Roorkee 247 667, India.

References
1. Korbahti, BK, Tanyolac, A: Continuous electrochemical treatment of phenolic wastewater in a tubular reactor. Water Res. **37**, 1505–1514 (2003)
2. Dellinger, B, Pryor, WA, Cueto, R, Squadrito, GL, Hegde, V, Deutsch, WA: Role of free radicals in the toxicity of airborne fine particulate matter. Chem. Res. Toxicol. **14**, 1371–1377 (2001)
3. Merck: The Merck Index, 11th edn. Merk, Rahway (1989)
4. Milligan, PW, Haggblom, MM: Biodegradation of resorcinol and catechol by denitrifying enrichment cultures. Environ. Toxicol. Chem. **17**, 1456–1461 (1998)
5. Hays, MD, Fine, PM, Geron, CD, Kleeman, MJ, Gullett, BK: Open burning of agricultural biomass: physical and chemical properties of particle-phase emissions. Atmos. Environ. **39**, 6747–6764 (2005)
6. Kumar, A, Kumar, S, Kumar, S: Adsorption of resorcinol and catechol on activated carbon: equilibrium and kinetics. Carbon **41**, 3015–3025 (2003)
7. Schweigert, N, Zehnder, AJB, Eggen, RIL: Chemical properties of catechols and their molecular modes of toxic action in cells, from microorganisms to mammals. Environ. Microbiol. **3**, 81–91 (2001)
8. Phutdhawong, W, Chowwanapoonpohn, S, Buddhasukh, D: Electro coagulation and subsequent recovery of phenolic compounds. Anal. Sci. **16**, 1083–1084 (2000)
9. Fiegel, H, Voges, HW, Hamamoto, T, Umemura, S, Iwata, T, Miki, H, Fujita, Y, Buysch, HJ, Garbe, D, Paulus, W: Phenol Derivatives in Ullmann's Encyclopedia of Industrial Chemistry. Wiley, New York (2002)
10. Meyer, J: Article for resorcinol: 10% hydrochloric acid on 1,3-diaminobenzene. Ber. **30**, 2569 (1897)
11. Raff, R, Ettling, BV: Hydroquinone, resorcinol and pyrocatechol. In: Kirk, RE, Othmer, DF (eds.) Encyclopedia of Chemical Technology, 2nd edn, pp. 462–492. Wiley, New York (1966)
12. IARC: Some fumigants, the herbicides 2,4-D and 2,4,5-T, chlorinated dibenzodioxins and miscellaneous industrial chemicals. In: IARC Monographs on the Evaluation of the Carcinogenic Risk of Chemicals to Humans, pp. 155–l75. IARC, Lyon (1977)
13. Zheng, LT, Ryu, GM, Kwon, BM, Lee, WH, Suk, K: Anti-inflammatory effects of catechols in lipopolysaccharide-stimulated microglia cells: inhibition of microglial neurotoxicity. Eur. J. Pharmacol. **588**, 106–113 (2008)
14. Smale, BC, Keil, HL: A biochemical study of the intervarietal resistance of *Pyrus communis* to fire blight. Phytochem. **5**, 1113–1120 (1966)
15. Ahtardjieff, C: Uber das Vorkommen von Arbutin und Gerbstoffen in einheimischen Vertretem der Familie der Ericaceen. Pharmazie. **21**, 59–60 (1966)
16. Sun, YG, Cui, H, Li, YH, Lin, X: Determination of some catechol derivatives by a flow injection electrochemiluminescent inhibition method. Talanta **53**, 661–666 (2000)
17. Wynder, EL, Hoffmann, D: Certain constituents of tobacco products: phenolic compounds and carboxylic acids. In: Tobacco and Tobacco Smoke, pp. 381–410. Academic, New York (1967)
18. Izard, C, Lacharpagne, J, Testa, P: Sur l'activite biologique de divers condensats de fumee de cigarettes, lee par les tests auxiniques. C.R. Acad. Sci. Paris Ser D **262**, 1859–1861 (1966)
19. Lazar, P, Izard, C, Morse-testa, P, Chouroulinkov, I: Interaction entre l'hydroquinone et le condensat de fumee de cigarette dans les tests cutanes i court terme du pouvoir carcinogene. C.R. Acad. Sci. Paris Ser D **274**, 496–499 (1972)
20. Dagon, TJ: Biological treatment of photo processing effluents. J. Water Poll. Control Fed. **45**, 2123–2135 (1973)
21. Umpelev, VL, Kogan, LA, Gagarinova, LM: Thin-layer chromatographic separation of polyhydric phenols in effluents. Zh. Anal. Khim. **29**, 179–180 (1974)
22. Jacquemain, R, Remy, F, Guinchard, C: Etudes et comparaisons des determinations des phenols dans les eaux; application a l'examen dun rejet de papeterie. J. Fr. Hydrol. **16**, 25–31 (1975)
23. Fritz, H, Reineke, W, Schmidt, E: Toxicity of chlorobenzene on *Pseudomonas* sp. strain RHO1, a chlorobenzene-degrading strain. Biodegradation **2**, 165–170 (1992)
24. Capasso, R, Evidente, A, Schivo, L, Orru, G, Marcialis, MA, Cristinzio, G: Antibacterial polyphenols from olive oil mill waste waters. J. Appl. Bacteriol. **79**, 393–398 (1995)
25. Boyd, GE, Adamson, AW, Meyers, LS: The exchange adsorption of ions from aqueous solution by organic zeolites. II Kinetics. J. Am. Chem. Soc. **69**, 2836–2848 (1947)
26. Rahouti, M, Steiman, R, Seigle-Murandi, F, Chritov, LP: Growth of 1044 strains and species of fungi on 7 phenolic lignin model compounds. Chemosphere **38**, 2549–2559 (1999)
27. Prager, JC: Environmental Contaminant Reference Data Book. Van Nostrand Reinhold, New York (1997)
28. Van Duursen, MBM, Sanderson, JT, de Jong, PC, Kraaij, M, van den Berg, M: Phytochemicals inhibit catechol-o-methyltransferase activity in cytosolic fractions from healthy human mammary tissues: implications for catechol estrogen-induced DNA damage. Toxicol. Sci. **81**, 316–324 (2004)
29. Irons, RD, Sawahata, R: Phenols, catechols, and quinones. In: Anders, MW (ed.) Bioactivation of Foreign Compounds, pp. 259–279. Academic, San Diego (1985)
30. Cavalieri, EL, Li, K-M, Balu, N, Saeed, M, Devanesan, P, Higginbotham, S, Zhao, J, Gross, ML, Rogan, EG: Catechol ortho-quinones: the electrophilic

compounds that form depurinating DNA adducts and could initiate cancer and other diseases. Carcinogenesis **23**, 1071–1077 (2002)
31. Bukowska, B, Kowalaska, S: Phenol and catechol induce prehemolytic and hemolytic changes in human erythrocytes. Toxicol. Lett. **152**, 73–84 (2004)
32. Moreno-Castilla, C: Adsorption of organic molecules from aqueous solutions on carbon materials. Carbon **42**, 83–94 (2004)
33. Kumar, A, Kumar, S, Kumar, S: Biodegradation kinetics of phenol and resorcinol using *Pseudomonas putida* MTCC 1194. Biochem. Eng. J. **22**, 151–159 (2005)
34. Stoilova, I, Krastanov, A, Stanchev, V, Daniel, D, Gerginova, M, Alexieva, Z: Biodegradation of high amounts of phenol, resorcinol, 2,4-dichlorophenol and 2,6-dimethoxyphenol by *Aspergillus awamori* cells. Enzyme Microb. Tech. **39**, 1036–1041 (2006)
35. Latkar, M, Swaminathan, K, Chakrabarti, T: Kinetics of anaerobic biodegradation of resorcinol and hydroquinone in upflow fixed film–fixed bed reactors. Bioresour. Technol. **88**, 69–74 (2003)
36. Subramanyam, R, Mishra, IM: Biodegradation of catechol (2-hydroxy phenol) bearing wastewater in an UASB reactor. Chemosphere **69**, 816–824 (2007)
37. Subramanyam, R, Mishra, IM: Co-degradation of resorcinol and catechol in an UASB reactor. Bioresour. Technol. **99**, 4147–4157 (2008)
38. Nasr, B, Abdellatif, G, Canizares, P, Saez, C, Lobato, J, Rodrigo, MA: Electrochemical oxidation of hydroquinone, resorcinol, and catechol on boron doped diamond anodes. Environ. Sci. Technol. **39**, 7234–7239 (2005)
39. Chien, SWC, Chen, HL, Wang, MC, Seshaiah, K: Oxidative degradation and associated mineralization of R, hydroquinone and resorcinol catalyzed by birnessite. Chemosphere **74**, 1125–1133 (2009)
40. Arana, J, Rodriguez, CF, Diaz, OG, Melian, JAH, Pena, JP: Role of Cu in the Cu–TiO_2 photocatalytic degradation of dihydroxybenzenes. Catal. Today **101**, 261–266 (2005)
41. Ahn, MY, Martinez, CE, Archibald, DD, Zimmerman, AR, Bollag, J-M, Dec, J: Transformation of resorcinol in the presence of a laccase and birnessite. Soil Biol. Biochem. **38**, 1015–1020 (2006)
42. Mohamed, FS, Khater, WA, Mostafa, MR: Characterization and phenols sorptive properties of carbons activated by sulphuric acid. Chem. Eng. J. **116**, 47–52 (2006)
43. Richard, D, Delgado, ML, Schweich, D: Adsorption of complex phenolic compounds on activated charcoal: adsorption capacity and isotherms. Chem. Eng. J. **148**, 1–7 (2009)
44. Blanco-Martinez, DA, Giraldo, L, Moreno-Pirajan, JC: Effect of the pH in the adsorption and in the immersion enthalpy of monohydroxylated phenols from aqueous solutions on activated carbons. J. Hazard. Mater. **169**, 291–296 (2009)
45. Richard, D, Delgado, ML, Schweich, D: Adsorption of complex phenolic compounds on active charcoal: breakthrough curves. Chem. Eng. J. **158**, 213–219 (2010)
46. Suresh, S, Srivastava, VC, Mishra, IM: Isotherm, thermodynamics, desorption and disposal study for the adsorption of catechol and resorcinol onto granular activated carbon. J. Chem. Eng. Data **56**, 811–818 (2011)
47. Suresh, S, Srivastava, VC, Mishra, IM: Study of catechol and resorcinol adsorption mechanism through granular activated carbon characterization, pH and kinetic study. Sep. Sci. Technol. **46**, 1750–1766 (2011)
48. Suresh, S, Srivastava, VC, Mishra, IM: Adsorption of hydroquinone in aqueous solution by granular activated carbon. J. Environ. Eng. **137**, 1145–1157 (2011)
49. Suresh, S, Srivastava, VC, Mishra, IM: Adsorptive removal of aniline-phenol and 4-nitrophenol-phenol binaries from aqueous solution by granular activated carbon. Chem. Eng. J. **171**, 997–1003 (2011)
50. Suresh, S, Srivastava, VC, Mishra, IM: Adsorptive removal of aniline by granular activated carbon from binary solutions with catechol and resorcinol. Environ. Technol. **33**, 773–781 (2012)
51. Yildiz, N, Gonulsen, R, Koyuncu, H, Calimli, A: Adsorption of benzoic acid and hydroquinone by organically modified bentonites. Colloid. Surf. A: Physicochem. Eng. Aspects **260**, 87–94 (2005)
52. Ayranci, E, Duman, O: Adsorption behaviors of some phenolic compounds onto high specific area AC cloth. J. Hazard. Mater. B **124**, 125–132 (2005)
53. Namasivayam, C, Sumithra, S: Adsorptive removal of catechol on waste Fe (III)/Cr(III) hydroxide: equilibrium and kinetics study. Ind. Eng. Chem. Res. **43**, 7581–7587 (2004)
54. Sun, Y, Chen, J, Li, A, Liu, F, Zhang, Q: Adsorption of resorcinol and catechol from aqueous solution by aminated hypercrosslinked polymers. React. Funct. Polym. **64**, 63–73 (2005)
55. Huang, J, Huang, K, Yan, C: Application of an easily water-compatible hyper crosslinked polymeric adsorbent for efficient removal of catechol and resorcinol in aqueous solution. J. Hazard. Mater. **167**, 69–74 (2009)
56. Vasudevan, D, Stone, TA: Adsorption of catechols, 2-aminophenols, and 1, 2-phenylenediamines at the metal (hydr) oxide/water interface: effect of ring substituents on the adsorption onto TiO_2. Environ. Sci. Technol. **30**, 1604–1613 (1996)
57. Arana, J, Melian, EP, Lopez, VMR, Alonso, AP, Rodriguez, JMD, Diaz, OG, Pena, JP: Photocatalytic degradation of phenol and phenolic compounds Part I. Adsorption and FTIR study. J. Hazard. Mater. **146**, 520–528 (2007)
58. Shakir, K, Ghoneimy, HF, Elkafrawy, AF, Beheir, SG, Refaat, M: Removal of catechol from aqueous solutions by adsorption onto organophilic-bentonite. J. Hazard. Mater. **150**, 765–773 (2008)
59. Srivastava, VC, Swamy, MM, Mall, ID, Prasad, B, Mishra, IM: Adsorptive removal of phenol by bagasse fly ash and activated carbon: equilibrium, kinetics and thermodynamics. Colloid. Surf. A: Physicochem. Eng. Aspects **272**, 89–104 (2006)
60. Suresh, S, Vijayalakshmi, G, Rajmohan, B, Subbaramaiah, V: Adsorption of benzene vapor onto activated biomass from cashew nut shell: batch and column study. Recent Patents Chem. Eng. **5**(2), 116–133 (2012)
61. Lewis Sr, RJ: Sax's Dangerous Properties of Industrial Materials (11th edn), 728th edn. Wiley, Hoboken (2004)
62. Yang, K, Wu, W, Jing, Q, Zhu, L: Aqueous adsorption of aniline, phenol, and their substitutes by multi-walled carbon nanotubes. Environ. Sci. Technol. **42**, 7931–7936 (2008)
63. International Programme on Chemical Safety (IPCS): International Chemical Safety Cards. http://www.cdc.gov/niosh/ipcsneng/neng0411.html (1994). Accessed 04 July 1997
64. Yogesh, R, Gupta, S, Javiya, S, Paul, P, Basu, S, Singh, K, Ganguly, B, Bhattacharya, A: Studies of performances by the interchanging of the sequence of the photomodified layer in the thin film composite (TFC) membrane. J. Appl. Polymer Sci. **108**, 2611–2616 (2008)
65. Kamlet, MJ, Doherty, RM, Abraham, MH, Marcus, Y, Taft, RW: Linear solvation energy relationships. 46. An improved equation for correlation and prediction of octanol-water partition coefficients of organic nonelectrolytes (including strong hydrogen bond donor solutes). J. Phys. Chem **92**, 5244–5255 (1988)
66. Rajkumar, D, Palanivelu, K, Balasubramanian, N: Combined electrochemical degradation and activated carbon adsorption treatments for wastewater containing mixed phenolic compounds. J. Environ. Eng. Sci. **4**, 1–9 (2005)
67. Rodriguez, E, Encinas, A, Masa, FJ, Beltran, FJ: Influence of resorcinol chemical oxidation on the removal of resulting organic carbon by activated carbon adsorption. Chemosphere **70**, 1366–1374 (2008)
68. Mondal, P, Balomajumder, C: Treatment of resorcinol and phenol bearing wastewater by simultaneous adsorption biodegradation (SAB): optimization of process parameters. Int. J. Chem. React. Eng. **5**, S1 (2007)
69. Bayram, E, Hoda, N, Ayranci, E: Adsorption/electrosorption of catechol and resorcinol onto high area activated carbon cloth. J. Hazard. Mater. **168**, 1459–1466 (2009)
70. Liao, Q, Sun, J, Gao, L: Adsorption of chlorophenols by multiwalled carbon nanotubes treated with HNO_3 and NH_3. Carbon **46**, 553–555 (2008)
71. ISI Web of Science Database.
72. Derbyshire, F, Jagtoyen, M, Andrews, R, Rao, A, Martin-Gullon, I, Grulke, EA: Carbon materials in environmental applications. In: Radovic, LK (ed.) Chemistry and Physics of Carbon, vol. 27. Dekker, New York (2001)
73. Figueiredo, JL, Pereira, MFR, Freitas, MMA, Orfao, JJM: Modification of the surface chemistry of activated carbons. Carbon **37**, 1379–1389 (1999)
74. Suresh, S, Srivastava, VC, Mishra, IM: Studies of adsorption kinetics and regeneration of Aniline, Phenol, 4-Chlorophenol and 4-Nitrophenol by Activated Carbon. Chem Ind. Chem. Eng. Q. doi:10.2298/CICEQ111225054S (2012)
75. Lin, D, Xing, B: Adsorption of phenolic compounds by carbon nanotubes: role of aromaticity and substitution of hydroxyl groups. Environ. Sci. Technol. **42**, 7254–7259 (2008)
76. Lin, S-H, Juang, R-S: Adsorption of phenol and its derivatives from water using synthetic resins and low-cost natural adsorbents: a review. J. Environ. Manage. **90**, 1336–1349 (2009)
77. Dabrowski, A: Adsorption—from theory to practice. Adv. Colloid Interface Sci. **93**, 135–224 (2001)

78. Dobbs, RA, Cohen, JM: Carbon Adsorption Isotherm for Toxic Organics, Tech. Re EPA-600/8-80-023. Environment Protection Agency, Cincinnati (1980)
79. Ozkaya, B: Adsorption and desorption of phenol on activated carbon and a comparison of isotherm models. J. Hazard. Mater. B **129**, 158–163 (2006)
80. Garcia-Araya, J, Beltrn De Heredia, J, Alvarez, P, Masa, FJ: Activated carbon adsorption of some phenolic compounds present in agroindustrial wastewater. Adsorption **9**, 107–115 (2003)
81. Franz, M, Arafat, HA, Pinto, NG: Effect of chemical surface heterogeneity on the adsorption mechanism of dissolved aromatics on activated carbon. Carbon **38**, 1807–1819 (2000)
82. Haghseresht, F, Nouri, S, Lu, GQ: Effects of the solute ionization on the adsorption of aromatic compounds from dilute aqueous solutions by activated carbon. Langmuir **18**, 1574–1579 (2002)
83. Streat, M, Patrick, JW, Camporro Perez, MJ: Sorption of phenol and parachlorophenol from water using conventional and novel ACs. Water Res **29**, 467–472 (1995)
84. Zhang, T, Walawender, WP, Fan, LT, Fan, M, Daugaard, D, Brown, RC: Preparation of activated carbon from forest and agricultural residues through CO_2 activation. Chem. Eng. J. **105**, 53–59 (2004)
85. Rodriguez-Reinoso, F: Controlled Gasification of Carbon and Pore Structure Development. Kluwer, Dordrecht (1991)
86. Rodriguez-Reinoso, F, Molina-Sabio, M, Gonzalez, MT: The use of steam and CO_2 as activating agents in the preparation of activated carbons. Carbon **33**, 15–23 (1995)
87. Tseng, RL, Tseng, SK, Wu, FC: Preparation of high surface area carbons from corncob with KOH etching plus CO_2 gasification for the adsorption of dyes and phenols from water. Colloid. Surf. A: Physicochem. Eng. Aspects **279**, 69–78 (2006)
88. Hameed, BH, Ahmad, AL, Latiff, KNA: Adsorption of basic dye (methylene blue) onto AC prepared from rattan sawdust. Dyes Pigments **75**, 143–149 (2007)
89. Johns, MM, Toles, CA, Marshall, WE: ACs from low-density agricultural waste. US Patent **834051**, 24 (March 2003)
90. Karagozoglu, B, Tasdemir, M, Demirbas, E, Kobya, M: The adsorption of basic dye (Astrazon Blue FGRL) from aqueous solutions onto sepiolite, fly ash and apricot shell AC: kinetic and equilibrium studies. J. Hazard. Mater. **147**, 297–306 (2007)
91. Kumar, S, Varadarajan, PR, Porkodi, K, Subbhuraam, CV: Adsorption of methylene blue onto jute fiber carbon: kinetics and equilibrium studies. J. Colloid Interface Sci. **284**, 78–82 (2005)
92. Kalavathy, MH, Karthikeyan, T, Rajgopal, S, Miranda, LR: Kinetic and isotherm studies of Cu(II) adsorption onto H3PO4-activated rubber wood sawdust. J. Colloid Interface Sci. **292**, 354–362 (2005)
93. Tans, IAW, Hameed, BH, Ahmad, AL: Equilibrium and kinetic studies on basic dye adsorption by oil palm fibre AC. Chem. Eng. J. **127**, 111–119 (2007)
94. Ayranci, E, Conway, BE: Adsorption and electrosorption at high-area carbon felt electrodes for waste-water purification: systems evaluation with inorganic S-containing anion. J. Appl. Electrochem. **31**, 257–266 (2001)
95. Niu, J, Conway, BE: Development of techniques for purification of wastewaters: removal of pyridine from aqueous solution by adsorption at high area AC-cloth electrodes using in situ optical spectrometry. J. Electroanal. Chem. **521**, 16–28 (2002)
96. Han, Y, Quan, X, Chen, S, Zhao, H, Cui, C, Zhao, Y: Electrochemically enhanced adsorption of aniline on activated carbon fibers. Sep. Purf. Technol. **50**, 365–372 (2006)
97. Han, Y, Quan, X, Ruan, X, Zhang, W: Integrated electrochemically enhanced adsorption with electrochemical regeneration for removal of acid orange 7 using activated carbon fibers. Sep. Purif. Technol. **59**, 43–49 (2008)
98. Baughman, RH, de Zakhidov, AA, Heer, WA: Carbon nanotubes-the route toward applications. Science **297**, 787 (2002)
99. Dillon, AC, Jones, KM, Bekkedahl, TA, Kiang, CH, Bethune, DS, Heben, MJ: Storage of hydrogen in single-walled carbon nanotubes. Nature **386**, 377–379 (1997)
100. Kong, J, Franklin, NR, Zhou, C, Chapline, MG, Peng, S, Cho, K, Dai, H: Nanotube molecular wires as chemical sensors. Science **287**, 622–625 (2000)
101. Cinke, M, Li, J, Bauschlicher Jr, CW, Ricca, A, Meyyappan, M: CO_2 adsorption in single-walled carbon nanotubes. Chem. Phys. Lett. **376**, 761–766 (2003)
102. Lu, C, Chiu, H, Liu, C: Removal of zinc(II) from aqueous solution by purified carbon nanotubes: kinetics and equilibrium studies. Ind. Eng. Chem. Res. **45**, 2850–2855 (2006)
103. Wang, HJ, Zhou, AL, Peng, F, Yu, H, Chen, LF: Adsorption characteristic of acidified carbon nanotubes for heavy metal Pb(II) in aqueous solution. Mater. Sci. Eng. A **466**, 201–206 (2007)
104. Lu, C, Chung, Y-L, Chang, K-F: Adsorption of trihalomethanes from water with carbon nanotubes. Water Res. **39**, 1183–1189 (2005)
105. Long, RQ, Yang, RT: Carbon nanotubes as superior sorbent for dioxin removal. J. Am. Chem. Soc. **123**, 2058–2059 (2001)
106. Cai, YQ, Jiang, GB, Liu, JF, Zhou, QX: Multiwalled carbon nanotubes as a solid-phase extraction adsorbent for the determination of bisphenol A, 4-n-nonylphenol, and 4-tert-octylphenol. Anal. Chem. **75**, 2517–2521 (2003)
107. Fagan, SB, Souza Filho, AG, Lima, JOG, Mendes Filho, J, Ferreira, OP, Mazail, IO, Alves, OL, Dresselhaus, MS: 1,2-Dichlorobenzene interacting with carbon nanotubes. Nano. Lett **4**, 1285 (2004)
108. Efremenko, I, Sheintuch, M: Enthalpy and entropy effects in hydrogen adsorption on carbon nanotubes. Langmuir **21**, 6282–6288 (2005)
109. Hilding, JM, Grulke, EA: Heat of adsorption of butane on multiwalled carbon nanotubes. J. Phys. Chem. B **108**, 13688–13695 (2004)
110. Peng, XJ, Luan, ZK, Di, ZC, Zhang, ZG, Zhu, CL: Carbon nanotubes iron oxides magnetic composites as adsorbent for removal of Pb(II) and Cu(II) from water. Carbon **43**, 880–883 (2005)
111. Coleman, JN, Khan, U, Gunko, YK: Mechanical reinforcement of polymers using carbon nanotubes. Adv. Mater. **18**, 689–706 (2006)
112. Masciangioli, T, Zhang, WX: Environmental technologies at the nanoscale. Environ. Sci. Technol. **37**, 102A–108A (2003)
113. Helland, A, Wick, P, Koehler, A, Schmid, K, Som, C: Reviewing the environmental and human health knowledge base of carbon nanotubes. Environ. Health Perspect. **115**, 1125–1131 (2007)
114. Nowack, B, Bucheli, TD: Occurrence, behavior and effects of nanoparticles in the environment. Environ. Pollut. **150**, 5–22 (2007)
115. Yang, K, Zhu, LZ, Xing, BS: Adsorption of polycyclic aromatic hydrocarbons by carbon nanomaterials. Environ. Sci. Technol. **40**, 1855–1861 (2006)
116. Yang, K, Xing, B: Desorption of polycyclic aromatic hydrocarbons from carbon nanomaterials in water. Environ. Pollut. **145**, 529–537 (2007)
117. Peng, X, Li, Y, Luan, Z, Di, Z, Wang, H, Tian, B, Jia, Z: Adsorption of 1,2-dichlorobenzene from water to carbon nanotubes. Chem. Phys. Lett. **376**, 154–158 (2003)
118. Daifullah, AAM, Girgis, BS: Removal of some substituted phenols by activated carbon obtained from agricultural waste. Water Res. **32**, 1169–1177 (1998)
119. Li, HT, Xu, MC, Shi, ZQ, He, BL: Isotherm analysis of phenol adsorption on polymeric adsorbents from nonaqueous solutions. J. Colloid Interface Sci. **271**, 47–54 (2004)
120. Zhu, DQ, Pignatello, JJ: Characterization of aromatic compound sorptive interactions with black carbon (charcoal) assisted by graphite as a model. Environ. Sci. Technol. **39**, 2033–2041 (2005)
121. Chen, W, Duan, L, Zhu, DQ: Adsorption of polar and non-polar organic chemicals to carbon nanotubes. Environ. Sci. Technol. **41**, 8295–8300 (2007)
122. He, BL, Huang, WQ: Ion Exchange and Adsorption Resins. The Science and Education, Shanghai (1992)
123. Davankov, VA, Tsyurupa, MP: Structure and properties of porous hypercrosslinked polystyrene sorbents Styrosorb. Pure Appl. Chem. **61**, 1881 (1989)
124. Davankov, VA, Tsyurupa, MP: Structure and properties of hypercrosslinked polystyrene—the first representative of a new class of polymer networks. React. Polym. **13**, 27–42 (1990)
125. Leon-Gonzalez, ME, Perez-Arribas, LV: Chemically modified polymeric sorbents for sample preconcentration. J. Chromatogr. A. **902**, 3–16 (2003)
126. Li, AM, Zhang, QX, Chen, JL: Adsorption of phenolic compounds on Amberlite XAD-4 and its acetylated derivative MX-4. React. Funct. Polym. **49**, 225–233 (2001)
127. Pan, BC, Xiong, Y, Su, Q, Li, AM, Chen, JL, Zhang, QX: Role of amination of a polymeric adsorbent on phenol adsorption from aqueous solution. Chemosphere **51**, 953 (2003)
128. Cai, JG, Li, AM, Shi, HY: Equilibrium and kinetic studies on the adsorption of aniline compounds from aqueous phase onto bifunctional polymeric adsorbent with sulfuric groups. Chemosphere **61**, 502–509 (2005)
129. Xu, ZY, Zhang, QX, Fang, HP: Applications of porous resin sorbents in industrial wastewater treatment and resource recovery. Crit. Rev. Environ. Sci. Technol. **33**, 363 (2003)

130. Pan, BC, Zhang, WM, Zhang, QX, Xu, ZY, Chen, JL: Application of resin adsorbents in pollution control of toxic compounds in water systems. In: International Symposium on Nanotechnology in Environmental Protection and Pollution (ISNEPP), pp. 18–21 Hongkong (June 2006)
131. He, BL, Ma, JB: Polymeric adsorbents for blood purification. Chem. J. Chin. U. **18**, 1212 (1997)
132. Stumm, W, Morgan, JM: Aquatic Chemistry. Wiley, New York (1981)
133. Giese, RF, Van Oss, CJ: Organophilicity and hydrophobicity of organo-clays. In: Yariv, S, Cross, H (eds.) Organo-Clay Complexes and Interactions, pp. 175–192. Marcel Dekker, New York (2002)
134. Zhu, L, Ren, X, Yu, S: Use of cetyltrimethylammonium bromide-bentonite to remove organic contaminants of varying polar character from water. Environ. Sci. Technol. **32**, 3374–3378 (1998)
135. Smith, JA, Grum, A: Sorption of nonionic organic contaminants to single and dual organication bentonite from water. Environ. Sci. Technol. **29**, 685–692 (1995)
136. Boyd, SA, Mortland, MM, Chiou, CT: Sorption characteristics of organic compounds on hexadecyltrimethylammonium-smectite. Soil Sci. Soc. Am. J. **52**, 652–657 (1988)
137. Dentel, SK, Jamrah, AI, Sparks, DL: Sorption and co-sorption of 1,2,4-trichlorobenzene and tannic acid by organoclays. Water Res. **32**, 3689–3697 (1998)
138. Mortland, MM: Clay-organic complexes and interactions. Adv. Agron. **22**, 75–117 (1970)
139. Janes, WF, Boyd, SA: Clay mineral type and organic compound sorption by hexadecyltrimethyl ammonium-exchanged clays. Soil Sci. Soc. Am. J. **55**, 43–48 (1991)
140. Chern, JM, Chien, YW: Adsorption of nitrophenol onto activated carbon beds: isotherms and breakthrough curves. Water Res. **36**, 647–655 (2002)
141. Yariv, S: Organo-clay complexes and interactions. In: Yariv, S, Cross, H (eds.) Introduction to Organo-Clay Complexes and Interactions, pp. 39–112. Marcel Dekker, New York (2002)
142. Kim, YS, Song, DI, Jecn, YW, Choi, SJ: Adsorption of organic phenols onto hexadecyltrimethylammonium-treated montmorillonite. Sep. Sci. Technol. **31**, 2815–2830 (1996)
143. McCarthy, JF, Zachara, JM: Subsurface transport of contaminants. Environ. Sci. Technol. **23**, 496–502 (1989)
144. Zachara, JM, Resch, CT, Smith, SC: Influence of humic substances on CO_2^+ sorption by a subsurface mineral separate and its mineralogic components. Geochim. Cosmochim. Acta. **58**, 553–566 (1994)
145. Cao, Y, Yang, W, Chen, Y, Du, H, Yue, P: Effect of chemisorbed surface species on the photocatalytic activity of TiO_2 nanoparticulate films. Appl. Surf. Sci. **236**, 223–230 (2004)
146. Piera, E, Ayllon, JA, Domenech, X, Peral, J: TiO_2 deactivation during gas-phase photocatalytic oxidation of ethanol. Catal. Today **76**, 259–270 (2002)
147. Regazzoni, AE, Blesa, MA, Marotao, AJG: Electrophoretic behavior of the hematite/complexing monovalent anion solution interface. J. Colloid Interface Sci. **122**, 315–325 (1988)
148. Rubio, J, Matijevic, E: Interactions of metal hydrous oxides with chelating agents: I -FeOOH-EDTA. J. Colloid Interface Sci. **68**, 408–421 (1979)
149. Jain, AK, Gupta, VK, Jain, S, Suhas, A: Removal of chlorophenols using industrial wastes. Environ. Sci. Technol. **38**, 1195–1200 (2004)
150. Ramakrishnan, SR, Subramanian, G: A continuous process for the purification of cooling tower blow down-water or the effluent containing hexavalent chromium. (1986). Indian Patent 158019
151. Namasivayam, C, Senthilkumar, S: Removal of arsenic (V) from aqueous solution using industrial solid waste: adsorption rates and equilibrium studies. Ind. Eng. Chem. Res. **37**, 4816–4822 (1998)
152. Namasivayam, C, Jayakumar, R, Yamuna, RT: Dye removal from wastewater by adsorption on waste Fe(III)/Cr(III) hydroxide. Waste Mangt. **14**, 643–648 (1994)
153. Perry, RH, Chilton, CH: Chemical Engineers' Handbook, 5th edn. McGraw-Hill, Kogakusha (1973)
154. Chen, GR: In: Chemical Industry Encyclopedia, vol. 1. Chemical Industry, Beijing (1990)
155. Ho, YS, McKay, G: Pseudo-second order model for sorption processes. Process Biochem. **34**, 451–465 (1999)
156. Ho, YS: Adsorption of Heavy Metals from Waste Streams by Peat. PhD Thesis, University of Birmingham, (1995)
157. Weber, WJ, Morris, JC: Kinetics of adsorption on carbon from solution. J. Sanitary Engg. Div. ASCE **89**, 31–60 (1963)
158. Wu, F-C, Tseng, R-L, Juang, R-S: Initial behavior of intraparticle diffusion model used in the description of adsorption kinetics. Chem. Eng. J. **153**, 1–8 (2009)
159. Shapiro, A, Stenby, EH: Multicomponent adsorption approaches to modeling adsorption equilibria. In: Encyclopedia of Surface and Colloid Science, pp. 4180–4189. Taylor & Francis, New York (2006)
160. Giles, CH, Mac Ewan, TH, Nakhwa, SN, Smith, D: Studies in adsorption. Part XI. A system of classification of solution adsorption isotherms, and its use in diagnosis of adsorption mechanisms and in measurement of specific surface area of solids. J. Am. Chem. Soc **11**, 3973–3993 (1960)
161. Nicoud, RM, Seidel-Morgenstern, A: Adsorption isotherms: experimental determination and application to preparative chromatography. Isolat. Purif. **2**, 165–200 (1996)
162. Ruthven, DM: Principles of Adsorption and Adsorption Processes. Wiley, New York (1984)
163. Brougton, DB, Gerhold, CG: Continuous sorption process employing fixed beds of sorbent moving inlets and outlets. (23 May 1961). US Patent 2,985,589
164. Mazzotti, M, Storti, G, Morbidelli, M: Operation of simulated moving bed units for nonlinear chromatographic separations. J. Chromatogr. A. **769**, 3–24 (1997)
165. Migliorini, C, Mazzotti, M, Morbidelli, M: Continuous chromatographic separation through simulated moving beds under linear and nonlinear conditions. J. Chromatogr. A. **827**, 161–173 (1998)
166. Kniep, H, Mann, G, Vogel, C, Seidel-Morgenstern, A: Separation of enantiomers through simulated moving-bed chromatography. Chem. Eng. Technol. **23**, 853–857 (2000)
167. Kaspereit, M, Jandera, P, Skavrada, M, Seidel-Morgenstern, A: Impact of adsorption isotherm parameters on the performance of enantioseparation using simulated moving bed chromatography. J. Chromatogr. A. **944**, 249 (2002)
168. Vijayalakshmi, PR, Raksh, VJ, Thirumaleswara, SGB: Adsorption of phenol, cresol isomers and benzyl alcohol from aqueous solution on activated carbon at 278, 298 and 323 K. J. Chem. Technol. Biotechnol. **71**, 173–179 (1998)
169. Banat, F, Sameer, AA, Leema, AM: Utilization of raw and activated date pits for the removal of phenol from aqueous solutions. Chem. Eng. Technol **27**, 80–86 (2004)
170. Costa, E, Calleja, G, Marijuan, L: Comparative adsorption of phenol, p-nitrophenol and p-hydroxybenzoic acid on activated carbon. Adsorpt. Sci. Technol. **5**, 213–228 (1988)
171. Terzyk, AP: Further insights into the role of carbon surface functionalities in the mechanism of phenol adsorption. J. Colloid Interface Sci. **268**, 301–329 (2003)
172. Arana, J, Rodriguez, JMD, Diaz, OG, Melian, JAH, Rodriguez, CF, Pena, JP: The effect of acetic acid on the photocatalytic degradation of catechol and resorcinol. Appl. Catal. A: Gen. **299**, 274–284 (2006)
173. Bansal, RC, Goyalo, M: Activated Carbon Adsorption. CRC, Boca Raton (2005)
174. Suzuki, M, Fujii, T: Concentration dependence of surface diffusion coefficient of propionic acid in activated carbon particles. AICHE J **28**, 380–385 (1982)
175. Srivastava, VC, Mall, ID, Mishra, IM: Adsorption thermodynamics and isosteric heat of adsorption of toxic metal ions onto bagasse fly ash (BFA) and rice husk ash (RHA). Chem. Eng. J. **132**, 267–278 (2007)
176. Faust, SD, Aly, OM: Adsorption Processes for Water. Treatment Butterwort, Boston (1987)
177. Sabah, E, Cinar, M, Celik, MS: Decolorization of vegetable oils: adsorption mechanism of β-carotene on acid-activated sepiolite. Food Chem. **100**, 1661–1668 (2007)
178. Mattson, J, Mark, H: Activated Carbon: Surface Chemistry and Adsorption from Solution. Marcel Dekker, New York (1971)
179. Murzin, D, Salami, T: Chemical Kinetics. Elsevier, Amsterdam (2005)
180. Raymon, C: Chemistry: Thermodynamic. McGraw-Hill, Boston (1998)
181. Scorecard: Pollution Information Site. http://www.scorecard.org/chemical-profiles (2011). Accessed 16 May 2012

3

Screening and evaluation of innate coagulants for water treatment: a sustainable approach

Saritha Vara

Abstract

Access to safe drinking water is important as a health and development issue at national, regional, and local levels. About one billion people do not have healthy drinking water. More than six million people (about two million children) die because of diarrhea which is caused by polluted water. Developing countries pay a high cost to import chemicals including polyaluminium chloride and alum. This is the reason why these countries need low-cost methods requiring low maintenance and skill. The use of synthetic coagulants is not regarded as suitable due to health and economic considerations. The present study was aimed to investigate the effects of alum as coagulant in conjunction with bean, sago, and chitin as coagulants on the removal of color, turbidity, hardness, and *Escherichia coli* from water. A conventional jar test apparatus was employed for the tests. The study was taken up in three stages, initially with synthetic waters, followed by testing of the efficiency of coagulants individually on surface waters and, lastly, testing of blended coagu ants. The experiment was conducted at three different pH conditions of 6, 7, and 8. The dosages chosen were 0.5, 1, 1.5, and 2 mg/l. The results showed that turbidity decrease provided also a primary *E. coli* reduction. Hardness removal efficiency was observed to be 93% at pH 7 with 1-mg/l concentration by alum, whereas chitin was stable at all the pH ranges showing the highest removal at 1 and 1.5mg/l with pH 7. In conclusion, using natural coagulants results in considerable savings in chemicals and sludge handling cost may be achieved.

Keywords: Alum, Chitin, Sago, Bean, Coagulation, Turbidity

Background

The explosive growth of the world's human population and subsequent water and energy demands have led to an expansion of standing surface water [1]. Nowadays, the concern about contamination of aquatic environments has increased, especially when water is used for human consumption. About one billion people do not have healthy drinking water. More than six million people (about two million children) die because of diarrhea which is caused by polluted water [2,3].

In most of the cases, surface water turbidity is caused by the clay particles, and the color is due to the decayed natural organic matter. Generally, the particles that determine the turbidity are not separated by settling or through traditional filtration. Colloidal suspension stability in surface water is also due to the electric charge of particle surface. Thus, there is great importance in either the development of more sophisticated treatments or the improvement of the current ones [4].

The production of potable water from most raw water sources usually entails the use of a coagulation flocculation stage to remove turbidity in the form of suspended and colloidal material. This process plays a major role in surface water treatment by reducing turbidity, bacteria, algae, color, organic compounds, and clay particles. The presence of suspended particles would clog filters or impair disinfection process, thereby dramatically minimizing the risk of waterborne diseases [5,6].

Many coagulants are widely used in conventional water treatment processes, based on their chemical characteristics. These coagulants are classified into inorganic, synthetic organic polymers, and natural coagulants [4]. Alum has been the most widely used coagulant because of its proven performance, cost effectiveness, relatively easy handling, and availability. Recently, much attention has been drawn on the extensive use of alum. Aluminum is regarded as an important poisoning factor in dialysis

Correspondence: vsjr08@gmail.com
Department of Environmental Studies, GITAM Institute of Science, GITAM University, Visakhapatnam, Andhra Pradesh 530045, India

encephalopathy. Aluminum is one of the factors which might contribute to Alzheimer's disease [7-9]. Alum reaction with water alkalinity reduces water pH and its efficiency in cold water [10,11]. However, some synthetic organic polymers such as acrylamide have neurotoxicity and strong carcinogenic effect [8,12].

In addition, the use of alum salts is inappropriate in some developing countries because of the high costs of imported chemicals and low availability of chemical coagulants [3]. This is the reason why these countries need low-cost methods requiring low maintenance and skill.

For these reasons, and also due to other advantages of natural coagulants/flocculants over chemicals, some countries such as Japan, China, India, and the United States have adopted the use of natural polymers in the treatment of surface water for the production of drinking water [13]. A number of studies have pointed out that the introduction of natural coagulants as a substitute for metal salts may ease the problems associated with chemical coagulants.

Natural macromolecular coagulants are promising and have attracted the attention of many researchers because of their abundant source, low price, multi-purposeness, and biodegradation [11,14,15]. Okra, rice, and chitosan are natural compounds which have been used in turbidity removal [16-18]. The extract of the seeds has been mentioned for drastically reducing the amount of sludge and bacteria in sewage [19].

In view of the above discussion, the present work has been taken up to evaluate the efficiency of various natural coagulants on the physico-chemical contaminant removal of water. To date, most of the research has been concentrated on the coagulant efficiencies in synthetic water, but in this study, we move ahead making an attempt to test the efficiency of the natural coagulants on surface water. The efficiencies of the coagulants as stated by [20] might alter depending on many factors: nature of organic matter, structure, dimension, functional groups, chemical species, and others.

Methods

Natural coagulants and their preparation

Sago is a product prepared from the milk of tapioca root. Its botanical name is '*Manihot esculenta* Crantz *syn. M. utilissima*'. Hyacinth bean with botanical name *Dolichos lablab* is chosen as another coagulant. Both the coagulants were used in the form of powders (starches). Starch consists mainly of a homopolymer of α-D-glucopyranosyl units that comes in two molecular forms, linear and branched. The former is referred to as amylose and the latter as amylopectin [21]. These have the general structure as per [22] (Figure 1).

The third coagulant was chitin ($[C_8H_{13}O_5N]_n$), which is a non-toxic, biodegradable polymer of high molecular weight.

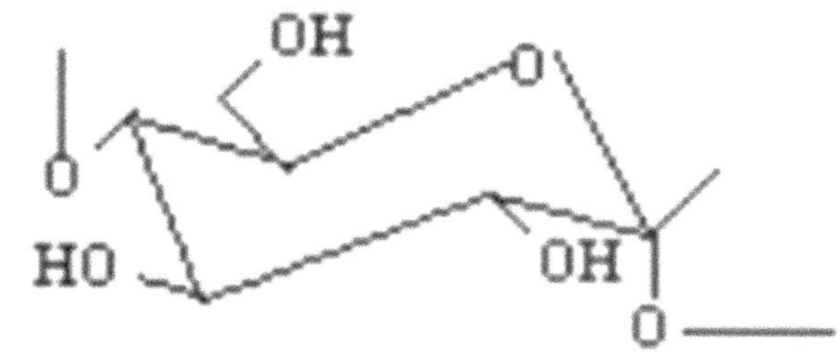

Figure 1 General structure of amylose and amylopectin.

Like cellulose, chitin is a fiber, and in addition, it presents exceptional chemical and biological qualities that can be used in many industrial and medical applications. The two plant originated coagulants were taken in the form of powder or starch. Chitin was commercially procured.

Stage I

The first stage included testing the efficiency of the four coagulants on the synthetic waters. Synthetic waters with turbidity of 70 and 100 nephelometric turbidity units (NTU) were prepared with fuller's earth in the laboratory and were used in this part of the study. The experiment was carried out using a jar test apparatus. The experiments were conducted in duplicates to eliminate any kind of error. Efficiency was evaluated by determination of reduction in turbidity of both the synthetic samples.

Stage II

In the second stage of the experiment, the individual coagulants were evaluated for their efficiency on the surface waters. The water samples for this stage and the preceding stage were collected from the surface reservoir, Mudasarlova, located at a distance of 5 km from the Environmental Monitoring Laboratory, GITAM University, where the experiments were carried out. This is the reservoir which serves as a source of domestic water for the nearby residents.

Care was taken while collecting the samples so that a representative sample is obtained. All samples were collected in sterile plastic containers. The samples were transported to the laboratory, and all the experiments were conducted within a duration of 24 h. The physical parameters like temperature and color were noted at the point of sample collection. The water samples were analyzed for the following parameters pre- and post-treatment with the coagulants (Table 1).

The coagulants were tested at various concentrations like 0.5, 1, 1.5, and 2 mg/l at three pH ranges of 6, 7, and 8.

Stage III

The results obtained from the second stage of the study have encouraged us to further extend the study in terms of blended coagulants. The blending of coagulants was

Table 1 Physico-chemical parameters tested (stage II)

Parameter	Measurement	Unit
Turbidity	Nepheloturbidimeter	NTU
Total hardness	Titrimetric analysis	mg/l
Calcium hardness	Titrimetric analysis	mg/l
Chlorides	Argentometric method	mg/l
E. coli	H_2S strip test	+/−

taken up from the fact that alum was the most widely used coagulant, and hence, it was taken as one part. The remaining combinations were 2, 3, 4, and 5 parts of the natural coagulants, i.e., 1:2, 1:3, 1:4, and 1:5.

Testing of the following parameters was adopted for evaluating the efficiency of the blended coagulants (pre- and post-coagulation) (Table 2). All the analysis has been performed as per the standard methods given by APHA, 2005 [23].

E. coli presence

The *E. coli* bacterial presence and absence were determined in the pre- and post-coagulated water using H_2S strip bottle. The water sample was filled into the bottle and allowed to stand for 24 h at room temperature. After 24 h, the water sample was observed for color change; black color change indicates the presence of *E. coli*.

Results

Coagulant actions onto colloidal particles take place through charge neutralization of negatively charged particles. If charge neutralization is the predominant mechanism, a stochiometric relation can be established between the particles' concentration and coagulant optimal dose.

In the initial stage of the experiment, the coagulants were tested against synthetic turbid samples with 70 and 100 NTU. According to Figure 2a,b, the optimum dosage of alum was observed to be 1mg/l for both the turbid samples, and the optimum pH is observed to be 7.

It is understood from Figure 3a,b that the optimum dosage for chitin as coagulant is 1.5 mg/l (turbidity to 40 NTU) for 100 NTU, whereas not much difference was observed between pH 7 and 8 for both the turbid samples. The optimum pH is observed to be 7 for both 70 and 100 NTU samples.

Table 2 Physico-chemical parameters tested (stage III)

Parameter	Measurement	Unit
Color	Spectrophotometry	Pt-Co scale
Electrical conductivity	Conductivity meter	mmhos
Total hardness	Titrimetric analysis	mg/l
Calcium hardness	Titrimetric analysis	mg/l
Chlorides	Argentometric method	mg/l
E. coli	H_2S strip test	+/−

Figure 4a,b exemplifies the trends of sago on the turbidity removal of the synthetic solutions. It is observed that sago was effective at both 1 and 1.5 mg/l (turbidity reduced to 50 and 45 NTU, respectively) for 100 NTU solution, and the efficiency was stable at pH 7 and 8.

Figure 5a,b illustrates the effect of bean on the synthetic turbid samples and turbidity removal. It is observed that bean was effective at 1mg/l (turbidity reduced to 55 NTU) for 100 NTU solution, and the efficiency was stable at pH 7 and 8.

Implications from the stage 1 experiment articulate that the coagulants are quite stable at the pH ranges tested; hence, in the proceeding experiments, all the three pH ranges were considered. In the second stage of experiment, the environmental samples from the surface water source were collected and tested for the removal of turbidity and other chemical parameters. The dosages were the same as the previous stage. The results are graphically represented as shown in Figures 6,7,8,9.

The turbidity removal efficiencies of the individual coagulants are depicted in Figure 6 wherein there was a broad variation among the pH ranges. The maximum reduction was observed with 1 mg/l (87%) of bean at pH 6 followed by 1 mg/l (82%) sago at the same pH. At pH 7, the maximum efficiency was shown by bean with 1.5 mg/l dosage (85.37%) followed by bean and sago with 1 (82.49%) and 1.5 mg/l (82.49%), respectively. Removal efficiencies of 41.46% and 36.59% were reported by 1 mg/l of bean and sago, respectively, at pH 8. The minimum reductions are not reported as there was a negative competence of the coagulants at different doses and pH variations. It can be observed from the graph that there was an increase in the turbidity of the water at these dosages like with 2 g of chitin the turbidity removal was −19.51. In the entire study, the best results were obtained with total hardness removal wherein no negative competence was reported as shown in Figure 7. The utmost removal was observed with 0.5-mg/l (97.67%) sago at pH 7. At pH 6, it was (90.70%) with 1.5 mg/l of bean. At pH 8, the reduction was (93.02%) with 0.5 mg/l of alum. Apart from these, the general observation was that all the coagulants were effective in an average removal of 65% total hardness at all pH variations and doses. The tracking for the least efficiency has showed chitin at pH 6 with 2-mg/l dose (34.88%).

The calcium hardness removal efficiencies are directly proportional with the total hardness removal; the highest removal was recorded by chitin (93.33%) at pH 7 with 1.5-mg/l dose as shown in Figure 8. Removal of 90% is at pH 8 and 7 with 0.5-mg/l alum and 1-mg/l chitin, respectively. Minimum effectiveness was observed by chitin (6.67%) at pH 6 with 2-mg/l dose. On an average, the removal competence was more than 60% with all coagulants at doses at all the pH conditions.

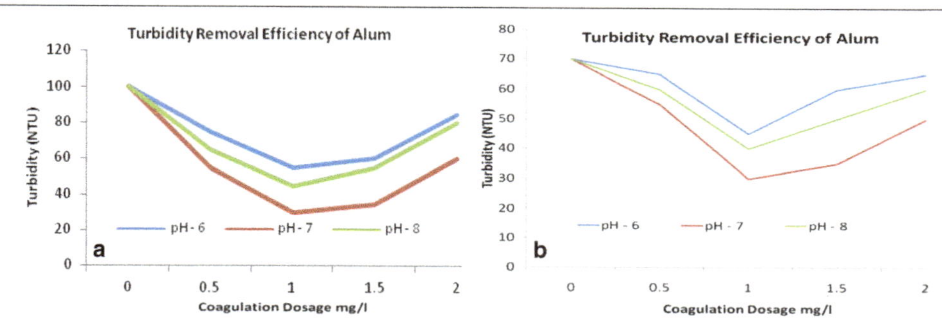

Figure 2 Turbidity removal efficiency of alum with initial turbidities of (a) 100 and (b) 70 NTU.

Figure 9 illustrates the chloride removal efficiency of the coagulants tested. The average competence was observed to be 40%. The maximum competence was noted at pH 7 by chitin (83.64%) at 1.5 mg/l followed by sago (81.82%) at 1 mg/l. Indeed at pH 7, the removal was observed to be superior as a whole. Similarly, pH has shown inferior effectiveness in the amputation of chloride. The remarkable point that was noted is that at pH 8, where the removal was superior, the increase in doses of sago and bean (1.5 and 2 mg/l) has shown a depressing outcome.

With the results obtained from the second stage experimentation, the study was carried forward for the evaluation of blended coagulants. From the literature, it was understood that blended coagulants show improved competence than that of the individual ones.

The regular test of turbidity was substituted with conductivity to establish a relation and test the difference with these parameters. The conductivity diminution was observed to be preeminent at the ratio of 1:2 of all the blended coagulants 26.12%, 26.00%, and 21.35% with alum/bean, alum/chitin, and alum/sago, respectively. The highest reduction was observed with alum/sago at pH 8 with 1:2 ratio (32.28%) (Figure 10).

The total hardness reduction trend of the blended coagulants was recorded as follows: at pH 7, all combinations of alum/bean have resulted in negative competence. Amputation of 100% was observed with alum/chitin and alum/sago at 1:2 and 1:4 and 1:5 doses, respectively (Figure 11). The overall competence of the alum/chitin and alum/sago were registered to be more than 80%. The calcium hardness efficiencies of the blended coagulants were similar to that of the total hardness. The highest removal efficiency was shown by alum/chitin with 1:5 ratio at pH 7 (Figure 12).

As said earlier, the turbidity was replaced by color determination taking into account the fact that turbidity is directly related to the color. pH 7 has been remarkably effective in the highest removal of color from the water. The blended coagulant alum/sago was found to be very effective with 98% to 100% reduction in color at all the ratios of dosage (Figure 13). The blended coagulants alum/chitin and alum/sago were relatively successful at an average rate of 80% decline in the color at almost all ratios of dosage at pH 7 and 8.

Alum/sago blend has a noteworthy effect on the removal of chloride from the water samples in which no negative result was noted. The highest reduction was observed with alum/chitin with dose of 1:5 (85.71%) at pH 7. Indeed, pH 7 can be optimized as perfect pH for this blend as all the ratios of dosages were quite efficient in the removal of chloride (Figure 14).

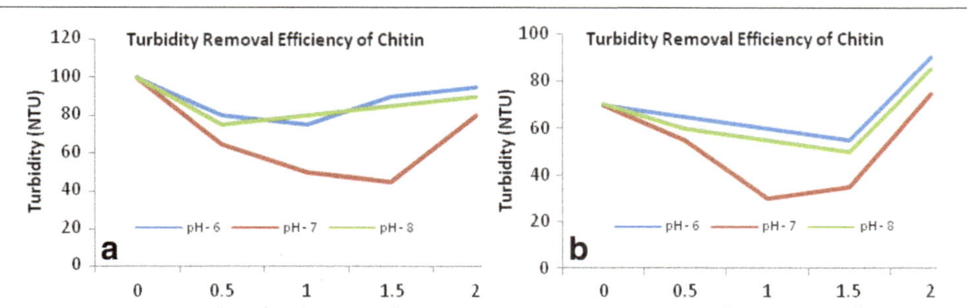

Figure 3 Turbidity removal efficiency of chitin with initial turbidities of (a) 100 and (b) 70 NTU.

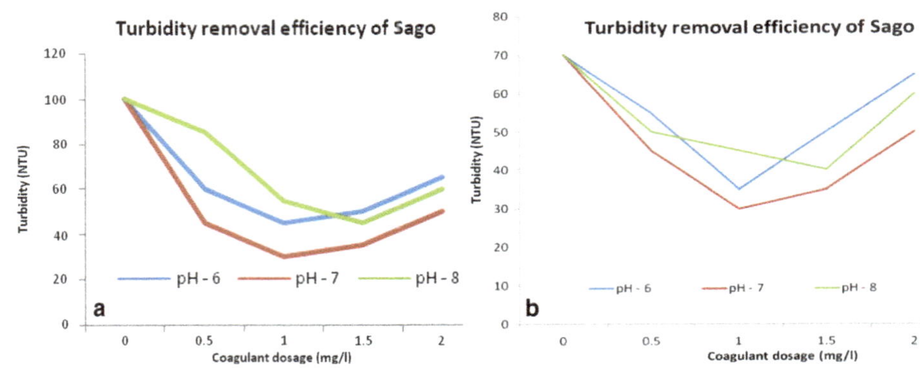

Figure 4 Turbidity removal efficiency of sago with initial turbidities of (a) 100 and (b) 70 NTU.

Discussion

Although many studies have used synthetic water in the experiments, this work chose to use raw water collected directly from the surface source. Therefore, it is important to consider that the natural compounds may cause variations in their composition, which interfere in the treatment process. All those factors are taken into account when evaluating the obtained results.

The characteristics of the superficial water used in this study are observed as that the water used has apparent color, turbidity, solids, and amount of compounds with a relatively high absorption in UV (254 nm). It is noticeable that the water has high turbidity and color.

The effectiveness of alum, commonly used as a coagulant, is severely affected by low or high pH. In optimum conditions, the white flocs were large and rigid and settled well in less than 10 min. This finding is in agreement with other studies at optimum pH [24,25]. The optimum pH was 7 and was similar to the obtained results by Divakaran [26]. At high turbidity, a significant improvement in residual water turbidity was observed. The supernatant was clear after about 20-min settling. Flocs were larger and settling time was lower. The results showed that above optimum dosage, the suspensions showed a tendency to restabilize.

The effectiveness of the chitin in the present study in the removal of various contaminants with varied pH individually and also in blended form can be traced to the explanation from the literature that chitin has been studied as biosorbent to a lesser extent than chitosan; however, the natural greater resistance of the former compared to the last, due to its greater crystallinity, could mean a great advantage. Besides, the possibility to control the degree of acetylation of chitin permits to enhance its adsorption potential by increasing its primary amine group density. Recent studies regarding the production of chitin-based biocomposites and its application as fluoride biosorbents have demonstrated the potential of these materials to be used in continuous adsorption processes. Moreover, these biocomposites could remove many different contaminants, including cations, organic compounds, and anions [27].

Chitosan has high affinity with the residual oil and excellent properties such as biodegradability, hydrophilicity, biocompability, adsorption property, flocculating ability, polyelectrolisity, antibacterial property, and its capacity of regeneration in many applications [28]. It has been used as non-toxic floccules in the treatment of organically polluted wastewater [29].

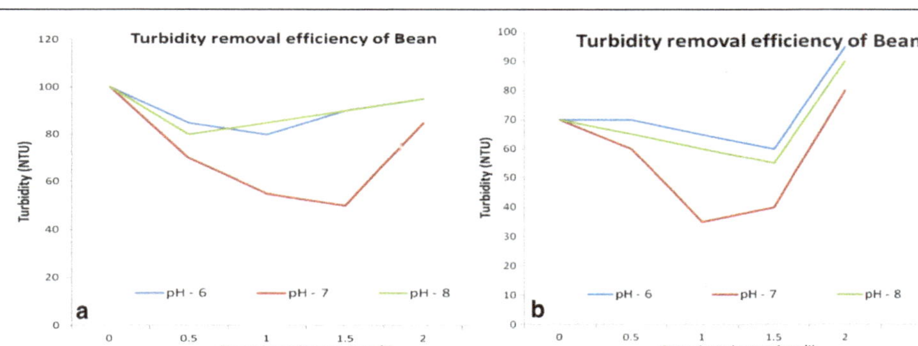

Figure 5 Turbidity removal efficiency of bean with initial turbidities of (a) 100 and (b) 70 NTU.

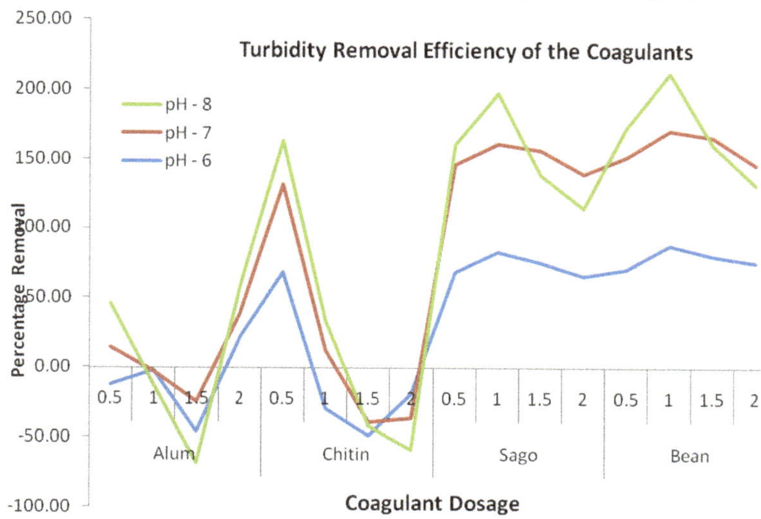

Figure 6 **Turbidity removal efficiency of individual coagulants.**

The effects of coagulation process on hardness are observed for varying levels of hardness, which resulted in significant decrease of hardness removal. The study correlates with the results obtained by [27], wherein they had a maximum hardness removal of 84.3% by chitosan in low turbid water with initial hardness of about 204 mg/l as $CaCO_3$.

Several experiments were carried out to determine the comparative performance of chitosan on *E. coli* in different turbidities. *E. coli* negative is present in the chitin-treated waters in all of the turbidities. The conclusive evidence was found for the negative influence of chitosan on *E. coli.* The regrowth of *E. coli* was not observed in the experiments after 24 h, which was similar to the observations by [27].

As far as sago is considered, the starch was effective both individually and as blended coagulant. Unlike polyaluminium chloride, the efficiency of the natural coagulants is not affected by pH. The pH increased their efficiency, which is one of the advantages of natural coagulants. The principle behind the efficiency of the sago from the literature can be stated as follows: Sago starch is a natural polymer that is categorized as polyelectrolyte and can act as coagulant aid. Coagulant aid can be

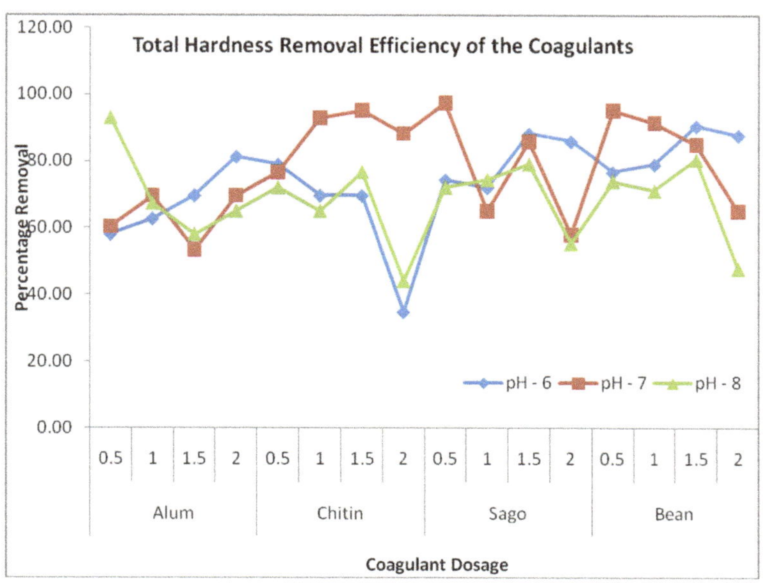

Figure 7 **Total hardness removal efficiency of individual coagulants.**

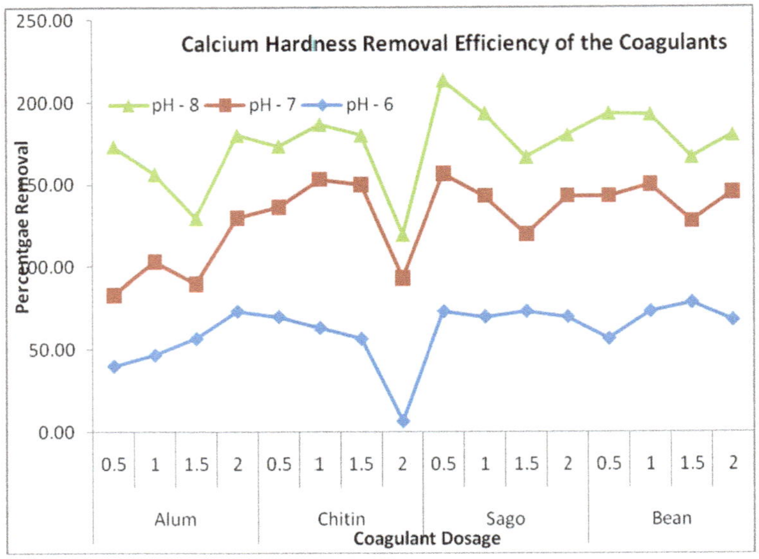

Figure 8 Calcium hardness removal efficiency of individual coagulants.

classified according to the ionization traits, which are the anions, cations, and amphoteric (with dual charges). Bratskaya et al. [30] mentioned that among the three groups, cation polymer is normally used to remove adsorbed negatively charged particles by attracting the adsorbed particles through electrostatic force. They discovered that anion polymer and those non-ionized cannot be used to coagulate negatively charged particles.

The chemical oxygen demand (COD) reduction is influenced by the concentration of sago used; the lower the concentration the better the removal of the COD. Using less than 1.50 g L-1, better COD reduction is observed. At this low concentration, settling time did not influence the COD reduction. Similarly, concentration of sago used at lower than 1.50 g L-1 reduced the turbidity in less than 15 min of settling time. Sago concentration higher than 1.50 g L-1 increased the turbidity; however, settling time has an influence on the turbidity reduction at higher sago concentrations. This pattern is congruent with the COD removal [31].

The sago starch-graft-polyacrylamide (SS-g-PAm) coagulants were found to achieve water turbidity removal up to 96.6%. The results of this study suggest that SS-g-PAm copolymer is a potential coagulant for reducing turbidity during water treatment [32].

At its optimum concentration, *D. lablab* seed powder does not affect the pH of the water. Total and calcium

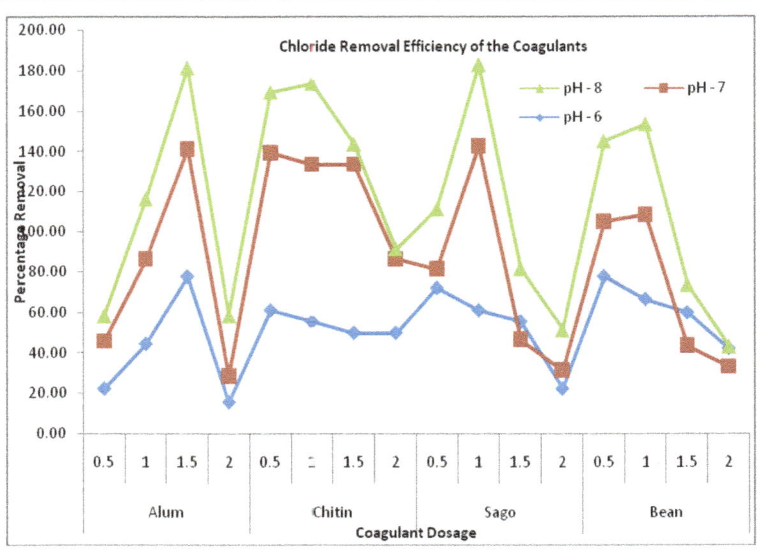

Figure 9 Chloride removal efficiency of coagulants.

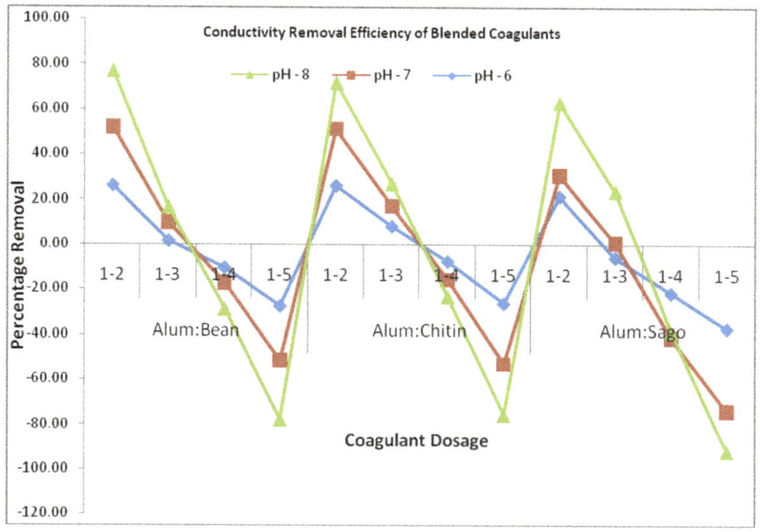

Figure 10 Conductivity removal efficiency of blended coagulants.

hardness remained almost constant and were within acceptable levels according to World Health Organization standards for drinking water. Moreover, coagulation of medium to high turbidity water with *D. lablab* seed powder with the finest grain size reduced turbidity further. The best performance of the finest seed powder could be due to its large total surface area, whereby most of the water-soluble proteins are at the solid–liquid interface during the extraction process as stated by Gassenschmidtet al. [33]. This might have increased the concentration of active coagulation polymer in the extract, which improved the coagulation process. The coagulant extract from seeds has shown antimicrobial activity in the comparative culture test, which was also observed in the study of Tandonet al. [34].

D. lablab demonstrated the best performance with turbid water, in which a turbidity removal efficiency of 87% was observed. The restabilization of destabilized colloidal particles, which was associated with higher residual turbidities, occurred at dosages above the optimum. It is commonly observed that particles are destabilized by small amounts of hydrolyzing metal salts and that optimum destabilization corresponds with neutralization of the particles' charge. Larger amounts of coagulants cause charge reversal so that the particles become positively charged

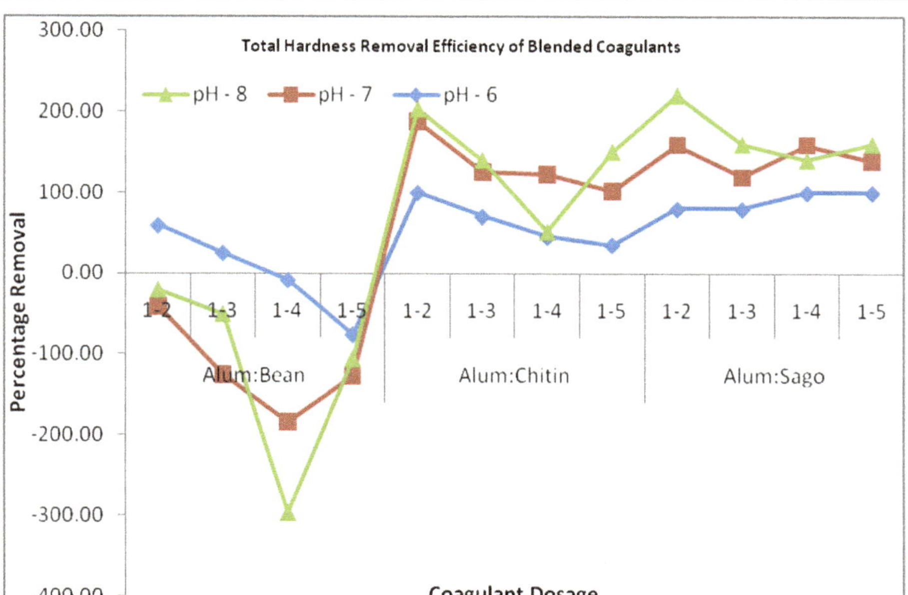

Figure 11 Total hardness removal efficiency of blended coagulants.

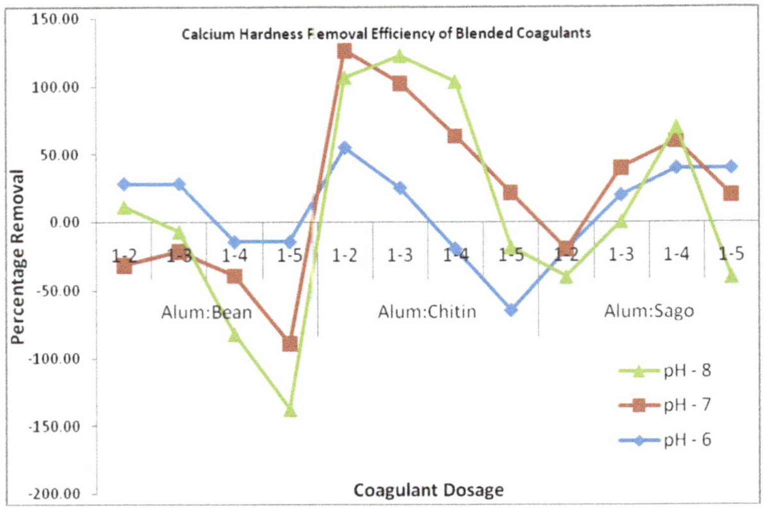

Figure 12 Calcium hardness removal efficiency of blended coagulants.

and, thus, restabilization occurs, which results in elevated turbidity levels [35]. It has also been observed that the reduction in turbidity is associated with significant improvements in bacteriological quality. The effect of natural coagulants on turbidity removal and the antimicrobial properties against microorganisms may render them applicable for simultaneous coagulation and disinfection of water for rural and peri-urban people in developing countries [36].

It is observed that blended coagulants gave utmost efficiency as compared to the traditional alum coagulants. Here in this blending process, we reduce the alum dose up to 80%; thus, we reduce the drawbacks of the alum. Also, we can reduce the cost of the treatment using the natural coagulants instead of the traditional coagulant.

E. coli is the best coliform indicator of fecal contamination from human and animal wastes. *E. coli* presence is more representative of fecal pollution because it is present in higher numbers in fecal material and generally not elsewhere in the environment [37]. Results showed the absence of *E. coli* increases with increasing time. A greater percentage of *E. coli* was eliminated in higher turbidities. The aggregation and, thus, removal of *E. coli* was directly proportional to the concentration of particles in the suspension. Chitosan and other natural coagulants showed antibacterial effects of 2 to 4 log reductions.

Antimicrobial effects of water-insoluble chitin and coagulants were attributed to both its flocculation and bactericidal activities. A bridging mechanism has been reported for bacterial coagulation by chitosan [38]. Especially with

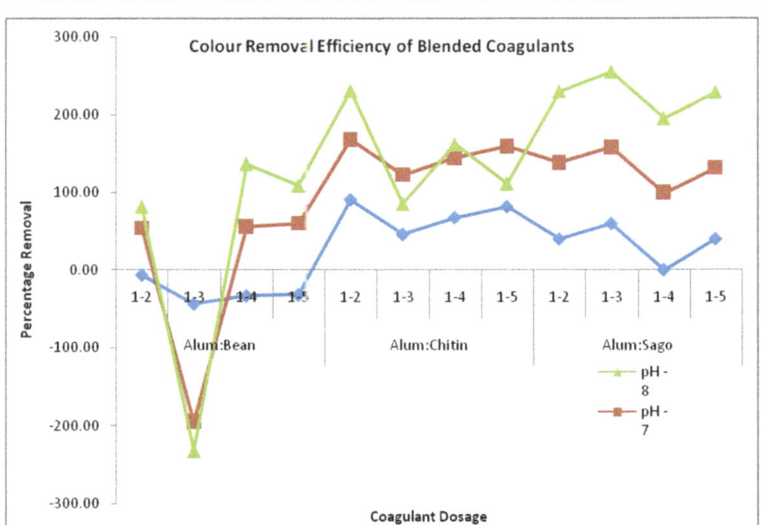

Figure 13 Color removal efficiency of blended coagulants.

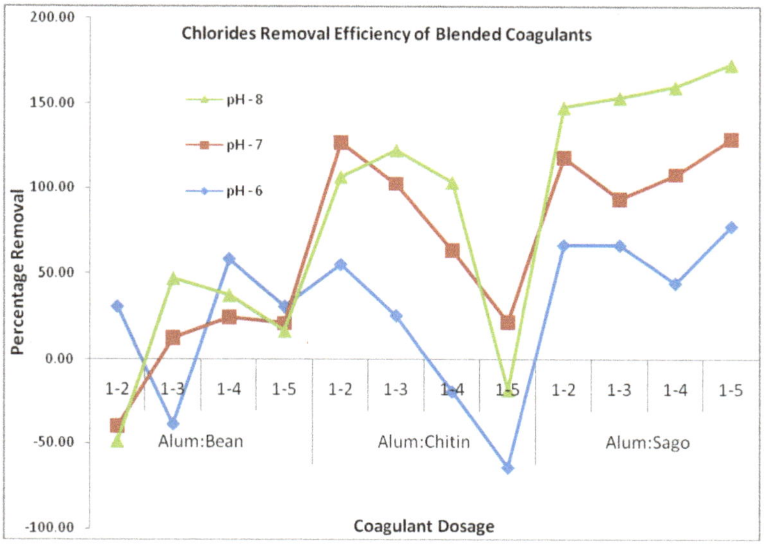

Figure 14 Chloride removal efficiency of blended coagulants.

reference to chitosan, molecules can stack on the microbial cell surface, thereby forming an impervious layer around the cell that blocks the channels, which are crucial for living cells [39]. On the other hand, cell reduction in microorganisms, such as *E. coli*, occurred without noticeable cell aggregation by chitosan.

This indicates that flocculation was not the only mechanism by which microbial reduction occurred. It was found that when samples were stored during 24 h, regrowth of *E. coli* was not observed for all turbidities. It should be noted that the test water contained no nutrient to support regrowth of *E. coli*, and chitosan is not a nutrient source for it. Another experiment was designed to check the effect of alum alone. Regrowth of *E. coli* was not observed for unaided alum after 24 h. The number of *E. coli* after resuspension of sediment reached to the initial numbers after 24 h and showed that it cannot be inactivated by alum. Such findings have been previously reported by Bina [40].

Conclusion

Access to clean and safe drinking water is difficult in rural areas of India. Water is generally available during the rainy season, but it is muddy and full of sediments. Because of a lack of purifying agents, communities drink water that is no doubt contaminated by sediment and human feces. Thus, the use of natural coagulants that are locally available in combination with solar radiation, which is abundant and inexhaustible, provides a solution to the need for clean and safe drinking water in the rural communities of India. Use of this technology can reduce poverty, decrease excess morbidity and mortality from waterborne diseases, and improve overall quality of life in rural areas.

The application of coagulation treatment using natural coagulants on surface water was examined in this study. The surface water was characterized by a high concentration of suspended particles with a high turbidity. At a varied range of pH, the suspended particles easily dissolved and settled along with the coagulants added. Research has been undertaken to evaluate the performance of natural starches of sago flour, bean powder, and chitin to act as coagulants individually and in blended form. In all three cases, the main variable was the dosage of the coagulant. The study shows that natural characteristics of starch and other coagulants can be an efficient coagulant for surface water but would need further study in modifying it to be efficient to the maximum. Thus, it can be concluded that the blended coagulants are the best which give maximum removal efficiency in minimum time.

It is chitin and chitosan which can readily be derivatized by utilizing the reactivity of the primary amino group and the primary and secondary hydroxyl groups to find applications in diversified areas. In this work, an attempt has been made to increase the understanding of the importance and effects of chitin at various doses and pH conditions, upon the chemical and biological properties of water. In view of this, this study will attract the attention of academicians and environmentalists.

Competing interests
The author declares that she has no competing interests.

Authors' information
SV is an assistant professor at the Department of Environmental Studies, GITAM Institute of Science, GITAM University, India.

Acknowledgments
The author thanks GITAM University for providing necessary facilities for carrying out this work.

References

1. Rosenberg, DM, McCully, P, Pringle, CM: Global-scale environmental effects of hydrological alterations: introduction. BioScience **50**, 746–751 (2000)
2. Choi, KJS, Kim, G, Kim, CW, Park, JK: Removal efficiencies of endocrine disrupting chemicals by coagulation/flocculation, ozonation, powdered/granular activated carbon adsorption, and chlorination. Korean J. Chem. Eng **23**, 399–408 (2006)
3. Rook, JJ: The formation of haloforms during chlorination of natural waters. Wat. Treatment & Exam **23**, 234–243 (1974)
4. GrasieleScaramal, M, GeovannaBordini, S, AngélicaMarquetottiSalcedo, V, Letícia, N, Karina Cordeiro, C, Rosângela, B: Study of the effect of saline solution on the extraction of the *Moringa oleifera* seed's active component for water treatment. Water Air Soil Pollut **211**, 409–415 (2010)
5. Mackenzie, LD, Cornwell, DA: Introduction to Environmental Engineering, 2nd edn, pp. 157–163. McGraw Hill, New York (1991)
6. Fatoki, O, Ogunfowokan, A: Effect of coagulant treatment on the metal composition of raw water. Water SA **28**(3), 293–298 (2002)
7. McLachlan, DRC: Aluminum and the risk for Alzheimer's disease. Environmetrics **6**, 233–275 (1995)
8. Ko, YS, Lee, YJ, Nam, SH: Evaluation of a pilot scale dual media biological activated carbon process for drinking water. Korean J. Chem. Eng **24**, 253–260 (2007)
9. Moon, CH, Lee, YJ, Ko, YS, Nam, SH: J. of KSSE. **25**, 227 (2003)
10. Merlet, N, Prevost, M, Merlet, Y, Coallier, J: Enlevement de la matiere organique dans les filters CAB. Sciences de l'Eau **5**, 143–164 (1991)
11. Rook, JJ: Haloforms in drinking water. AWWA **68**, 168–172 (1976)
12. Joret, JC, Levi, Y, Volk, C: Biodegradable dissolved organic carbon (BDOC) content of drinking water and potential regrowth of bacteria. Water Sci Technol **24**, 95–101 (1991)
13. Kawamura, S: Effectiveness of natural polyelectrolytes in water treatment. Journal Awa, Japan **79**(6), 88–91 (1991)
14. Singer, PC, Chang, SD: Correlations between trihalomethanes and total organic halides formed during water treatment. J. AWWA **81**, 61–65 (1989)
15. Edzwald, JK, Becker, WC, Wattier, KL: Surrogate parameters for Monitoring Organic Matter and THM precursors. J. AWWA **77**, 122–132 (1985)
16. Reckhow, DA, Singer, PC, Malcolm, RL: Chlorination of humic materials: byproduct formation and chemical interpretations. Environ Sci Technol **24**, 1655–1664 (1990)
17. Camp, TR, Root, DA: Effect of temperature on rate of floc formation. J. AWWA **1940**, 32 (1913)
18. Kawamura, SJ: AWWA **70**, 328 (1976)
19. Muyibi, SA, Evison, LM: *Moringa oleifera* seeds for softening hard water. Water Res **29**(4), 1099–1105 (1995)
20. Driscoll, CT, Letterman, RD: Factors regulating residual aluminium concentrations in treated waters. Environmetrics **6**, 287–309 (1995)
21. Taggart, P: Starch as an ingredient: manufacture and applications. In: Eliasson, AC (ed.) Starch in Food: Structure, Function and Applications. Woodhead Publishing Limited and CRC Press LLC, New York (2004)
22. Dzulkefly, K, See Yaw, K, Anuar, K, Atan, S, AbdHalim, A: Chemical modification of sago starch by solventless esterification with fatty acid chlorides. The Malaysian Journal of Analytical Sciences **11**(2), 395–399 (2007)
23. American Public Health Association: Standard Methods for the Examination of Water and Wastewater. APHA, Washington, DC (2005)
24. Lin, SD, Evans, RL, Beuscher, DB: Algal Removal by Alum Coagulation. Illinois State Water Survey Division: Urbana, pp. 1–19. Report of Investigation 68. Illinois State Water Survey, Illinois (1971)
25. Ebeling, JM, Sibrell, P, Ogden, S, Summerfelt, S: Evaluation of chemical coagulation–flocculation aids for the removal of suspended solids and phosphorus from intensive recirculating aquaculture effluent discharge. AquacEng **29**, 23–42 (2003)
26. Divakaran, R, Pillai, VN: Flocculation of river silt using chitosan. Water Res **36**, 2414–2418 (2002)
27. Jose Rene, R-M, Vladimir Alonso Escobar, B, Jose Luis, D-R: Chitin based biocomposites for removal of contaminants from water: a case study of fluoride adsorption. In: Elnashar, M (ed.) Biopolymers, pp. 163–180. Sciyo, Croatia (2010)
28. Ravi Kumar, MNV: A review of chitin and chitosan applications. React Funct Polym **46**(1), 1–27 (2000)
29. An, HK, Park, BY, Kim, DS: Crab shell for the removal of heavy metals from aqueous solution. Water Res **35**(15), 3551–3556 (2001)
30. Bratskaya, S, Schwarz, S, Liebert, T, Heinze, T: Starch derivatives of high degree of functionalization 10. Flocculation of kaolin dispersions. *Colloids and Surfaces. A*. Physicochemical and Engineering Aspects **254**, 75–80 (2005)
31. FatehahMohd, O, NikNorulainiNik Abdul, R, Anees, A: COD reduction in semiconductor wastewater by natural and commercialized coagulants using response surface methodology. Water Air Soil Pollut **195**, 345–352 (2008)
32. Qudsieh, IY, Fakhru'l-Razi, A, Kabbashi, NA, Mirghani, MES, Fandi, KG, Alam, MZ, Muyibi, SA, Nasef, MM: Preparation and characterization of a new coagulant based on the sago starch biopolymer and its application in water turbidity removal. J Appl Polym Sci **109**(5), 3140–3147 (2008)
33. Gassenschmidt, U, Jany, KD, Tauscher, B, Niebergall, H: Isolation and characterization of a flocculating protein from *Moringa oleifera* Lam. Biochem Biophys Acta **1243**(3), 477–481 (1995)
34. Tandon, P, Chhibber, S, Reed, RH: Inactivation of *Escherichia coli* and coliform bacteria in traditional brass and earthenware water storage vessels. Ant van Leeuw **88**, 35–48 (2005)
35. Ghebremichael, KA, Gunaratna, KR, Henriksson, H, Brumer, H, Dalhammar, G: A simple purification and activity assay of the coagulant protein from *Moringa oleifera* seed. Water Res **39**(11), 2338–2344 (2005)
36. Özacar, M, Sengil, A: Evaluation of tannin biopolymer as a coagulant aid for coagulation of colloidal particles. Colloids and Surfaces A: Physicochem Eng Aspects **229**(1–3), 85–96 (2003)
37. Tyagi, VK, Chopra, AK, Kazmi, AA, Kumar, A: Alternative microbial indicators of faecalpollution:current perspective. Iran. J. Environ. Health. Sci. Eng **3**(3), 205–216 (2006)
38. Roussy, J, Van Vooren, M, Dempsey, B, Guibal, E: Influence of chitosan characteristics on the coagulation and the flocculation of bentonite suspensions. Water Res **39**, 3247–3258 (2005)
39. Qin, C, Li, H, Xiao, Q, Liu, Y, Zhu, J, Du, Y: Water-solubility of chitosan and its antimicrobial activity. Carbohydr Polym **63**, 367–374 (2006)
40. Bina, B: The use of *Moringa oleifera* seed as natural plant coagulant in removal of clay particles and E.coli. Water & Sewage **14**, 4–12 (1995)

4

Integrated approach for the optimal selection of environmentally friendly drilling systems

Ok-Youn Yu[1*], Zenon Medina-Cetina[2], Seth D Guikema[3], Jean-Louis Briaud[2] and David Burnett[2]

Abstract

There is a pressing need in the energy industry to develop technologies capable of reducing the environmental impact during oil and gas drilling operations. However, these technologies have not been fully integrated into a decision-making system that can reflect a quantitative effort toward this goal. This paper introduces two quantitative decision methods for the selection of environmentally friendly drilling systems. One is based on a multi-attribute utility approach and the other one is based on the analysis of interventions or causal approach. To illustrate the applicability of the proposed methods and to contract their benefits and limitations, a case study is presented using data collected from Green Lake at McFaddin, TX, USA.

Keywords: Comparative analysis, Environmentally friendly drilling, Decision-making, Multi-attribute utility, System selection, Intervention, Causal approach

Background

One of the current goals of the oil and gas industry is to minimize the environmental impact during drilling operations. This is because an effective management of the environmental impacts during drilling operation has proven to lead to a greater access of reserves in environmentally sensitive areas, particularly those classified as 'off-limits' [1-3]. As a consequence, a significant number of Environmentally Friendly Drilling (EFD) technologies continue to emerge, but these have not been integrated into a decision-making method capable of combining them to define an optimal drilling system for specific conditions on a given site. In practice, the major challenge is to select the best combination of EFD technologies based on a set of competing evaluation criteria. In this paper, a 'system' will be defined as a set of EFD technologies.

From an engineering perspective, the civil infrastructure needed to complete a drilling operation may strongly condition its success (e.g., access and maintenance of roads, power supply, water availability and management of residuals, traffic and noise control). This interaction is exacerbated when the drilling operations expand on large areas, and at a rapid pace, threatening the sustainability of the inherent civil infrastructure.

A number of studies have introduced decision support systems for the selection of drilling well locations [4-7]. A few studies on the best practices on the use of EFD technologies are also available such as in the case of drilling waste discharge [8] and in the design of cementing [1]. However, to the best knowledge of the authors, there are few precedents on a quantitative decision-making method for the integral selection of standard drilling systems.

This work aims at introducing a decision-making evaluation protocol to find the optimal EFD system for a given drilling site and also discusses the sensitivity of the inherent input parameters with respect to the expected outcomes. A search algorithm is proposed as the basis of a multi-attribute utility model combined with an exhaustive enumeration of all available technology combinations. This work hypothesizes that optimal decision-making on EFD technology selection can be achieved by an integrated approach, which allows decision-makers to minimize the environmental impact, to maximize the expected profits, to account for the influence of public perception, and, most importantly, to guarantee the operation's safety [9,10].

To support this hypothesis, two competing methods for the selection of EFD systems are presented in this

* Correspondence: yuo@appstate.edu
[1]Appalachian State University, Boone, NC 28608, USA
Full list of author information is available at the end of the article

paper. One follows a simple system selection approach where no formal consideration on the dependencies between the system components was considered (non-causal approach). The other takes into account the same drilling system components as in the previous approach but, in addition, introduces the effect of the dependencies between events taking place during the drilling operation (causal approach). As it is expected, the non-causal approach permits to reduce the computation time for the selection of the optimal drilling system, becoming a very good reference for preliminary analyses. On the other hand, the causal approach requires further knowledge on the dependencies taking place during the decision-making process, which adds computational effort to the optimal selection of EFD systems. Since the decision-making process is inherently conditioned in a drilling sequence (i.e., reservoir assessment dictates the drilling technologies to use, the area where the drilling is located dictates the technologies used to reduce emission, etc.), a comparison between both approaches seems convenient to address the relevance of interventions as the decision-making for the EFD system progresses. Irrespective of the method used, results show consistent support to the hypothesis that an integrated approach for the selection of EFD systems is needed to maximize the benefits of all stakeholders participating in an oil and gas related drilling operation.

To illustrate the benefits and limitations of the proposed methods (non-causal and causal), a case study is presented based on prescribed EFD system selection criteria for a drilling site located in Green Lake at McFaddin, TX, USA. The aim is to help decision-makers select an optimal drilling system for the site by minimizing environmental impact, maximizing profit, and at the same time accounting for perceptions and safety. In addition to showing the need for the use of an integrated decision-making approach on this case, the differences between each approach are discussed. Results of this comparative analysis show the relevance of introducing causal dependencies between system components during the search for the optimal selection of EFD systems.

Methods

Optimal drilling system selection

This section summarizes an optimal drilling system selection procedure for a given site [9]. The proposed system evaluation protocol defines a decision-making process that ensures the selection of an optimal drilling system according to given criteria.

The basis for an EFD system selection includes four main subsystems (Access, Drill Site, Rig, and Operation) and 13 subsets, which have been previously identified through standard EFD operations (see Figure 1). Design of the decision models has been undertaken as part of a comprehensive academic-industry collaboration funded by the US Department of Energy (DOE) and the Research Partnership to Secure Energy for America (RPSEA), which integrates the key drilling phases [2].

Once the subsystems and subsets are defined, available technologies within each subset need to be fully characterized. A drilling technology selection example is also presented in Figure 1, where technologies, indicated within circles for each subset box, represent one possible combination defining a drilling system. Further combinations of technologies are required to evaluate all possible systems and, consequently, to find the optimal drilling system for a given site [11].

For the proposed selection criteria, an attribute is defined as one of the parameters considered in the evaluation of the system (e.g., cost, footprint, emission, perceptions, and safety). Each attribute has an attribute scale used to score the technology on how well it meets the objective for this attribute (e.g., minimization of cost, footprint, and emissions and maximization of positive perceptions and safety value). To evaluate available technologies against each attribute, it is required to introduce attribute scales that explicitly reflect their possible impacts on the system selection process [12]. In this case, nine attributes and the corresponding scales are considered for the selection of the EFD system as shown in Figure 2. These attributes should describe accurately their corresponding technologies. Notice that the proposed attributes are measurable (i.e., dollars, hectares, and decibels) or constructed (i.e., perceptions and safety values) [9].

The attributes considered in this study are as follows:

- Total cost (x_1) = the total expenditure in dollars during the drilling operation.
- Footprint (x_2) = the total used land area in hectares.
- Emissions of air pollutants (x_3) = emissions of three air contaminants (i.e., carbon monoxide (CO), nitrogen oxides (NO_x), and particulate matter (PM)). The relative importance of these contaminants is CO (20%), NO_x (40%), and PM (40%) as shown in Table 1. Table 1 shows an example of how to calculate the air emission score for each technology: First, estimate the three contaminants' real value for each technology in pounds per operating hour; second, in order to get an overall air emission score for each technology, transform each contaminant's score into a non-dimensional score (normalization) between 0 and 1 using the proportional scoring

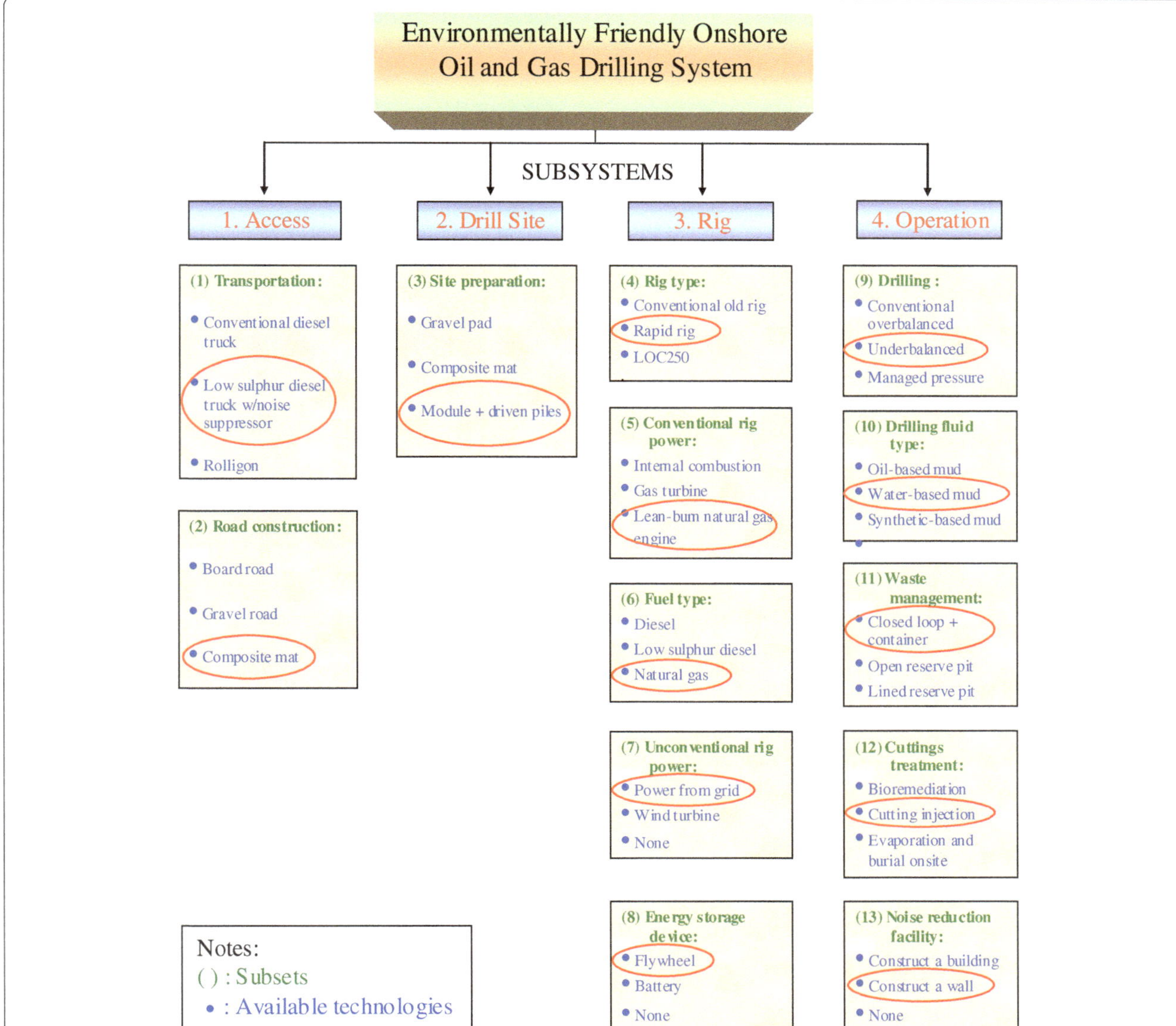

Figure 1 Example of one possible EFD system selection (adapted from [9]).

approach, (x – Worst score) / (Best score – Worst score); and third, calculate the overall air emission score of a technology as $\sum k_i u_i$ (where k_i is a weight factor for each air contaminant and u_i is a non-dimensional score for each contaminant).

- Emissions of solid and liquid pollutants (x_4) =? >thinsp;the ordinal scale as constructed in Table 2.
- Emissions of noise pollutants (x_5) = the 8-h time-weighted average (TWA) sound level given in decibels.
- Perception of government, as regulator, (x_6) = the ordinal scale as constructed in Table 3.
- Perception of industry, as decision maker, (x_7).
- Perception of the general public (x_8).
- Safety value (x_9).
- Notice that the ordinal scales of x_7 through x_9 are similar to that of x_6 [9].

To estimate the impact of available technologies with respect to the proposed nine attributes (i.e., x_1 through x_9), the decision method relies on experts' beliefs and on factual computations generated from available information sources. For instance, 'Composite mat (rent)', a selected technology for subset (2) 'Road construction', is estimated in terms of the required attributes based on the engineering calculations (cost, footprint, and emissions) and experts' beliefs (perceptions and safety) as shown in Table 4. It is noted that attribute scores are not evaluated for the empty cells because

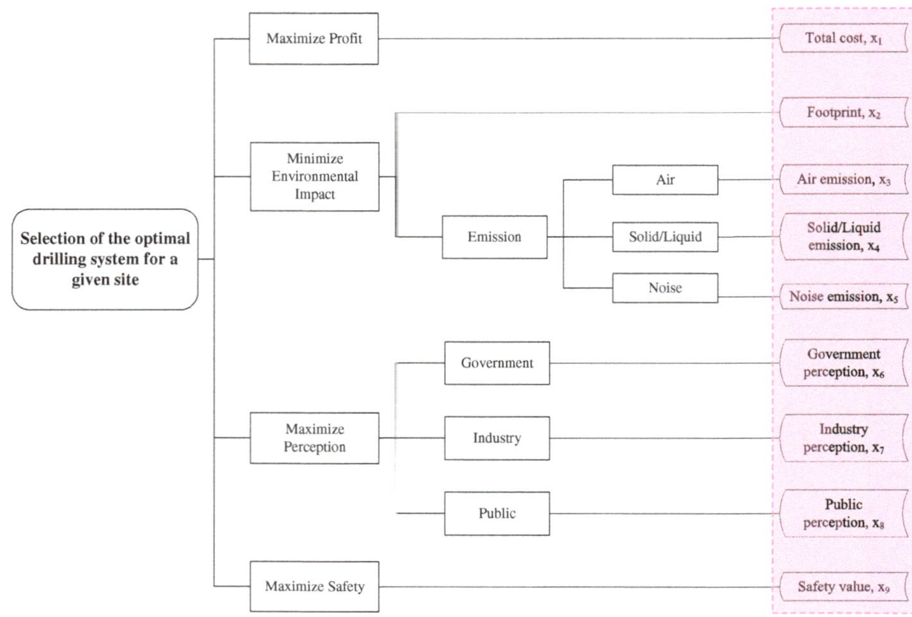

Figure 2 Hierarchy of objectives and attributes ($x_1 \sim x_9$) for a given EFD project.

those attribute scores are not relevant to the particular subsets or because these are already included in technologies within other subsets.

After each technology is evaluated in terms of the attributes (i.e., x_1 through x_9), for each attribute, the overall attribute score of a system is calculated by adding the technology scores of the system or by selecting the minimum technology score of the system. The addition of individual scores is used for attributes such as cost, footprint, and emissions as indicated in Equation 1, while the minimum score is selected for attributes such as perceptions and safety as indicated in Equation 2. Accordingly, the overall score on the ith attribute (X_i) is as follows:

$$X_i = \sum_{n=1}^{N} x_{i,n} y_n \quad \text{for attributes } x_1 \text{ through } x_5 (\text{i.e., } i = 1 \text{ to } 5), \tag{1}$$

$$X_i = \text{Min}\left[x_{i,n} y_n\right] \text{for attributes } x_6 \text{ through } x_9 (\text{i.e., } i = 6 \text{ to } 9), \tag{2}$$

where n is the index for possible technologies, N is the number of possible technologies, i is the index for the attributes, $x_{i,n}$ is the score of the nth technology on the ith attribute, and y_n is a binary decision variable that is one if the nth technology is selected and zero if it is not.

The following constraint is required:

$$\sum_{n=1}^{M} y_n = 1 \text{ for each subset except subsets } (7), (8), \text{ and } (13), \tag{3}$$

where n is the index for possible technologies, M is the number of possible technologies within each subset, and y_n is a binary decision variable.

One technology must be selected for each subset, except for subsets (7), (8), and (13) since subsets (7), (8), and (13) are optional (see Figure 1). Table 4 shows the overall attribute score for each attribute of a system. It is observed from this table that the overall scores of cost (x_1), footprint (x_2), and emissions (x_3 through x_5) are calculated by summing the scores of technologies selected within each subset. The overall scores of perceptions (x_6 through x_8) and safety (x_9), on the other hand, are calculated by choosing the worst score among technologies selected within each subset for a given system because it is suggested that perception and safety values should be included on the systems level, and not on the individual technology level.

Once the overall attribute score for each attribute of a system is calculated in terms of the nine attributes (i.e., x_1 through x_9), for each of these, and in order to homogenize the scores, a utility function (u_i) needs to be introduced to convert the overall dimensional score of a system into a non-dimensional utility value of the

Table 1 Example of air emission score calculation

Technologies	Unit	20% CO	40% NO_x	40% PM	Overall score
Gravel road: diesel truck + dust	g/hp-h	15.5	4	0.1	0.566
	lb/hp-h	0.03418	0.00882	0.00022	
	(lb/h)/unit	10.253	2.646	0.216	
	lb/operation	3,250.280	838.782	68.520	
	U-value	0.000	0.822	0.593	
Composite mat: low-sulphur diesel truck with noise suppressor	g/hp-h	15.5	0.2	0.01	0.976
	lb/hp-h	0.03418	0.00044	0.00002	
	(lb/h)/unit	10.253	0.132	0.007	
	lb/operation	369.117	4.763	0.238	
	U-value	0.886	0.999	0.999	
Internal combustion engine	lb/MWh	6.2	21.8	0.78	0.118
	(lb/h)/unit	6.200	21.800	0.780	
	(lb/h) × portion	6.200	21.800	0.780	
	lb/operation	1,339.200	4,708.800	168.480	
	U-value	0.588	0.000	0.000	
Internal combustion engine with SCR and with noise suppressor	(lb/MWh)	6.2	4.7	0.78	0.431
	(lb/h)/unit	6.200	4.700	0.780	
	(lb/h) × portion	6.200	4.700	0.780	
	lb/operation	1,339.200	1,015.200	168.480	
	U-value	0.588	0.784	0.000	
Lean-burn natural gas engines with noise suppressor	lb/MWh	5	2.2	0.03	0.878
	(lb/h)/unit	5.000	2.200	0.030	
	(lb/h) × portion	5.000	2.200	0.030	
	lb/operation	1,080.000	475.200	6.480	
	U-value	0.668	0.899	0.962	
Power from grid	lb/MWh	0	0	0	1.000
	(lb/h)/unit	0.000	0.000	0.000	
	(lb/h) × portion	0.000	0.000	0.000	
	lb/operation	0.000	0.000	0.000	
	U-value	1.000	1.000	1.000	

SCR, *selective catalytic reduction.*

Table 2 Proposed attribute scale for solid and liquid emissions

Waste management technologies	Cuttings treatment	Solid/liquid emission score
Closed loop	Cuttings injection	1.00
-	Bioremediation, composting, *in situ* vitrification, land spreading, plasma arc, microwave technology	0.75
Lined reserve pit	Thermal desorption	0.50
-	Chemical fixation and solidification	0.25
Open reserve pit	Evaporation and burial on-site	0.00

Table 3 Proposed attribute scale for government perception

Scale	Description	Perception score
Strong support	All parties will encourage its use and are willing to appropriate funds for the cause	1.00
Moderate support	There is interest from a majority. Its use will be encouraged, but funds will not be appropriated	0.75
Neutrality	All parties are indifferent. There is no resistance, but there is also no help	0.50
Moderate opposition	Some resistance from the majority. Its use may be discouraged, but fines or restrictions will not be imposed	0.25
Strong opposition	Strong resistance to its use from all parties. Restrictions or fines will be set up to eliminate this option	0.00

system, which ranges between 0 and 1. This allows the decision-makers to make the overall attribute score for each attribute uniform and comparable. Although Keeney and Raiffa [12] introduced the distinction between a value function and a utility function, the authors believe that this is still largely a personal choice within the decision analysis community. Herein, it is called a utility function since it is based on assessed preference parameters of a decision-maker and it represents their utility. Also, there are different approaches to develop single-attribute functions (i.e., linear and nonlinear). The proportional scoring approach (i.e., linear approach) is mainly used in this paper because of the limited expert assessment. This can and should be revisited for future applications as needed based on interactions with EFD subject matter experts, mainly due to changes in technology. A general formula for the proportional scoring approach is given by [11]:

$$u_i(X_i) = \frac{X_i - \text{Worst score}}{\text{Best score} - \text{Worst score}}, \tag{4}$$

where X_i is the overall score on the ith attribute of a system.

Figure 3 shows an example of utility function curves used in this study. As can be seen in Figure 3, the maximum and minimum values of each attribute need to be

Table 4 Example of a system matrix

Selected technologies in each subset	Weights (∑=100% ∴ O.K!)								
	40%	25%	5%	5%	5%	5%	5%	5%	5%
	Total cost, x_1, ($)	Footprint, x_2, (ha)	Emissions (x_3 ~ x_5)			Perceptions (x_6 ~ x_8)			Safety value (x_9)
			Air	Solid and liquid	Noise (TWA)	Government	Industry	Public	
(1) Transportation: conventional diesel truck						0.250	1.000	0.250	0.750
(2) Road construction: composite mat (rent)	132,000	0.615	0.964		82.870	1.000	0.500	1.000	1.000
(3) Site preparation: composite mat (rent)	90,000	0.420	0.976		79.945	0.750	0.750	0.750	1.000
(4) Rig type: LOC250 (CWD)	182,000		0.977		77.458	1.000	0.500	1.000	1.000
(5) Rig power (conventional): internal combustion engine with SCR and with noise suppressor	73,369		0.574		90.259	0.750	0.750	0.750	0.750
(6) Fuel type: low-sulphur diesel	47,628					0.750	0.750	1.000	0.750
(7) Rig power (unconventional): electric power from grid (10%)	5,918	0.000	1.000		0.000	0.500	1.000	1.000	1.000
(8) Energy storage: flywheels	30,000	0.000				0.500	1.000	1.000	0.750
(9) Drilling technology: conventional overbalanced drilling	170,000				115.385	1.000	0.500	0.500	0.500
(10) Fluid type: water-based muds	47,940					1.000	1.000	1.000	1.000
(11) Waste management: closed loop + containers + solid control equipment[a]	30,000	0.000		1.000		1.000	0.500	1.000	0.750
(12) Cuttings management: cuttings injection	54,000			1.000		1.000	0.500	1.000	0.750
(13) Noise reduction facility: N/A									
Overall attribute scores (∑ or minimum value)	862,855	1.035	4.490	2.000	445.917	0.250	0.500	0.250	0.500
Single-attribute utility values (U_i)	0.883	0.764	0.757	1.000	0.655	0.250	0.500	0.250	0.500

∴ Multi-attribute utility value = *0.740*; [a]*solid control equipment includes shakers, possibly cone centrifuge, desander, desilter, cuttings dryer, and perhaps decanting centrifuge;* CWD, *casing-while-drilling;* SCR, *selective catalytic reduction;* N/A, *not applicable.*

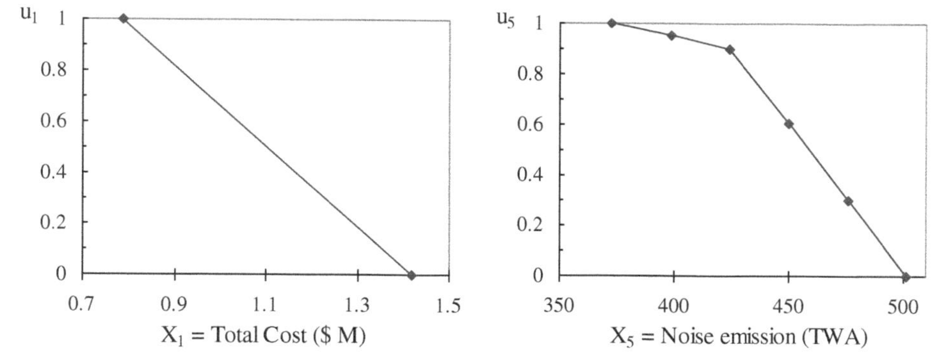

Figure 3 Examples of single-attribute utility function curves.

computed in order to generate the utility function curves. For example, it is known that the range of the cost attribute (X_1) values in Figure 3 is from \$0.79 million to \$1.42 million, where the minimum total costs are preferred over the maximum ones. Thus, to remain consistent with the scaling rule where the utility functions ranged from 0 to 1, u_1 (\$0.79 million) = 1 and u_1 (\$1.42 million) = 0 are defined. Simple linear curves are used for attributes x_1 through x_9 except x_5. The noise attribute (x_5) utility function discussed in Yu et al. [9] is modified in this paper in order to better satisfy the noise level in normal working conditions, as described by the Occupational Safety and Health Administration [13]:

$$u_5(X_5) = \frac{X_5 - \text{Worst score}}{425 - \text{Worst score}} \times 0.9 \text{ for } X_5 > 425, \quad (5)$$

$$u_5(X_5) = \frac{X_5 - 425}{\text{Best score} - 425} \times 0.1 + 0.9 \text{ for } X_5 \leq 425, \quad (6)$$

where X_5 is the noise attribute score, an 8-h TWA sound level in decibels, and u_5 is the noise attribute utility value of the system.

Once each single-attribute utility function $u_i(X_i)$ is derived for its attribute measure, these individual utility values are combined into a final utility value. If mutual preferential and utility independence are satisfied in this study, it is possible to define the multi-attribute utility function with the additive form [12]:

$$\begin{aligned} U(X_1, X_2, \ldots, X_9) &= U\{u_1(X_1), u_2(X_2), \ldots, u_9(X_9)\} \\ &= k_1u_1(X_1) + \ldots + k_9u_9(X_9) = \sum_{i=1}^{9} k_iu_i(X_i), \end{aligned} \quad (7)$$

where $u_i(X_i)$ is the expected utility of the ith attribute scaled from 0 to 1, and k_i is the weighting constant for the ith attribute.

Since it is assumed that there is no interaction between each attribute (no cause-effect dependency), all of the weights are positive and they must sum to one [14]. The weight combination represents the trade-offs between the utility of the different attributes, and in practice, they are chosen by the stakeholders. One weight combination example assigned to a system is shown in Table 4. It is suggested to examine the appropriateness of any independence condition before adopting the additive utility function to see if violations of the independence condition are found [15]. In this study, for example, it was verified that the trade-offs for attributes x_1 (total cost) and x_2 (footprint), keeping the levels of the other attributes (x_3 through x_9) fixed, do not depend on the particular values of these fixed levels and so on for each pair of attributes.

Since the exhaustive search optimization is a simple, practical, and very robust method given the speed of modern computers [16], it is proposed to evaluate all possible systems according to nine attributes, with the relative importance of the different attributes (i.e., weight combination) defined by the stakeholders. Then, it follows to find the 'best' available system that should be particularly suitable for a specific site. Once all possible systems have been evaluated, the system containing the highest overall utility score is the best system with given weighting factors.

After the optimization scheme has determined the best system, it is suggested to conduct a sensitivity analysis to examine the impact of possible changes in the attribute scores, weight factors, and utility functions on the best system. For example, the weight assigned to the cost attribute shown in Table 4 could be changed from the initially assigned value of 0.40. Since the weighting factors must sum to one in this study, the weights assigned to other attributes are known once a weight assigned to the cost attribute is decided. Note that the final result needs not be a single system but that a few 'optimal' systems close to the best score. This may provide some flexibility

Figure 4 Satellite map of Green Lake on the McFaddin Ranch, TX, USA (adapted from [9]).

Table 5 Available technologies for each subset

Subset number	Alternative #1	Alternative #2	Alternative #3
(1) Transportation	Conventional diesel truck	Low-sulphur diesel truck with noise suppressor	-
(2) Road construction	Gravel road	Composite mat (rent)	-
(3) Site preparation	Gravel pad	Composite mat (rent)	Aluminum modules + driven piles
(4) Rig type	Conventional rig	Rapid rig	LOC250 (CWD)
(5) Conventional power	Internal combustion engine	Internal combustion engine with SCR and with noise suppressor	Lean-burn natural gas engines with noise suppressor
(6) Fuel	Conventional diesel	Low-sulphur diesel	Natural gas
(7) Unconventional power	None	Electric power from grid (10% to 30%)	-
(8) Energy storage	None	Flywheel (10% to 30%)	-
(9) Drilling technology	Conventional overbalanced drilling	Underbalanced drilling with noise suppressor	Managed pressure drilling with noise suppressor
(10) Fluid	Water-based muds	-	-
(11) Waste management	Lined reserve pit + solid control equipment	Closed loop + containers + solid control equipment	-
(12) Cuttings treatment	Cuttings injection	CFS	-
(13) Noise reduction facility	None	-	-

SCR, *selective catalytic reduction; CFS,* chemical fixation and solidification; CWD, *casing-while-drilling.*

for the person in charge of the overall decision-making about the drilling process [15,17].

Results and discussion

Applications

To illustrate the applicability of an integrated approach for the selection of EFD systems, a case study was conducted at a site in Green Lake at McFaddin, TX, USA. This assumed that an independent operator was to drill a well on their lease in South Texas, located in an environmentally sensitive wetland area. The lease extends to the center of Green Lake on the McFaddin Ranch as shown in Figure 4. The formation target belongs to the upper Frio sand at a vertical depth of approximately 8,500 ft [18]. The available technologies within each subset for this case study are shown in Table 5. This table shows that some subsets (i.e., (3), (4), (5), and (9)) have three available technologies while subset (10) has only one available technology. It is also noticed that subset (13) is not considered in this case study.

Non-causal approach

An exhaustive search optimization generates all possible combinations of technologies for a given site. The corresponding influence diagram for this approach is shown in Figure 5. Notice that the configuration of the influence diagram for the non-causal system selection approach denotes no influences between the system's main components.

The influence diagram is a compact way for describing the dependencies among variables and decisions [19]. Normally, an arc in an influence diagram denotes an influence (i.e., the fact that the node at the tail of the arc influences the value of the node at the head of the arc). These arcs are drawn as solid lines. Arcs coming into decision nodes have different meanings. As decision nodes are under the decision maker's control, these arcs do not denote influences but rather temporal precedence (in the sense of flow of information). The outcomes of all nodes at the tail of informational arcs will be known before the decision will need to be made. In particular, if there are multiple decision nodes, they need to be all connected by information arcs. This reflects the fact that the decisions are made in a sequence and the outcome of each decision is known before the next decision is made. Informational arcs are drawn as dashed lines [20]. Figure 5 shows a drilling system comprised of 13 subsets and the corresponding dependencies between them. In a real application, a non-causal system selection approach is not recommended as the best practice. However, this can help to assess preliminary responses of the system selection due to its simple implementation and efficient computational effort. Moreover, it typically serves as the basis for the development of more complex system selections.

Only three basic dependencies were considered in this specific application as shown in Figure 6 (i.e., dependencies between subsets (5) and (7), between subsets (5) and (6), and between subsets (7) and (8)). For example, the number of possible fuel types for a conventional power generation engine varies by what kind of engine is selected and whether using an energy storage device or not should be dependent on whether an unconventional power generation method is used or not. If it is decided not to use an unconventional power generation method, an energy storage device is not considered as a possible subset in the 'Rig' subsystem. In this case study, the range of unconventional power usage is varied from 0% to 30% of total power usage (see Figure 6). The

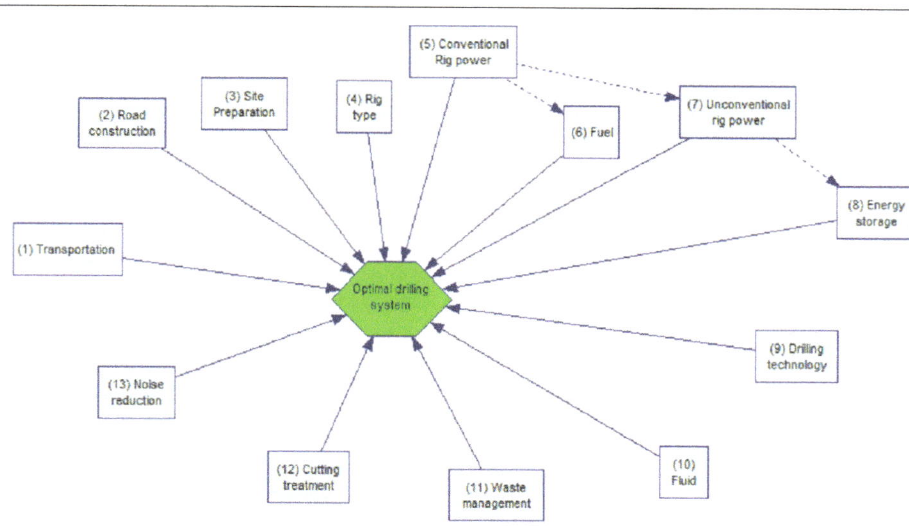

Figure 5 Influence diagram considered in the non-causal approach.

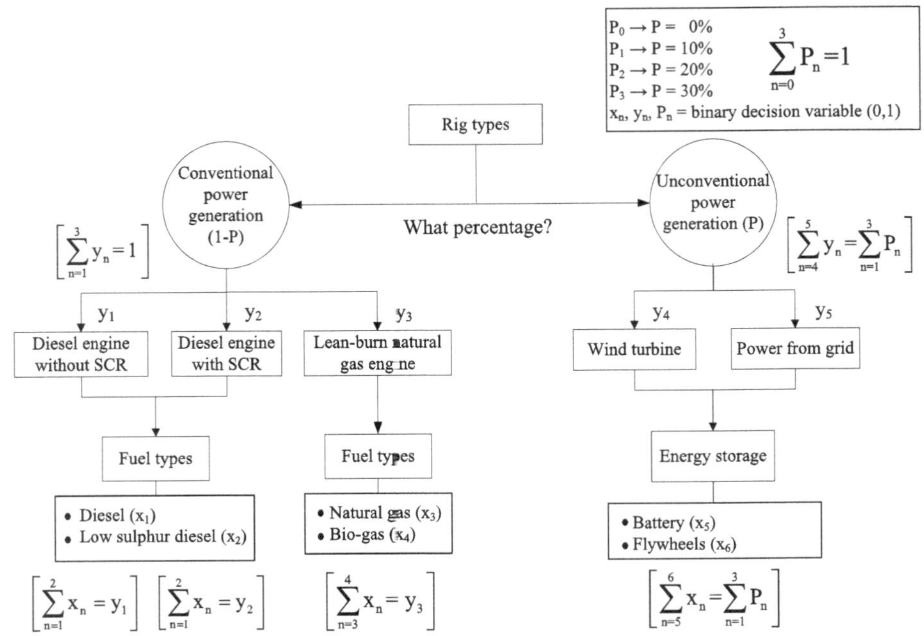

Figure 6 Selection procedure for the 'Rig' subsystem considered for the case study (adapted from [9]).

construction strategy and constraints for the 'Rig' subsystem are also specified in Figure 6 [9].

The basic assumptions and an example of input spreadsheet used to evaluate available technologies for this case study are shown in Tables 6 and 7, respectively. Since this application does not consider varying drilling time, the total drilling time was fixed to 16 days for the computation of total cost. Other attribute values such as emissions were also evaluated considering the same drilling time.

Causal approach

To take into account causal dependencies for the selection of an optimal drilling system, it is necessary to build a consistent influence diagram. In this case, the influence diagram for the given drilling site was developed through a series of meetings with EFD experts, and it is presented in Figure 7 [9]. Notice that the effort required to build the causal model is significantly higher than the

Table 6 Basic assumptions used in the formulation of the non-causal approach

Basic assumptions	Value
Power consumption (peak)	1 MW
Access road width	7.6 m (2 lanes)
Access road length	1.6 km
Width of drilling site	107 m
Length of drilling site	107 m
Total drilling time	16 days

effort required to build the model for the non-causal approach because it requires deeper knowledge between the dependencies of the system components. The influence diagram representing dependencies between the subsets should be considered before estimating attribute scores of technologies because attribute scores of a technology can be dependent on key influence variables as those presented in Table 8. This table includes the basic assumptions considered for the case study as well as key input variables that affect the proposed technologies.

To illustrate the impact of considering dependencies between the EFD subsets, it is suggested to first consider the influence of a selected rig type in subset (4) causing the estimate of total drilling time (i.e., 10 to 16 days) because the total drilling time causes the various estimates of attribute scores of technologies within many different subsets (i.e., total cost, emissions, etc.). For example, it is necessary to estimate different total costs for the same technology within subset (5) 'Conventional rig power' based on a selected rig type because each rig type has its own drilling speed, and thus, total drilling time should be varied by the selected rig type; total drilling time causes the variation of total cost of a technology. The total cost of the same technology within subset (2) 'Road construction' also varies by a selected transportation type in subset (1) because it is likely that the mobilization cost of a technology within subset (2) increases as a more expensive transportation type is selected. Moreover, the total footprint of the same technology within subset (3) 'Site

Table 7 Example of input scores used in the non-causal approach

Subsets	Technologies	Buy ($)	Resell value ($)	Rent/ day ($)	Daily rate ($)	Total cost ($)	Ecological footprint (ha)	Emissions			Perceptions			Safety value
								Air	Solid and liquid	Noise (TWA)	Government	Industry	Public	
1	Conventional diesel truck					20,000	0.000				0.250	1.000	0.250	0.750
	Low-sulphur diesel truck with tier III engine and with noise suppressor					28,000	0.000				1.000	0.500	1.000	0.750
2	Gravel roads					148,500	1.226	0.570		99.000	0.250	1.000	0.250	0.500
	Composite mat					132,000	0.613	0.960		83.000	1.000	0.500	1.000	1.000
3	Gravel pad					137,000	1.133	0.600		98.000	0.250	1.000	0.250	0.500
	Composite mat					120,000	0.567	0.970		82.000	0.750	0.750	0.750	1.000
	Aluminum modules + driven piles					220,000	0.003	0.970		97.600	1.000	0.500	1.000	0.500
4	Conventional order vintage rig					200,000		0.970		79.000	0.500	1.000	0.500	0.500
	Rapid rig					224,000		0.970		77.000	1.000	0.500	1.000	1.000
	LOC250 (CWD)					240,000		0.980		77.500	1.000	0.500	1.000	1.000
5	Internal combustion engine					56,000		0.384		107.000	0.500	1.000	0.500	0.750
	Internal combustion engine with SCR and with noise suppressor					83,020		0.601		90.000	0.750	0.750	0.750	0.750
6	Conventional diesel					65,800					0.500	1.000	0.500	0.500
	Low-sulphur diesel					69,160					0.750	0.750	1.000	0.750
7	Electronic power from grid					25,800.00	0.000	1.000		0.000	0.500	1.000	1.000	1.000
8	Flywheels	450,000	360,000			90,000	0.000				0.500	1.000	1.000	0.750
9	Conventional overbalanced drilling					204,000				117.000	1.000	0.500	0.500	0.500
	Underbalanced drilling with noise suppressor					221,400				101.000	0.750	0.750	0.750	0.750
	Managed pressure drilling with noise suppressor					232,200				99.000	0.750	0.750	1.000	1.000
10	Water-based muds					48,000					1.000	1.000	1.000	1.000
11	Lined reserve pit + solid control equipment[a]					24,000	0.015		0.500		0.750	0.750	0.750	0.750
	Closed loop + containers + solid control equipment[a]					36,000	0.000		1.000		1.000	0.500	1.000	0.750
12	Cuttings injection					54,000			1.000		1.000	0.500	1.000	0.750
	CFS					61,710			0.250		0.750	0.750	1.000	0.500
13	N/A													

[a]*Solid control equipment includes shakers, possibly cone centrifuge, desander, desilter, cuttings dryer, and perhaps decanting centrifuge;* CWD, *casing-while-drilling;* SCR, *selective catalytic reduction; CFS,* chemical fixation and solidification; *N/A, not applicable.*

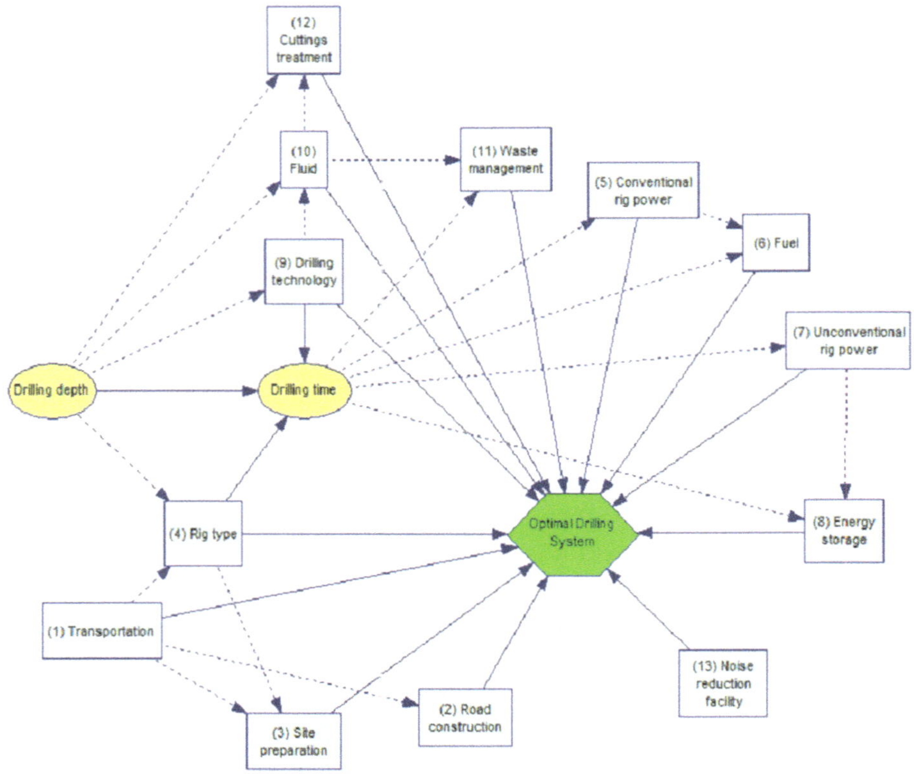

Figure 7 Influence diagram for the drilling site of the case study.

preparation' varies by a selected rig type in subset (4) because each rig type uses different land area. All the other causal dependencies can be easily established by following Figure 7.

An example of an input spreadsheet used to assign scores to different technologies is presented in Table 9. Notice that the cost, footprint, and emission scores of a technology in subset (1) 'Transportation' are already included in a mobilization part of technologies within other subsets. For example, the total cost of gravel road in subset (2) shown in Table 9 includes material, mobilization, and installation costs.

Analysis of results

A base-case weight combination is defined as shown in Table 10 to compare the optimal drilling systems selected by the non-causal and causal approaches. The resulting optimal drilling systems are presented in Tables 11 and 12.

Basic comparisons between these tables indicate, for instance, that in subset (4) 'Rig type', a conventional rig is always selected in terms of the cost attribute in the non-causal approach because the daily cost of LOC250 rig ($15,000) is more expensive than a conventional rig ($12,500). On the other hand, LOC250 rig is always selected in the causal approach because it takes into account total drilling time variation and LOC250 rig reduces the total drilling time by 5 days compared to a conventional rig. Therefore, the total costs of LOC250 rig and a conventional rig in the non-causal approach are $240,000 (16 days × $15,000) and $200,000 (16 days × $12,500), respectively, whereas $165,000 (11 days × $15,000) and $200,000 (16 days × $12,500), respectively, in the causal approach. This result shows that the causal approach yields a more systematic system selection - considering total drilling time variation. In addition, both findings support the main hypothesis of this work, stressing that unnecessary costs would be generated in the absence of an integrated decision-making model.

Other relevant inferences can be obtained by following an integrated decision-making approach for the selection of EFD systems. This can be illustrated by comparing the benefits and limitations between the proposed models when varying the weight on the cost attribute from zero to one, as shown in Tables 13 and 14. It is noted that since the weights must sum to one, as one weight increases, the others must decrease. For example, as the weight assigned to the cost attribute increases, the weights assigned to other attributes decrease by the ratio of the base-case weight combination given in Table 10. Underbalanced drilling (UBD) and managed pressure drilling (MPD) methods are mainly selected as an optimal drilling system's component in the causal approach because USB or MPD methods can reduce a system's

Table 8 Basic assumptions and key influence variables used in formulation of the causal approach

Basic assumptions and key influence variables	Value
Basic assumptions	
Power consumption (peak)	1 MW
Access road width	7.6 m (2 lanes)
Access road length	1.6 km
Width of drilling site	107 m (conventional rig + pad)
	91 m (conventional rig + modules + piles)
	61 m (compact rig + pad)
	46 m (compact rig + modules + piles)
Length of drilling site	107 m (conventional rig + pad)
	91 m (compact rig + pad)
	38 m (conventional rig + modules + piles)
	30 m (compact rig + modules + piles)
Key influence variables	
Transportation type	Conventional diesel truck
Rig type	LOC250 (CWD)
Engine type	Internal combustion engine
Drilling type	Conventional overbalanced drilling
Noise reduction type	N/A
Proportion of unconventional power	30.0%
Resale value	80.0%
Drilling time	9.0 days
Move/rig up	1.0 day
Number of wells	1 well

CWD, *casing-while-drilling; N/A, not applicable.*

total cost due to the decreased drilling time, compared to conventional overbalanced drilling (COBD) method. In the non-causal approach, however, once the weight factor of the cost attribute is greater than 33%, COBD is selected as an optimal drilling system's component because its daily cost is cheaper than that of UBD or MPD.

The technologies in subsets (2), (4), (10), (12), and (13) selected by the causal approach shown in Table 14 are always the same for all possible weights on the cost attribute, while selected technologies in other subsets can change. For example, set 3 is preferred over set 2 as the weight assigned to the cost attribute increases, and set 1, containing 30% of unconventional power usage, is only selected as the optimal system when the cost attribute has a very low weight ($w_1 < 2\%$). This is simply because currently developed unconventional power generation methods and energy storage devices are costly even though they significantly decrease emission rates. Moreover, 'Conventional diesel truck' is selected for subset (1) rather than 'Low-sulphur diesel truck with noise suppressor' when the weight assigned to the cost attribute is greater than 48% because 'Conventional diesel truck' is cheaper than 'Low-sulphur diesel truck with noise suppressor'. It is anticipated that further sensitivity analyses need to be conducted for other input variables, including attribute scores and the utility functions for each attribute, in addition to weighting factors for attributes to suggest more robust optimal systems for this case study.

In summary, LOC250 rig, UBD, and MPD, which can reduce the total drilling time causing the reduction of cost, footprint, and emissions, are most likely to be selected as components of optimal drilling systems when following the causal approach. It is reasonable to state that if one technology can cause to reduce other technologies' cost, footprint, and emissions, it can be selected as a component of an optimal drilling system even if its daily cost is more expensive than that of other alternatives. The causal approach thoroughly takes into account this kind of dependencies between each drilling component, as opposed to just reflecting the exhaustive search of optimal technologies (i.e., less expensive conventional technologies) in the non-causal approach. Therefore, results of this case study suggest that the causal approach generates improved inferences that can better guide the decision-making process for the optimal

Table 9 Example of input scores of the case study in the formulation of the causal approach

Subsets	Technologies	Buy ($)	Resell value ($)	Rent/day ($)	Daily rate ($)	Total cost ($)	Ecological footprint (ha)	Emissions			Perceptions			Safety value
								Air	Solid and liquid	Noise (TWA)	Government	Industry	Public	
1	Low-sulphur diesel truck with tier III engine and with noise suppressor										1.000	0.500	1.000	0.750
2	Gravel roads	198,396	0		12,400	198,396	1.226	0.679		77.250	0.250	1.000	0.250	0.500
	DURA-BASE from composite mat (rent)					147,840	0.613	0.976		64.696	1.000	0.500	1.000	1.000
3	Gravel pad	184,118	0		11,507	184,118	1.138	0.702		76.815	0.250	1.000	0.250	0.500
	DURA-BASE from composite mat (rent)					137,200	0.569	0.978		64.194	0.750	0.750	0.750	1.000
	Aluminum modules + driven piles	1,870,041	1,496,033		23,376	374,008	0.003	0.983		77.990	1.000	0.5000	1.0000	0.500
4	Traditional older vintage rig			12,500	12,500	228,000		0.982		61.304	0.500	1.000	0.500	0.500
5	Internal combustion engine			3,500	3,500	56,000		0.382		106.775	0.500	1.000	0.500	0.750
	Internal combustion engine with SCR and with noise suppressor			5,188	5,188	83,003		0.602		89.574	0.750	0.750	0.750	0.750
6	Conventional diesel			4,704	4,704	65,856					0.500	1.000	0.500	0.500
	Low-sulphur diesel			4,939	4,939	69,149					0.750	0.750	1.000	0.750
7	Electrical power from grid			1,614	1,614	25,824.00	0.000	1.000		0.000	0.500	1.000	1.000	1.000
8	Flywheels	450,000	360,000		5,625	90,000	0.000				0.500	1.000	1.000	0.750
9	Conventional over balanced drilling			17,000	17,000	204,000				116.700	1.000	0.500	0.500	0.500
10	Water-based muds					47,940					1.000	1.000	1.000	1.000
11	Lined reserve pit + solid control equipment[a]			2,000	2,000	24,000	0.015		0.500		0.750	0.750	0.750	0.500
	Closed loop + containers + solid control equipment[a]			3,000	3,000	36,000	0.000		1.000		1.000	0.500	1.000	0.750
12	Cuttings injection			5,000	5,000	54,000			1.000		1.000	0.500	1.000	0.750
	CFS					61,710			0.250		0.750	0.750	1.000	0.500
13	N/A													

[a]*Solid control equipment includes shakers, possibly cone centrifuge, desander, desilter, cuttings dryer, and perhaps decanting centrifuge;* SCR, *selective catalytic reduction; CFS,* chemical fixation and solidification; *N/A, not applicable.*

Table 10 Base-case weight combination

Attributes	Weights ($\sum = 1$)
Total cost (x_1)	0.40
Footprint (x_2)	0.20
Air emission (x_3)	0.20/3
Solid/liquid emission (x_4)	0.20/3
Noise emission (x_5)	0.20/3
Government perception (x_6)	0.05
Industry perception (x_7)	0.05
Public perception (x_8)	0.05
Safety (x_9)	0.05

selection of EFD systems for a given site. However, it is worth mentioning that the main limitation of the causal approach is associated to the effort needed to build the decision-making model (i.e., availability of knowledge base, time, and cost).

A limitation that applies to the use of an integrated decision-making approach for the selection of EFD systems relies on the use of multi-attribute utility values, which typically represent deterministic estimates. This may contain significant uncertainty components, such as total drilling time and drilling depth, which affect the rest of the decision-making process. This limitation represents a challenge in terms of future research, which should aim at providing a confidence metric on the predictions of the integrated approach models proposed above.

Nevertheless, what is consistently observed from the use of the integrated decision-making models discussed above is a significant increase in the number of inferences that becomes available to the stakeholders, contributing to a more informed decision-making process. In addition, the proposed models set the basis to 'connect' with the peripheral decision-making processes, such as those mentioned before, related to impacts to current and future civil infrastructure and their inherent environmental impacts. In particular, because of the imminent growth of the natural gas industry across the country, it becomes a matter of urgency to improve the understanding of the decision-making of these processes as well as of the collateral consequences stemmed from related disciplines.

Web-based decision optimization tools

Finally, to further illustrate the applicability of the proposed models, and its readability to connect to other concurrent disciplines, these have been put into the

Table 11 Example of an optimal drilling system selected by the non-causal approach

	Weights ($\sum = 100\%$ ∴ O.K!)								
	40%	20%	6.67%	6.67%	6.67%	5%	5%	5%	5%
Selected technologies in each subset	**Total cost, x_1, ($)**	**Footprint, x_2, (ha)**	**Emissions (x_3 ~ x_5)**			**Perceptions (x_6 ~ x_8)**			**Safety value (x_9)**
			Air	**Solid and liquid**	**Noise (TWA)**	**Government**	**Industry**	**Public**	
(1) Transportation: conventional diesel truck	20,000					0.250	1.000	0.250	0.750
(2) Road construction: composite mat (rent)	132,000	0.627	0.960		83.000	1.000	0.500	1.000	1.000
(3) Site preparation: composite mat (rent)	120,000	0.567	0.970		82.000	0.750	0.750	0.750	1.000
(4) Rig type: conventional older vintage rig	200,000		0.970		79.000	0.500	1.000	0.500	0.500
(5) Rig power (conventional): lean-burn natural gas engines with noise	125,000		0.880		93.000	0.250	1.000	0.250	1.000
(6) Fuel type: natural gas	58,800					1.000	0.500	1.000	0.750
(7) Rig power (unconventional): N/A (0%)	0	0.000	1.000		0.000	0.250	1.000	0.250	1.000
(8) Energy storage: N/A	0	0.000				0.250	1.000	0.250	1.000
(9) Drilling technology: conventional over balanced drilling	204,000				117.000	1.000	0.500	0.500	0.500
(10) Fluid type: water-based muds	48,000					1.000	1.000	1.000	1.000
(11) Waste management: closed loop + containers+ solid control equipment[a]	36,000	0.000		1.000		1.000	0.500	1.000	0.750
(12) Cuttings management: cuttings injection	54,000			1.000		1.000	0.500	1.000	0.750
(13) Noise reduction facility: N/A									
Overall attribute scores ($\sum$ or minimum value)	997,800	1.194	4.780	2.000	454.000	0.250	0.500	0.250	0.500
Single-attribute utility values(U_i)	0.929	0.679	0.971	1.000	0.565	0.250	0.500	0.250	0.500

∴ Multi-attribute utility value = *0.751; [a]solid control equipment includes shakers, possibly cone centrifuge, desander, desilter, cuttings dryer, and perhaps decanting centrifuge; N/A, not applicable.*

Table 12 Example of an optimal drilling system selected by the causal approach

	Weights ($\sum = 100\%$ ∴ O.K!)								
	40%	20%	6.67%	6.67%	6.67%	5%	5%	5%	5%
			Emissions ($x_3 \sim x_5$)			Perception ($x_6 \sim x_8$)			
Selected technologies in each subset	Total cost, x_1, ($)	Footprint, x_2, (ha)	Air	Solid and liquid	Noise (TWA)	Government	Industry	Public	Safety value (x_9)
(1) Transportation: low-sulphur diesel truck with noise suppressor						1.000	0.500	1.000	0.750
(2) Road construction: composite mat (rent)	147,840	0.613	0.976		64.696	1.000	0.500	1.000	1.000
(3) Site preparation: composite mat (rent)	100,800	0.418	0.984		62.356	0.750	0.750	0.750	1.000
(4) Rig type: LOC250 (CWD)	173,800		0.985		60.366	1.000	0.500	1.000	1.000
(5) Rig power (conventional): lean-burn natural gas engines with noise suppressor	70,353		0.918		89.499	1.000	0.500	1.000	0.750
(6) Fuel type: natural gas	25,650					1.000	0.500	1.000	0.750
(7) Rig power (unconventional): electric power from grid (10%)	5,380	0.000	1.000		0.000	0.500	1.000	1.000	1.000
(8) Energy storage: flywheels	30,000	0.000				0.500	1.000	1.000	0.750
(9) Drilling technology: underbalanced drilling with noise suppressor	184,500				99.624	0.750	0.750	0.750	0.750
(10) Fluid type: water-based muds	47,940					1.000	1.000	1.000	1.000
(11) Waste management: closed loop + containers + solid control equipment[a]	27,000			1.000		1.000	0.500	1.000	0.750
(12) Cuttings management: cuttings injection	54,000			1.000		1.000	0.500	1.000	0.750
(13) Noise reduction facility: N/A								1.000	
Overall attribute scores ($\sum$ or minimum value)	867,263	1.031	4.863	2.000	376.542	0.500	0.500		0.750
Single-attribute utility values (U_i)	0.876	0.764	0.986	1.000	0.993	0.500	0.500		0.750

∴ Multi-attribute utility value = *0.827*; [a]*solid control equipment includes shakers, possibly cone centrifuge, desander, desilter, cuttings dryer, and perhaps decanting centrifuge*; N/A, not applicable.

Table 13 Optimal systems selected by the non-causal approach

Subset number	Set 1 ($0\% \le w_1 < 20\%$)	Set 2 ($20\% \le w_1 < 33\%$)	Set 3 ($33\% \le w_1 < 64\%$)	Set 4 ($64\% \le w_1 \le 100\%$)
(1)	Low-sulphur diesel truck with noise suppressor	Low-sulphur diesel truck with noise suppressor	Conventional diesel truck	Conventional diesel truck
(2)	Composite mat (rent)	Composite mat (rent)	Composite mat (rent)	Composite mat (rent)
(3)	Aluminum modules + driven piles	Composite mat (rent)	Composite mat (rent)	Composite mat (rent)
(4)	Rapid rig	Rapid rig	Conventional rig	Conventional rig
(5)	Lean-burn natural gas engines with noise suppressor	Lean-burn natural gas engines with noise suppressor	Lean-burn natural gas engines with noise suppressor	Internal combustion engine
(6)	Natural gas	Natural gas	Natural gas	Conventional diesel
(7)	Electric power from grid (10%)	Electric power from grid (10%)	None	None
(8)	Flywheel	Flywheel	None	None
(9)	Managed pressure drilling with noise suppressor	Underbalanced drilling with noise suppressor	Conventional overbalanced drilling	Conventional overbalanced drilling
(10)	Water-based muds	Water-based muds	Water-based muds	Water-based muds
(11)	Closed loop + containers + solid control equipment[a]	Closed loop + containers + solid control equipment[a]	Closed loop + containers + solid control equipment[a]	Lined reserve pit + solid control equipment[a]
(12)	Cuttings injection	Cuttings injection	Cuttings injection	Cuttings injection
(13)	None	None	None	None

The weight on the cost attribute (w_1) is varied from 0 to 1; [a]*solid control equipment includes shakers, possibly cone centrifuge, desander, desilter, cuttings dryer, and perhaps decanting centrifuge.*

Table 14 Optimal systems selected by the causal approach

Subset number	Set 1	Set 2	Set 3	Set 4	Set 5
	(0% ≤ w_1 < 2%)	(2% ≤ w_1 < 27%)	(27% ≤ w_1 < 48%)	(48% ≤ w_1 < 81%)	(81% ≤ w_1 ≤ 100%)
(1)	Low-sulphur diesel truck with noise suppressor	Low-sulphur diesel truck with noise suppressor	Low-sulphur diesel truck with noise suppressor	Conventional diesel truck	Conventional diesel truck
(2)	Composite mat (rent)	Composite mat (rent)	Composite mat (rent)	Composite mat (rent)	Composite mat (rent)
(3)	Aluminum modules + driven piles	Aluminum modules + driven piles	Composite mat (rent)	Composite mat (rent)	Composite mat (rent)
(4)	LOC250 (CWD)	LOC250 (CWD)	LOC250 (CWD)	LOC250 (CWD)	LOC250 (CWD)
(5)	Lean-burn natural gas engines with noise suppressor	Lean-burn natural gas engines with noise suppressor	Lean-burn natural gas engines with noise suppressor	Lean-burn natural gas engines with noise suppressor	Internal combustion engine
(6)	Natural gas	Natural gas	Natural gas	Natural gas	Conventional diesel
(7)	Electric power from grid (30%)	Electric power from grid (10%)	Electric power from grid (10%)	None	None
(8)	Flywheel	Flywheel	Flywheel	None	None
(9)	Managed pressure drilling with noise suppressor	Managed pressure drilling with noise suppressor	Underbalanced drilling with noise suppressor	Underbalanced drilling with noise suppressor	Underbalanced drilling with noise suppressor
(10)	Water-based muds	Water-based muds	Water-based muds	Water-based muds	Water-based muds
(11)	Closed loop + containers + solid control equipment[a]	Closed loop + containers + solid control equipment[a]	Closed loop + containers + solid control equipment[a]	Closed loop + containers + solid control equipment[a]	Lined reserve pit + solid control equipment[a]
(12)	Cuttings injection	Cuttings injection	Cuttings injection	Cuttings injection	Cuttings injection
(13)	None	None	None	None	None

The weight on the cost attribute (w_1) is varied from 0 to 1; [a]*solid control equipment includes shakers, possibly cone centrifuge, desander, desilter, cuttings dryer, and perhaps decanting centrifuge; CWD, casing-while-drilling.*

form of two web-based decision optimization tools (non-causal and causal), replicating the system selection processes discussed above. These applications provide a quantitative basis for suggesting appropriate EFD systems, which explicitly evaluate the proposed selection criteria and allow for using the best available evidence, including expert's belief, data, and models.

The non-causal approach has been already used by students enrolled in the 'Drilling Engineering (PETE661)' class at Texas A&M University as an integrated decision-making tool for their well site design [21]. More recently, the causal application was completed and used for the same course, allowing for the generation of similar inferences and comparisons as discussed in this work. The key features of both applications are summarized in Table 15 and can be accessed at [22].

Conclusions

This work makes the case for introducing an integrated decision-making approach for the selection of optimal environmentally friendly drilling (EFD) systems to provide a more logical and comprehensive approach that maximized the economic and environmental goals of both the landowner and the natural gas industry. Two different drilling system selection approaches were discussed based on a combination of the multi-attribute utility theory and the exhaustive search optimization, which are formulated as a non-causal and a causal approach. The proposed models were implemented to a case study defined in an environmentally sensitive area in Green Lake at McFaddin, TX, USA.

Results showed the relevance of using an integrated decision-making approach for the selection of EFD

Table 15 Key features of the two web-based decision-making optimization tools

Features	Non-causal	Causal
Providing a quantitative basis for selecting EFD drilling systems	Included	Included
Evaluating different EFD drilling technologies	Included	Included
Optimizing selection of EFD systems for given conditions	Included	Included
Introducing drilling time variation effects	Not included	Included
Introducing causal dependencies between the system components	Not included	Included

systems. This was corroborated by assessing each model and by showing the type of inferences that can be retrieved from the definition of such a complex system. A comparison between the proposed models also helped to address the need for the use of an integrated decision-making approach because it was possible to replicate the real trade-off between technology costs and reduction of the environmental impacts.

The proposed decision-making methods aim at indicating a critical shift from non-causal to causal system selection, which the authors strongly believe can improve the finding of optimal components in such a complex system as of EFDs. By having access to such decision-making models, optimal scenarios can be suggested, contributing to a more informed decision-making process. This becomes significantly relevant in light of an imminent growth of the natural gas industry and the potential consequences this may impose in concurrent processes such as those related to civil infrastructure and its corresponding environmental impacts.

Competing interests

The authors declare that they have no competing interests.

Authors' contributions

OY carried out the optimization studies and drafted the manuscript. ZMC drafted the manuscript. SG participated in the optimization study and analyzed the results. JB and DB conceived of the study and participated in its design and coordination. All authors read and approved the final manuscript.

Acknowledgements

The authors would like to thank the EFD subject matter experts for their assistance with this research. They want to acknowledge in particular Ms. Carole Fleming, Dr. John Rogers, and Dr. Jerome Schubert. The information contained in this paper was provided in part from the research project 'Field Testing of Environmentally Friendly Drilling Systems' sponsored by the US DOE, from the project 'Environmentally Friendly Drilling Systems' sponsored by the RPSEA, and by the contribution of experts from oil and gas companies. The authors want to thank all the sponsors and subject experts for their support.

Author details

[1]Appalachian State University, Boone, NC 28608, USA. [2]Texas A&M University, College Station, TX 77843, USA. [3]Johns Hopkins University, Baltimore, MD 21218, USA.

References

1. Al-Yami, A, Schubert, J, Medina-Cetina, Z, Yu, O-Y: Drilling expert system for the optimal design and execution of successful cementing practices. In: Proceedings of IADC/SPE Asia Pacific Drilling Technology Conference and Exhibition, Ho Chi Minh City, 1–3 Nov 2010
2. Haut, R, Williams, T, Burnett, D, Theodori, G, Veil, J: The Environmentally Friendly Drilling Systems Program Report. Houston Advanced Research Center, The Woodlands (2010)
3. Rogers, JD, Knoll, B, Haut, R, McDole, B, Deskins, G: Assessments of Technologies for Environmentally Friendly Drilling Project: Land-Based Operations. Texas A&M University Environmentally Friendly Drilling Report, Houston Advanced Research Center, The Woodlands (2006)
4. Herbert, M, Pedersen, J, Pedersen, T: A step change in collaborative decision making - onshore drilling center as the New Work space. In: Proceedings of SPE Annual Technical Conference and Exhibition, Denver, 5–8 Oct 2003
5. King, WW: Iterative drilling simulation process for enhanced economic decision making. (22 Nov 2001). US Patent 6612382
6. Hui, Z, Deli, G: Screening the key techniques for oil & gas drilling in consideration of safety. Nat. Gas Ind. **25**(4), 77 (2005)
7. Simon, HA, Dantzig, GB, Hogarth, R, Plott, CR, Raiffa, H, Schelling, TC, Shepsle, KA, Thaler, R, Tversky, A, Winter, S: Decision making and problem solving. Interfaces **17**(5), 11–31 (1987)
8. Sadiq, R, Husain, T, Veitch, B, Bose, N: Risk-based decision-making for drilling waste discharges using a fuzzy synthetic evaluation technique. Ocean Eng. **31**(16), 1929–1953 (2004)
9. Yu, O-Y, Guikema, S, Briaud, J-L, Burnett, D: Quantitative decision tools for system selection in environmentally friendly drilling. Civ. Eng. Environ. Syst. **28**(3), 185–208 (2011)
10. Yu, O-Y, Guikema, S, Briaud, J-L, Burnett, D: Sensitivity analysis for multi-attribute system selection problems in onshore environmentally friendly drilling (EFD). Syst. Eng. **15**(2), 153–171 (2012)
11. Clemen, RT, Reilly, T: Making Hard Decisions with Decision Tools. Duxbury, Pacific Grove (2001)
12. Keeney, RL, Raiffa, H: Decisions with Multiple Objectives: Preferences and Value Tradeoffs. Cambridge University Press, New York (1976)
13. Occupational Safety & Health Adminstration: Noise exposure computation. http://www.osha.gov/dts/osta/otm/noise/hcp/index.html. Accessed Feb 2012
14. Hardaker, JB: Coping with Risk in Agriculture. CABI, Cambridge (2004)
15. Keeney, RL: Value-Focused Thinking: A Path to Creative Decisionmaking. Harvard University Press, Cambridge (1992)
16. Cover, KS, Verbunt, JPA, de Munck, JC, van Dijk, BW: Fitting a single equivalent current dipole model to MEG data with exhaustive search optimization is a simple, practical and very robust method given the speed of modern computers. Int. Congr. Ser. **1300**, 121–124 (2007)
17. Guikema, S, Milke, M: Quantitative decision tools for conservation programme planning: practice, theory and potential. Environ. Conserv. **26**(03), 179–189 (1999)
18. Hovorka, SD, Doughty, C, Knox, PR, Green, CT, Pruess, K, Benson, SM: Evaluation of brine-bearing sands of the Frio formation, upper Texas Gulf Coast for geological sequestration of CO_2. In: Proceedings of First National Conference on Carbon Sequestration, Washington, DC, 14–17 May 2001
19. Howard, RA, Matheson, JE: Influence diagrams. Decis. Anal. **2**(3), 127–143 (2005)
20. Decision Systems Laboratory: GeNIe Tutorial. http://genie.sis.pitt.edu/ (2005–2007). Accessed Sept 2011
21. Burnett, D, Yu, O-Y, Schubert, J: Well design for environmentally friendly drilling systems: using a graduate student drilling class team challenge to identify options for reducing impacts. In: Proceedings of the SPE/IADC Conference and Exhibition, Amsterdam, 17–19 March 2009
22. Medina-Cetina, Z, Yu, O-Y: Environmentally Friendly Drilling Systems. https://stochasticgeomechanics.civil.tamu.edu/efd/ (2009–2011). Accessed 18 Sept 2012

5

Influence of turbulent mixing intensity on the MILD combustion and the pollutant formation

Hamdi Mohamed* and Bentîcha Hmaeid

Abstract

Moderate or intense low-oxygen dilution (MILD) combustion has been acknowledged as one of the most interesting combustion technologies to meet both the targets of high process efficiency and low pollutant emissions. Therefore, the potential for its implementation in gas turbine combustors is investigated in recent research activities. In the MILD combustion regime, we find that the characteristic times of kinetics and turbulence become comparable and the two phenomena are coupled with each other. In this work, a partially stirred reactor is applied to investigate the influence of the turbulent mixing intensity on the combustion process as well as its impact on NO_x formation. The proposed model is based on a combination of simple turbulent mixing approach with a detailed chemistry sub-model for methane oxidation and NO_x formation. The computational results demonstrate that the combustion processes as well as the pollutant formation are very sensitive to the mixing intensity. However, the total NO_x emissions after complete combustion are seen to be only slightly influenced by the mixing intensity.

Keywords: MILD combustion, Turbulent mixing intensity, Gas turbine combustion, Pollutant formation, NO_x emissions

Background

Pollutant emissions from combustion systems have created increasing environmental concerns. Industrial gas turbine emissions have potentially many different climatic effects such as atmospheric ozone production and destruction, changes in composition of greenhouse gases, alterations of cloud properties and coverage, etc.

Hence, in order to overcome the conflict of interests between energy saving and pollutant emission reductions, alternative combustion concepts are needed. New modes of combustion have been recently introduced in gas turbines, including lean premixed combustion, staged combustion, catalytic combustion and rich-quench-lean combustion. In the last 10 years, scientific investigations and industrial applications have indicated that flameless combustion, elsewhere called flameless oxidation [1], diluted [2], or moderate or intense low-oxygen dilution (MILD) [3] combustion, is one of the most promising combustion technologies to meet both the targets of high process efficiency and low pollutant emissions. MILD combustion is based on large recirculation of the hot combustion products allowing stable combustion in vitiated air. In this regime, combustion occurs with spontaneous ignition, and it appears to develop without a visible and audible flame [4]. The chemical reaction zone is quite diffuse (volumetric combustion [5]) resulting in a heat release spread out and at the same time in a smoother temperature profile.

Plessing et al. [4] investigated the mixing process inside a laboratory-scale furnace fitted with a FLOX burner using laser techniques to simultaneously measure temperature and OH radical. By analyzing the reaction zone, they found that the reaction rate is slow compared with the mixing rate, giving a low Damköhler number. They concluded that MILD combustion could be modeled as a well-stirred reactor.

In conventional flames, the chemical kinetics occurs much faster than the mixing, whereas under MILD conditions, the combustion is controlled by both kinetics and the mixing. As an extension to this, in a furnace environment, MILD combustion may be likened to a well-stirred reactor (WSR) as the chemical time scales

* Correspondence: hamdimh@yahoo.fr
Laboratoire d'Etudes des Systèmes Thermiques et Energétique, Ecole Nationale d'Ingénieurs de Monastir, University of Monastir, Monastir 5000, Tunisia

becoming larger in relation to the turbulence time scales [4]. The link between MILD combustion and a well-stirred reactor may tend to indicate that this regime should be readily modeled, since a well-stirred reactor is well-defined. Recently, Hamdi et al. [6] studied the possibility of integration of the flameless combustion mode to an adiabatic combustor, typically used in gas turbine engines by means of a specific zero-dimensional loop reactor model. They illustrated clearly that the flameless combustion appears as a promising technique for industrial gas turbine combustor applications, even at high operating temperatures and pressures.

The interaction between the turbulence and the chemistry in describing the nature of MILD combustion is expected to play a significant role however [7], so the analogy to a WSR may not always be applicable. Kraft et al. [8] investigated numerically the influence of varying intensities of turbulent mixing on NO_x formation using a partially stirred plug flow reactor model describing the combustion process of methane in tubular flow reactor. By analyzing the sensitivity of reaction paths to various degrees of mixing intensities, they found that the formation pathways of NO formation are sensitive to the mixing intensities.

The main aim of this work is to deepen the influence of turbulence-chemistry interaction on MILD combustion process as well as its impact on the pollutant emissions. Throughout this study, we use the complex finite rate chemistry scheme GRI3.0 for methane and NO_x formation [9], and a simple partially stirred reactor (PaSR) model to account for the turbulence effect on the combustion.

Methods

MILD combustion model

To determine the interaction of the turbulent mixing and the MILD combustion process as well as its impact on pollutant formation in gas turbine combustor, the combustor was simulated by the flow model shown in Figure 1. The flow model proposed in this work stimulates the 'sequential combustor system' where a second combustion chamber is employed after the first stages of the turbine. This form of oxidation leads to the MILD combustion at a second chamber. The hot gases exit the first combustor chamber and enter the second combustor chamber. This second stage combustor, representing the MILD combustion process, is described as a 'PaSR,' where mixing and chemical reactions occur simultaneously.

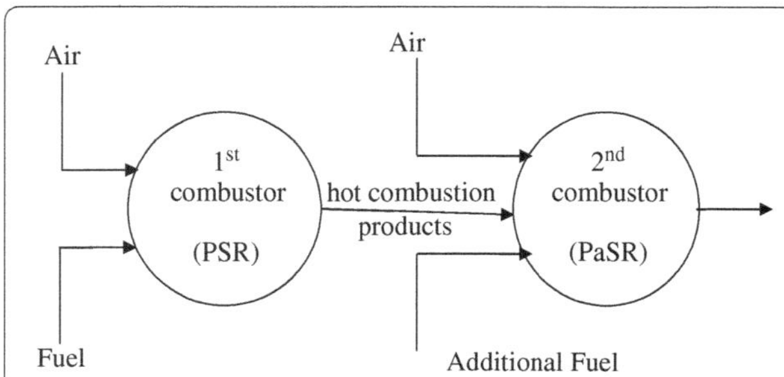

Figure 1 Principle of the MILD combustion model.

Numerical simulation approach

Modeling the interaction of a complex set of chemical reactions and high Reynolds number turbulent fluid flow typical for gas turbine combustion is a challenging task. If the emphasis is on the formation of pollutants such as NO_x, a full chemistry model is prohibitive. Direct numerical simulation is not feasible due to high computational costs and storage requirements. Therefore, severe simplifications have to be made to be able to study the interaction of turbulence and detailed chemistry with respect to NO_x formation. The first combustion chamber is assumed to be an ideal, turbulent, adiabatic, constant pressure, perfectly stirred reactor (PSR). It is assumed that there are no boundary effects, and the turbulence created at the inlet is homogeneous, isotropic, and stationary. Then, the second combustion chamber is described as a PaSR. The widely available CHEMKIN package [10] and specially its software application PaSR [11] are used to model the flow field that occurs in the second combustion chamber.

The PaSR is an extension of a continuously stirred tank reactor (CSTR) model that addresses the interaction between chemical reactions and turbulence [12,13]. The basic assumptions for the PaSR are similar to those of the CSTR or PSR. The major difference between a PSR and a PaSR lies in the treatment of the molecular mixing inside the reactor. In a PSR, the contents of the reactor are well-mixed by assuming high-intensity turbulent stirring action, and the only influence from fluid dynamics is controlled by the reactor residence time τ_R. Unlike the PSR, a PaSR allows fluid dynamics to control the extent of the mixing and, consequently, the chemical reactions by means of an additional parameter: the scalar mixing time, τ_{mix}. The turbulent mixing time scale is often considered to be proportional to the turbulent eddy turnover time as follows [11].

$$\tau_{mix} = C_\phi \frac{\tilde{k}}{\tilde{\epsilon}} \tag{1}$$

where C_ϕ usually treated as a constant, but its value varies for different flow configurations.

Lindstedt et al. [14], Cao et al. [15], and Christo and Dally [16] reported similar requirement for adjusting C_ϕ when using the joint PDF approach. Typically, C_ϕ values of no higher than 3 are reported in the literature, but they found that for MILD flames a higher value is required. Thus, it was necessary to increase the value of C_ϕ from the default value of 2, as suggested by Pope [17], to 3 to stabilize the flame.

The composition and temperature in the PaSR are described by a probability density function (PDF). This composition PDF is a subset of the joint velocity-composition PDF because the flow field in the PaSR is assumed to be spatially homogeneous. Velocity fluctuations are also ignored, that is, the PDF is over scalars only, but is not a delta function in scalar space because reactants, intermediates, and products are not mixed at the molecular level.

The PaSR consists of an adiabatic chamber having M inlet streams and one outlet. Steady flows of reactants are introduced through the inlets with given gas compositions and temperatures. The reactor pressure is assumed to be constant.

The overall mass balance for the gas mixture inside the PaSR is as follows:

$$\frac{d(\prec\rho\succ V)}{dt} = \sum_{i=1}^{M} \dot{m}_i - \dot{m}_0 = 0 \qquad (2)$$

where $\dot{m}_i$ is the mass flow rate of the ith inlet, and $\dot{m}_0$ is the through-flow mass flow rate.

The average properties of the PaSR are obtained from the ensemble of particles inside the reactor. Each particle is regarded as an independent PSR and interacts with others only through the molecular mixing process. Therefore, the conservation of energy and species is applied to an individual particle rather than to the reactor.

The species equation for a particle is then similar to that of a PSR:

$$\frac{dY_k^{(n)}}{dt} = \frac{1}{\dot{m}_0\tau_R}\sum_{i=1}^{M}\left\{\dot{m}_i\left(Y_{i,k} - \prec Y_k\succ\right)\right\} + \frac{W_k\dot{\omega}_k^{(n)}}{\rho^{(n)}} \qquad (3)$$

and so is the energy equation for a particle:

$$\frac{dT^{(n)}}{dt} = \frac{1}{C_p^{(n)}\dot{m}_0\tau_R}\sum_{i=1}^{M}\dot{m}_i\left(\sum_{k=1}^{k_g}Y_{i,k}\left(h_{i,k} - \prec h_k\succ\right)\right) - \sum_{k=1}^{k_g}\frac{W_k\dot{\omega}_k^{(n)}h_k^{(n)}}{\rho^{(n)}C_p^{(n)}} \qquad (4)$$

In the above equations, the angled bracket ($\prec\ \succ$)indicates the ensemble average that we use to approximate the density-weighted average in the simulation. The average residence time of the reactor, τ_R, is calculated as follows:

$$\tau_R = \frac{\prec\rho\succ V}{\dot{m}_0} \qquad (5)$$

Initial conditions for the numerical simulations

The inlet gas compositions and temperatures used in the present study are listed in Table 1. The pressure is fixed at 1 atm, and the residence time in the reactor is fixed to 100 ms. In all the computations, the combustion was supposed to proceed adiabatically (i.e., Q_{loss} = 0). A value of 1 ms was chosen for the time step of the Monte Carlo simulation. We used 1,000 stochastic particles in the PaSR calculations in order to correctly predict the statistics of the computed scalars. The only parameter varied in this study is the characteristic turbulent mixing time τ_{mix}. It corresponds to the degree of mixing intensity in the flow reactor. Numerical simulations are performed for τ_{mix} = 0, 25, and 50 ms for which the characteristic physical and chemical times are the same order of magnitude.

Results and discussion

In the following section, we discuss the influence of the turbulent mixing intensity on the chemical kinetics under MILD combustion conditions. In order to demonstrate the influence of turbulent mixing intensity on the chemical reactions, we vary the turbulent mixing time τ_{mix}, from the limiting case of the PSR (i.e., τ_{mix} = 0 s), which corresponds to very fast mixing; we move to realistic turbulent time scales τ_{mix} = 50 ms, which is slow mixing, and an intermediate mixing time τ_{mix} = 25 ms.

As we mentioned earlier, the turbulence times are about the same order as the characteristic time for chemical reactions to guarantee that both processes influence each other. This will be demonstrated with the time evolution of the statistical moments estimated from the approximation of the PDF for the important chemical species as well as two-dimensional marginal PDF. Additionally, the estimations of the overall production of NO from the chemical source term of each particle will give insight on how reaction paths will vary.

Figure 2 shows the time evolution of the estimated mean of OH mass fraction. We note that the mixing

Table 1 Inlet conditions for the numerical simulations

Combustor	Gas type	Flow rate (kg/s)	Temperature (K)	Composition (%vol)
First combustor (PSR model)	Natural gas	12.5	770	CH_4 94.5, C_2H_6 1.2, C_3H_8 0.4, N_2 2.3, O_2 0.6
	Air	545.5	770	O_2 20.7, N_2 77, H_2O 1.58, CO_2 0.0
Second combustor (FLOX) (PaSR model)	Natural gas	4.0	1,300	CH_4 94.5, C_2H_6 1.2, C_3H_8 0.4, N_2 2.3, O_2 0.6
	Air	558.0	1,300	O_2 12.5, N_2 74, H_2O 8.77, CO_2 3.7

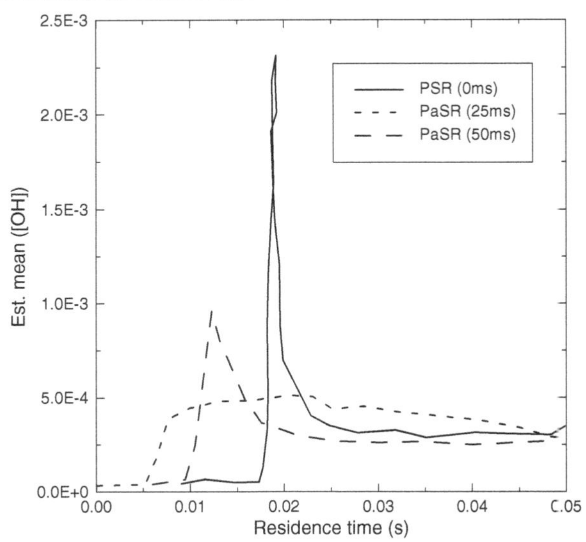

Figure 2 Time evolution of estimated mean of OH mass fraction.

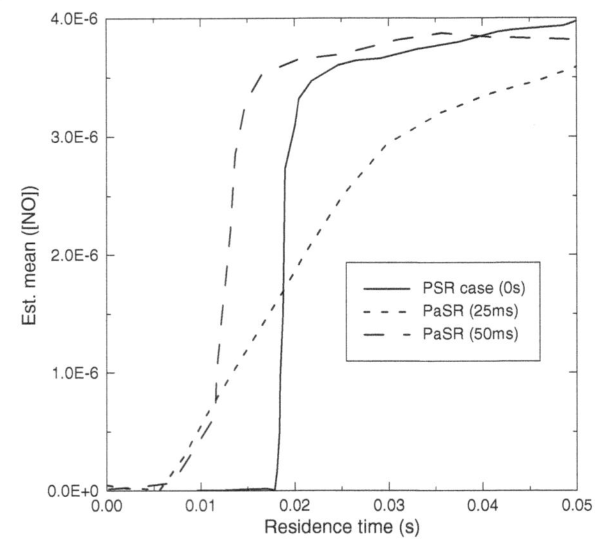

Figure 3 Time evolution of estimated mean of NO mass fraction.

time has a significant influence on the auto-ignition delay time: with increasing mixing time τ_{mix}, ignition delay times decrease significantly (the ignition delay time is defined as the time elapsed between τ_{res} = 0 s and the sudden increase in OH mass fraction). The distribution of the mean mass fraction of OH radical is strongly influenced by the mixing intensity. A rapid increase in the OH concentration signals that ignition takes place. Due to the reduction of the mixing intensity (i.e., higher mixing time), the presence of OH radical is shifted to an earlier point in time. Moreover, the sharp peak present in the diagram of the OH concentration for the PSR case (i.e., τ_{mix} =0 s) is broadened, meaning that the ignition process is distributed over a longer interval. The combustion process, which is very short in the PSR case, is stretched for less intense mixing, and the OH peak concentration is much lower. Our results agree qualitatively with those obtained by Correa and Dean [18] for the auto-ignition of lean n-heptane/air mixtures using an initial temperature of 1,000 K. They explained the decreasing of the auto-ignition delay with the increase of the mixing time arguing that the rich particles lead the ignition process because of their intrinsic shorter ignition delays, in agreement with the results of Chen [13]. These results are of capital importance for the understanding of the MILD combustion regime since it can be considered as a continuous auto-ignition [3].

Figures 3 and 4 plotted respectively the time evolution of estimated means of NO and NO_2 in PSR case and PaSR model for different turbulent mixing times. It can be seen that the ignition delay time is shorter, but the residence time, to achieve complete burnout, increases significantly. Nevertheless, total NO_x emissions after complete burnout are influenced only slightly or even tend to decrease with less intense mixing. This is in contrast to the common expectation that imperfect mixing in the combustion process leads to overall higher NO_x emissions. This result is similar to that obtained by Kraft et al. [8] for the combustion of methane in a tubular flow reactor.

We note that the ignition delay time can be defined as the time needed by the system to increase its temperature of 10 K in comparison with the inlet temperature [19]; therefore, the time evolution of mass fraction of OH radical at varying intensities of turbulent

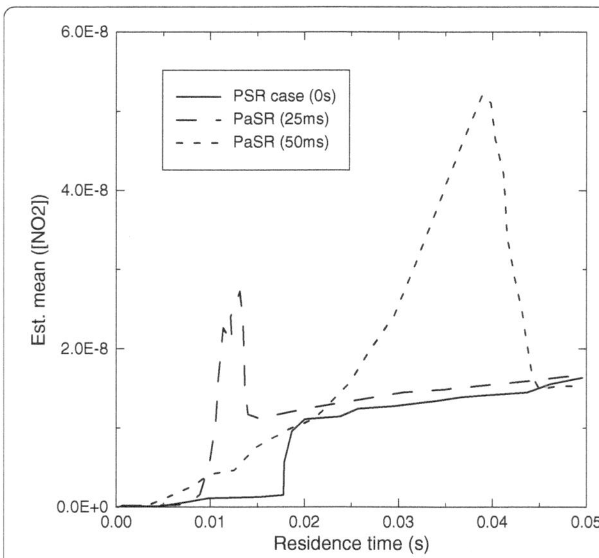

Figure 4 Time evolution of estimated mean of NO2 mass fraction.

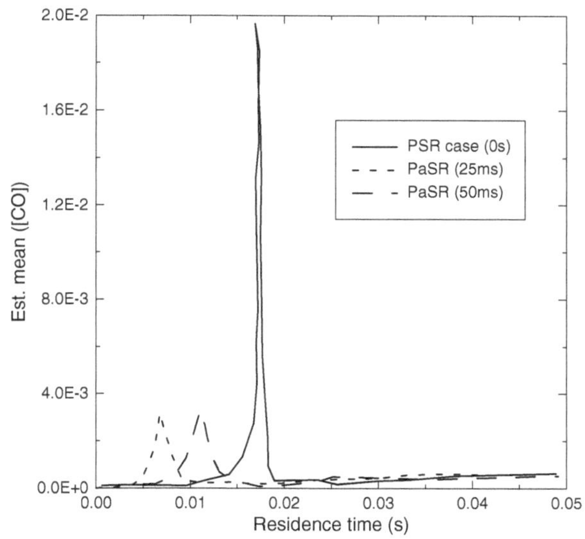

Figure 5 Time evolution of estimated mean of CO mass fraction.

mixing, given in Figure 2, can give an idea on the influence of the mixing intensity on the combustion temperature and, therefore, their influence on NO_x formation. Figure 2 shows that the effect of mixing times on the ignition delay (i.e., on the temperature) becomes less discernible at large mixing times.

Figures 5 and 6 represent respectively the time evolution of estimated means of carbon monoxide (CO) and carbon dioxide (CO_2) in PSR case and PaSR model for different turbulent mixing times. Figures 5 and 6 show that in all cases, the combustion process is fully completed within the residence time of 50 ms, and all radicals reached their equilibrium concentrations and mass fraction of CO_2 its maximum value. During ignition, carbon monoxide is formed and then oxidized to CO_2.

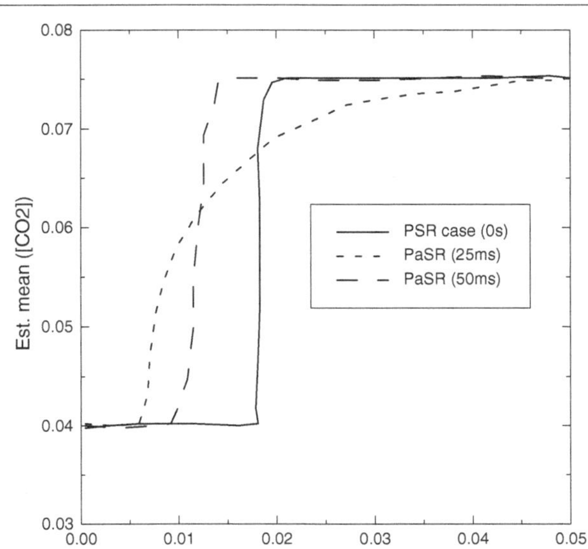

Figure 6 Time evolution of estimated mean of CO2 mass fraction.

Conclusion

In the present work, computational results are presented for natural gas combustion at gas turbine relevant MILD combustion conditions: A PaSR model has been developed to study the influence of turbulent mixing intensity on the combustion process and pollutant emissions. The computational results show that the combustion processes as well as the NO_x formations are very sensitive on the mixing intensity. With increasing turbulent mixing time (i.e., decreasing mixing intensity), the combustion process is stretched out. Total NO_x emissions after complete combustion depended only slightly on the mixing intensity. This is in contrast to the common expectation that imperfect mixing in the combustion process leads to overall higher NO_x emissions.

Competing interests

The authors declare that they have no competing interests.

Authors' contributions

HM carried out the simulation and calculation of data and drafted, interpreted, and corrected the manuscript. BH participated in the organization and presentation of the results and corrected the draft manuscript. All authors read and approved the final manuscript.

Authors' information

HM received his Diploma of Engineering (2000), his M.Sc. (2001), and Ph.D. (2005), all in Energetic Engineering from the National School of Engineering in Monastir, Tunisia. He is now an assistant professor in the Preparatory Institute for Engineering Studies of Monastir, Tunisia where he continues to work on his research in combustion pollutant formation and control and MILD combustion within his former research group in Monastir, Tunisia. BH received his B.Sc. (1979) in Physics from the Faculty of Sciences in Tunisia and his M.Sc. (1984) and Ph.D. (1986) from the University of Paris VII, France, both in Physics with a specialization in reactive flows. Since 1988, he has been with the National School of Engineering in Monastir, Tunisia where he is teaching and doing research in the areas of thermodynamics, combustion of gaseous premixed and diffusion flames, pulverized solid particle jet combustion with radiative transfer, and air pollutant formation and control.

References

1. Wünning, JA, Wünning, JG: Flameless oxidation to reduce thermal NO formation. Prog Energy Combust Sci **23**, 81–94 (1997)
2. Giammartini, S, Girardi, G, Cipriani, R, Cucco, F, Sica, M: Diluted Combustion with High Air Pre-Heating: Experimental Characterization of Laboratory Furnaces by Means of Advanced Diagnostics, pp. 21–24. INFUB, Porto (2000)
3. Cavaliere, A, De Joannon, M: Mild combustion. Prog. Energy Combust. Sci. **30**, 329–366 (2004)
4. Plessing, T, Peters, N, Wünning, JG: Laser optical investigation of highly preheated combustion with exhaust gas recirculation. Proc Combustion Inst **27**, 3197–3204 (1998)
5. Milani, A, Saponaro, A: Diluted combustion technologies. IFRF Combustion Journal (2001). http://www.journal.ifrf.net. Accessed 2008
6. Hamdi, M, Benticha, H, Sassi, M: Detailed chemical modeling of a flameless combustion turbine for pollution prevention. Int J Environ Pollut **32**(3), 410–421 (2008)

7. Katsuki, M, Hasegawa, T: The science and technology of combustion in highly preheated air combustion. Proc Combust Inst **27**, 3135–3146 (1998)
8. Kraft, M, Fey, H, Schlegel, A, Chen, JY, Bockhorn, H: A numerical study on the influence of mixing intensity on NOx formation, pp. 24–26. 3rd Workshop on Modelling of Chemical Reaction Systems Proceedings, Heidelberg (1997)
9. Smith, GP, Golden, DM, Frenklach, M, Moriarty, NW, Eiteneer, B, Goldenberg, M, Bowman, CT, Hanson, R, Song, S, Gardiner Jr, WC, Lissianski, V, Qin, Z:, Gri_mech 3.0. http://www.me.berkeley.edu/gri_mech. Accessed 2010
10. Miller, JA, Kee, RJ, Rupley, FM, Meeks, E: CHEMKIN III: A Fortran chemical kinetics package for the analysis of gas-phase chemical and plasma kinetics, pp. 96–2816. Sandia National Laboratories, SAND, Livermore (1996)
11. Kee, RJ, Rupley, FM, Miller, JA, Meeks, E, Lutz, AE, Warnatz, J, Miller, SF: PaSR application user manual: modeling the mixing and kinetics in partially stirred reactors. Chemkin Collection Release 3.7. Reaction Design, Inc, San Diego (2002)
12. Correa, SM: Turbulence-chemistry interactions in the intermediate regime of premixed combustion. Combust Flame **93**, 41–60 (1993)
13. Chen, JY: Stochastic modelling of partially stirred reactors. Combust Sci Technol **122**, 63–94 (1997)
14. Lindstedt, RP, Louloudi, SA, Vaos, EM: Joint scalar probability density function modeling of pollutant formation in piloted turbulent jet diffusion flames with comprehensive chemistry. Proc Combust Inst **28**, 149–156 (2000)
15. Cao, R, Pope, SB, Masri, AR: Turbulent lifted flames in a vitiated coflow investigated using joint PDF calculations. Combust Flame **142**, 438–453 (2005)
16. Christo, FC, Dally, BB: Modeling turbulent reacting jets issuing into a hot and diluted coflow. Combust. Flame **142**, 117–129 (2005)
17. Pope, SB: A Monte Carlo method for the PDF equations of turbulent reactive flow. Combust Sci Technol **25**, 159–174 (1981)
18. Correa, SM, Dean, AJ: Turbulent broadening of auto-ignition limits. Proc Combust Inst **25**, 1293–1299 (1994)
19. De Joannon, M, Cavaliere, A, Donnarumma, R, Ragucci, R: Dependence of auto-ignition delay on oxygen concentration in Mild combustion of high molecular weight paraffin. Proc Combus Inst **29**, 1139–1146 (2003)

A study conducted on the impact of effluent waste from machining process on the environment by water analysis

Punnose P Kovoor[1*], Mohd Razif Idris[1], Masjuki Haji Hassan[2] and Tengku Fazli Tengku Yahya[3]

Abstract

Ferrous block metals are used frequently in large quantities in various sectors of industry for making automotive, furniture, electrical and mechanical items, body parts for consumables, and so forth. During the manufacturing stage, the block metals are subjected to some form of material removal process either through turning, grinding, milling, or drilling operations to obtain the final product. Wastes are generated from the machining process in the form of effluent waste, solid waste, atmospheric emission, and energy emission. These wastes, if not recycled or treated properly before disposal, will have a detrimental impact on the environment through air, water, and soil pollution. The purpose of this paper is to determine the impact of the effluent waste from the machining process on the environment through water analysis. A twofold study is carried out to determine the impact of the effluent waste on the water stream. The preliminary study consists of a scenario analysis where five scenarios are drawn out using substances such as spent coolant, tramp oil, solvent, powdered chips, and sludge, which are commonly found in the effluent waste. The wastes are prepared according to the scenarios and are disposed through the Institute of Product Design and Manufacturing (IPROM) storm water drain. Samples of effluent waste are collected at specific locations according to the APHA method and are tested for parameters such as pH, ammoniacal nitrogen, dissolved oxygen, biochemical oxygen demand, chemical oxygen demand, and total suspended solids. A subsequent study is done by collecting 30 samples of the effluent waste from the machining operations from two small- and medium-scale enterprise locations and the IPROM workshop to test the quality of water. The results obtained from the tests showed high values of chemical oxygen demand, ammoniacal nitrogen, and total suspended solids when compared with the Standard B specification for inland water bodies as specified by the Department of Environment, Malaysia.

Keywords: Effluent waste, Ammoniacal nitrogen, Biochemical oxygen demand, Chemical oxygen demand, Total suspended solids, Dissolved oxygen, pH

Background

The rapid deterioration of the world's environment in the name of development is a major cause of concern for every human being on this planet. Both individuals and businesses have a social and environmental responsibility to keep this planet clean. With the rapid development in today's industrialized world, the issue of preserving and managing our environment has become crucial. The need for a good environmental management system is evidenced by the amendments made to the Environmental Quality Act 1974 and other environmental regulations, which serve to ensure stricter compliance of environmental standards in Malaysia. Malaysia also adopted the National Environment Policy in 2002 as a means of addressing environmental issues in an integrated manner, while more and more companies are striving to attain the MS14001 EMS certification [1].

Small- and medium-scale enterprises (SMEs) are the backbone of industrialization in many developed and developing countries and play a crucial role in the growth of the country's economy. While most SMEs are in the service sector, approximately one quarter or so that are

* Correspondence: punnose@iprom.unikl.edu.my
[1]Institute of Product Design & Manufacturing, IPROM, Universiti Kuala Lumpur, Kuala Lumpur 56100, Malaysia
Full list of author information is available at the end of the article

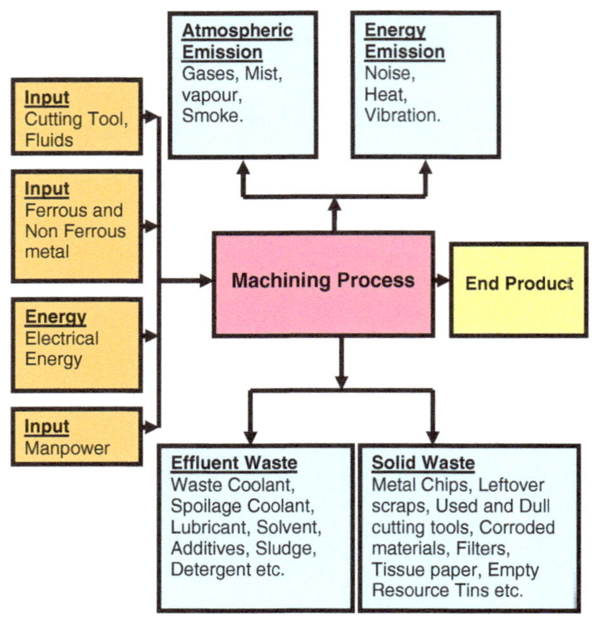

Figure 1 Schematic representation of machining process.

engaged in manufacturing contributes to a major portion of industrial waste [2]. Most of the SMEs that are in the metal industry and engaged in the manufacturing sector utilize one form of machining or another for material removal process. During the machining operation, any material or energy generated in addition to the final product is termed as waste. The waste generated due to machining process is a major environmental concern for manufacturers. With the emergence of new manufacturing industries, a large amount of block metals and other ferrous metals are subjected to machining process, resulting in a considerable amount of machining waste generated in the form of effluent and solid waste, and atmospheric and energy emission, as shown in Figure 1.

The cutting oil, coolant, lubricants, solvents, and additives used in the metal working processes helps to improve machinability, increase productivity, and extend tool life. They are also used to reduce friction and wear, to flush away the chips from the cutting zone, and to protect the machined surface from environmental corrosion. When performing these functions, they become contaminated with foreign organisms, causing coolants to lose effectiveness and develop foul odors. Hence, coolants are used only for a short time and then discarded. The practice of discarding and recycling coolants can be costly and troublesome. Many small- and medium-scale enterprises often dispose the used coolant and cutting fluids in an improper manner into the environment, often neglecting the consequences of such behavior. The wastes, if not recycled or treated properly before disposal, have negative impacts on the environment in terms of water, soil, and sediment pollution [3]. People, the flora and fauna, and the living environment in general are therefore affected adversely. The effluent wastes from the industries are classified as scheduled wastes and are hazardous to nature. In view of the problem above, the Department of Environment (DOE) has made it mandatory for the manufacturers to comply with the Malaysian Environmental Quality (Industrial Effluent) Regulations 2009 (PU (A) 434).

The hazardous waste in Malaysia has been gradually increasing over the years, as shown in Figure 2, a response to a survey from United Nations Statistical Division (UNSD) and United Nations Environment Programme (UNEP).This paper investigates the potential impact of effluent waste from the machining processes on the environment through water analysis.

Numerous studies have been carried out in Malaysia and abroad on the environmental performance assessment among the SMEs. The impact of machining waste on the environment is an area that has to be considered in view of the machines used widely among the manufacturing companies in Malaysia. The machines under consideration are the lathe machines, drilling machines, boring machines, and the milling machines. Work was conducted on field exposures of grade 304 and 316

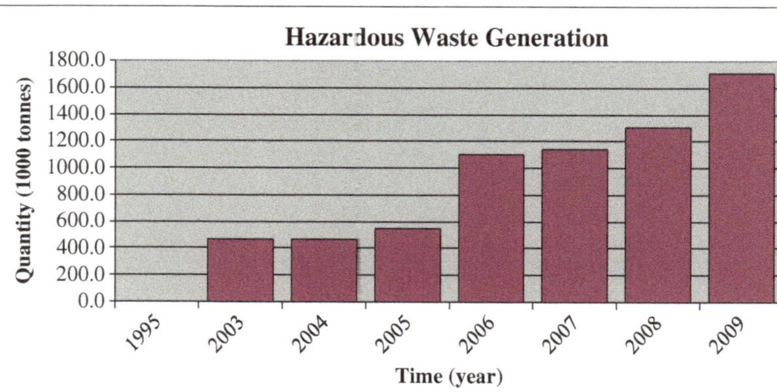

Figure 2 Hazardous waste generation in Malaysia. Source: UNSD/UNEP Questionnaires on Environment Statistics [4].

stainless steels and a laboratory percolation study simulating 20 to 25 years of chromium- and nickel-containing runoff water interaction with soil [5]. Total metal annual release rates varied between 0.2 and 0.7 mg/m^2/year for Cr, between 0.1 and 0.8 mg/m^2/year for Ni, and between 10 and 200 mg/m^2/year for Fe. Soil extractions showed Cr and Ni to be very strongly retained within the soil. This confirms that the corrosion from the metals can enter the nearby field and pollute the environment through the runoff water [5].

Continuous monitoring and assessment of the physical, chemical, and biological parameters of water is an essential part of the water quality control program. The disposal of machining waste through the water stream affects the aquatic organisms living in the water. Most countries practice the water quality index (WQI) method, which is similar to the existing DOE index [6]. The concept of WQI is based on the comparison of the water quality parameter with respective regulatory standards [7].

According to the DOE, the WQI is defined by six parameters, namely the biochemical oxygen demand (BOD), chemical oxygen demand (COD), dissolved oxygen (DO), pH, total suspended solids (TSS), and ammoniacal nitrogen. BOD is the most commonly used parameter for determining the oxygen demand on the receiving water of a municipal or industrial discharge. High BOD is an indication of poor water quality, while COD is an indicator of organics in the water, usually used in conjunction with BOD. COD test is commonly used to measure the amount of organic and inorganic oxydizable compounds in the water [8]. Generally, the BOD/COD ratio should be equal to or less than 1.0 [9]. The BOD/COD ratios for industrial waste water can be lower than 0.5 [10]. Oxygen is essential for all plants and animals to survive, whether they live on land or in water. Aquatic organisms rely on oxygen that is dissolved in the water. A low DO (less than 2 mg/l) indicates poor water quality and, thus, is not conducive for sustaining many sensitive aquatic lives. DO also depends on physiochemical parameters such as temperature, salinity, and pH [11]. The ammonia molecule is a nutrient required for life, but excess ammonia accumulates in the organism and causes alteration of metabolism or increase in body pH. TSS refers to the particles in the water that are larger than 0.45 μm. TSS includes pollutants such as minute metal swarfs, filings, and powdered chips [12], high concentrations of which can damage the habitats of fish and other aquatic organisms [13]. The presence of synthetic organic chemicals like solvents, paints, and detergents can impart offensive tastes, odors, and colors to fish and aquatic plants even if present in low concentrations [14]. Oil in the water also affects the lives of the aquatic beings by inhibiting the atmospheric oxygen from dissolving in the water. Free oil found in the effluent waste rises to the surface of the water and prevents the sunlight from reaching the depths of the water. Oil and grease also become chemically emulsified, primarily through the use of detergents and other alkalis [15]. pH is an indicator of the existence of biological life as most of them thrive in a quite narrow and critical pH range. The discharge of industrial effluents and urban wastes into the river water causes an increase in BOD, COD, and TSS [14,15].

Methods

A twofold study was conducted to determine the impact of the effluent waste on the water stream. The preliminary study consisted of a scenario analysis where five scenarios were drawn out using substances which included spent coolant, tramp oil, solvent, powdered chips, and sludge [16]. These substances are used for preparing the scenarios as they are commonly found in the effluent wastes which are disposed from the manufacturer's premises. Of all the life cycle stages directly influenced by metal working fluids, the disposal stage has the largest impact on the environment as it impacts the environment through hazardous metal carry-off, hazardous chemical constituents, oxygen depletion, oil content, and nutrient loading [16]. The reaction of the coolant between the fluids, additives, and the metals being machined may cause spent fluid to become hazardous even if the 'clean' product was not hazardous [17]. Table 1 shows the most commonly found combinations of effluent waste confined to machining process, taken from the preliminary survey among the SMEs in Malaysia.

A scenario analysis was chosen because the effluent waste from the manufacturers may contain different combinations of the above mentioned substances depending upon the usage during the machining process. The variations in the effluent waste can cause varying impact on the water body if the waste reaches the water stream without any prior treatment or separation. The purpose of conducting this study is to determine the scenario of effluent waste that can cause the greatest impact on the water stream if it were to be disposed in an inappropriate manner.

Scenario 1 consists of the basic coolant obtained from a lathe machine after 6 months of usage. Scenario 2 consists of the basic coolant mixed with 10% of tramp oil.

Table 1 Scenarios of waste for simulation

Number	Different scenarios
1	Coolant
2	Coolant, oil
3	Coolant, oil, chip
4	Coolant, sludge
5	Coolant, sludge, solvent

The amount of tramp oil in the effluent waste was found to be between 1% and 16% [18]. Tramp oil originates as lubrication oil seeping out from the slideways and washing into the coolant mixture, as the protective film with which the steel supplier coats the bar stock to prevent rusting, or as hydraulic oil leaks. In extreme cases, it can be seen as a film or skin on the surface of the coolant or as floating specks of oil. A tramp oil level above 2% could cause health issues due to emulsification [18]. Scenario 3 consists of the basic coolant mixed with 10% oil and around 10 g of ferrous material chips in near-powdered form [19]. Generally the effluent waste from a grinding machine includes minute particles of iron and worn-out or burnt-out abrasives from the wheel, which are difficult to segregate from the used coolant. The complexity of the fluid is compounded by contamination from a combination of substances from the manufacturing process, such as tramp oils, hydraulic fluids, and particulate matter from the machining operations [19]. Scenario 4 consists of 10% sludge mixed with the basic coolant. Sludge is a heavy residue that lies at the bottom of the sump or in a drum where the used coolants are stored before disposal. Microorganisms such as bacteria tend to multiply and grow in the sludge that is present in the effluent waste. Scenario 5 consists of the basic coolant mixed with 10% sludge and 2% solvent. Solvents are used to clean and degrease the finished goods and even certain machine parts after the machining operation. The effluents, in combination with oil and grease, can form a toxic substance for the aquatic organisms [8].

These wastes are then diluted with 20% of water from the Institute of Product Design and Manufacturing (IPROM) workshop to simulate the type of effluent waste which is disposed by manufacturers into the water stream. These wastes are disposed into the storm water drain in the IPROM; from there, the effluent waste flows into a bigger drain which runs along the periphery of the campus.

The experiments are conducted on three different occasions, and on each occasion, the experiments are repeated thrice for reliability. Samples of effluent waste are collected for each of the simulated scenarios from the IPROM storm water drain according to APHA methods. A bar chart is plotted using the mean values to assess the WQI parameters against the Standard B specification from the DOE, Malaysia.

The second part of the research consists of a comparative study where 30 samples of effluent wastes were collected from the machining workshops of two SMEs and the IPROM. The samples were diluted and tested for the WQI parameters. A statistical analysis was conducted using the SPSS software to determine the homogeneity of variance, the mean between the samples, and the effect of covariance among the samples of effluent waste collected from the SMEs and from the IPROM.

Results and discussion

The effluent quality of any discharge from a point of source to an inland body of water should meet the minimum requirements of the Environmental Quality Act 1974, and the limits set down by the Environmental Quality Act are the Standard A and Standard B specification in accordance with the Sewage and Industrial Effluent Regulations, 1979. Standard A criteria applies to catchment areas located upstream of drinking water supply offtakes, while Standard B specification refers to areas downstream from the point of discharge.

Scenario analysis

Figure 3 shows the BOD values for different scenarios of effluent waste. The maximum BOD is between 15 to 16 mg/l, which is below the Standard B specification for

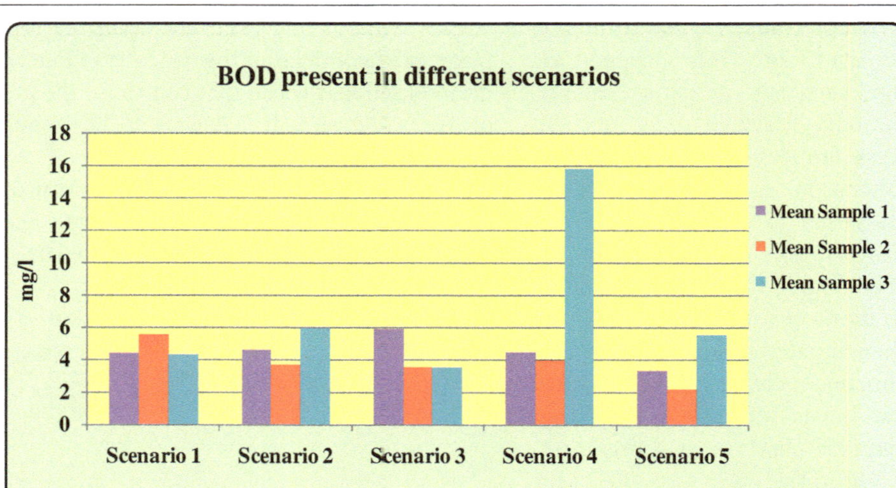

Figure 3 BOD for different scenarios.

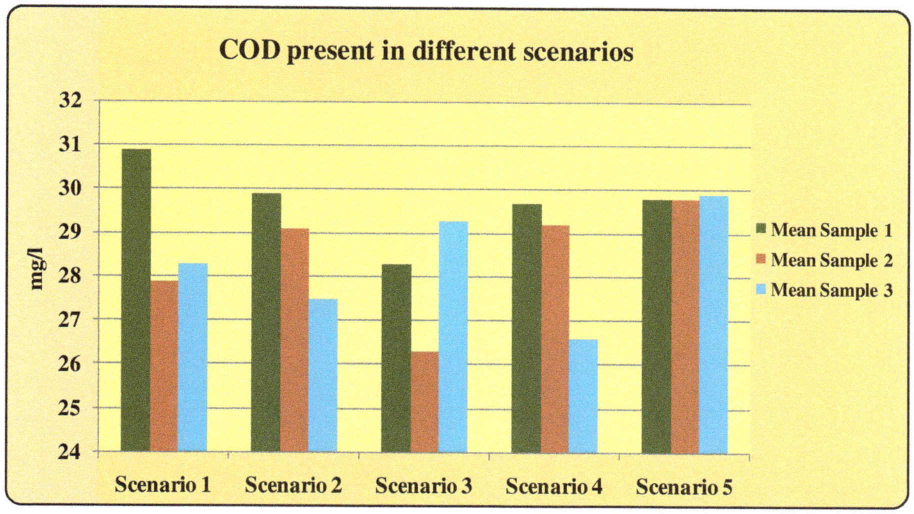

Figure 4 COD for different scenarios.

inland bodies of water and is within the permissible range by the DOE. However, this would place the BOD limit above category 4 according to the Interim National Water Quality Standard (INWQS). The increase in values of the BOD for scenario 4 can be due to the presence of microorganisms in the sludge generated during the storage of the waste coolant before disposal. The untreated or partially treated sewage from the outlets into the drain can also cause the BOD to go high [15].

Figure 4 shows the COD values for different scenarios of effluent waste. The maximum COD is around 31 mg/l, which is below the Standard B specification, and it can be placed in category 3 according to the INWQS. Sample1 shows a higher COD limit due to the presence of anaerobic microbes generated in the stored effluent waste before disposal. A higher COD results in a reduced oxygen level in the water.

Figure 5 shows the DO values for different scenarios of effluent waste. The minimum DO is around 1 mg/l, which is very low for aquatic beings and indicates poor water quality according to Standard B specification. This would place the DO limit to category 4 according to the INWQS. Scenarios 1, 2, and 3 show the presence of greater amount of oxygen in the water. However, for scenarios 4 and 5, the amount of DO is below 2 mg/l in samples 2 and 3. Scenarios 4 and 5 constitute one part of sludge, which is a thick viscous liquid where microorganisms such as bacteria can easily multiply and grow. These microorganisms take in the dissolved oxygen, consequently reducing the overall level of DO in the water. The low water level in the drain, together with high atmospheric temperature, can also reduce the level of DO in water.

Figure 6 shows the values of ammoniacal nitrogen (AN) for different scenarios of effluent waste. The

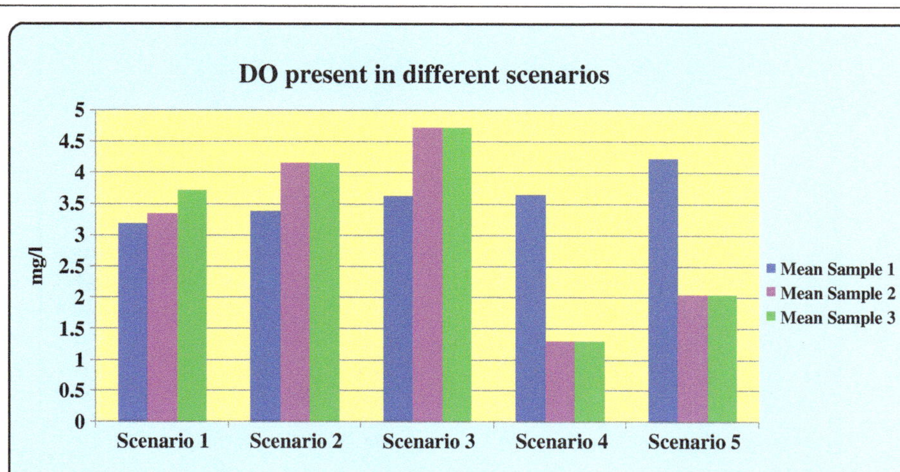

Figure 5 DO for different scenarios.

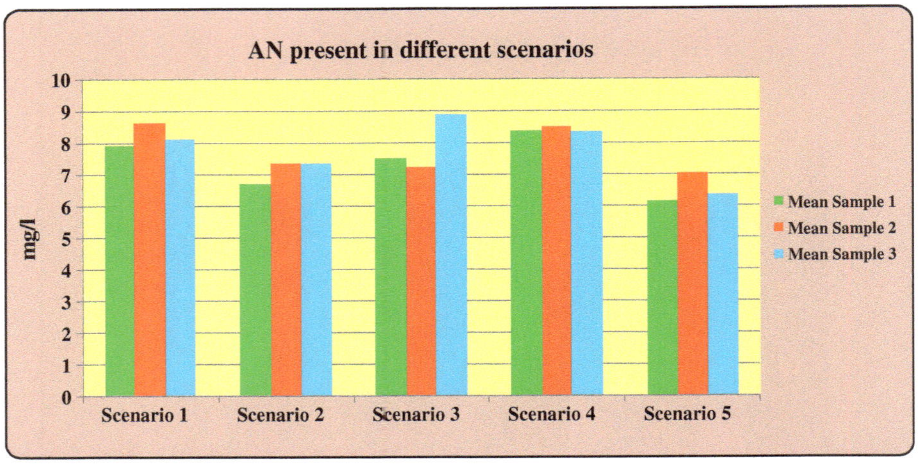

Figure 6 Ammoniacal nitrogen for different scenarios.

maximum AN is between 8 and 9 mg/l, which is higher than the Standard B specification limit and places the AN limit in category 5 according to the INWQS. Ammonia is produced from the decay of organic matter and animal waste. The bacteria break down the organic matter into nitrate. The sources of AN can also be from the domestic sewage that is resident in the drain. Excess ammonia accumulates in the organism and causes alteration of metabolism or increase in body pH.

Figure 7 shows the pH values for different scenarios of effluent waste. The samples containing the effluent waste appear to be mildly alkaline as they exceed 7 on the pH scale. The levels are within the range of Standard B specification and can be placed in category 2 according to the INWQS.

Figure 8 shows the values for TSS for different scenarios of effluent waste. The maximum limit of TSS obtained from the samples is between 50,000 and 60,000 mg/l.

This is much higher than the Standard B specification limit. It places the TSS limit to category 5 as specified by the INWQS. The TSS limits are very high during the third sampling conducted with scenario 4 of the effluent waste. The effluent waste from the IPROM containing sludge and small particles of iron chips contributes to this effect. High suspended solids also prevent the sunlight from penetrating into the water [8].

The Water Quality Index for the five scenarios is shown in Figure 9. According to the DOE standards, a range from 0 to 59 indicates a high pollution level, while a range from 60 to 80 indicates slight pollution, and any value higher than 81 indicates no pollution. The results of the experiment show that when the effluent waste is disposed into the water stream, the water can be polluted. However,

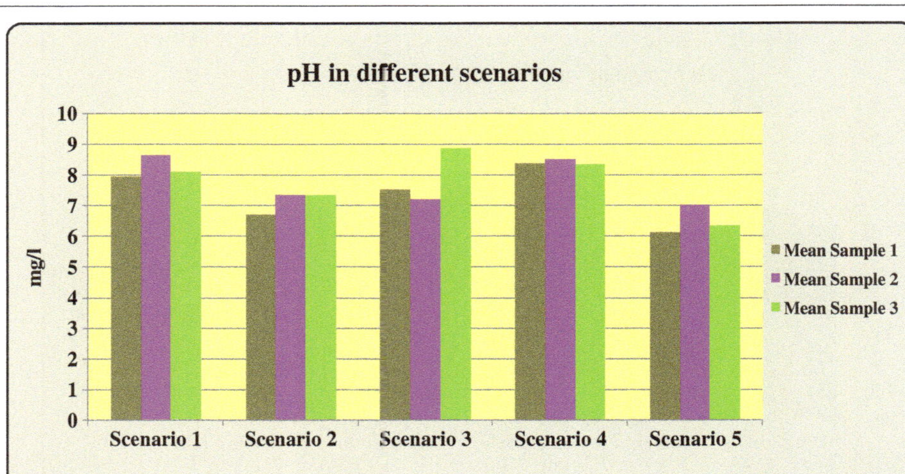

Figure 7 pH values for different scenarios.

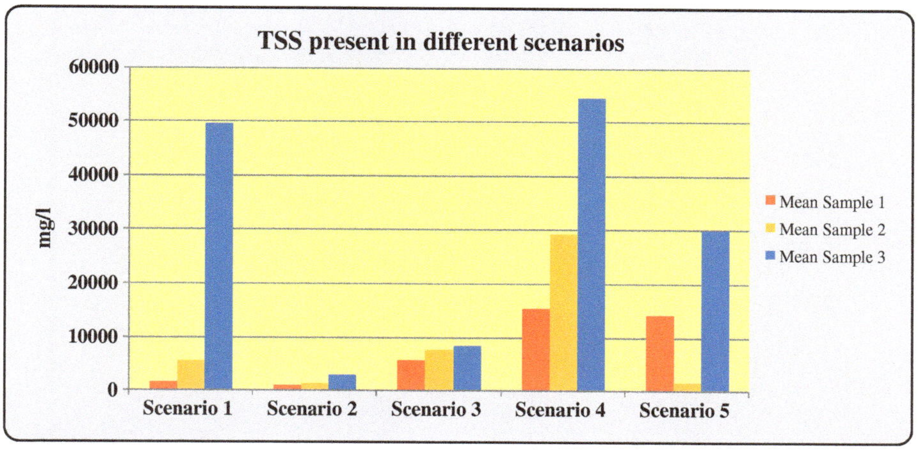

Figure 8 TSS values for different scenarios.

the level of pollution will be minimized when there is a heavy flow of water in the drain.

Comparative study

A comparative study through water analysis was also conducted among the SMEs on the impact of effluent waste on the environment. Thirty samples of effluent waste derived from the machining process were collected from two SMEs and the IPROM for the comparative study. A statistical analysis was done on the data obtained through experimentation.

The data from the samples indicate a positive correlation to that obtained from the IPROM during the scenario analysis. An ANOVA was performed on the data obtained from the two SME premises and the IPROM. The BOD analysis shows that the test of homogeneity of variance is not significant at the 5% significance level according to Levene's statistic (as P value > 0.05). Thus, the null hypothesis of homogeneity of variances cannot be rejected. The estimated F statistics is 9.413 and significant at the 5% level (as P value $0.00 < 0.05$). Hence, the decision was to reject the null hypothesis of equal means at the 5% level of significance. This shows that the mean of the BOD between the groups of samples are different from each other in a significant manner. The post hoc comparison test shows that the SME1 samples indicate a higher BOD compared to the SME2 and IPROM samples, as shown in Figure 10.

The COD analysis shows that the test of homogeneity of variance is not significant at the 5% significance level (as P value > 0.05). Thus, the null hypothesis of homogeneity of variances cannot be rejected. The estimated F statistics is 0.538 and is not significant at the 5% level of significance (as P value $0.586 > 0.05$). Hence, the decision was not to reject the null hypothesis of equal means at the 5% level of significance. This proves that the mean

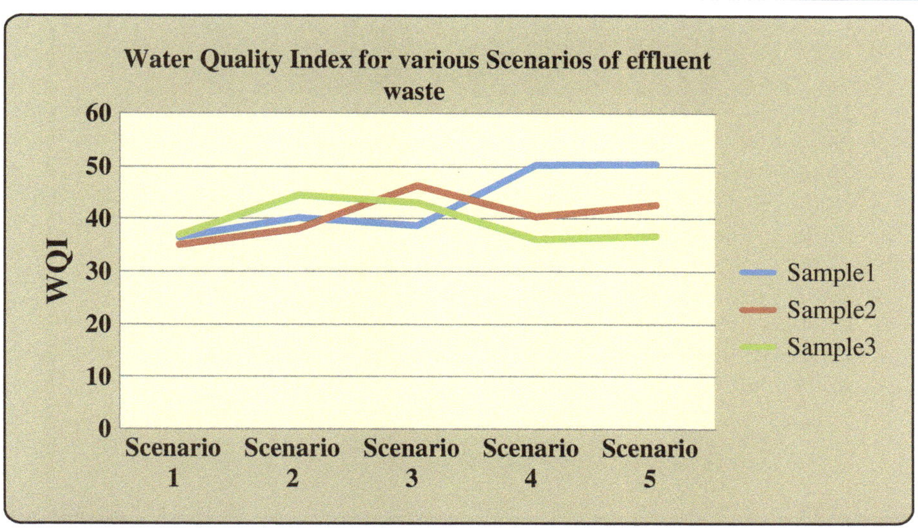

Figure 9 WQI for different scenarios.

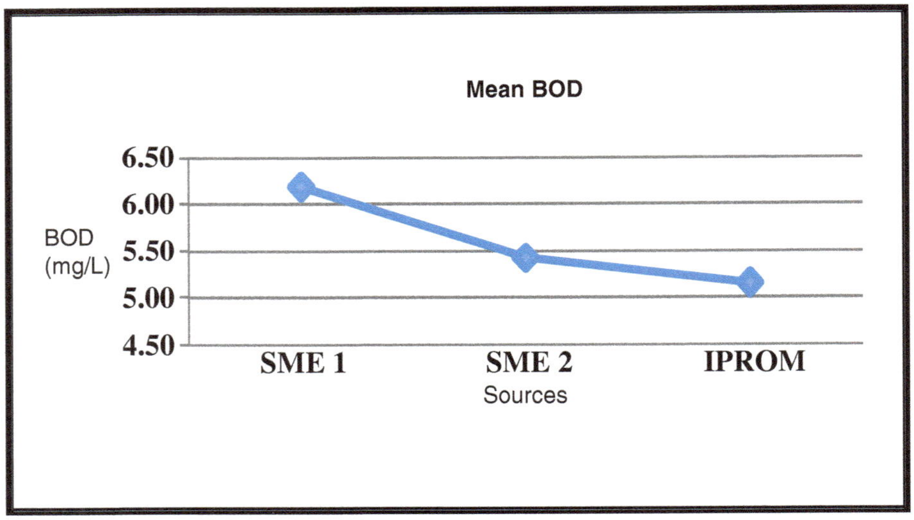

Figure 10 Mean value of BOD between samples.

COD between the groups of samples is not significantly different from each other. The post hoc comparison test shows that the SME1 samples indicate a higher COD compared to the SME2 and IPROM samples, as shown in Figure 11.

The DO present among the three samples was also subjected to an ANOVA test. The test of homogeneity of variance shows that Levene's statistic is not significant at the 5% significance level (as P value > 0.05). Thus, the null hypothesis of homogeneity of variances is not rejected. The estimated F statistics is 15.970 and significant at the 5% level of significance (as P value 0.00 < 0.05). Hence, the decision was to reject the null hypothesis of equal means at the 5% level of significance. This proves that the mean of the DO between the groups of samples are different from each other in a significant manner. The post hoc comparison test with honestly significant difference (HSD) test shows that the SME1 samples indicate a higher DO compared to the SME 2 and IPROM samples, as shown in Figure 12.

In the analysis for TSS, the test of homogeneity of variance shows that Levene's statistic is not significant at the 5% significance level (as P value > 0.05). Thus, the null hypothesis of homogeneity of variances cannot be rejected. The estimated F statistics is 0.021 and is not significant at the 5% level of significance (as P value 0.979 > 0.05). Hence, the decision was not to reject the null hypothesis of equal means at the 5% level of significance. This proves that the mean TSS values of SME1, SME2, and the IPROM are not significantly different from each other. The post hoc comparison test shows that the SME1 samples indicate a higher TSS

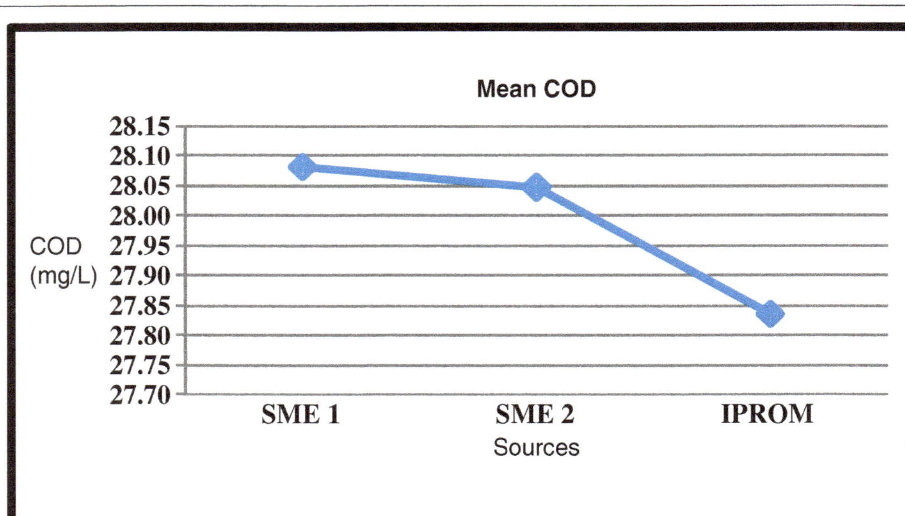

Figure 11 Mean value of COD between samples.

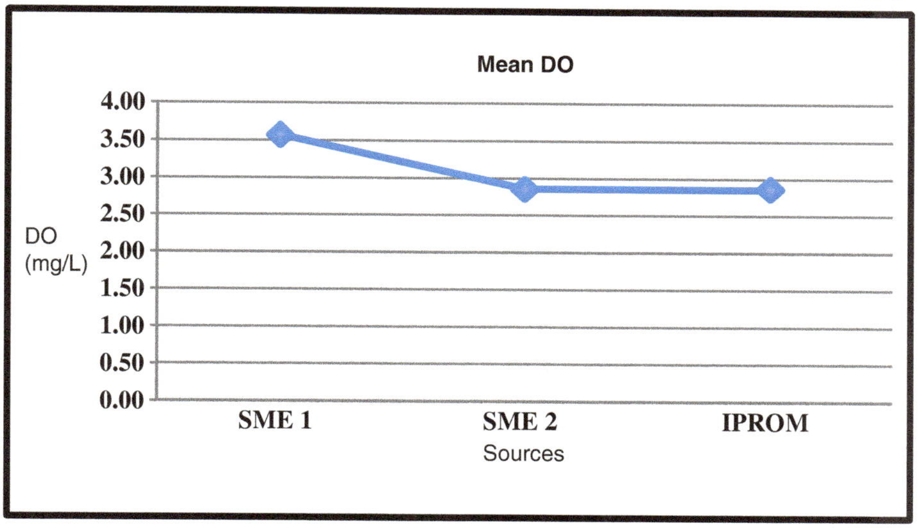

Figure 12 Mean value of DO between samples.

compared to the SME2 and IPROM samples, as shown in Figure 13.

The AN present among the three samples was also subjected to an ANOVA test. The test of homogeneity of variance shows that Levene's statistic is not significant at the 5% significance level (as P value > 0.05). Thus, the null hypothesis of homogeneity of variances cannot be rejected. The estimated F statistics is 1.657 and not significant at the 5% level of significance (as P value 0.197 < 0.05). Hence, the decision was not to reject the null hypothesis of equal means at the 5% level of significance. This proves that the mean AN values of the SME1, SME2, and IPROM samples are not significantly different from each other. The post hoc comparison test with HSD shows that the SME2 samples indicate a higher AN compared to the SME1 and IPROM samples, as show in Figure 14.

The statistical analysis indicate that the mean of the samples may or may not be different from each other, depending upon several factors such as the age of the effluent waste that was disposed, the nature of the effluent waste, the scenario of waste disposal, and the amount of waste disposed.

An analysis of covariance (ANCOVA) was also conducted based on the samples to determine the effect of the independent variables on the dependent variables sub-index value of BOD (QBOD) and sub-index value of COD (QCOD) without the existence of any extraneous variables. The possible effect of any extraneous variables such as temperature and turbidity (covariates) on the dependent variable is statistically controlled in the analysis of covariance.

When an ANCOVA is performed on QBOD with turbidity as the covariate, Levene's test of equality of error

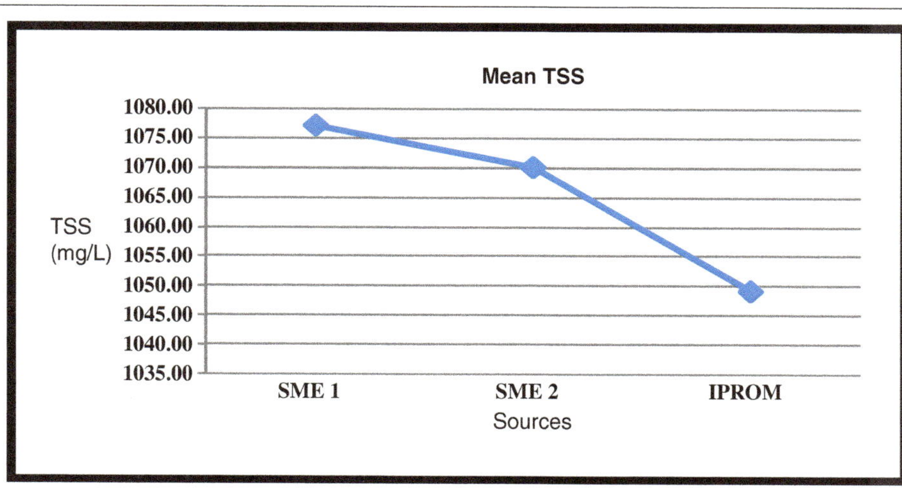

Figure 13 Mean value of TSS between samples.

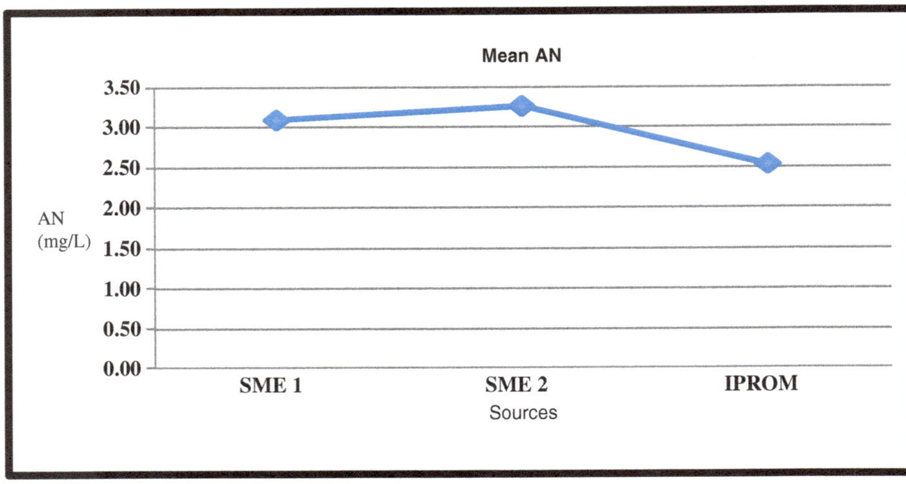

Figure 14 Mean value of AN between samples.

variances shows that there is no significant difference between the variances of the QBOD values of the three samples (as P value $0.174 > 0.05$). As the probability value obtained from SPSS (0.662 for the sample comprised of the IPROM, SME1, and SME2) is higher than the predetermined value (0.05), the null hypothesis was not rejected. Hence, there exists adequate evidence to show that there is no significant difference in the mean score of QBOD between the different samples when turbidity is statistically controlled. As the probability value obtained from SPSS (0.730 for turbidity) is higher than the predetermined value (0.05), the null hypothesis was not rejected. There exists adequate evidence to show that there is no significant difference in the mean score of QBOD due to turbidity. When the ANCOVA was performed on QBOD with temperature as the covariate, Levene's test of equality of error variances shows that there is no significant difference between the variances of the QBOD values of the three samples (as P value $0.153 > 0.05$). As the probability value obtained from SPSS (0.006 for the samples comprised of the IPROM, SME1, and SME2) is lower than the predetermined value (0.05), the null hypothesis is rejected. Hence, there exists adequate evidence to show that there is a significant difference in the mean score of QBOD between the different samples when temperature is statistically controlled.

The ANCOVA was also conducted on QCOD with turbidity as the covariate. Levene's test of equality of error variances shows that there is no significant difference between the variances of the QCOD values of the three samples (as P value $0.365 > 0.05$). As the probability value obtained from SPSS (0.388 for the sample comprised of the IPROM, SME1, and SME2) is higher than the predetermined value (0.05), the null hypothesis was not rejected. Hence, there exists adequate evidence to show that there is no significant difference in the mean score of QCOD between the different samples when turbidity is statistically controlled. As the probability value obtained from SPSS (0.294 for turbidity) is higher than the predetermined value (0.05), the null hypothesis was not rejected. There exists adequate evidence to show that there is no significant difference in the mean score of QCOD due to turbidity. When the ANCOVA was conducted with temperature as the covariance, Levene's test of equality of error variances shows that there is no significant difference between the variances of the QCOD values of the three samples (as P value $0.311 > 0.05$). As the probability value obtained from SPSS (0.576 for the sample comprised of the IPROM, SME1, and SME2) is higher than the predetermined value (0.05), the null hypothesis was not rejected. Hence, there exists adequate evidence to show that there is no significant difference in the mean score of QCOD between the different samples when temperature is statistically controlled. As the probability value obtained from SPSS (0.591 for Temperature) is higher than the predetermined value (0.05), the null hypothesis was not rejected. There exists adequate evidence to show that there is no significant difference in the mean score of QCOD due to temperature.

The ANCOVA is used to minimize the possible errors by individual differences in the samples. The results for the ANCOVA show that the covariates considered in this study, turbidity and temperature, do not have significant influence on the dependent variable, apart from the case when temperature causes a fluctuation in the QBOD. The temperature of the water causes the DO to fluctuate, and colder water can hold more oxygen compared to warmer waters [20].

Conclusions

Numerous studies through water analysis have been made on the impact of effluent waste on the environment, a few

Table 2 Summary of environmental assessment from past researches

Number	Title	Experimental observation
1	Assessment of the impact of industrial effluents on water quality of receiving rivers in urban areas of Malawi [20]	During dry season, a low DO, high BOD, and SS were observed. During rainy season, the DO is higher with the BOD and SS going lower.
2	Comparative Analysis of the Impact of Industrial and Agricultural Effluent on Ona Stream in Ibadan, Nigeria [21]	Lower DO and higher BOD and TSS at the point of discharge compared to upstream conditions.
3	Industrial effluents and their impact on water quality of receiving rivers in Nigeria [22]	High levels of BOD, COD, and suspended solids.
4	Impact of Industrial Effluents on Water Quality of Behgul River at Bareilly [23]	The physicochemical parameters show high levels of BOD and COD and a low level of DO during the experiment.

of which are compiled in Table 2. Most of these experiments show high amounts of BOD and COD in combination with low DO in the presence of effluent waste. The resident waste in the channel, wastes flowing down from upstream, wastes from sewages, and waste due to the influx of the effluents from the industries can be the source of contamination. The current study showed that the BOD and COD are within the Standard B specification, while TSS and ammoniacal nitrogen are high compared to the Standard B specification for inland waters. The WQI parameters may show a different result, depending upon several factors. These factors can be broadly classified as internal factors and external factors. The internal factors consist of the nature of the effluent waste that is disposed from the SME, the amount of waste disposed on one occasion, the treatment done on the effluent waste before disposal, the frequency of disposal, and the time of disposal. These factors are within the control of the company, while the external factors lie outside the control of the company. The external factors consist of the amount of water flowing down the drain at the point of disposal, the flow of water in the drain, the slope of the drain towards the merging river, and upstream pollution. Adequate measures must be taken on the internal factors to comply with the Malaysian environmental law before the effluent waste is released into the drain. Subsequent to this research, an acute test with LC_{50} as the end point is recommended to determine the degree of toxicity of the effluent waste.

Abbreviations

AN: Ammoniacal nitrogen; BOD: Biochemical oxygen demand; COD: Chemical oxygen demand; DO: Dissolved oxygen; DOE: Department of Environment; INWQS: Interim National Water Quality Standard; IPROM: Institute of Product Design and Manufacturing; QBOD: Sub-index value of BOD; QCOD: Sub-index value of COD; SME: Small- and medium-scale enterprises; TSS: Total suspended solids; UNSD: United Nations Statistical Division; WQI: Water quality index.

Competing interests

The authors declare that they have no competing interests.

Authors' contributions

PPK was involved in the design of the study, carried out the experimentations, collected the samples, performed the graphical and statistical analyses, and drafted the manuscript. MRI guided the whole study. MHH was involved in the design of the study. TFTY carried out the water analysis test at the MICET environmental laboratory. All authors read and approved the final manuscript.

Authors' information

PPK is a senior lecturer at Universiti Kuala Lumpur, Malaysia, currently doing his PhD study with IPROM, Universiti Kuala Lumpur. This research paper is a part of his PhD program. MRI is an associate professor and Deputy Dean for IPROM and is also the PhD supervisor for PPK. MHH is a senior professor of University Malaya and Co-supervisor for PPK. TFTY is a Senior Lecturer with MICET, Universiti Kuala Lumpur, and is also the Head of Section for the Environmental Engineering Department.

Acknowledgments

The current research is funded by the Short Term Research Grant (STRG) from Universiti Kuala Lumpur to carry out studies in manufacturing and on the environment.

Author details

[1]Institute of Product Design & Manufacturing, IPROM, Universiti Kuala Lumpur, Kuala Lumpur 56100, Malaysia. [2]Department of Mechanical Engineering, University of Malaya, Kuala Lumpur 50603, Malaysia. [3]Malaysian Institute of Chemical & Bio Engineering Technology, MICET, Universiti Kuala Lumpur, Alor Gajah, Melaka 78000, Malaysia.

References

1. Ensearch: Environmental management for SMIs/SMEs. SMI/SME Business Directory, SMI/SME Editorial. http://www.smibusinessdirectory.com.my/smisme-editorial/more/safetyhealthenvironmentalissues/389-environmental-management-for-smissmes.html (2008)
2. UNEP: Industry and Environment. **26** (2003). No. 4
3. Amin, A, Ahmad, T, Ehsanullah, M, Irfanullah, Khatak, Khan, MA: Evaluation of industrial and city effluent quality using physicochemical and biological parameters. EJEAFChe **9**(5), 931–939 (2010)
4. The United Nations Statistics Division: UNSD/UNEP questionnaires on environment statistics, waste section. (). http://unstats.un.org/unsd/environment/hazardous.htm. (2011)
5. Wallinder, IO, Bertling, S, Kleja, B, Leygraf, C: Corrosion-induced release and environmental interaction of chromium, nickel and iron from stainless steel. Water, Air, and Soil Pollution **170**, 17–35 (2006)
6. Department of Environment: Classification of Malaysian rivers, final report on development of water quality criteria and standards of Malaysia, (Phase IV-River classification). (1994). Department of Environment Malaysia, Ministry of Science, Technology and the Environment
7. Khan, S, Noor, M: Investigation of pollutants in waste water of Hayatabad Industrial Estate Peshawar. Pakistan Journal of Agriculture Sciences **2**, 457–461 (2002)
8. Ministry of Natural Resources and Environment Malaysia: Study on the river water quality trends and indexes in peninsular Malaysia. Water Resources Publication No. 21, Water Resources Management and Hydrology Division, Department of Irrigation and Drainage. (2009)

9. Samudro, G, Mangkoedihardjo, S: Review on BOD, COD and BOD/COD ratio: a triangle zone for toxic, biodegradable and stable levels. International Journal of Academic Research. **2**, 4 (2010)
10. Mangkoedihardjo, S: Biodegradability improvement of industrial wastewater using hyacinth. Journal of Applied Sciences. **6**(6), 1409–1414 (2006)
11. Jack, PJ, Abdsalam, AT, Khalifa, NS: Assessment of dissolved oxygen in coastal waters of Benghazi. Libya. J. Black Sea/Mediterranean Environment **15**, 135–150 (2009)
12. Said, A, Stevens, DK, Sehlke, G: An innovative index for evaluating water quality in streams. Environ Manage **34**(3), 406–414 (2004)
13. Avvannavar, SM, Shrihari, S: Evaluation of water quality index for drinking purposes for river Netravathi. South India, Environmental Monitoring and Assessment, Mangalore (2007)
14. Sisodia, R, Chaturbhuj, M: Assessment of water quality index of wetland Kalakho Lake, Rajasthan. India. Journal of Environmental Hydrology. **14**, 23 (2006)
15. Tariq, M, Ali, M, Shah, Z: Characteristics of industrial effluents and their possible impacts on quality of underground water. Soil & Environ. **25**(1), 64–69 (2006)
16. Handbook of Environmentally Conscious Mechanical Design: Environmentally Conscious Manufacturing. Prevention of Metalworking Fluid Pollution: Environmentally Conscious Manufacturing at the Machine Tool, 2nd edn. Wiley, New York (2000)
17. USEPA: Pollution prevention in machining and metal fabrication: a manual for technical assistance providers. NEWMOA (2001). http://www.newmoa.org/prevention/topichub/23/newmoamanual.pdf
18. Gauthier, SL: Metalworking fluids: oil mist and beyond. Appl Occup Environ Hyg **18**, 818–824 (2003)
19. National Institute for Occupational Safety and Health (NIOSH): Metalworking Fluids. Centers for Disease Control and Prevention. Atlanta. http://www.cdcinfo@cdc.gov. Accessed January 2012
20. Phiri, O: Assessment of the impact of industrial effluents on water quality of receiving rivers in urban areas of Malawi. Int. J. Environ. Sci. Tech. **2**(3), 237–244 (2005)
21. Ogedengbe, K, Akinbile, CO: Comparative analysis of the impact of industrial and agricultural effluent on Ona Stream in Ibadan. Nigeria. New York Science Journal. **3**, 7 (2010)
22. Kanu, I, Jeoma, P, Achi, OK: Industrial effluents and their impact on water quality of receiving rivers in Nigeria. Journal of Applied Technology in Environmental Sanitation **1**(1), 75–86 (2011)
23. Kumara, P, Kumar, S, Agarwal, A: Impact of industrial effluents on water quality of Behgul River at Bareilly. Advances in Bioresearch. **1**(2), 127–130 (2010)

7

Review of underground coal gasification technologies and carbon capture

Stuart J Self, Bale V Reddy and Marc A Rosen*

Abstract

It is thought that the world coal reserve is close to 150 years, which only includes recoverable reserves using conventional techniques. Mining is the typical method of extracting coal, but it has been estimated that only 15% to 20% of the total coal resources can be recovered in this manner. If unrecoverable coal is considered in the reserves, the lifetime of this resource would be greatly extended, by perhaps a couple hundred years. Mining involves a large amount of time, resources, and personnel and contains many challenges such as drastic changes in landscapes, high machinery costs, elevated risk to personnel, and post-extraction transport. A new type of coal extraction method, known as underground coal gasification (UCG), that addresses most of the problems of coal mining is being investigated and implemented globally. UCG is a gasification process applied to *in situ* coal seams. UCG is very similar to aboveground gasification where syngas is produced through the same chemical reactions that occur in surface gasifiers. UCG has a large potential for providing a clean energy source through carbon capture and storage techniques and offers a unique option for CO_2 storage. This paper reviews key concepts and technologies of underground coal gasification, providing insights into this developing coal conversion method.

Keywords: underground gasification, *in situ* coal gasification, UCG, clean energy, coal technology, carbon capture

Review

Introduction

The global energy supply is comprised of many different sources, including fossil fuels, uranium, and various alternative and renewable sources. Currently, over 85% of the global energy supply is derived from fossil fuels, and a high fossil fuel dependency appears likely to remain in the immediate future [1]. The amount of energy required globally is projected to increase due to growing population and industrialization [2,3]. Some feel that the total primary energy demand will double or even triple by the year 2050 relative to levels today, and as the energy demand continues to increase, future fossil fuel shortages are predicted [4,5].

Hammond [6] argues that fossil fuel depletion is a significant factor when considering sustainable energy systems for the future. Fossil fuel resources are finite and being consumed rapidly, beginning with the most economically attractive resources [7]. In the future, fossil fuel resource extraction and production rates are expected to peak and begin to decline [7]. Oil production is predicted to peak in 5 to 15 years and gas production within 40 years, with significant exhaustion of oil and gas reserves by the years 2050 and 2070, respectively [4,8]. As fossil fuel demand approaches supply levels, the cost of energy is anticipated to increase drastically, prompting research and technological developments for improved ways to convert more fossil fuel resources into useable reserves [9].

Currently, coal generates 41.5% of the world's electricity and provides 26.5% of global primary energy needs [1]. Coal has the largest reserves in the world of the fossil fuels and is abundant in many countries. It is thought that the world's recoverable coal reserve is close to 150 years at current production rates, but this only represents 15% to 20% of the entire resource [8]. Remaining global coal resources have recently been estimated to be 18 trillion tonnes [10]. This contrasts significantly with the typical figure of tens of billions of tonnes for recoverable reserves [4]. If unrecoverable coal is considered in the recoverable reserves, the lifetime of the resource could be extended by a couple hundred

* Correspondence: marc.rosen@uoit.ca
Faculty of Engineering and Applied Science, University of Ontario Institute of Technology, 2000 Simcoe Street North, Oshawa, ON L1H 7K4, Canada

years. For this to be realized, new, economic extraction techniques need to be implemented.

Coal is conventionally extracted by mining, both underground and open pit. Mining operations require much time, personnel, and natural resources; typically, coal reserves lie too deep underground, or are otherwise too costly, to exploit using conventional mining methods. Conventional mining also has other issues including land subsidence, high machinery costs, hazardous work environments, coal transport requirements, localized flooding, and methane buildup in cellars of nearby homes [11].

Underground coal gasification (UCG) is a newer type of coal extraction that is being investigated and implemented around the world and that avoids most of the problems of mining coal. UCG involves the conversion of unmined coal, where coal seams are gasified, without mining, and synthetic gas (syngas) is produced for use in power generation or as chemical feedstock [12]. UCG limits the amount of underground work required by personnel, lowering risks of harm relative to conventional mining. Power generation and chemical processing plants can be built directly above a coal resource and use syngas produced through UCG, avoiding coal transport. UCG has the ability to significantly widen the resource base, where the energy contained within inaccessible coal reserves, considered uneconomical for recovery, could be recovered using UCG [11]. It has been estimated by the Underground Coal Gasification Partnership that around 4 trillion tonnes of otherwise unusable coal could be suitable for UCG [9].

UCG is appealing for expanding recoverable coal reserves, but as with the combustion of all fossil fuels, there are associated greenhouse gas emissions. Coal is the most carbon-intensive of all fossil fuels and has high associated CO_2 emissions [4]. The calorific value of fossil fuel sources varies, with typical values of 50 GJ/tonne for natural gas, 45 GJ/tonne for crude oil, and 30 GJ/tonne for coal [4]. Hence, coal has the highest CO_2 emissions per unit of thermal energy produced [13]. If coal is to become a major contributor in the future global energy supply, CO_2 capture and storage techniques would need to be incorporated in the process. UCG has good potential for CO_2 reduction. During gasification, CO_2 is produced, which can be captured from the syngas and stored for long terms. If UCG is successfully linked to such carbon capture and storage (CCS), a method will be available for exploiting the energy in previously unrecoverable coal reserves while satisfying standards for reducing CO_2 emissions.

The aim of this paper is to review key areas and technologies for underground coal gasification so as to provide insights into this developing coal conversion method.

Underground coal gasification

Brief UCG history

The concept of coal gasification has been recognized for more than 200 years and was first used during the late 1800 s to produce town gas fuel for heating and lighting applications [14]. Today, coal gasification is primarily used to provide fuel for advanced power plants and chemical feedstocks for use in the chemical industry [9]. Conventionally, coal is extracted from the ground through mining, processed, transported, and then gasified in a surface gasifier unit to produce syngas. Underground coal gasification is a combined extraction and conversion process.

Experimentation on UCG was first performed in 1912 by Sir William Ramsay in England [4]. The experiments demonstrated the potential of UCG, but Ramsay's work was interrupted by the First World War. After the war, further UCG research did not continue, since coal was relatively inexpensive and available through conventional mining techniques in Western Europe [4]. The former Soviet Union was the first to begin considerable research and development programs with respect to large-scale UCG systems in the 1930s [15,16]. UCG technology was first utilized within commercial operations by the former Soviet Union for heating and power generation applications, which has continued to implement these systems for over 50 years [17]. Even though UCG has the appearance of being commercially mature, the technology from the former Soviet Union has been gaining interest only recently, with a rapid increase in the number of pilot plants throughout the rest of the world over the last decade. There are many commercial projects entering pilot plant phase and undergoing study, in Australia, New Zealand, the USA, India, Pakistan, Canada, and Italy. National research programs are being reconsidered in the USA and the UK, after preliminary systems failed to reach commercial maturity. As of 2008, the number of UCG trials includes 200 in the former Soviet Union, 33 in the USA, and approximately 40 distributed between South Africa, China, Australia, Canada, New Zealand, India, Pakistan, and Europe [17,18].

UCG concepts and technology

UCG is similar to surface gasification [19], with syngas produced through the same chemical reactions [12]. The main difference is that surface gasification occurs in a manufactured reactor whereas the reactor for a UCG system is a natural geological formation containing unmined coal [19,20]. UCG also has similarities to *in situ* combustion processes applied in heavy-oil recovery and oil shale retorting, with such common operational parameters as roof/floor stability, seam continuity and permeability, and ground water influx [19,21].

The basic UCG concept is illustrated in Figure 1. UCG involves an arrangement of injection and production wells drilled into coal seams. The coal is ignited and compressed gasification agents are fed into the coal seam through injection wells which triggers and controls an *in situ* sub-stoichiometric combustion process, producing syngas [22]. Syngas is extracted using production wells and is processed for use [17]. Suitable gasifying agents include air, oxygen, steam/air, and steam/oxygen [23]. The main difference between using oxygen and atmospheric air is that utilization of oxygen increases the heating value of the syngas [24], but producing pure oxygen requires additional energy and resources.

Coal ignition is initiated through the use of an electric coil or gas firing near the face of the coal seam. Continuous oxidant flow through the injection well allows for gasification to be sustained [1]. The temperature of the gasification process is maintained through varying the oxidant flow to the reactor [23]. In UCG systems, the temperature of the coal face can reach temperatures in excess of 1,500 K [24,25].

Various chemical reactions, temperatures, pressures, and gas compositions exist at different locations within a UCG gasifier. The gasification channel is normally divided into three zones: oxidization, reduction, and dry distillation and pyrolysis [26,27]. In the oxidization zone, multiphase chemical reactions occur involving the oxygen in the gasification agents and the carbon in the coal. The highest temperatures in the gasifier occur in the oxidation zone, due to the large release of energy during the initial reactions [28]. The following reactions occur in the oxidation zone:

$$C + O_2 \rightarrow CO_2 + 393.8\ \text{kJ} \tag{1}$$

$$2C + O_2 \rightarrow 2CO + 231.4\ \text{kJ} \tag{2}$$

$$2CO + O_2 \rightarrow 2CO_2 + 571.2\ \text{kJ}. \tag{3}$$

In the reduction zone, the main reactions involve the reduction of $H_2O(g)$ and CO_2 into H_2 and CO at high temperatures within the oxidation zone [26]. The following endothermic reactions occur in the reduction zone [21,28]:

$$C + CO_2 \rightarrow 2CO - 162.4\ \text{kJ} \tag{4}$$

$$C + H_2O(g) \rightarrow CO + H_2 - 131.5\ \text{kJ}. \tag{5}$$

Under the catalytic action of coal ash and metallic oxides, a methanation reaction occurs:

$$C + 2H_2 \rightarrow CH_4 + 74.9\ \text{kJ}. \tag{6}$$

The energy terms, within the above equations, represent the amount of energy released or consumed during each reaction with the stoichiometric coefficients in equations representing moles. Equations 1 to 6 are taken from [21] and [28]. The endothermic reactions in the reduction zone decrease the temperature in the gasification channel to below that required for the reduction reactions.

Within the distillation (pyrolysis) zone, the coal seam is decomposed into multiple volatiles including H_2O, CO_2, CO, C_2H_6, CH_4, H_2, tar, and char [21,28]. At the exit of the gasification channel, the volatile composition of the syngas consists mostly of CO, H_2, and CH_4. The UCG process can also have other products, including H_2S, As, Hg, Pb, and ash [27,29,30]. The composition of syngas is highly dependent on the gasification agent, air injection method, and coal composition [31,32]. During

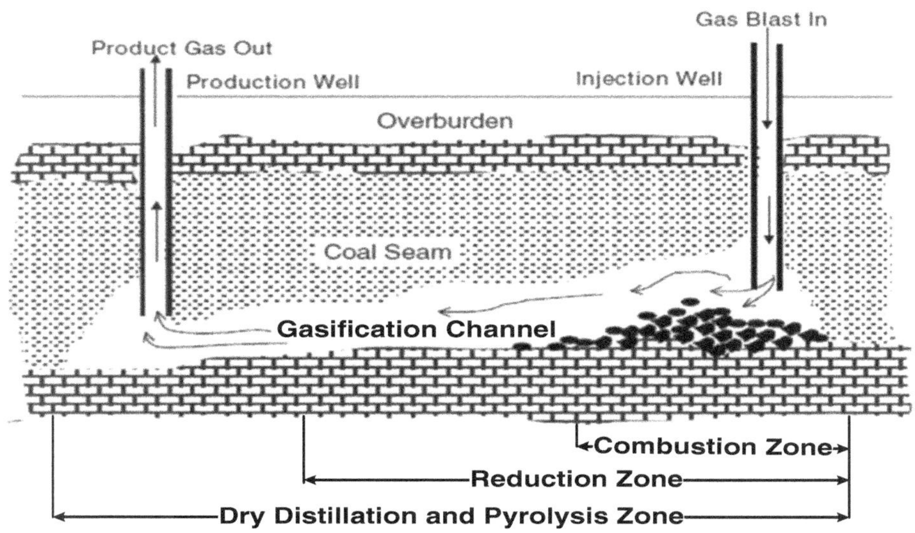

Figure 1 Schematic of *in situ* underground coal gasification process (modified from [21]).

operation, the three gasification zones move along the direction of the air flow, ensuring continuous gasification reactions [21]. A distinguishing feature of UCG, compared to surface gasification, is that drying, pyrolysis, and char gasification occur simultaneously within the coal [26].

By-products of the UCG process pose an environmental hazard to the local surroundings through leaching of organic and inorganic materials into groundwater. Environmental data were first made available after later trials in the USA, including Hanna and Hoe Creek UCG trials, for which groundwater contamination monitoring was conducted before, during, and after gasification. The results illustrated that at shallow depths, UCG can pose a significant risk to groundwater in adjacent strata [30].

Groundwater pollution around UCG zones is mainly caused by one of the following mechanisms: dispersion and penetration of the pyrolysis products of the coal seam to the surrounding rock layers, the emission and dispersion of high contaminants with gas products after gasification, and migration of residue by leaching and penetration of groundwater [30]. In addition, the escaped gases such as carbon dioxide, ammonia, and sulfide may change the pH value of the local strata if dissolved.

The entire process is confined to the space of the coal seam and is sealed from the surface by natural geological formations or man-made barriers; the coal seam and strata serve, to some extent, as a natural groundwater cleaning system. In general, systems have active pressure control, in which the cavity pressure is held in equilibrium or below that of the surrounding strata [17,30]. The pressure difference induces flow into the reactor space, which inhibits gasification products from leaking away from the cavity [33,34].

The quality of the product gas is influenced by several parameters - such as the pressure inside the coal seam, coal properties, feed conditions, kinetics, and heat and mass transport within the coal seam - and the product of the UCG process is a multi-compound, high-temperature, and high-pressure syngas [1]. When the syngas reaches the surface, it is cleaned and undesired by-products are removed from the product stream [24]. Removal techniques are similar to those used with surface gasifiers. Once the by-products are removed, they can be disposed of safely, or used for other chemical processes [16]. The degree of cleaning required is dependent on the use of the syngas; syngas is cleaned either to meet the specification for input into a gas turbine (for electricity generation) or to be of sufficient purity for use as a chemical feedstock for conversion to synthetic fuels [20].

Methods for UCG. Two standard methods of preparing a coal seam for gasification have been utilized successfully: shaft and shaftless. The method implemented is dependent on parameters such as the natural permeability of the coal seam, geochemistry of the coal, seam thickness, depth, width and inclination, proximity of urban developments, and the amount of mining desired [34].

1. *Shaft UCG methods.* Shaft methods use coal mine galleries and shafts to transport gasification reagents and products, which sometimes entail the creation of shafts and the drilling of large-diameter openings through underground labor [34]. The shaft method was the first technique utilized within UCG systems. Currently, the shaft method is only employed in closed coal mines due to economic and safety reasons [34]. The following are examples of common UCG shaft methods:

   - *Chamber or warehouse method.* This method utilizes constructed underground galleries with brick walls separating coal panels. Gasification agents are supplied to a previously ignited coal face on one side of the wall, and the syngas is removed from a gallery on the other side. The chamber method strongly relies on the natural permeability of the coal seam to allow for sufficient oxidant flow through the system. The syngas composition may vary during operation, and the gas production rates are often low. To improve system output, coal seams are often outfitted with explosives for rubblization prior to the reaction zone [21].
   - *Borehole producer method.* For this method, parallel underground galleries are created within a coal seam with sufficient distance between them. The galleries are connected by drilling boreholes from one gallery to the other [34]. Remote electric ignition of the coal in each borehole is used to initiate the gasification process. This method is designed to gasify considerably flat-lying seams. Some variations exist where linking of the galleries is accomplished through hydraulic and electric linking [21,34].
   - *Stream method.* This method is designed for sharply inclined coal beds. Parallel pitched galleries following the contour of the coal seam are constructed and are connected at the bottom of the seam by a horizontal gallery also known as a fire-drift. To initiate gasification, fire is introduced within the horizontal gallery. The hot coal face moves up the seam slope with oxidant fed through one inclined gallery and syngas leaving through the other. The main advantage of this method is that the ash and roof material drop down to fill the void space created during the

process, which prevents suffocating the gasification process at the coal front [21].

- *LLT gasification method.* This method utilizes mined tunnels or constructed roadways to connect the injection well to the production well [4]. Typical long and large tunnel (LLT) systems consist of a gasification channel, two auxiliary holes, and two auxiliary tunnels (Figure 2). The auxiliary holes are arranged between the injection and production wells and are used as malfunction holes for the injection of air and water vapor, or to discharge gas for added gasifier control. LLT also includes an auxiliary tunnel constructed of bricks, which is an auxiliary installation for air injection that prevents blockage in the gasification channel. The mined tunnels are isolated by sealing walls to prevent leakage of combustible gases from the gasifier [35]. The location and height of the oxidant injection points and gas outlet points can be adjusted, allowing for two-dimensional control of oxidant injection and gas production [28].

2. *Shaftless UCG methods.* Recently, most of the focus of research has been on the shaftless methods, which employ directional drilling techniques [6]. The preparation of a reactor for the directional drilling technique consists of the creation of dedicated in-seam boreholes for oxidant injection and product collection using drilling and completion technology that has been adapted from oil and gas production.

With shaftless methods, all preparation and operational processes are carried out through a series of boreholes drilled from the surface into a coal seam and do not require underground labor. Preparation of a shaftless reactor consists of the creation of dedicated in-seam boreholes for oxidant injection and product collection using drilling and completion technology that has been adapted from oil and gas production [34]; the approach generally includes drilling inlet and outlet boreholes into a coal seam, increasing the coal permeability between the inlet and outlet boreholes, igniting the coal seam, introducing an oxidant for gasification, and extraction of the product gas from the outlet well [21]. Currently, there are two main classifications of shaftless UCG methods: linked vertical well (LVW) and controlled retractable injection point (CRIP).

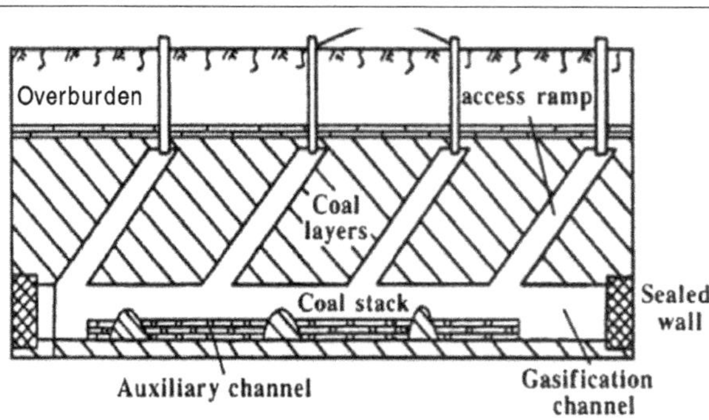

Figure 2 Schematic of the structure of a LLT underground gasifier (modified from [28]).

- *LVW method.* The LVW method is one of the oldest methods for UCG and is derived from technology developed in the former Soviet Union [16]. Vertical wells are drilled into a coal seam, and internal pathways in the coal are utilized to direct the oxidant and product gas flow from the inlet to the outlet borehole. Internal pathways can be naturally occurring or constructed [35]. In its simplest form, the LVW method has inlet and outlet borehole locations that are static for the life of the system. During operation, the coal face migrates and it is found that system control, performance, and syngas quality are affected negatively as the distance from the coal face to the oxidant injection point increases [4]; this factor greatly reduces the feasibility of simple LVW systems.

A more advanced LVW approach involves a series of dedicated injection boreholes located along the length of a coal seam [21]. Over the life of a UCG reactor, the coal face, being gasified, travels as localized coal is exhausted [4]. Having multiple boreholes for injection allows for improved static operating conditions. A more complex variation of the LVW method also exists where multiple inlet and outlet boreholes are drilled into a coal seam, forming inlet and outlet borehole pairs. Parallel inlet and outlet manifolds are connected to the boreholes to provide a path for oxidant and syngas flows, respectively. Coal between each pair of inlet and outlet boreholes forms a zone. When the coal in a zone has been exhausted, new boreholes are drilled in a location of fresh coal, forming new zones [21].

Low-rank coals, such as lignites, have considerable natural permeability and can be exploited for UCG without the need for linking technologies. However, high-rank coals, such as anthracites, are far less permeable, making the gas production rate more limited if UCG is employed [35]. For the use of high-rank coals in UCG, a method of linking must be employed to increase the permeability and fracture the coal seam [36]. The boreholes in traditional LVW gasifiers are linked by special methods including forward combustion, reverse combustion, fire linkage, electric linkage, hydrofracturing, and directional drilling to create sizable gasification channels [35,37].

- *Controlled retractable injection point.* Over the span of a coal seam, the geometry may change, resulting

> in variable UCG operation and system performance [38]. In the past, this problem was solved by having multiple injection and/or production wells so that static operating conditions could be accomplished through moving the gasifier zones to fresh coal [16]. CRIP offers an alternative approach where the vertical injection well is not moved, but the injection point is moved within the coal seam to fresh coal when necessary [39].

The CRIP method relies on a combination of conventional drilling and directional drilling to access the coal seam and physically form a link between the injection and production wells, without the use of linking technologies utilized in LVW methods [38]. A vertical section of injection well is drilled to a predetermined depth, after which directional drilling is used to expand the hole and drill along the bottom of the coal seam creating a horizontal injection well [40]. At the end of the injection well, a gasification cavity is initiated in a horizontal section of the coal seam, creating a localized reactor. The CRIP system utilizes a burner attached to retractable coiled tubing which is used to ignite the coal [39]. The burner burns through the borehole casing to ignite the coal. The ignition point can be moved to any desired location along the horizontal injection well for the creation of a new gasification cavity after a deteriorating reactor has been deserted [38]. Typically, the injection point is retracted using a gas burner, which burns a section of the liner at a desired location [39]. In this manner, accurate control of the gasification process can be obtained. This UCG method has gained popularity in Europe and the USA, but the use of the CRIP method for UCG is fairly new and currently has not become commonly employed [4].

UCG with CO_2 capture and storage

All fossil fuels emit CO_2 when combusted. Currently, coal has the highest CO_2 emissions, per unit energy produced, of the fossil fuels used in combustion [13,41]. To maintain and expand the use of coal, implementation of CCS technologies is becoming imperative.

CO_2 capture can be performed in three main fashions: pre-combustion, post-combustion, and oxy-firing [42]. A broad range of technology options are available for capturing CO_2 including physical absorption, chemical absorption, membrane separation, and cryogenic separation [42,43]. Within UCG, the syngas compositions, temperatures, and pressures of production streams at the exit of a production well are comparable to those of surface gasifiers, which allow similar methods of CO_2 capture. Due to similarities, it is believed that UCG syngas could take advantage of separation using physical sorbents, within a pre-combustion arrangement, which has costs comparable to capture technologies commonly utilized in integrated gasification combined cycles [4,12]. Post-combustion methods are also applicable and would be directly comparable in terms of cost and performance to typical post-combustion systems utilized in power plants. Oxy-firing options are possible for UCG as well, and within a power generating scenario, an air separation unit can generate O_2 streams for injection into the UCG and for use in an oxy-fired plant utilizing the syngas [12].

The spatial coincidence of geological carbon storage (GCS) options with UCG opportunities suggests that designers could colocate and combine UCG and GCS systems with high potential for effective CO_2 storage [18]. In general, these storage options would be the same for conventional carbon sequestration operations, including saline formations and mature oil and gas fields [44]. For UCG-CCS utilizing conventional sequestration options, there could exist common interests in site characterization and monitoring between UCG and CCS projects, where work performed during the design and implementation of one project could be used within the other. Coordinating UCG and CCS designs would improve economics for both projects.

If UCG and CCS are coupled, there is an attractive carbon management scheme associated, where most of the expected CO_2 emissions are sequestered back into a coal seam void that has been recently created by spent subsurface reactors through existing injection and production wells [13,21]. When voids are created, they typically collapse, similar to voids produced during longwall coal mining, leaving zones of artificial breccias with high permeability. Suitable containment zones prevent vertical flow of CO_2 to the surface, where storage locations are isolated from the surface by low-permeability strata (known as seals or caprocks, often shales or evaporites) [4,45]. For a spent UCG system to accommodate CO_2 storage, the void must be at depths below approximately 800 to 1,000 m [44-46]. These depths are required so that supercritical pressures and temperatures exist that allow the CO_2 density to be high enough (approximately 500 to 700 kg/m^3) to limit the storage volume required [45].

The UCG-CCS approach, if successful, could offer an integrated energy recovery and CO_2 storage system, which exploits a new sequestration resource created during operation. A significant challenge with CCS is its large energy requirement [47], of which a considerable portion is consumed during CO_2 capture and compression [48]. The pressure after compression is generally high enough to allow for a reduction in pressure during transport while allowing the fluid to be in a liquid state [9]. If CO_2 storage is accommodated in spent UCG reactors, CO_2 transport and compression requirements decline. CO_2 transport accounts for 5% to 15% of a conventional CCS financial budget, which can be lowered with a self-contained UCG-CCS project, through reduced piping and shipping requirements associated with long-distance transport [18]. A large portion of the budget for a CCS project is allotted for CO_2 storage, typically 10% to 30%, most of which is used for geological and geophysical studies and drilling injection wells [18,48]. These tasks are commonly completed during UCG construction and would not need to be repeated for the implementation of CCS, thus reducing system cost relative to conventional storage methods [18].

As of 2009, it remains unclear if CCS using UCG-produced voids is viable [44]. Until recently, this alternative has received little attention, and there remains substantial scientific uncertainty associated with the technological challenges and environmental risks of storing CO_2 in this manner [13,44]. For full-scale commercialization, extensive research and development is needed to alleviate the uncertainties. Currently, CO_2 sequestration is under development internationally by such organizations as the Intergovernmental Panel on Climate Change and Carbon Sequestration Leadership Forum [13].

Conclusions

Although the earth is an abundant source of coal, a significant amount is currently unrecoverable. With the introduction of UCG, recoverable coal reserves can be expanded by possibly a couple hundred years. Coal is likely to remain used in many countries, increasing the needs for new technologies that permit more environmentally benign extraction and utilization. Wide-scale use of UCG is such a technology option, with the syngas it produces usable as a fuel. Fossil fuels typically utilized in power production could then be used for other purposes, which would result in large reductions in their consumption rates.

UCG offers a coal extraction and conversion method in a single process that avoids many of the challenges associated with conventional mining practices. UCG has a high potential for integration with CCS using conventional methods utilized in power production due to similarities with surface gasifier units. UCG also has the potential to store CO_2 within voids created during its operation, which reduces the need for transport and storage site identification. In essence, UCG could provide a cost-effective, near-zero-carbon, energy source through the use of a self-contained system with a closed carbon loop.

Abbreviations

CCS: carbon capture and storage; CRIP: controlled retractable injection point; GCS: geological carbon storage; LLT: long and large tunnel; LVW: linked vertical well; UCG: underground coal gasification.

Competing interests

The authors declare that they have no competing interests.

Authors' contributions

SS carried out the review and drafted the manuscript. BR and MR co-supervised the investigation, conceived of the study, reviewed the manuscript, and helped draft parts. All authors read and approved the final manuscript.

Authors' information

SS is a PhD candidate at the University of Ontario Institute of Technology studying Mechanical Engineering. His fields of research include energy production and use, thermal energy storage, advanced energy systems, thermofluids, and exergy analysis. BR is a professor in the Faculty of Engineering and Applied Science at the University of Ontario Institute of Technology in Canada. His research interests are energy systems, including heat transfer in fluidized bed combustors, coal, biomass, natural gas, waste heat recovery energy systems analysis, hydrogen, and solar energy. BR has supervised graduate students and postdoctoral fellows, and is actively involved in the organization of conferences. He has delivered invited keynote presentations on energy systems and developments in international conferences, and has published over 120 papers in various technical journals and conferences and contributed book chapters in the energy area. He has received the Best Professor Award for Teaching Excellence three times. MR is a professor in the Faculty of Engineering and Applied Science at the University of Ontario Institute of Technology in Canada, where he was the founding dean of the faculty from 2002 to 2008. He has served as the President of the Engineering Institute of Canada and the Canadian Society for Mechanical Engineering. With over 70 research grants and contracts and 700 technical publications, MR is an active teacher and researcher in thermodynamics, energy technology, sustainable energy, and the environmental impact of energy systems. He has worked for such organizations as Imatra Power Company in Finland, Argonne National Laboratory near Chicago, and the Institute for Hydrogen Systems near Toronto. MR has received numerous awards and honors, including the Engineering Institute of Canada's Smith Medal for achievement in the development of Canada.

Acknowledgments

The authors acknowledge the financial support of the Natural Sciences and Engineering Research Council of Canada.

References

1. Daggupati, S., Mandapati, R.N., Mahajani, S.M., Ganesh, A., Pal, A.K., Sharma, R.K., Aghalayam, P.: Compartment modeling for flow characterization of underground coal gasification cavity. Ind. Eng. Chem. Res. **50**, 277–290 (2011)
2. Tanaka, N.(e.): World Energy Outlook 2009. International Energy Agency, Paris (2009)
3. Energy Information Administration: International Energy Outlook 2010. Department of Energy, U.S. Government, Washington (2010)
4. Roddy, D.J., Younger, P.L.: Underground coal gasification with CCS: a pathway to decarbonising industry. Energy Environ. Sci. **3**, 400–407 (2010)
5. Ediger, V.S., Hosgor, E., Surmeli, A.N., Tatlidil, H.: Fossil fuel sustainability index: an application of resource management. Energy Policy **35**, 2969–2977 (2007)
6. Hammond, G.P.: Energy, environment and sustainable development: a UK perspective. Transactions of the Institution of Chemical Engineers, Part B **78**, 304–323 (2000)
7. Aleklett, K., Hook, M., Jakobsson, K., Lardelli, M., Snowden, S., Soderbergh, B.: The peak of the oil age–analyzing the world oil production reference scenario in World Energy Outlook 2008. Energy Policy **38**, 1398–1414 (2010)
8. World Energy, C.: Survey of Energy Resources 2007. London, World Energy Council (2007)
9. Ghose, M.K., Paul, B.: Underground coal gasification: a neglected option. Int. J. Environ. Stud. **64**(6), 777–783 (2007)
10. Couch, G.: Underground Coal Gasification. IEA Clean Coal Centre, London (2009)

11. Shackley, S., Mander, S., Reiche, A.: Public perceptions of underground coal gasification in the United Kingdom. Energy Policy **34**, 3423–3433 (2006)
12. Burton, E., Friedmann, J., Upadhye, R.: Best Practices in Underground Coal Gasification. Lawrence Livermore National Laboratory, Livermore (2006)
13. Khadse, A., Qayyumi, M., Mahajani, S., Aghalayam, P.: Underground coal gasification: a new clean coal utilization technique for India. Energy **32**, 2061–2071 (2007)
14. Breault, R.W.: Gasification processes old and new: a basic review of the major technologies. Energies **3**, 216–240 (2010)
15. Gregg, D.W., Hill, R.W., Olness, D.U.: An Overview of the Soviet Effort in Underground Gasification of Coal. Lawrence Livermore National Laboratory, Livermore (1976)
16. Shafirovich, E., Varma, A.: Underground coal gasification: a brief review of current status. Ind. Eng. Chem. Res. **48**, 7865–7875 (2009)
17. Van der Riet, M.: Underground coal gasification. In: Proceedings of the SAIEE Generation Conference. Eskom College, Midrand (19 Feb 2008)
18. Roddy, D., Gonzalez, G.: Underground coal gasification (UCG) with carbon capture and storage (CCS). In: Hester, R.E., Harrison, R.M. (eds.) Issues in Environmental Science and Technology, vol. 29, pp. 102–125. Royal Society of Chemistry, Cambridge (2010)
19. Pana, C.: Review of Underground Coal Gasification with Reference to Alberta's Potential. Energy Resources Conservation Board, Edmonton (2009)
20. Walker, L.: Underground coal gasification: a clean coal technology ready for development. The Australian Coal Review **8**, 19–21 (1999)
21. Lee, S., Speight, J.G., Loyalka, S.K.: Handbook of Alternative Fuel Technologies. CRC, Boca Raton (2007)
22. Kempka, T., Plötz, M.L., Schlüter, R., Hamann, J., Deowan, S.A., Azzam, R. Carbon dioxide utilisation for carbamide production by application of the coupled UCG-urea process. Energy Procedia **4**, 2200–2205 (2011)
23. Perkins, G., Sahajwallaa, V.: Modelling of heat and mass transport phenomena and chemical reaction in underground coal gasification. Chem. Eng. Res. Des. **85**(3), 329–343 (2007)
24. Perkins, G., Sahajwalla, V.: Steady-state model for estimating gas production from underground coal gasification. Energy Fuel **22**, 3902–3914 (2008)
25. Peng, F.F., Lee, I.C., Yang, R.Y.K.: Reactivities of in situ and ex situ coal chars during gasification in steam at 1000-1400 °C. Fuel Process. Technol. **41** 233–251 (1995)
26. Perkins, G., Sahajwallaa, V.: A mathematical model for the chemical reaction of a semi-infinite block of coal in underground coal gasification. Energy Fuel **19**, 1679–1692 (2005)
27. Yang, L.H., Pang, X.L., Liu, S.Q., Chen, F.: Temperature and gas pressure features in the temperature-control blasting underground coal gasification. Energy Sources: Part A **32**, 1737–1746 (2010)
28. Yang, L., Liang, J., Yu, L.: Clean coal technology—study on the pilot project experiment of underground coal gasification. Energy **28**, 1445–1460 (2003)
29. Liu, S., Wang, Y., Yu, L., Oakey, J.: Thermodynamic equilibrium study of trace element transformation during underground coal gasification. Fuel Process. Technol. **87**, 209–215 (2006)
30. Shu-qin, L., Jing-gang, L., Mei, M., Dong-lin, D.: Groundwater pollution from underground coal gasification. J. China Univ. Mining & Technol. **17**(4), 0467–0472 (2007)
31. Stańczyk, K., Howaniec, N., Smoliński, A., Świadrowski, J., Kapusta, K., Wiatowski, M., Grabowski, J., Rogut, J.: Gasification of lignite and hard coal with air and oxygen enriched air in a pilot scale ex situ reactor for underground gasification. Fuel **90**, 1953–1962 (2011)
32. Prabu, V., Jayanti, S.: Integration of underground coal gasification with a solid oxide fuel cell system for clean coal utilization. Int. Journal of Hydrogen Energy **37**, 1677–1688 (2012)
33. Yang, L.H.: A review of the factors influencing the physicochemical characteristics of underground coal gasification. Energy Sources, Part A **30** (11), 1038–1049 (2008)
34. Wiatowski, M., Stańczyk, K., Świądrowski, J., Kapusta, K., Cybulski, K., Krause, E., Grabowski, J., Rogut, J., Howaniec, N., Smoliński, A.: Semi-technical underground coal gasification (UCG) using the shaft method in Experimental Mine "Barbara". Fuel **99**, 170–179 (2012)
35. Liang, J., Liu, S., Yu, L.: Trial study on underground coal gasification of abandoned coal resource. In: Xie, H., Golosinki, T.S. (eds.) Proceedings of the '99 International Symposium on Mining Science and Technology, Beijing, August 1999. Mining Science and Technology 99, pp. 271–275. A.A. Balkema, Rotterdam (1999)
36. Blinderman, M.S., Klimenko, A.Y.: Theory of reverse combustion linking. Combustion and Flame **150**, 232–245 (2007)
37. Blinderman, M.S., Saulov, D.N., Klimenko, A.Y.: Forward and reverse combustion linking in underground coal gasification. Energy **33**, 446–454 (2008)
38. Nourozieh, H., Kariznovi, M., Chen, Z., Abedi, J.: Simulation study of underground coal gasification in Alberta reservoirs: geological structure and process modeling. Energy Fuel **24**, 3540–3550 (2010)
39. Klimenko, A.Y.: Early ideas in underground coal gasification and their evolution. Energies **2**, 456–476 (2009)
40. Wang, G.X., Wang, Z.T., Feng, B., Rudolph, V., Jiao, J.L.: Semi-industrial tests on enhanced underground coal gasification at Zhong-Liang-Shan coal mine. Asia-Pac. J. Chem. Eng. **4**, 771–779 (2009)
41. Nag, B., Parikh, J.: Indicators of carbon emission intensity from commercial energy use in India. Energy Economics **22**, 441–461 (2000)
42. Göttlicher, G., Pruschek, R.: Comparison of CO2 removal systems for fossil-fuelled power plant processes. Energy Convers. Mgmt. **38**, 173–178 (1997)
43. Ho, M.T., Allinson, G., Wiley, D.E.: Comparison of CO_2 separation options for geo-sequestration: are membranes competitive? Desalination **192**, 288–295 (2006)
44. Friedmann, S.J., Upadhye, R., Kong, F.M.: Prospects for underground coal gasification in carbon-constrained world. Energy Procedia **1**(1), 4551–4557 (2009)
45. Orr, F.M.: Onshore geologic storage of CO_2. Science **325**(5948), 1656–1658 (2009)
46. Budzianowski, W.M.: Value-added carbon management technologies for low CO2 intensive carbon-based energy vectors. Energy **41**, 280–297 (2012)
47. Steinberg, M.: Fossil fuel decarbonization technology for mitigating global warming. International Journal of Hydrogen Energy **24**, 771–777 (1999)
48. Gibbins, J., Chalmers, H.: Carbon capture and storage. Energy Policy **36**, 4317–4322 (2008)

8

Techno-economic analysis of a hybrid mini-grid system for Fiji islands

Sandeep Lal and Atul Raturi*

Abstract

The Pacific Island Countries constantly struggle with the challenges of high petroleum dependence for their electricity production and lack of adequate energy services. It is possible to alleviate the energy poverty by utilizing abundant renewable energy resources available in the region. The objective of this work is to investigate the feasibility of a wind/solar photovoltaic/diesel generator-based hybrid power system in a remote location in Fiji islands. We used the Hybrid Optimisation Model for Electric Renewables (HOMER) software to simulate the system and perform system optimisation analysis. The system characteristics were studied in terms of optimal configuration, net present cost and the cost of energy. An entirely renewable energy-based configuration is feasible if 10% annual capacity shortage is allowed, while for a scenario with no capacity shortage, addition of a diesel generator is necessary. The addition of renewable energy components results in greenhouse gas reduction which could be used for carbon trading.

Keywords: Hybrid mini-grid system, HOMER, Fiji islands

Background

The Pacific Island Countries (PICs) suffer from the dual challenge of lack of access to modern energy services and high dependence on imported fossil fuels. Some of the countries like Papua New Guinea and Solomon Islands have less than 20% of their population connected to any kind of electricity system whereas 85% of total electricity production in the region is through diesel-based generators [1]. Figure 1 shows the access to electricity in the PICs.

Fiji is reasonably better off than its larger neighbours with almost 80% of its population (92% in urban areas) having access to electricity. In 2008, average electricity generation mix was composed of 62.1% hydro, 33.8% diesel, 0.6% wind and 3.5% biomass components. The energy consumption in Fiji increased from 202.8 million kWh in 1980 to 715.3 million kWh in 2009 [2]. Since the hydro component is highly variable depending on the weather conditions, the increasing demand has to be met through imported diesel-based generation. The Fiji Electricity Authority has an ambitious plan to produce 90% of its electricity using renewable energy resources (mainly hydro) by 2015.

Renewable energy technologies can help mitigate the twin problems mentioned above significantly as there are abundant natural resources like solar, wind, biomass and small hydro in most of the islands. The dispersed population and remoteness of islands make large grid-connected systems unviable, and mini-grid or stand-alone systems are the only options available for expanding electrification. There is a growing interest among PIC governments to utilise renewable energy resources for electricity generation in the region [3]. One of the major challenges in using renewable energy sources for electricity generation is their intermittent and weather-dependent nature. For reliability of electricity supply, it is better to have a combination of technologies. The so-called hybrid system can have different configurations involving various renewable and conventional technologies.

A wind and solar photovoltaic (PV)-based hybrid system holds the best promise for remote area applications [4–7]. Normally, a diesel generator is added to the system to make it more reliable. The many advantages of replacing a completely diesel generator-based system with a hybrid system include reduced fuel bills, reduced greenhouse gas (GHG) emission and lower maintenance

* Correspondence: raturi_a@usp.ac.fj
School of Engineering and Physics, The University of the South Pacific, Suva, Fiji Islands

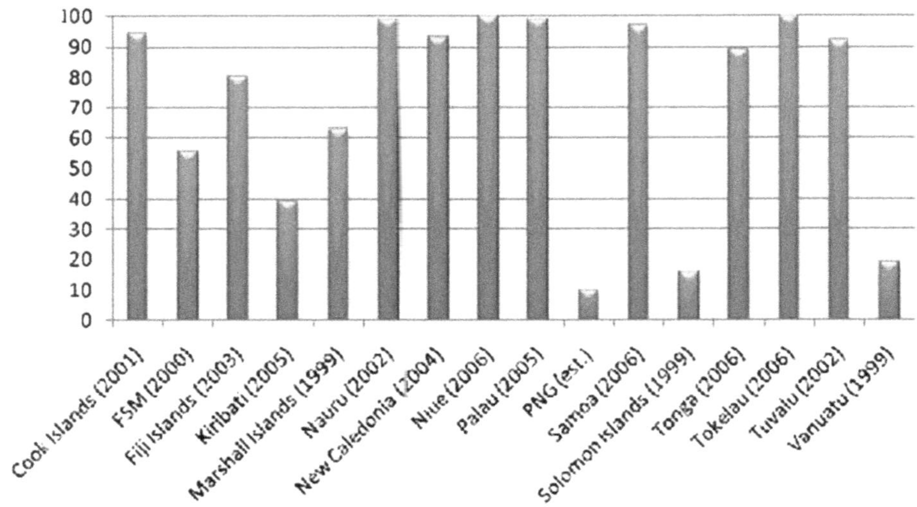

Figure 1 Electricity access in the PICs [1].

costs. The complementary nature of solar and wind resources combined with a storage system significantly increases the reliability of electricity supply [8].

A number of resource assessments have been done in the Pacific region to gauge the possibility of hybrid system development. Table 1 shows wind and solar data for some locations. It can be seen that there are reasonably good prospects for developing such systems.

One of the major barriers in developing renewable energy (RE)-based hybrid systems is the lack of available resource data for the Pacific Islands region. The Asia Sustainable and Alternative Energy Program of the World Bank has produced a wind resource atlas for the region [10], which is very useful in locating sites with good wind energy potential. The NASA Surface Meteorology and Solar Energy website provides global radiation data based on satellite measurements [11]. At the University of the South Pacific, a South Korean-funded project is helping to collect solar and wind resource data for 12 PICs.

Table 1 Wind and solar radiation data for some locations in the PICs [9]

Location	Average wind speed (m/s)	Average wind energy flux (W/m^2)	Average daily solar radiation (Wh/m^2)
Ngatangiia, Rarotonga, Cook Islands	5.5	179	4,300
Vunatovau (10 m), Viti Levu, Fiji	5.5	188	>4,800
Vunatovou (21 m), Viti Levu, Fiji	5.7	204	4,800
Hakupu, Niue	5.9	219	4,600
Tungi Estate, Tongatapu, Tonga	4.2	77	>4,500
White Sands, Efate, Vanuatu	4.7	116	>4,600

A typical hybrid system will comprise a PV array, wind turbines, possibly a diesel generator, inverters and a battery bank with associated control devices. Due to the variability of RE resources, correct sizing of the system components is of paramount importance. Different approaches have been applied to determine the optimal size of the system components and are discussed in detail by Zhou et al. [8] and Bajpai and Dash [12]. Software tools such as Hybrid Optimisation Model for Electric Renewables (HOMER) and HYBRID2 are available to perform simulation and optimisation of hybrid systems.

The HOMER is designed to simulate long-term operation of a combination of micro-power system configurations that could include components like solar PV, wind, small hydro, diesel generators and storage devices like battery banks. It can also model grid-connected systems. The three main tasks performed by HOMER are simulation, optimisation and sensitivity analysis [13]. After simulating a number of combinations, HOMER suggests the optimal configuration based on the lowest net present cost (NPC). Using sensitivity variables (e.g. wind speed and diesel price), it is possible to do a sensitivity analysis that enables the designer to determine the best combination of system components under different conditions.

HOMER has been used to conduct feasibility study of hybrid systems in many locations around the world [14–18]. Bekele and Palm determined the optimal system for supplying electricity to a community of 200 families in Ethiopia [14]. They found that in the 2009 diesel price, the diesel generator/battery/converter set-up was the most cost-effective. A 51% RE-based system

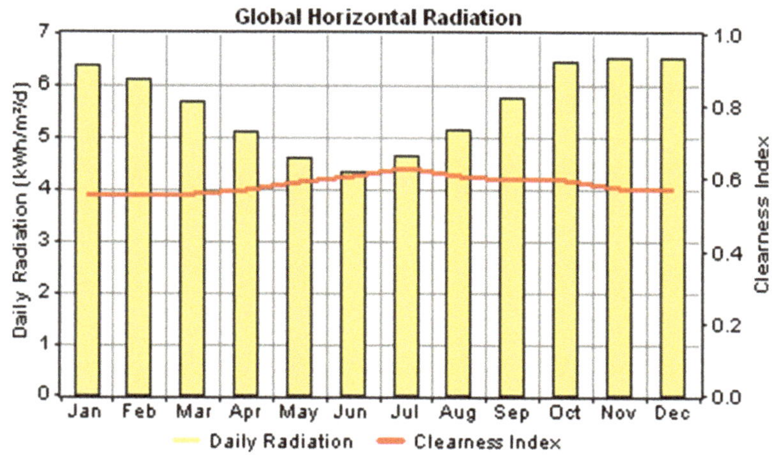

Figure 2 Solar radiation data.

was 19% more expensive but with half the GHG emissions. With the ever-increasing diesel price and continued decrease in solar PV module prices, RE-based systems are becoming more competitive. Al-Karaghouli and Kazmerski applied HOMER to study the life cycle cost of a hybrid system for a rural health clinic in Iraq [5]. A system comprising PV/battery/inverter emerged as the most economic system. Shaahid and El-Amin performed a techno-economic evaluation of PV/diesel/battery systems for rural electrification in Saudi Arabia [15]. They examined the effect of the increase in PV/battery on the cost of energy (COE), operational hours of diesel generators and reduction in GHG emissions. Van Alphen et al. used HOMER to create optimal RE system designs in the Maldives [7]. This study showed that 10% of the total electricity needs of Maldives can be supplied using RE sources.

Methods

Location of the project

This study is done in Nabouwalu (16°59′S, 178°42′E), which is located on the island of Vanua Levu, Fiji. Owing to its distance from the two main town centres on the island, grid electricity is not available at this location. In 1998, a wind/solar/diesel-based hybrid mini-grid system was established at this location, but the hybrid system has fallen into a state of disrepair due to a number of reasons; now, the electricity is supplied using the diesel generators. Considering the high costs involved in the operation and maintenance of these generators, there is a renewed interest in bringing back the hybrid system. This work looks at the feasibility of an optimised wind/solar/diesel hybrid system to replace the existing system.

The following sections describe the main parameters used in the simulation and analysis of the proposed hybrid system.

Solar resource

The Fiji Meteorological Service has been monitoring solar radiation data at Nabouwalu for the last many years. Figure 2 shows the average monthly solar radiation and the clearness index for this location. The annual scaled average value is found to be 5.60 kWh/m^2/day, with the lowest radiation received in the month of June.

Wind resource

The monthly average wind speeds as measured by the meteorological office are shown in Figure 3. The wind

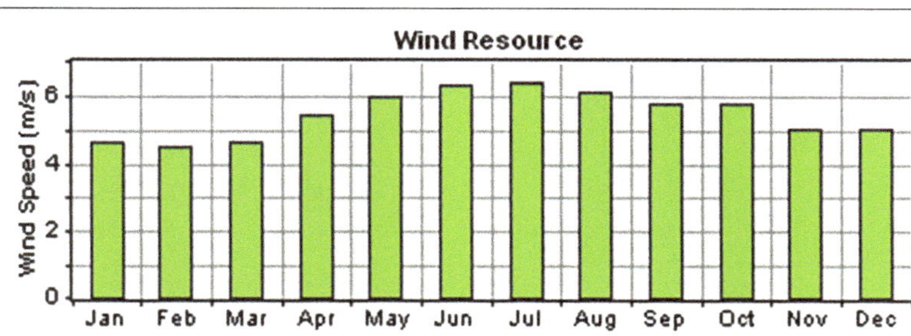

Figure 3 Wind speed data.

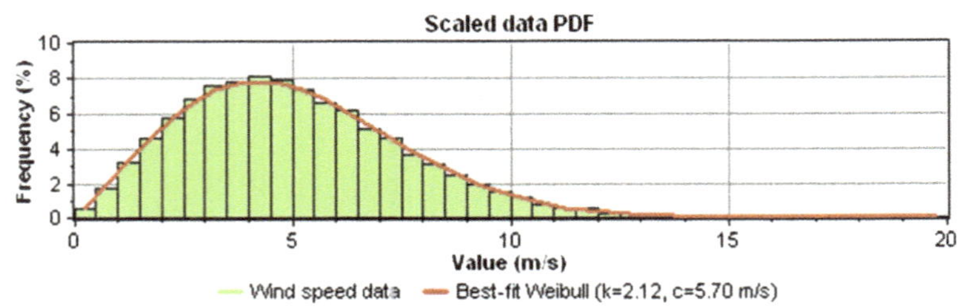

Figure 4 Wind speed distribution.

speed varies between 6.40 m/s in July and 4.60 m/s in January. Figure 4 shows the Weibull distribution for the Nabouwalu wind data with a Weibull factor of 2.1 and peak wind speeds appearing in the afternoon. This would be useful in a solar-wind hybrid system - the wind is picking up as the sun goes down.

Load profile

The Nabouwalu hybrid power system will supply electrical energy to the government hospital, the post office, a primary school, the provincial council building, the Agriculture and Fisheries Department, the police station and staff quarters, the meteorological station, the Public Works Department depot and staff quarters, shops, bakeries, a fisheries ice plant, and 180 residents of the village. Hourly load profile was not available for this location, and the load data were synthesized using daily load profiles. A random variability of daily 10% and hourly 10% noise was added to obtain a peak load of 120 kW. Figure 5 shows the load profile for the proposed system.

Hybrid system design

Considering the solar and wind resources available at this location, the proposed hybrid system will be based on a combination of PV panels, wind turbines, diesel generators and storage batteries. Figure 6 shows the HOMER schematic diagram of the system showing all possible components. The final system configuration is decided after performing the optimisation.

In order to arrive at optimal number of system components, HOMER allows multiple values to be entered. These user-supplied values are then shown in a table depicting the so-called 'search space'. Table 2 shows the different sizes of system components considered. A total of 18,816 system configurations are possible. During simulation, HOMER searches through this space to determine the optimal configuration.

Solar PV

In HOMER analysis, the PV output is taken as directly proportional to the incident radiation. As shown in Table 2, a number of options are available from no solar panels (0 kW) to 400 kW. No restriction was placed for the minimum renewable energy fraction of the total energy mix. The panel costs were taken as Fijian $8,000/kW[a]. The lifetime was taken as 25 years.

Wind turbine

BWC Excel-R wind turbines manufactured from the Bergey WindPower Co. (Norman, OK, USA) were considered in this simulation. Each turbine was rated at 8.1 kW DC and

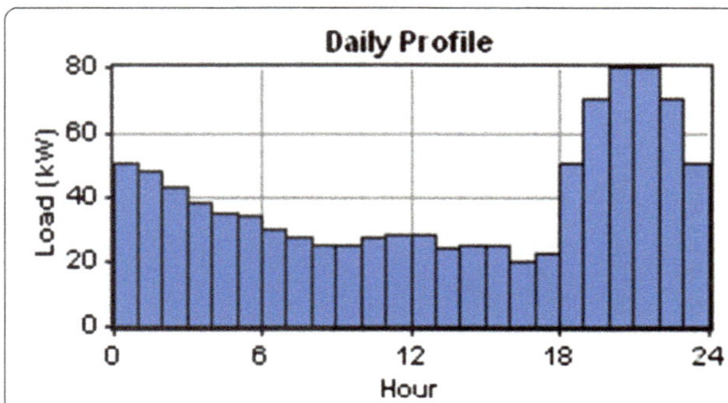

Figure 5 Load profile.

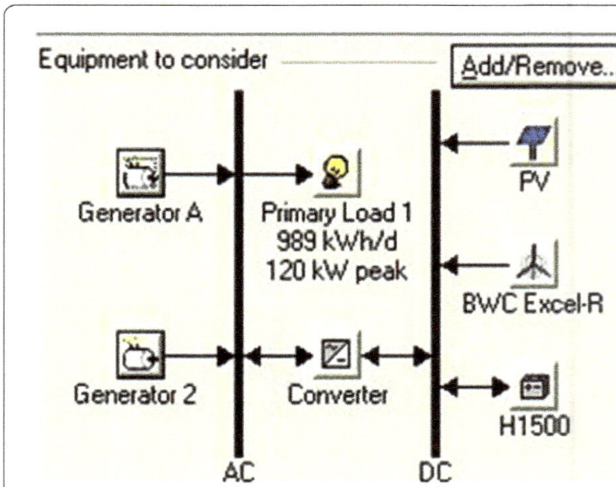

Figure 6 Schematic of the proposed system.

Table 2 System search space

PV array (kW)	XLR (quantity)	Gen A (kW)	Gen 2 (kW)	H1500 (quantity)	Converter (kW)
10	8	85	85	250	150
20	10	150	150	300	200
30	20	200	200	400	
40	25			500	
60	30			550	
80				600	
100					
150					
200					
300					
350					
400					

The components considered include XLR (BWC Excel-R) turbines, Gen A and Gen 2 (85 kW) diesel generators, H1500 (Hoppecke 12 OpzS 1500) batteries and converters.

costs Fijian $76,000. A hub height of 25 m was considered, and turbine lifetime was taken as 25 years. Figure 7 shows the turbine power curve with a cut-in speed of 3 m/s. The number of turbines considered for simulation is shown in Table 2.

Diesel generator

Two diesel generators with a number of size options (Table 2) were considered. The cost of the generators was taken as Fijian $5.00/kW with a lifetime of 40,000 h.

Battery bank

Hoppecke 12 OpzS 1500 batteries manufactured from HOPPECKE Batterien GmbH & Co. KG (Brilon, Germany) were considered as the storage devices. The nominal battery capacity was 1,500 A h with a nominal voltage of 2 V. The initial cost was taken as Fijian $1,200 per battery.

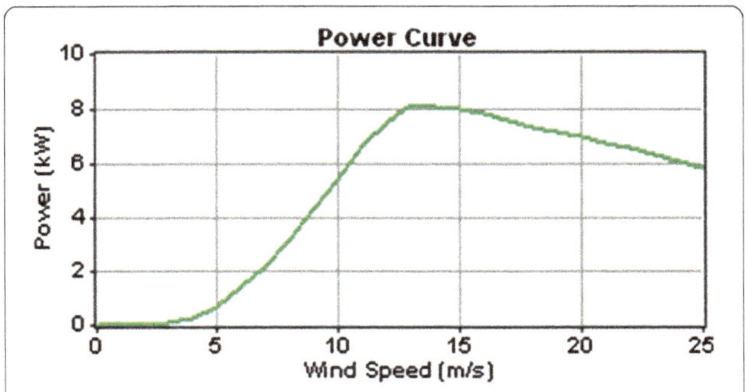

Figure 7 Power curve for the BWC Excel-R wind turbine.

Converter

Two converter sizes were considered: 150 and 200 kW. HOMER suggests an optimal size of 150 kW costing Fijian $50,000.

Hybrid system control parameters and constraints

Two types of dispatch strategies are available in HOMER. In the 'load following' strategy, the generators supply just enough power to service the loads whenever there is insufficient renewable energy contribution. In the 'cycle charging' strategy, the generator (if present) runs at full power and excess electricity is used for charging the batteries. In the present work, an 80% state of charge was set for this strategy.

No minimum renewable energy contribution to the electricity production was set, and simulations were done for 0% and 10% annual capacity shortages. A 6% annual real interest rate was applied

Results and discussion

The input parameters and system constraints, as described above, were used to simulate hybrid systems and perform optimisation and sensitivity analysis. HOMER determines the optimal system by choosing suitable system components (system configuration) depending on sensitivity parameters like wind speed, diesel price and maximum annual capacity shortage. The feasibility of a configuration is based on the NPC and hourly performance.

Sensitivity variables are those over which the designer/modeller does not have any control *viz.* wind speed, solar radiation and diesel price. The HOMER sensitivity analysis assesses the effect of such changes on the configuration rankings. The results are presented in graphical and tabular forms.

Optimisation results

Two scenarios are presented below: (I) with 100% capacity and (II) with 10% annual capacity shortage.

Scenario I: Constraints: Maximum capacity shortage - 0%

Under this scenario, 100% annual load requirements are supplied. According to the sensitivity graph (Figure 8), PV/generator A/generator 2 with battery storage appears as the optimal configuration at lower wind speeds. At higher wind speeds, wind turbine-based systems become more feasible.

Figure 9 shows the categorized list of the most feasible systems for a fixed diesel price of Fijian $2.2/L and an average wind speed of 5.04 m/s. According to this table, the most feasible configuration is made up of 200-kW PV panels and two 85-kW diesel generators with battery storage. The dispatch strategy is 'load following', i.e. the generators operate only to supply power to the loads and the PV system charges the batteries. The NPC of

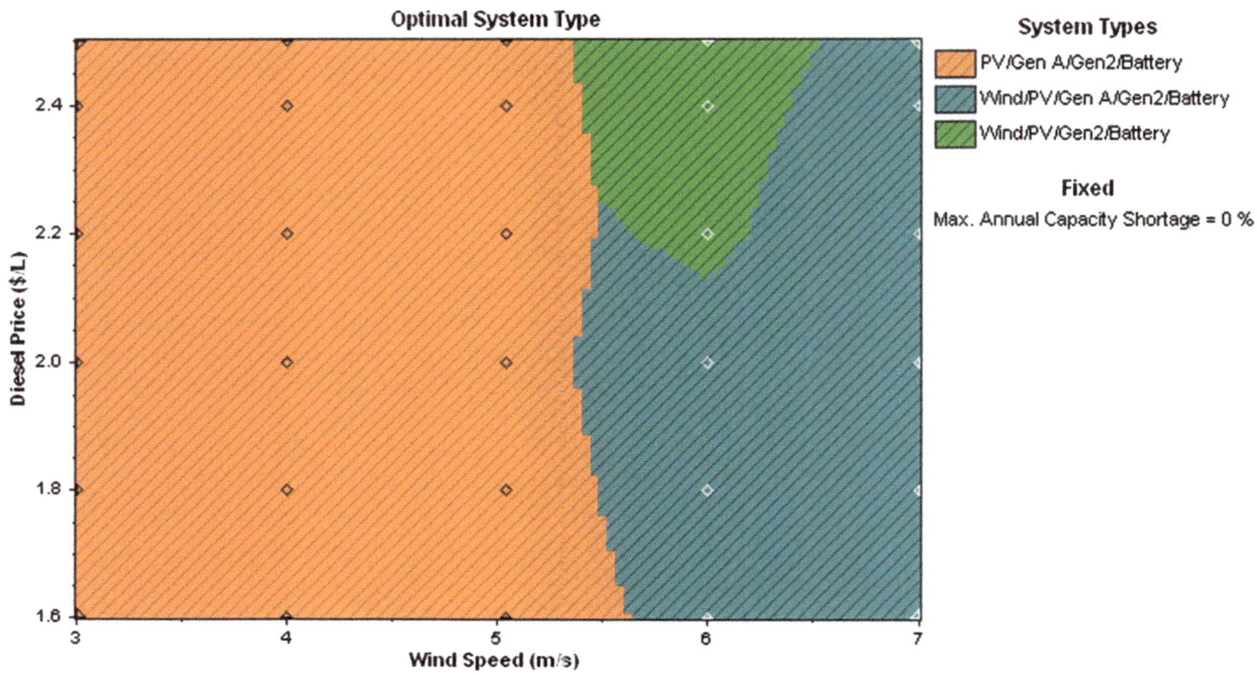

Figure 8 Optimal system graph (0% capacity shortage).

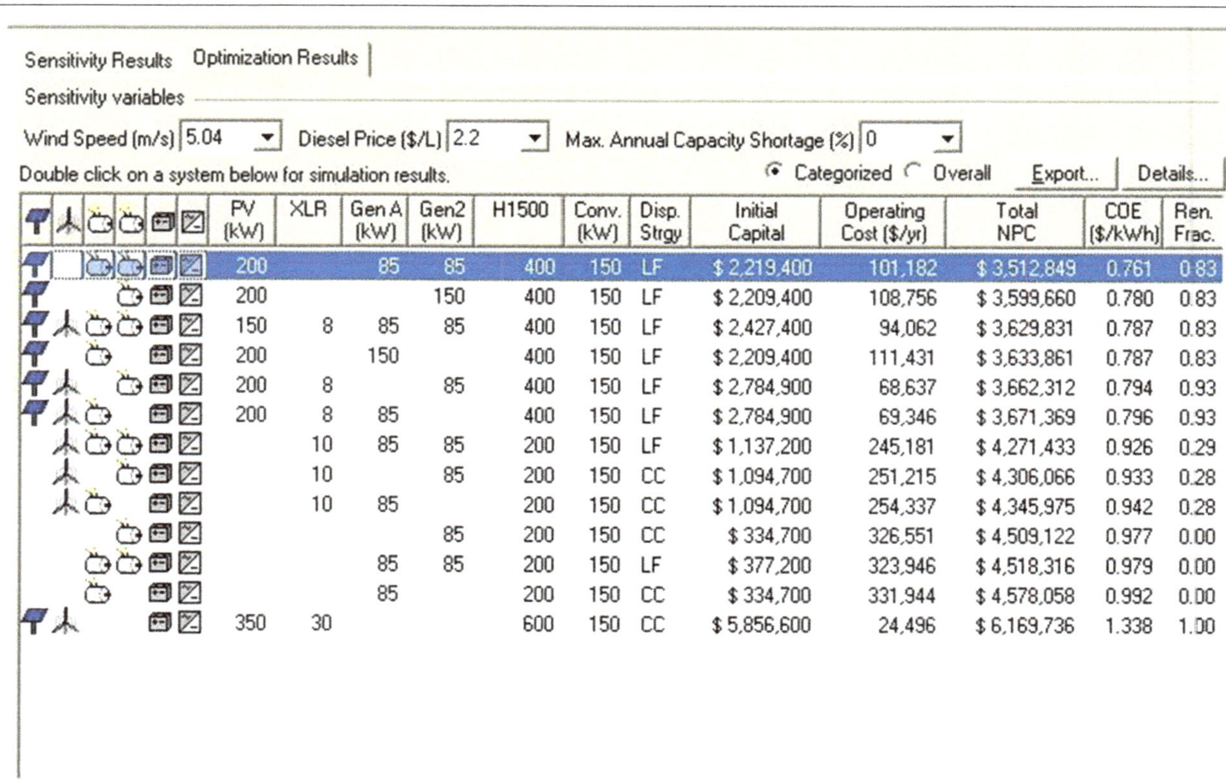

	PV (kW)	XLR	Gen A (kW)	Gen2 (kW)	H1500	Conv. (kW)	Disp. Strgy	Initial Capital	Operating Cost ($/yr)	Total NPC	COE ($/kWh)	Ren. Frac.
	200		85	85	400	150	LF	$ 2,219,400	101,182	$ 3,512,849	0.761	0.83
	200			150	400	150	LF	$ 2,209,400	108,756	$ 3,599,660	0.780	0.83
	150	8	85	85	400	150	LF	$ 2,427,400	94,062	$ 3,629,831	0.787	0.83
	200		150		400	150	LF	$ 2,209,400	111,431	$ 3,633,861	0.787	0.83
	200	8		85	400	150	LF	$ 2,784,900	68,637	$ 3,662,312	0.794	0.93
	200	8	85		400	150	LF	$ 2,784,900	69,346	$ 3,671,369	0.796	0.93
		10	85	85	200	150	LF	$ 1,137,200	245,181	$ 4,271,433	0.926	0.29
		10		85	200	150	CC	$ 1,094,700	251,215	$ 4,306,066	0.933	0.28
		10	85		200	150	CC	$ 1,094,700	254,337	$ 4,345,975	0.942	0.28
				85	200	150	CC	$ 334,700	326,551	$ 4,509,122	0.977	0.00
			85	85	200	150	LF	$ 377,200	323,946	$ 4,518,316	0.979	0.00
			85		200	150	CC	$ 334,700	331,944	$ 4,578,058	0.992	0.00
	350	30			600	150	CC	$ 5,856,600	24,496	$ 6,169,736	1.338	1.00

Figure 9 Optimisation table for feasible configurations (0% capacity shortage).

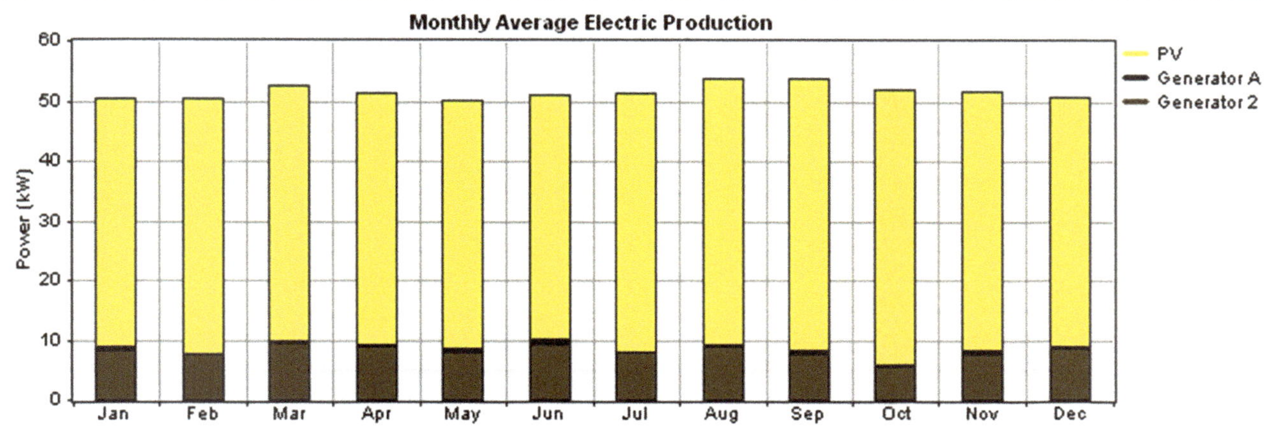

Figure 10 Electricity production (0% capacity shortage).

this configuration is Fijian $3,512,849 and the COE is Fijian $0.761/kWh.

Figure 10 shows the contribution to the total electricity generated by individual system components. The PV system has a capacity factor of 21.5% and supplies 83% of the annual electricity production. Generator A has an overall efficiency of 27%, operates for 1,896 h/annum and has a capacity factor of 9.6%. With the increase in future diesel prices and lower PV module costs, a higher contribution from the RE component can be expected.

Scenario II: Constraints: Maximum capacity shortage - 10%

Under this scenario, 10% of annual capacity shortage is allowed. Figure 11 shows the feasible configuration ranking in this situation. Now, a 100% RE system consisting of a 200-kW PV and eight XLR wind turbines with battery storage turns out to be the optimal configuration. The NPC is Fijian $3,312,476 and the COE is Fijian $0.78/kWh.

Figure 12 depicts the sensitivity analysis for this scenario. At very low wind speeds and low diesel prices, the PV/Gen/battery system is the most feasible configuration, but at higher wind speeds and prevailing diesel price, the wind/PV/battery system becomes the optimal one. The monthly electricity production is shown in Figure 13. The PV system supplies 81% of the total electricity, while the wind turbines contribute 19% with a capacity factor of 15.6%. The 10% capacity shortage might look unacceptable from a utility point of view, but in most rural areas of Fiji, diesel-based electricity is normally available only for 4 h/day. The mini-grid system would be a major step towards an improved rural energy supply. The transportation costs for diesel are

Sensitivity Results | Optimization Results

Sensitivity variables

Wind Speed (m/s) 5.04 | Diesel Price ($/L) 2.2 | Max. Annual Capacity Shortage (%) 10

Double click on a system below for simulation results. Categorized / Overall | Export... | Details...

PV (kW)	XLR	Gen A (kW)	Gen2 (kW)	H1500	Conv. (kW)	Disp. Strgy	Initial Capital	Operating Cost ($/yr)	Total NPC	COE ($/kWh)	Ren. Frac.
200	8			500	150	CC	$ 2,863,500	35,122	$ 3,312,476	0.780	1.00
200			85	400	150	LF	$ 2,176,900	101,683	$ 3,476,756	0.754	0.83
200		85		400	150	LF	$ 2,176,900	103,048	$ 3,494,195	0.758	0.83
200		85	85	400	150	LF	$ 2,219,400	101,182	$ 3,512,849	0.761	0.83
300				400	150	CC	$ 2,934,400	49,512	$ 3,567,329	0.836	1.00
150	8		85	300	150	LF	$ 2,263,800	104,024	$ 3,593,577	0.779	0.81
150	8	85		400	150	LF	$ 2,384,900	96,078	$ 3,613,099	0.783	0.83
150	8	85	85	400	150	LF	$ 2,427,400	94,062	$ 3,629,831	0.787	0.83
	10		85	200	150	LF	$ 1,094,700	243,146	$ 4,202,927	0.915	0.29
	10	85		200	150	LF	$ 1,094,700	248,204	$ 4,267,577	0.929	0.29
	10	85	85	200	150	LF	$ 1,137,200	245,181	$ 4,271,433	0.926	0.29
			85	200	150	LF	$ 334,700	319,628	$ 4,420,614	0.966	0.00
		85		200	150	LF	$ 334,700	325,641	$ 4,497,486	0.983	0.00
		85	85	200	150	LF	$ 377,200	323,946	$ 4,518,316	0.979	0.00

Figure 11 Optimisation table for feasible configurations (10% capacity shortage).

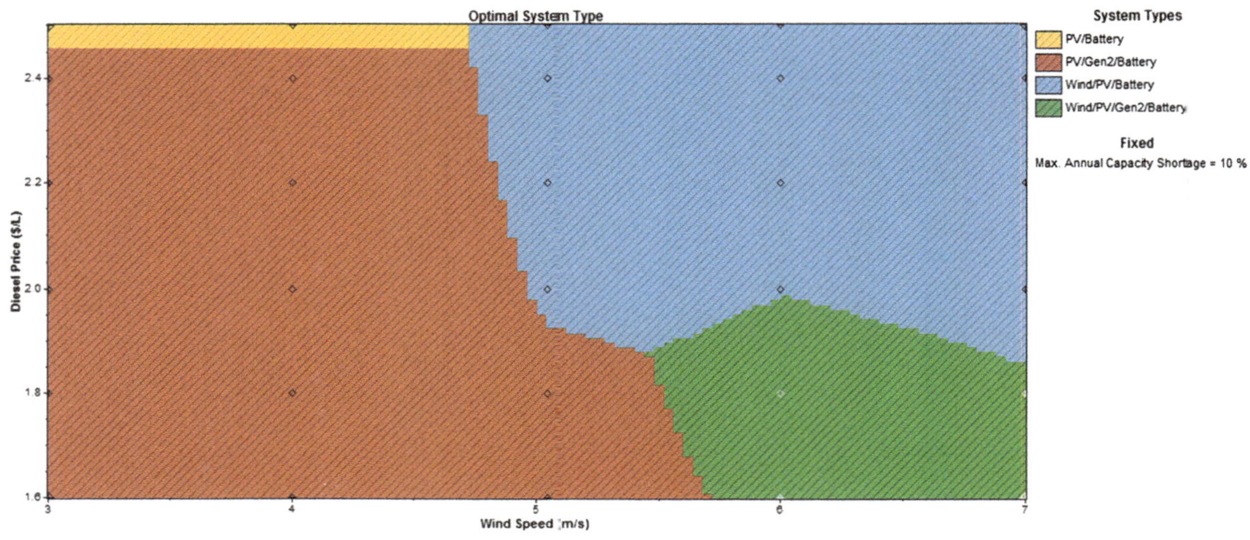

Figure 12 Optimal system graph (10% capacity shortage).

prohibitively high in most of the outer islands in the Pacific region, and idle diesel generators are a common sight in many locations. The operation and maintenance issues are critical for any electricity system, and RE-based hybrid systems have an obvious edge over the conventional diesel generators.

GHG emission reduction

Figure14 shows CO_2 emission from a hybrid system as a function of wind speed and the diesel price. As discussed above, at higher wind speeds and/or higher diesel prices, RE-based configurations become more feasible, resulting in less CO_2 emission. A diesel generator/battery-based system will generate approximately 250 tonnes of CO_2 annually, whereas a wind/solar hybrid system (scenario II) will have zero emissions. This emission reduction can bring in additional income to the project developers through carbon trading. PICs have not really benefited (only two projects) from the Clean Development Mechanism (CDM) activity under the Kyoto Protocol. The main hindrance is the low capacity for the CER generation - the total power generation capacity in the region is only around 1,100 MW. The recently introduced programmatic CDM (p-CDM) might be able to make regional carbon trading projects plausible. A number of solar and wind-based mini-grid systems can be developed as programme activities to be part of a regional p-CDM project. Obviously, a number of institutional and regulatory interventions will be required for this scheme to be successful.

Fiji's electricity sector and the role of renewable energy systems

The Fiji Electricity Authority (FEA) is a corporate body responsible for electricity generation and transmission in

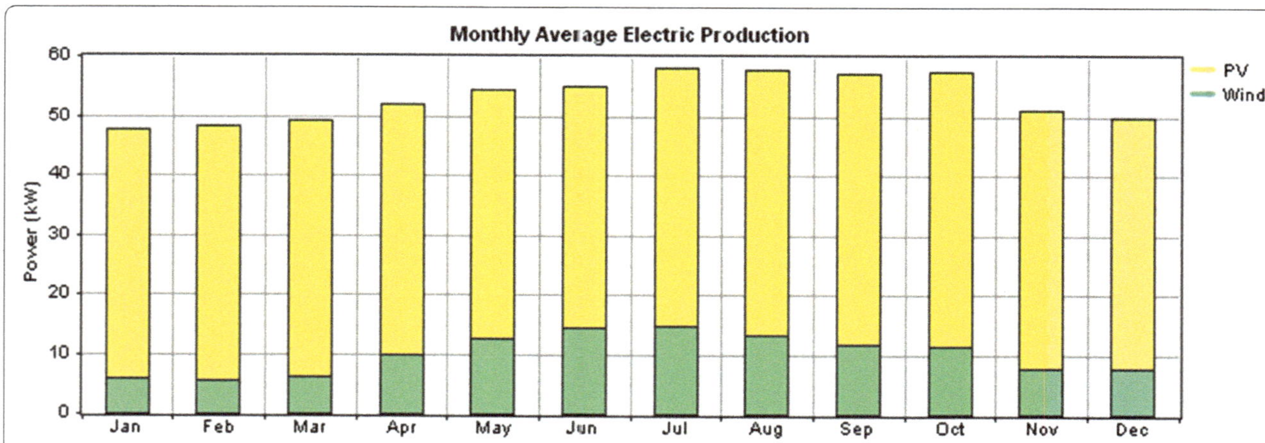

Figure 13 Electricity production (10% capacity shortage).

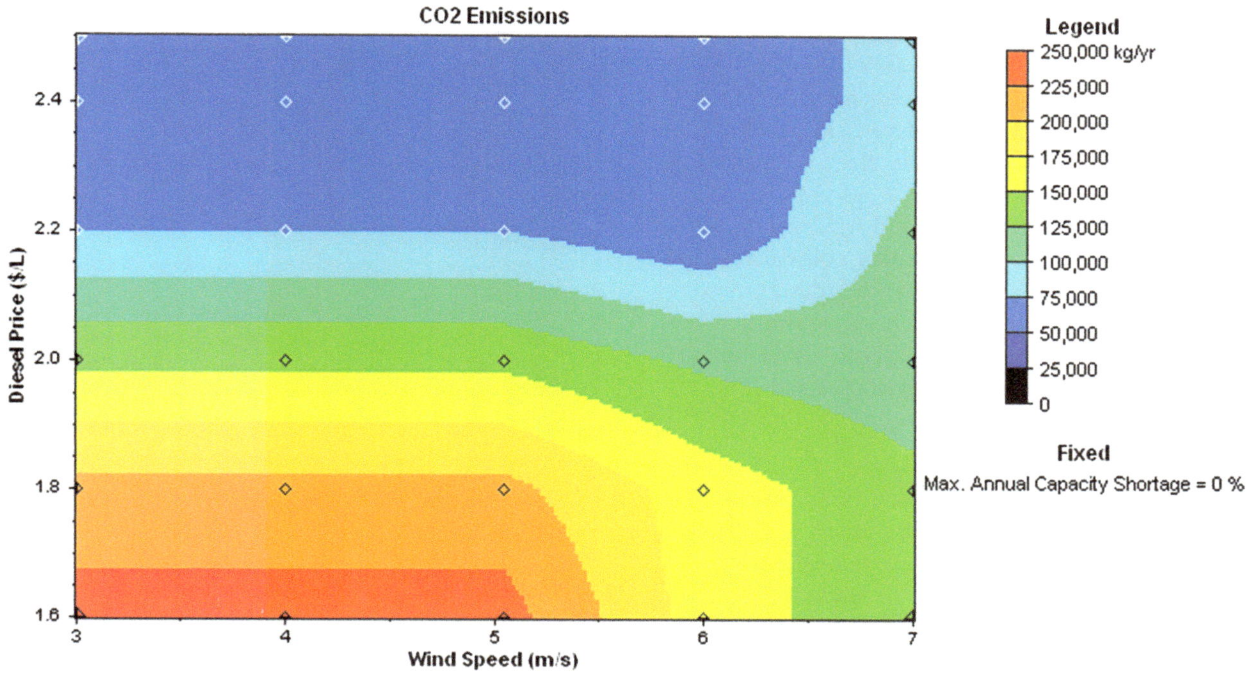

Figure 14 CO_2 **emissions under various scenarios.**

Fiji. Its mission is 'to provide clean and affordable energy solutions to Fiji and the Pacific' [2]. Under the Fiji Commerce Act, the Commerce Commission controls the electricity price and any changes in the tariff rates have to be authorized by the commission. Fiji's tariff rates are the lowest among the Pacific Islands region, and until recently, the rates had stagnated despite a marked increase in demand and resulting diesel usage. One of the major implications of lower tariffs is the lack of interest from the independent power producers (IPPs) as FEA was unable to pay them a suitable buying rate. As Figure 2 shows, at present, FEA is able to supply only about 54% of the demand from its hydropower stations and the rest is from thermal generation. This will change when a new 42-MW hydropower system becomes operational in 2012. According to FEA estimates, diesel-based electricity generation in Fiji is 15% costlier than hydro-based generation, and in 2010 alone, FEA reported a loss of Fijian $17 million ascribed mainly to higher diesel consumption.

The FEA grid is only available on two main islands: Viti Levu and Vanua Levu, while a smaller network is available in Ovalau. The Fiji government has enacted a rural electrification policy under which a rural community can request for access to electricity. This policy provides three options for rural electrification [19]:

1. Grid extension by FEA. The government subsidizes grid extension by providing 95% of the costs, and the community contributes the remaining 5%. The number of communities benefiting from this scheme depends on the rural electrification budget. FEA also implements a grid extension policy based on the internal rate of return (IRR). For projects with an IRR greater than 15%, the grid is extended without any cost to the community; for those with a lower IRR, a contribution from the community is required. This option is available only on the two main islands and in Ovalau.
2. Diesel generator-based mini-grid systems. The Fiji Department of Energy (FDOE), through its Rural Electrification Unit, is responsible for bringing electricity to rural communities spread all over Fiji islands. Presently, the department operates 600 diesel generator-based systems. Due to the costs of involved in operation and maintenance of these systems, electricity is usually available only for 4 h in the evening.
3. Renewable energy-based systems (mainly PV systems). FDOE has supplied more than 1,200 homes with solar home systems (SHSs) in Vanua Levu under its rural electrification scheme. This scheme is run on a Renewable Energy Service Company model where the consumers pay for the service but do not own the systems.

It is obvious that FEA will not be able to extend its grid network to rural communities living on dispersed islands and SHSs are useful only for providing lighting. Diesel-based generation is very expensive and unreliable due to maintenance issues. This is where renewable energy-based hybrid mini-grid systems become important and help in the sustainable development of rural communities. With the new tariffs in place, IPPs will be attracted to develop renewable energy-based systems.

A number of grid-connected PV and wind energy systems have been established or are under construction in the PIC region. Utilities are actively considering newer tariff regimes like feed-in tariff and net metering for renewable energy-based generation. This development augurs well for renewable energy developers and IPPs.

Capacity building

For any RE-based programme to be successful, availability of a sustained cadre of trained professionals is essential. The University of the South Pacific, a regional university, provides RE education and training to students from its 12 member countries. A number of demonstration projects that include a 45-kW grid-connected PV system and stand-alone and wind/solar hybrid systems are part of the training infrastructure. Plans are afoot to develop a certification scheme and competency standards for RE installers/designers. This is being done in cooperation with the Sustainable Energy Industry Association of Pacific Island (SEIAPI).

Conclusions

Renewable energy-based mini-grid systems can play a vital role in bringing sustainable energy to rural communities in the Pacific Islands. In this work, an optimisation and sensitivity analysis of a solar PV/wind/diesel hybrid mini-grid system in Fiji islands has been presented. This study indicates that for the chosen location, the most feasible system consists of a 200-kW PV, 170-kW diesel generators and battery storage if no capacity shortage is demanded. Allowing for 10% capacity shortage, a fully renewable energy-based system becomes feasible.

Endnote

[a]Fijian $1 = US $0.54.

Competing interests
The authors declare that they have no competing interests.

Acknowledgments
The authors are thankful to Fiji Meteorological Service for the solar resource data and Fiji Department of Energy for the load data.

Authors' contributions
SL carried out the site survey and helped in HOMER modeling. AR performed the analysis and drafted the manuscript. Both authors read and approved the final manuscript.

Authors' information
SL received his BETech degree from the School of Engineering and Physics, University of the South Pacific. AR is an associate professor in Physics at the same school.

References

1. SPC: Pacific regional information system (PRISM). http://www.spc.int/prism/ (2012). Accessed 1 Feb 2012
2. Fiji Electricity Authority: Fiji electricity authority. www.fea.com.fj (2012). Accessed 5 Jan 2012
3. CROP Energy Working Group: Ministers' communiqué, Proceedings of the Pacific Energy Minister's Meeting and Regional Energy Officials' Meeting, Nuku'alofa, Tonga. http://ict.sopac.org/VirLib/JC0200.pdf (2009). Accessed 10 Oct 2011
4. Khan, MJ, Iqbal, MT: Feasibility study of standalone hybrid systems for applications in Newfoundland. Renew. Energy **30**, 835–854 (2005)
5. Al-Karaghouli, A, Kazmerski, L: Optimization and life-cycle cost of health clinic PV system for a rural area in southern Iraq using HOMER software. Solar Energy **84**, 710–714 (2010)
6. Mondal, AH, Denish, M: Hybrid systems for decentralized power generation in Bangladesh. Energy Sustain. Dev. **14**, 48–55 (2010)
7. Van Alphen, K, Wilfried, GJHM, Hekkert, MP: Renewable energy technologies in the Maldives—determining the potential. Renew. Sustain. Energy Rev. **11**, 1650–1674 (2007)
8. Zhou, W, Lou, C, Li, Z, Lu, L, Yang, H: Current status of research on optimum sizing of stand-alone hybrid solar–wind power generation systems. Applied Energy **87**, 380–389 (2010)
9. SOPAC: SOPAC Report ID JCO 121. www.sopac.org (1995). Accessed 10 Oct 2011
10. AWS Truewind/ASTAE: The 2009 Pacific Islands Wind Energy Resource Atlas. http://siteresources.worldbank.org/INTEAPASTAE/Resources/FarEastIslandsWindResourceMap_2009.pdf (2009). Accessed 1 June 2012
11. NASA: Surface Meteorology and Solar Energy. http://eosweb.larc.nasa.gov/sse/ (2004). Accessed 12 Oct 2011
12. Bajpai, P, Dash, V: Hybrid renewable energy systems for power generation in stand-alone applications: a review. Renew. Sustain. Energy Rev. **16**(1), 2926–2939 (2012)
13. HOMER Energy LLC: Hybrid optimization model for electric renewables. http://www.homerenergy.com/ (2009). Accessed 5 Aug 2011
14. Bekele, G, Palm, B: Feasibility study for a standalone solar–wind-based hybrid energy system for application in Ethiopia. Applied Energy **87**, 487–495 (2010)
15. Shaahid, SM, El-Amin, I: Techno-economic evaluation of off-grid hybrid based hybrid energy system for photovoltaic–diesel–battery power systems for rural electrification in Saudi Arabia—a way forward for sustainable development. Renew. Sustain. Energy Rev. **13**(3), 625–633 (2009)
16. Hassan, K, Faitma, K, Mahmood, HS: Feasibility of hybrid power generation over wind and solar standalone system. 5th Power Engineering and Optimization Conference (PEOCO), Selangor **7**, 6–7 (2011)
17. Nandi, SK, Ghosh, HR: Prospect of wind–PV-battery hybrid power system as an alternative to grid extension in Bangladesh. Energy **35**, 3040–3047 (2010)
18. Rehman, S, Al-Hadhrami, ML: Study of a solar PV–diesel–battery hybrid power system for a remotely located population near Rafha Saudi Arabia. Energy **36**, 4986–4995 (2010)
19. Fiji Department of Energy: www.energy.gov.fj (2012). Accessed 5 Jan 2012

Analysis of wind speed data and wind energy potential in three selected locations in south-east Nigeria

Sunday O Oyedepo[1], Muyiwa S Adaramola[2*] and Samuel S Paul[3]

Abstract

In this study, the wind speed characteristics and energy potential in three selected locations in the southeastern part of Nigeria were investigated using wind speed data that span between 24 and 37 years and measured at a height of 10 m. It was shown that the annual mean wind speed at a height of 10 m for Enugu, Owerri and Onitsha are 5.42, 3.36 and 3.59 m/s, respectively, while the annual mean power densities are 96.98, 23.23 and 28.34 W/m^2, respectively. It was further shown that the mean annual value of the most probable wind speed are 5.47, 3.72 and 3.50 m/s for Enugu, Owerri and Onitsha, respectively, while the respective annual value of the wind speed carrying maximum energy are 6.48, 4.33 and 3.90 m/s. The performance of selected commercial wind turbine models (with rated power between 50 and 1,000 kW) designed for electricity generation and a windmill (rated power, 0.36 kW) for water pumping located in these sites was examined. The annual energy output and capacity factor for these turbines, as well as the water produced by the windmill, were determined. The minimum required design parameters for a wind turbine to be a viable option for electricity generation in each location are also suggested.

Keywords: Mean wind speed, Wind power density, South-east Nigeria, Electricity, Water pumping

Background

The quest to reduce environmental impacts of conventional energy resources and, more importantly, to meet the growing energy demand of the global population had motivated considerable research attention in a wide range of environmental and engineering application of renewable form of energy. It is recognized that wind energy, as a renewable energy source, has stood out as the most valuable and promising choice. Wind energy by nature is clean, abundant, affordable, inexhaustible and environmentally preferable. Due to its many advantages, wind energy has also become the fastest growing renewable source of energy in both developed and developing countries. For example, wind energy is widely used to produce electricity in countries like Denmark, Spain, Germany, USA, China and India. Interestingly, the global cumulative installed capacity of wind power had increased sharply from 6,100 MW in 1996 to about 237,669 MW in 2011 [1]. In Africa, for example, Egypt, Morocco and Tunisia are the leading countries with installed capacities of 550, 291 and 114 MW, respectively, at the end of 2011 [1].

The increasing energy demand, the rapidly depleting fossil fuel reserves and the environmental problems associated with the use of fossil fuel have necessitated the development of alternative energy sources like wind energy for electricity generation in Nigeria. It is reported that the electricity production in Nigeria as of the end of 2010 is less than 4,000 MW due to fluctuations in the availability and maintenance of production sources, leading to a shortfall in supply [2]. However, analyses of available wind data for selected cities have confirmed a high prospect of wind energy resources in Nigeria. Several studies on renewable sources of energy have also been performed. A detailed review and discussion of these studies can be found in [2-6] and are not repeated here. Worthy of mention here from these studies, however, is that the effective utilization of wind energy at a typical location requires sound knowledge of the wind characteristics and accurate wind data analysis. For

* Correspondence: muyiwa.adaramola@usask.ca
[2]Department of Energy and Process Engineering, Norwegian University of Science and Technology, Trondheim, NO-7491, Norway
Full list of author information is available at the end of the article

example, the choice of wind turbine design must be based on the average wind velocity at a selected wind turbine installation site [7]. Prior studies have also shown that the wind flow patterns are influenced by terrains, vegetation and water bodies.

Although several studies have been performed to investigate the characteristics and pattern of wind speed across Nigeria, less attention has been given to sites in the south-east region. According to [8,9], the few reported studies on wind speed in this part of the country were limited to wind speed distributions, while less attention was paid to the wind energy potential evaluation. The focus of this study is, therefore, to evaluate the wind energy potential in three selected locations (Enugu, Owerri and Onitsha) in the south-east region and to assess the performance of selected small- to medium-size commercial wind turbines. It is the authors' view that this information will be helpful to the government and any organization in making an informed decision with regard to investment in wind energy resource in this part of Nigeria.

Methods

The wind data used in this study were obtained from the Nigerian Meteorological Agency, Oshodi, Lagos. The geographical coordinates of the meteorological stations where the wind speed data were captured at a height of 10 m by a cup-generator anemometer are given in Table 1. There are many sources of measurement uncertainty in cup-anemometer measurements. The guidelines and steps necessary to minimize these errors are outlined in Manwell et al. [10]. Following the methodologies proposed and explained in the ISO guide [11] to the expression of uncertainty in measurement, the uncertainty in the mean velocities at 95% confidence level was determined to be ±2%. Monthly wind data that span between 24 and 37 years were obtained for Enugu, Owerri and Onitsha. The recorded wind speeds were computed as the mean of the speed for each month. It should be noted that using monthly wind speed has some limitations such as loosing extremely low or high wind speeds within the month as well as inability to observe diurnal variations in the wind speed. However, using monthly mean wind speed, which is mostly available for most locations, can be used to study the seasonal changes in wind speed and facilitates wind data analysis.

Table 1 The geographical location of the selected stations

Station	Latitude (N)	Longitude (E)	Altitude (m)	Measurement period
Enugu	6° 26′	7° 29′	304.70	1971 to 2007
Owerri	5° 29′	7° 02′	186.05	1977 to 2002
Onitsha	6° 10′	6° 47′	63.14	1978 to 2003

N, North; E, East.

Frequency distribution and site wind speed parameters

Several mathematical models such as normal and lognormal have been used for wind data analysis. Prior studies have also shown that statistical methods such as the Weibull and Rayleigh distribution models can equally be used [12]. According to [13-15], the two-parameter Weibull probability distribution function is the most appropriate, accepted and recommended distribution function for wind speed data analysis. This is because it gives a better fit for measured monthly probability density distributions than other statistical functions [12,15]. In addition, the Weibull parameters at known height can be used to estimate wind parameters at another height [13]. Therefore, the two-parameter Weibull probability density function was used in this study. In Weibull distribution, the variation in wind velocity is characterized by two parameter functions: the probability density function and the cumulative distribution. The probability density function $f(V)$ indicates the probability of the wind at a given velocity V, while the corresponding cumulative distribution function of the velocity V gives the probability that the wind velocity is equal to or lower than V, or within a given wind speed range. The Weibull probability density function is given as, e.g., [12,16]:

$$f(V) = \left(\frac{k}{c}\right)\left(\frac{V}{c}\right)^{k-1} \exp\left[-\left(\frac{V}{c}\right)^{k}\right] \tag{1}$$

where $f(V)$ = the probability of observing wind speed (V), k = dimensionless Weibull parameter and c = the Weibull scale parameter (in meter per second). The scale factor could be related to the mean wind speed through the shape factor, which determines the uniformity of the wind speed in a given site. The cumulative distribution $F(V)$ is the integral of the probability density function, and it is expressed as, e.g., [12,16]:

$$F(V) = 1 - e^{-\left(V/c\right)^{k}} \tag{2}$$

The monthly and annual values of Weibull parameters were calculated using standard deviation method. This method is useful where only the mean wind speed and standard deviation are available. In addition, it gives better results than graphical method and has relatively simple expressions when compared with other methods [13,17,18]. Moreover, it is unlike most of the other methods that may require more detailed wind data (which, in some cases, are not readily available) for the determination of the Weibull distribution shape and scale parameters. The shape and scale factors are thus computed from Equations 3 and 4 given by [13,19]:

$$k = \left(\frac{\sigma}{V_m}\right)^{-1.086} \quad (3)$$

$$c = \frac{V_m}{\Gamma(1 + \frac{1}{k})} \quad (4)$$

where σ is the standard deviation, V_m is the mean wind speed (in meter per second) and $\Gamma(x)$ is the gamma function, which is defined as [17,19]:

$$\Gamma(x) = \int_0^\infty t^{x-1} e^{-t} dt. \quad (5)$$

Alternatively, scale factor can be determined from the following expressions given by [20]:

$$c = \frac{V_m k^{2.6674}}{0.184 + 0.816 k^{2.73855}} \quad (6)$$

Equation 6 is used in this study to estimate the monthly and annual scale factors.

In addition to the mean wind speed, the other two significant wind speeds for wind energy estimation are the most probable wind speed (V_F) and the wind speed carrying maximum energy (V_E). They can be expressed respectively as [12,21]:

$$V_F = c\left(\frac{k-1}{k}\right)^{1/k} \quad (7)$$

$$V_E = c\left(\frac{k+2}{k}\right)^{1/k} \quad (8)$$

The most probable wind speed corresponds to the peak of the probability density function, while the wind speed carrying maximum energy can be used to estimate the wind turbine design or rated wind speed. Prior studies have shown that wind turbine system operates most efficiently at its rated wind speed. Therefore, it is required that the rated wind speed and the wind speed carrying maximum energy should be as close as possible [16].

Extrapolation of wind speed at different hub height

In most cases, the available wind data are measured at a height different from the wind turbine hub height. It is noted that it is the wind speed at the hub height that is of interest for wind power application; therefore, the available wind speeds are adjusted to the wind turbine hub height using the following power law expression, e.g., [12]:

$$\frac{V}{V_o} = \left(\frac{h}{h_o}\right)^{\alpha} \quad (9)$$

where V is the wind speed at the hub height h, V_o is wind speed at the original height h_o and α is the surface roughness coefficient and is assumed to be 0.143 (or 1/7) in most cases. The surface roughness coefficient α can be determined from the following expression [22]:

$$\alpha = [0.37\text{–}0.088\ln(V_o)]/\left[1\text{–}0.088\ln\left(\frac{h_o}{10}\right)\right] \quad (10)$$

Alternatively, the Weibull probability density function can be used to obtain the extrapolated values of wind speed at different heights. Since the boundary layer development and the effect of the ground are non-linear with respect to wind speed, the scale factor c and form factor k of the Weibull distribution will change as a function of height by the following expressions [13]:

$$c(h) = c_o\left(\frac{h}{h_o}\right)^{n} \quad (11)$$

$$k(h) = k_o\left[1 - 0.088\ln\left(\frac{h_o}{10}\right)\right]/\left[1 - 0.088\ln\left(\frac{h}{10}\right)\right] \quad (12)$$

where c_o and k_o are the scale factor and shape parameter, respectively, at the measurement height h_o. The exponent n is defined as:

$$n = [0.37 - 0.088\ln(c_o)]/\left[1 - 0.088\ln\left(\frac{h}{10}\right)\right] \quad (13)$$

Mean wind power density and energy density

The mean wind power density can be estimated by using the following equation:

$$P_D = \frac{P(V)}{A} = \frac{1}{2}\rho V_m^3 \quad (14)$$

where $P(V)$ = the wind power (in watt), P_D = the wind power density (watt per square meter), ρ = the air density at the site (assumed to be 1.225 kg/m^3 in this study) and A = the swept area of the rotor blades (in square meter). Both the mean wind speed and power density are generally used to classify the wind energy resource (e.g., Pacific Northwest Laboratory (PNL) wind power classification scheme, Illica et al. [23]). However, the wind power density (wind power per unit area) based on the Weibull probability density function can be calculated using the following equation [24]:

$$P_D = \frac{P(V)}{A} = \frac{1}{2}\rho c^3 \Gamma\left(1 + \frac{3}{k}\right) \quad (15)$$

The mean energy density (E_D) over a period of time T is the product of the mean power density and the time T, and it is expressed as:

$$E_D = \frac{1}{2}\rho c^3 \Gamma\left(1 + \frac{3}{k}\right) T \quad (16)$$

Wind turbine energy output and capacity factor

A wind energy conversion system can operate at its maximum efficiency only if it is designed for a particular site because the rated power and cut-in and cut-off wind speeds must be defined based on the site wind characteristics [12]. It is essential that these parameters are selected so that energy output from the conversion system is maximized. The performance of a wind turbine installed in a given site can be examined by the amount of mean power output over a period of time ($P_{e,ave}$) and the conversion efficiency or capacity factor of the turbine. The capacity factor C_f is defined as the ratio of the mean power output to the rated electrical power (P_{eR}) of the wind turbine [12,20].

The mean power output $P_{e,ave}$ and capacity factor C_f of a wind turbine can be estimated using the following expressions based on Weibull distribution function [12]:

$$P_{e,ave} = P_{eR}\left(\frac{e^{-\left(\frac{v_c}{c}\right)^k} - e^{-\left(\frac{v_r}{c}\right)^k}}{\left(\frac{v_r}{c}\right)^k - \left(\frac{v_c}{c}\right)^k} - e^{-\left(\frac{v_f}{c}\right)^k}\right) \quad (17)$$

$$C_f = \frac{P_{e,ave}}{P_{eR}} \quad (18)$$

where v_c, v_r and v_f are the cut-in wind speed, rated wind speed and cut-off wind speed, respectively. For an investment in wind power to be cost effective, it is suggested that the capacity factor should be greater than 0.25 [25].

Wind-driven rotodynamic pumps

There are three types of wind-powered pumping systems. They are the mechanical-piston pump, the mechanical-air lift (rotodynamic) pump and the electrical pump. In general, the volume of water produced by rotodynamic and electrical pumps are considered to be more than that of a piston pump at the same wind speed regime. This is because there is a better match between the rotodynamic and electrical pumps and the wind rotor than for a piston pump [26,27]. In this study, the performance of a rotodynamic pump is simulated. For a rotodynamic pump driven by a wind turbine with a given cut-in wind speed and a cut-out wind speed, the water produced over a period of time T can be determined from:

$$Q = T\int_{V_i}^{V_o} Q(V) f(V)\, dV \quad (19)$$

where $Q(V)$ is the discharge of the pump at any wind speed, and it is given by [28]:

$$Q(V) = \frac{1}{8} C_P \eta V D\left(\frac{\rho_a}{\rho_w}\right)\left(\frac{V_d^2}{gH}\right)\left(\frac{G\lambda_D}{N_{PD}}\right) \quad (20)$$

where V_d is design wind speed, G is the gear ratio, N_{PD} is the speed of the pump at design condition, D is the wind turbine rotor diameter and λ_D is the design tip speed ratio of the wind turbine. For a water-pumping application, a tip speed ratio between 1 and 3 is recommended [10]. By substituting Equation 20 into Equation 19 and assuming Rayleigh probability density function $f(V)$, the total water produced over a given time is expressed as [26]:

$$Q = \frac{\pi}{16 V_m^2} C_{P_D} \eta_{PD} T D\left(\frac{\rho_a}{\rho_w}\right)\left(\frac{V_d^2}{gH}\right)\left(\frac{G\lambda_D}{N_{PD}}\right)\int_{V_i}^{V_o} V^2 \exp^{-\left(\frac{\pi}{4}\left(\frac{V}{V_m}\right)^2\right)} dV \quad (21)$$

Results and discussion

Wind speed frequency distribution

The annual probability density frequency and cumulative distributions of wind speed for the three locations obtained using the Weibull distribution function are shown in Figure 1. The probability density function is used to illustrate the fraction of time for which given wind speed possibly prevails at a location. As expected, the peak of the density function frequencies of all the sites skewed towards the higher values of mean wind speed (Figure 1a). It should be remarked that the peak of the probability density function curve indicates the most frequent velocity. It can be observed from Figure 1a that the most frequent wind speed expected in Enugu, Owerri and Onitsha are about 5.5, 3.5 and 3.5 m/s, respectively. It can be further observed that Enugu has the highest spread of wind speed toward high wind speed among the locations.

The cumulative probability distributions of the wind speed at all the study locations (Figure 1b) show a similar trend. The cumulative distribution function can be used for estimating the time for which wind speed is within a certain speed interval. For wind speeds greater or equal to 2.5 m/s cut-in wind speed, Enugu, Owerri and Onitsha have frequencies of about 96.9%, 86.5% and

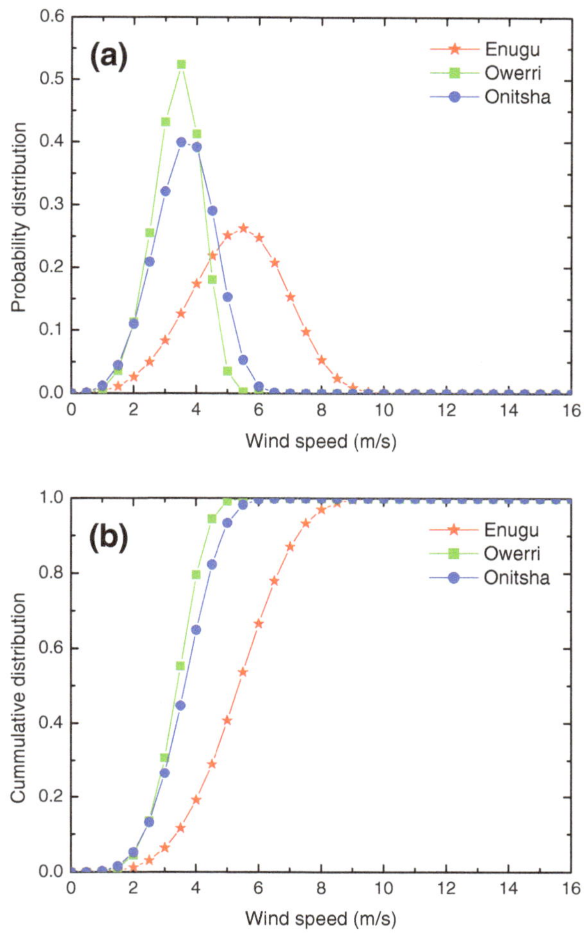

Figure 1 Annual wind speed distribution. **(a)** Probability density function and **(b)** cumulative density function.

86.9%, respectively, while the same locations respectively have frequencies of about 88.4%, 44.7% and 55.3% for wind speed of 3.5 m/s cut-in wind speed. According to Ojosu and Salawu [3], if a wind turbine system with a design cut-in wind speed of 2.2 m/s is used in these sites for wind energy resource for electricity generation, all the sites will have frequencies of more than 92%.

Mean wind speed and mean power density

The monthly variation of the mean wind speed characteristics (V_m, V_F and V_E), mean power density and mean energy density as well as the annual values of these parameters at a height of 10 m are presented in Tables 2, 3 and 4.

The monthly mean wind speed varies between 4.13 m/s in November and 6.30 m/s in March for Enugu site (Table 2). The monthly mean power density varies between 43.15 W/m^2 in November and 153.16 W/m^2 in March. Therefore, based on PNL wind power classification scheme [23], the monthly mean power density mostly falls into class 1 ($P_D \leq 100$) except in January, February, April and July, when it falls into class 2 ($100 < P_D \leq 150$), and in March, when it falls into class 3 ($150 < P_D \leq 200$). However, the annual mean power density for this site is 96.98 W/m^2 (class 1). For Owerri (Table 3), the minimum and maximum values of the monthly mean wind speeds are 2.72 and 3.70 m/s, respectively, while the annual mean wind speed for this site is 3.36 m/s. The monthly mean power density varies between 11.66 W/m^2 in November and 31.02 W/m^2 in January. The monthly mean power density falls into class 1 wind resource category ($P_D \leq 100$)

Table 2 Characteristic speeds and mean power density in Enugu at a height of 10 m

	V_m (m/s)	k	c (m/s)	V_F (m/s)	V_E (m/s)	P_D (W/m^2)	E_D (kWh/m^2)
Jan	5.62	3.49	6.25	5.68	7.12	108.72	78.279
Feb	5.67	4.86	6.19	5.90	6.64	111.65	80.387
Mar	6.30	5.75	6.81	6.58	7.17	153.15	110.271
Apr	6.22	5.11	6.77	6.49	7.22	147.39	106.123
May	5.35	4.44	5.87	5.54	6.39	93.79	67.530
Jun	5.21	4.82	5.69	5.42	6.11	86.62	62.367
Jul	5.48	5.45	5.94	5.72	6.29	100.80	72.574
Aug	5.44	3.93	6.01	5.58	6.68	98.61	70.996
Sept	4.85	4.79	5.30	5.04	5.70	69.88	50.311
Oct	4.56	5.24	4.96	4.76	5.27	58.08	41.815
Nov	4.13	3.75	4.58	4.22	5.13	43.15	31.066
Dec	4.95	3.34	5.52	4.96	6.36	74.29	53.488
Annual	5.42	4.05	5.87	5.47	6.48	96.98	717.619

V_m, mean wind speed; k, dimensionless Weibull shape parameter; c, Weibull scale parameter; V_F, most probable wind speed; V_E, wind speed carrying maximum energy; P_D, wind power density; E_D, mean energy density.

Table 3 Characteristic speeds and mean power density in Owerri at a height of 10 m

	V_m (m/s)	k	c (m/s)	V_F (m/s)	V_E (m/s)	P_D (W/m^2)	E_D (kWh/m^2)
Jan	3.52	2.61	3.96	3.29	4.92	26.71	19.234
Feb	3.54	8.11	3.74	3.68	3.84	27.17	19.564
Mar	3.64	7.00	3.87	3.79	4.02	29.54	21.269
Apr	3.70	6.61	3.96	3.87	4.12	31.02	22.338
May	3.45	8.48	3.63	3.58	3.72	25.15	18.109
Jun	3.55	8.46	3.74	3.68	3.83	27.40	19.730
Jul	3.32	6.30	3.56	3.47	3.72	22.41	16.138
Aug	3.42	5.30	3.72	3.57	3.95	24.50	17.641
Sept	3.31	7.01	3.53	3.45	3.66	22.21	15.993
Oct	3.04	5.41	3.30	3.17	3.49	17.21	12.390
Nov	2.67	6.28	2.97	2.88	3.10	11.66	8.394
Dec	3.00	4.80	3.28	3.12	3.53	16.54	11.907
Annual	3.36	5.10	3.65	3.50	3.90	23.23	203.530

V_m, mean wind speed; k, dimensionless Weibull shape parameter; c, Weibull scale parameter; V_F, most probable wind speed; V_E, wind speed carrying maximum energy; P_D, wind power density; E_D, mean energy density.

Table 4 Characteristic speeds and mean power density in Onitsha at a height of 10 m

	V_m (m/s)	k	c (m/s)	V_F (m/s)	V_E (m/s)	P_D (W/m^2)	E_D (kWh/m^2)
Jan	3.59	4.91	3.92	3.74	4.20	28.34	20.404
Feb	3.73	4.11	4.11	3.84	4.53	31.79	22.886
Mar	4.23	4.86	4.62	4.41	4.96	46.36	33.378
Apr	4.07	5.00	4.44	4.24	4.75	41.29	29.732
May	3.76	4.71	4.11	3.91	4.43	32.56	23.442
Jun	3.72	4.55	4.08	3.86	4.42	31.53	22.702
Jul	3.58	4.51	3.92	3.71	4.26	28.10	20.234
Aug	3.61	4.77	3.95	3.76	4.25	28.82	20.747
Sept	3.52	4.74	3.85	3.66	4.15	26.71	19.234
Oct	3.28	4.23	3.60	3.37	3.94	21.61	15.562
Nov	3.01	4.78	3.29	3.13	3.54	16.70	12.026
Dec	3.05	3.52	3.39	3.08	3.85	17.38	12.512
Annual	3.59	4.27	3.96	3.72	4.33	28.34	248.252

V_m, mean wind speed; k, dimensionless Weibull shape parameter; c, Weibull scale parameter; V_F, most probable wind speed; V_E, wind speed carrying maximum energy; P_D, mean wind power density; E_D, mean energy density.

in all the months, and the annual mean power density for this site is 23.23 W/m^2 (class 1). In the case of Onitsha (Table 4), the minimum and maximum values of the monthly mean wind speeds are 3.01 m/s (in November) and 4.23 m/s (in March), respectively. The monthly mean power density varies between 16.70 W/m^2 in November and 46.36 W/m^2 in March. The monthly mean power density falls into class 1 wind resource category ($P_D \leq 100$) in all the months, and the annual mean power density for this site is 28.34 W/m^2 (class 1). Detailed information about these sites' wind speed characteristics (mean wind speed, most probable wind speed (V_F) and the wind speed carrying maximum energy (V_E)) and mean power density are illustrated in Tables 2, 3 and 4.

The least monthly value of the Weibull shape parameter k for Owerri is 2.61 in January and reached the highest value of 8.48 in the month of May. Therefore, the wind speed is most uniform in Owerri in May, while it is least uniform in December. The annual shape factors for Enugu, Owerri and Onitsha are 4.05, 5.10 and 4.27, respectively. The least monthly value of Weibull scale parameter c is obtained as 2.97 m/s in the month of November in Owerri, and the highest value of 6.81 m/s in the month of March in Enugu. The annual shape factors for Enugu, Owerri and Onitsha are 5.87, 3.65 and 3.96 m/s, respectively.

In summary, Enugu has the highest annual mean wind speed among the sites considered in this study. Also, this site has the highest values of annual power density. Even though the most probable wind speed (V_F) is a statistical characteristic, which may not be directly connected to wind energy [21], it does not necessarily mean that Enugu has much higher wind potentials than the other locations considered. However, as mentioned earlier, the efficiency of a wind turbine is closely related to these parameters, especially V_E, which should be as close as possible to the design or rated wind speed of the system. Therefore, the proposed wind turbine, if installed in Enugu, would likely produce more power than other locations. Moreover, it can be considered as the best site for wind energy development in southern Nigeria (based on the three sites considered in this study). Furthermore, the monthly mean wind speeds in south Nigeria ranges from 2.72 to 6.30 m/s. The monthly mean power density varies between 11.66 and 153.15 W/m^2, while the annual mean power density is in the range of 23.23 to 96.98 W/m^2. It can be inferred from this analysis that the wind resource in this part of Nigeria can be classified mostly into class 2 or less category. Furthermore, the annual mean energy density varies between 203.53 and 717.62 kWh/m^2.

Even though the wind resource in these locations falls into class 2 or less, which is considered as marginally or unsuitable for wind power development, the wind power can be used for water pumping and small-scale electricity generation, providing intermittent power requirements for a variety of purposes that need low-energy capacity, slow-running high-torque wind turbines with multi-blade, e.g., [3,4,6,29]. For a modern wind turbine, the cut-in wind speed required by it to start generating electricity is generally between 3 to 5 m/s. Depending on the size of the turbine, the peak power output can be attained when the wind speed (rated wind speed) is in the range of 10 to 15 m/s [6]. For water pumping, wind turbine can be operated at a lower wind speed; however, they can function effectively when the wind speed is more than 3 m/s. Based on the required quantity of water, a site with a mean wind speed around 2.0 m/s can be considered for wind-powered pump application [30]. Similarly, depending on the end use of the generated power, it can be concluded that these locations may be suitable for utilization of wind energy.

Performance of selected wind turbines

Seven small- to medium-size commercial wind turbine models with rated power range from 50 to 1,000 kW [31-33] were selected to simulate their performance at Enugu, Owerri and Onitsha. These are P15-50, P19-100, P50-500 and P62-1000 models (Polaris America LLC, Lakewood, NJ, USA); WES30 model (Wind Energy Solutions BV, The Netherlands); WWD-1-60 model (Winwind, Espoo, Finland) and BONUS 1000–54 (Siemens AG, Erlangen, Germany). The selected wind turbine models and their characteristic properties are given in Table 5. For each location, the annual energy output and capacity factor based on Weibull distribution

Table 5 Characteristics of the selected wind turbines

	POLARIS P15-50	POLARIS P19-100	WES30	POLARIS P50-500	POLARIS P62-1000	WWD-1-60	BONUS-1000-54
Rated power (kW)	50	100	250	500	1,000	1,000	1,000
Hub height (m)	30	30	36	50	60	70	45
Rotor diameter (m)	15.2	19.1	30	50	62	60	54
Cut-in wind speed (m/s)	2.5	2.5	2.7	2.5	2.5	3.6	3
Rated wind speed (m/s)	10	12	12.5	12	12	12.5	14
Cut-out wind speed (m/s)	25	25	25	25	25	25	25

[12,31-33].

function parameters at their respective hub height are determined using Equations 17 and 18, respectively.

The performance of the selected wind turbine models at all the locations is presented in Figure 2. The figure clearly reveals that irrespective of the wind turbine model, Enugu seems to be the best site for wind power development for electricity generation. This is expected because, when compared with other sites, Enugu has the highest annual mean wind speed and highest value of V_E at the hub height for each turbine. The annual energy output for Enugu ranges from about 113 MWh/year using POLARIS 19–100 model to 2,444 MWh using WWD-1-60 model. Among the 1,000-kW model turbines, the WWD-1-60 model produced the highest power output, closely followed by the POLARIS 62–1000 (2,431 MWh/year), while BONUS 1000–54 produced the least power (849 MWh/year). This observed trend is related to the hub height and rotor diameter (which are lowest for the BONUS model), the design or rated wind speed (highest for the BONUS model, 14 m/s) and the cut-in wind speed (highest for the WWD-1-60 model, 3.6 m/s). Even though the hub height of the WWD-1-60 model is higher than that of the POLARIS 62 model, the POLARIS 62 model still produced almost the same amount of power (2,431 MWh/year) due to its lower cut-in wind speed and rated wind speed. Furthermore, it is observed that the energy output from the POLARIS 15–50 model is slightly more than the power output from the POLARIS 19–100 model. This is because both models have the same hub height, and the POLARIS 19–100 model has higher rated wind speed than the POLARIS 15–50 model.

The minimum annual energy outputs of 6.05 and 18.83 MWh/year are observed for Owerri and Onitsha, respectively, using the P19-100 model. While the maximum annual energy outputs are 178.58 and 461.32 MWh/year, respectively, for Owerri and Onitsha with the WWD-1-60 model. For each of these sites, the power generated by each wind turbine follows the same trend observed in Enugu. Regardless of the location, the WWD-1-60 wind turbine model produced the highest quantity of annual energy output. For example, if 1,000-kW turbines are to be operated at the same hub height, the POLARIS 62–1000 will likely perform better than WWD-1-60 and BONUS-1000-54 because of its low cut-in wind speed and rated wind speed as well as its bigger rotor diameter compared with other models.

The cost effectiveness of a wind turbine can be roughly estimated by the capacity factor of the turbine. This factor is a useful parameter for both consumer and manufacturer of the wind turbine system [24]. The WWD-1-60 model

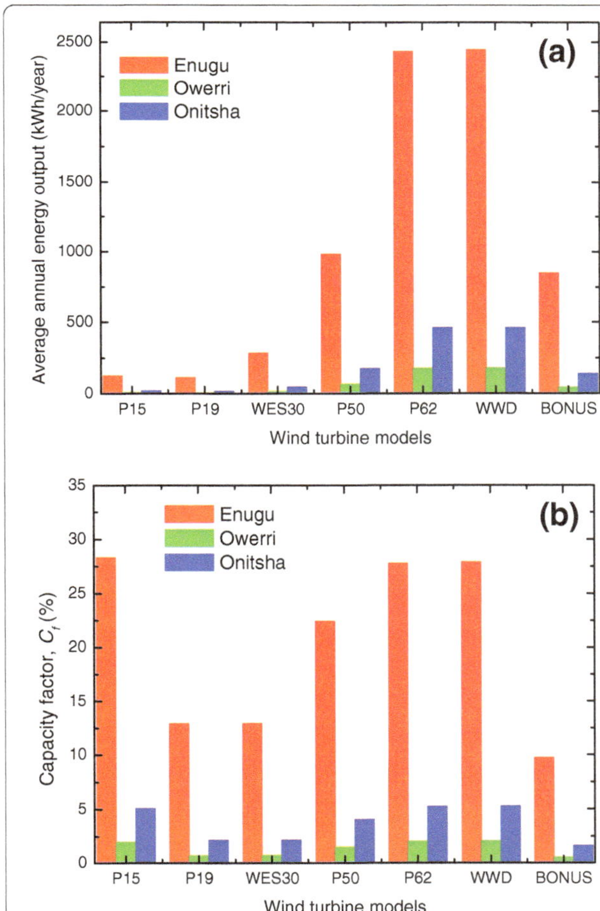

Figure 2 The performance of the selected wind turbine models for all the locations. (a) Average annual energy output and **(b)** capacity factor.

has the highest value among the models considered for all the sites. The C_f values for this model are 27.90%, 2.04% and 5.27% for Enugu, Owerri and Onitsha, respectively. The C_f values for Enugu for POLARIS 15–50, POLARIS 62–1000 and WWD-1-60 are 28.30%, 27.75% and 27.90%, respectively. These values are greater than the suggested recommended value before an investment can be considered worthwhile. Hence, these wind turbines or similar turbine model will be good for electricity generation in Enugu. In Owerri, however, the capacity factor for other wind turbine models ranges from 0.49% for the BONUS-1000-54 model to 2.04% for the WWD-1-60 model. Similarly, the minimum and maximum capacity factors at Onitsha are 1.62% for the BONUS-1000-54 model and 5.27% for the WWD-1-60 model, respectively. Therefore, these two sites may be considered for wind energy development for small-scale applications such as water pumping (see 'Wind-powered pumps performance'). It should be noted that the cost of generating electricity using wind turbine is inversely proportional to the capacity factor. The higher the capacity factor (or higher wind speed regime), the lower the cost of generated electricity, assuming that all factors remain the same (see, e.g., Paul et al. [34]).

Based on the annual energy output and the capacity factor, the POLARIS 62–1000 and WWD-1-60 models or wind turbines with similar designed characteristics will be best suited for electricity generation at Enugu and small-scale application in other locations. However, by redesigning the selected wind turbine models to operate at lower cut-in wind speed (especially the WWD-1-60 and BONUS-100-54 models), lower rated wind speed (especially BONUS-100-54) and higher hub height compared with their current design parameters (cut-in and rated wind speeds and hub height), both the annual energy output and capacity factor could significantly be improved. For instance, if POLARIS 62–1000 is to be operated at a hub height of 70 m and rated wind speed of 10 m/s, the capacity factor for Enugu, Owerri and Onitsha will be 59.17%, 8.20% and 16.99%, respectively. However, increasing the hub height may increase the overall capital cost of the wind turbines. But this is compensated for by increased in capacity factor and, hence, the energy output from the wind turbines.

In order to meet the minimum recommended capacity factor (25%) for electricity generation, the following design parameters are suggested: wind turbine model with a minimum hub height of 55 m, cut-in wind speed of less than 3.5 m/s, rated wind speed of around 12 m/s and cut-out wind speed of 25 m/s are recommended for Enugu; for Owerri, wind turbine with a minimum height of 75 m, cut-in wind speed of less than 3.5 m/s, rated wind speed of around 8.5 m/s and cut-out wind speed of 20 m/s are recommended; while a system with a minimum hub height of 65 m, cut-in wind speed of less than 3.5 m/s, rated wind speed of around 9 m/s and cut-out wind speed of 20 m/s are recommended for Onitsha.

Wind-powered pumps performance

In this section, the performance of wind-powered pumps assumed to be located in each of the locations considered in this study is presented and discussed. For the performance assessment, a wind turbine model with a rated power of about 0.36 kW is chosen. The characteristic properties of this wind turbine model and the specifications of the pumps are given in Table 6. The wind turbine characteristics are similar to the FT-2.5 wind turbine model (produced in China, according to [35]) [35,36] for water pumping. The hub height of the wind turbine is assumed to be 15 m, which is within the hub height range for common commercial wind turbines for water-pumping application [30]. The pump head of 10 m is used for the analysis. This pump head was chosen because it is within the range of the water table level of 3 to 20 m in this part of Nigeria [37] and within the range of the design lifting height of 5 to 10 m for the FT-2.5 windmill [35]. The monthly mean wind speeds at hub height at each selected site are computed using Equations 9 and 10. The quantity of water pumped per month is computed using Equation 23. This equation was solved using Wind Energy Resource Analysis software developed by Mathew [25]. The software is based on Rayleigh distribution function and requires the mean wind speed, wind turbine and pump parameters as inputs.

The comparison between water produced per month by rotodynamic pump at the three locations is presented in Table 7. As expected from the wind speed frequency distribution, the amount of water output from the pump at Enugu is highest among the three locations. In Enugu, the monthly water output varies between about 3,136 and 3,512 m^3, and the average volume discharge is about

Table 6 Wind turbine parameters and rotodynamic pump specifications

Parameters/specifications	Value
Wind turbine	
Diameter (m)	2.5
Rated power (W)	360
Rated speed (m/s)	7
Cut-in speed (m/s)	2.5
Cut-out speed (m/s)	10
Design speed ratio	2
Design power coefficient	0.35
Pump	
Efficiency (pump and transmission)	0.6
Gear ratio	10

Table 7 Monthly water produced and the number of habitants that can be served per month

Month	Enugu				Owerri				Onitsha			
	V_m (m/s)	Q (m^3)	Habitants 50 L/day	Habitants 36 L/day	V_m (m/s)	Q (m^3)	Habitants 50 L/day	Habitants 36 L/day	V_m (m/s)	Q (m^3)	Habitants 50 L/day	Habitants 36 L/day
Jan	6.16	3,511.51	2,341	3,251	3.92	2,687.47	1,792	2,488	4.00	2,752.48	1,835	2,549
Feb	6.21	3,510.02	2,340	3,250	3.95	2,710.74	1,807	2,510	4.15	2,859.98	1,907	2,648
Mar	6.88	3,436.48	2,291	3,182	4.05	2,787.89	1,859	2,581	4.69	3,194.29	2,130	2,958
Apr	6.79	3,450.62	2,300	3,195	4.12	2,843.95	1,896	2,633	4.52	3,100.75	2,067	2,871
May	5.87	3,507.62	2,338	3,248	3.85	2,633.46	1,756	2,438	4.18	2,885.82	1,924	2,672
Jun	5.72	3,495.80	2,331	3,237	3.96	2,721.07	1,814	2,520	4.14	2,857.08	1,905	2,645
Jul	6.01	3,512.36	2,342	3,252	3.71	2,510.67	1,674	2,325	3.98	2,738.42	1,826	2,536
Aug	5.97	3,511.42	2,341	3,251	3.82	2,605.90	1,737	2,413	4.03	2,770.28	1,847	2,565
Sept	5.34	3,432.44	2,288	3,178	3.70	2,502.78	1,669	2,317	3.93	2,692.31	1,795	2,493
Oct	5.04	3,343.72	2,229	3,096	3.41	2,232.75	1,489	2,067	3.65	2,460.31	1,640	2,278
Nov	4.58	3,136.10	2,091	2,904	3.11	1,936.35	1,291	1,793	3.38	2,205.90	1,471	2,043
Dec	5.45	3,456.25	2,304	3,200	3.37	2,192.40	1,462	2,030	3.42	2,240.39	1,494	2,074

Hub height is assumed as 15 m. V_m, the mean wind speed; Q, water produced at a given time.

3,442 m^3/month. In Owerri, where the pump has the least performance among the three sites, the monthly water output varies between 1,936 and 2,844 m^3 with an average discharge volume of about 2,530 m^3/month. In the case of Onitsha, the monthly water output varies between about 2,206 and 3,194 m^3, and the average volume discharge is about 2,730 m^3/month. Therefore, all the sites have strong potential for wind energy development for water pumping. In fact, wind resource in these sites could also be used for electricity generation, as previously shown.

The total numbers of habitants that can be served by water discharged from these sites are also shown in Table 7. Based on water usage of 36 L/capita/day in Nigeria as of 2006 [38], the water output at Enugu can serve between 2,900 and 3,250 habitants depending on the month. The average number of people that can be served per month is estimated to be around 3,190. However, if the estimate is based on the minimum recommended water usage of 50 L/capita/day [38], the water produced can only serve about 2,290 habitants per month on average. In Owerri, the water produced can serve about 2,340 and 1,690 habitants per month on average based on 36 and 50 L/capita/day, respectively, while the total numbers of habitants that can be served by the water produced in Onitsha are 2,530 and 1,820/month on average based on water usage of 36 and 50 L/capita/day, respectively. Therefore, for small rural communities scattered across the southeastern part of Nigeria where access to good water and unreliable supply of water is a regular problem, wind resource development for water pumping will be a good option. For a larger community, the performance of a pump can be increased if a wind turbine with higher rated power (or diameter) is used instead of the small size used in this study. In addition, two or more wind turbines can be installed in these sites in order to increase the quantity of water produced.

Conclusions

In this study, the wind speed and wind energy potential in selected three locations in the southeastern part of Nigeria were investigated. In addition, the performance of selected commercial wind turbine models designed for both electricity generation and water pumping located in these sites was investigated. The findings from this study can be summarized as follows:

1. The annual mean wind speeds for Enugu, Owerri and Onitsha are 5.42, 3.36 and 3.59 m/s, respectively. The annual values of the wind speed carrying maximum energy for these locations are respectively 6.48, 4.33 and 3.90 m/s.
2. The mean annual value of Weibull shape parameter k is between 4.05 and 5.10, while the annual value of scale parameter c is between 3.96 and 5.87 m/s.
3. The annual mean power density for Enugu, Owerri and Onitsha are 96.98, 23.23 and 28.34 W/m^2, respectively. Therefore, based on the wind data used in this study, the wind energy resource in south-east Nigeria may generally be classified into class 1. However, based on monthly mean power density, the wind resource may fall into higher class category in some cases.
4. Based on the capacity factor, the POLARIS 15–50, POLARIS 62–1000 and WWD-1-60 models or wind turbines with similar designed characteristics will be best suited for electricity generation in Enugu. However, in order to meet the minimum

recommended capacity factor (25%) for electricity generation, wind turbine models with cut-in wind speed of less than 3.5 m/s minimum and hub height of 55, 75 and 65 m, as well as rated wind speed of about 12, 8.5 and 9 m/s, respectively, are recommended for Enugu, Owerri and Onitsha.

5. Using a 0.36-kW wind turbine, the average monthly water produced by a rotodynamic pump assumed to be installed in Enugu, Owerri and Onitsha is determined as 3,442, 2,530 and 2,730 m^3, respectively. The quantity of water can serve about 2,290, 1,690 and 1,820 habitants in respective locations.

Abbreviations

A: wind power rotor swept area; *c*: Weibull scale parameter; C_f: capacity factor; C_O: scale factor at the height h_o; C_P: wind turbine power coefficient; C_{PD}: wind turbine design power coefficient; *D*: wind turbine rotor diameter; E_D: mean energy density; *f*(*V*): probability of observing wind speed (*V*); *F* (*V*): cumulative of observing wind speed (*V*); *G*: pump gear ratio; *g*: acceleration due to gravity; *H*: water head; *H*: wind turbine hub height; h_o: cup-generator anemometer height; *k*: dimensionless Weibull shape parameter; k_o: shape parameter height h_o; N_{PD}: pump speed at design condition; *n*: exponent; P_D: mean wind power density; $P_{e,ave}$: mean power output; P_{eR}: rated wind turbine power; *P*(*V*): wind power; *Q*: water produced at a given time; *Q*(*V*): volume discharge of the pump at any wind speed *V*; *T*: period or time; *V*: wind speed; V_E: wind speed carrying maximum energy; V_F: most probable wind speed; V_m: mean wind speed; V_o: wind speed at the cup-generator anemometer height; v_c: cut-in wind speed; V_d: v_r, rated wind speed; v_f: cut-off wind speed.

Symbols

α, surface roughness coefficient; *σ*, standard deviation; *Γ*(*x*), gamma function; *λD*, design tipspeed ratio of the wind turbine; *ρ*, air density; ρ_w, water density; *η*, overall pump efficiency; η_{PD}, overall design pump efficiency.

Competing interests

The authors declare that they have no competing interests.

Acknowledgments

The authors are grateful to the Nigerian Meteorological Agency (NIMET) Oshodi, Lagos, Nigeria, for providing data for this study.

Author details

[1]Mechanical Engineering Department, Covenant University, Ota, Ogun State 112101, Nigeria. [2]Department of Energy and Process Engineering, Norwegian University of Science and Technology, Trondheim, NO-7491, Norway. [3]Department of Mechanical and Manufacturing Engineering, University of Manitoba, Winnipeg, Manitoba R3T 5V6, Canada.

Authors' contributions

SOO collected the wind data and carried out initial analysis. Wind turbine calculations were performed by MSA. 'Background' section was drafted by SSP. 'Results and discussion' section was jointly drafted by MSA and SSP. All authors read, edited and eventually approved the manuscript to be submitted.

References

1. Global Wind Energy Council: Global wind report: annual market updates 2011. http://www.gwec.net/ (2011). Accessed 30 April 2012
2. Fagbenle, RO, Katende, J, Ajayi, OO, Okeniyi, JO: Assessment of wind energy potential of two sites in North-East, Nigeria. Renew. Energy **36**, 1277–1293 (2011)
3. Ojosu, JO, Salawu, RI: A survey of wind energy potential in Nigeria. Solar and Wind Technology **7**, 155–67 (1990)
4. Adekoya, LO, Adewale, AA: Wind energy potential of Nigeria. Renew. Energy **2**, 35–39 (1992)
5. Fadare, DA: The application of artificial neural networks to mapping of wind speed profile for energy application in Nigeria. Applied Energy **87**, 934–942 (2010)
6. Adaramola, MS, Oyewola, OM: On wind speed pattern and energy potential in Nigeria. Energy Policy **39**, 2501–2506 (2011)
7. Marcius-Kaitis, M, Katnals, V, Karaliauskas, A. Wind power usage and prediction prospects in Lithuania. Renew. Sustain. Energy Rev. **12**, 265–277 (2008)
8. Anyanwa, EE, Iwuagwu, CJ: Wind characteristics and energy potentials for Owerri Nigeria. Renew. Energy **6**, 125–128 (1995)
9. Asiegbu, AD, Iwuoha, GS: Studies of wind resources in Umudike, south east Nigeria-an assessment of economic viability. J. Eng. Appl. Sci. **2**, 1539–1541 (2007)
10. Manwell, JF, McGowan, JG, Rogers, AL: Wind Energy Explained: Theory, Design and Application, 2nd edn. Wiley, Wiltshire (2010)
11. International Standards Organisation: Guide to the Expression of Uncertainty in Measurement, 1st edn. (1992)
12. Akpinar, EK, Akpinar, S: An assessment on seasonal analysis of wind energy characteristics and wind turbine characteristics. Energy Conversion and Management **46**, 1848–67 (2005)
13. Justus, CG, Hargraves, WR, Mikhail, A, Graber, D: Methods for estimating wind speed frequency distributions. J. Appl. Meteorol. **17**, 350–353 (1978)
14. Akdag, SA, Dinler, A: A new method to estimate Weibull parameters for wind energy applications. Energy Conversion and Management **50**, 1761–1766 (2009)
15. Akdag, SA, Bagiorgas, HS, Mihalakakou, G: Use of two-component Weibull mixtures in the analysis of wind speed in the Eastern Mediterranean. Applied Energy **87**, 2566–2573 (2010)
16. Mathew, S, Pandey, KP, Kumar, AV: Analysis of wind regimes for energy estimation. Renew. Energy **25**, 381–399 (2002)
17. Jowder, FAL: Wind power analysis and site matching of wind turbine generators in Kingdom of Bahrain. Applied Energy **86**, 538–545 (2009)
18. Kwon, SD: Uncertainty analysis of wind energy potential assessment. Applied Energy **87**, 856–865 (2010)
19. Ouammi, A, Dagdougui, H, Sacile, R, Mimet, A: Monthly and seasonal assessment of wind energy characteristics at four monitored locations in Liguria region (Italy). Renew. Sustain. Energy Rev. **14**, 1959–1968 (2010)
20. Balouktsis, A, Chassapis, D, Karapantsios, TD: A nomogram method for estimating the energy produced by wind turbine generators. Solar Energy **72**, 251–259 (2002)
21. Bagiorgas, HS, Assimakopoulos, MN, Theoharopoulos, D, Matthopoulos, D, Mihalakakou, GK: Electricity generation using wind energy conversion systems in the area of Western Greece. Energy Conversion and Management **48**, 1640–1655 (2007)
22. Ucar, A, Balo, F: Evaluation of wind energy potential and electricity generation at six locations in Turkey. Applied Energy **86**, 1864–1872 (2009)
23. Ilinca, A, McCarthy, E, Chaumel, J-L, Retiveau, J-L: Wind potential assessment of Quebec Province. Renew. Energy 28, 1881–1897 (2003)
24. Celik, AN: A statistical analysis of wind power density based on the Weibull and Rayleigh models at the southern region of Turkey. Renew. Energy **29**, 593–604 (2004)
25. Mathew, S: Wind Energy: Fundamentals, Resource Analysis and Economics. Springer, Heidelberg (2006)
26. Mathew, S, Pandey, KP: Modelling the integrated output of wind-driven roto-dynamic pumps. Renew. Energy **28**, 1143–1155 (2003)
27. Hau, E, Renouard, R: Wind turbines: Fundamentals, technologies, applications, economics, 2nd edn. Springer, Heidelberg (2006)
28. Mathew, S, Pandey, KP: Modelling the integrated output of mechanical wind pumps. J. Solar Energy Engineering **122**, 203–206 (2000)
29. Fagbenle, RL, Karayiannis, TG: On the wind energy resource of Nigeria. Intern. J. Energy Res. **18**, 493–508 (1994)
30. Meel, J, Smulders, P: Wind Pumping: A Handbook (World Bank Technical Paper No. 101). World Bank, Washington (1989)
31. Polaris: http://www.polarisamerica.com/ (2011). Accessed 04 April 2011
32. Wind Energy Solutions: http://www.wes30.com/ (2011). Accessed 04 April 2011
33. WinWinD: http://www.winwind.com/Documents/Press%20Kit/wwd1.pdf (2011). Accessed 04 April 2011

34. Paul, SS, Oyedepo, SO, Adaramola, MS: Economic assessment of water pumping systems using wind energy conversion systems in the southern part of Nigeria. Energy Exploration and Exploitation **30**, 1–18 (2012)
35. Shi, J, Shen ,D, Wei, J: The development of wind pumping technology in China. Biomass **20**, 13–23 (1989)
36. Bobs Harries Engineering Ltd: Kijito wind pumps – Pumping water with wind: Information package. Brochure, Bob Harries Engineering Limited (2009)
37. Adelana, SMA, Olashinde, PI, Bale, RB: An overview of the geology and hydrogeology of Nigeria. In: Adelana, S, MacDonald, A (Eds) Applied Groundwater Studies in Africa, pp. 171–197. Taylor and Francis, London (2008)
38. United Nations Development Programme: Human Development Report. United Nations Development Programme, New York (2006)

10

Thermal and mechanical aspect of entropy-exergy relationship

Pierfrancesco Palazzo

Abstract

The mechanical aspect of entropy-exergy relationship, together with the thermal aspect usually considered, leads to a formulation of physical exergy based on both useful work and useful heat that are the outcomes of available energy of a thermodynamic system with respect to a reservoir. This approach suggests that a mechanical entropy contribution can be defined, in addition to the already used thermal entropy contribution, with respect to work interaction due to pressure and volume variations. The mechanical entropy is related to energy transfer by means of work interaction and it is complementary to the thermal entropy that accounts energy transfer by means of heat interaction. Furthermore, the study proposes a definition of exergy based on Carnot cycle that is reconsidered in the case the inverse cycle is adopted and, as a consequence, the concept that work depends on pressure similarly as heat depends on temperature, is pointed out. Then, the logical sequence to get mechanical exergy expression to evaluate useful work withdrawn from available energy is demonstrated. Based on mechanical exergy expression, the mechanical entropy set forth is deduced in a general form valid for any process. Finally, the formulation of physical exergy is proposed that summarizes the contribution of either heat or work interactions and related thermal exergy as well as mechanical exergy that both result as the outcome from the available energy of the composite of the system interacting with a reservoir. This formulation contains an additional term that takes into account the volume and, consequently, the pressure that allow to evaluate exergy with respect to the reservoir characterized by constant pressure other than constant temperature. The basis and related conclusions of this paper are not in contrast with principles and theoretical framework of thermodynamics and highlight a more extended approach to exergy definitions already reported in literature that remain the reference ground of present analysis.

Keywords: Thermal entropy, Mechanical entropy, Total entropy, Thermal exergy, Mechanical exergy, Physical exergy

Background

The first purpose of the present paper is to highlight the mechanical aspect of entropy property, complementary to the thermal aspect, and the definitions already existing in classical thermodynamics literature. The second purpose, on the basis of entropy property structure and the thermal and mechanical aspects of entropy-exergy relationship, is to propose a formulation of physical exergy based on both useful work and useful heat. This formulation would be suitable to evaluate either the net useful work or the net useful heat of a system being both the outcome of available energy, considering the role of pressure other than temperature, in the evaluation of work and heat interactions in combination with a reservoir as defined in the study of Gyftopoulos and Beretta [1]. The formulation, already established in literature, is based on mutual stable equilibrium state between system and reservoir due to temperature equality only. The tentative of the paper is to adopt the equality of pressure as a further condition of mutual stable equilibrium and to consider the pressure, too, in the formulation of the physical exergy as well as in the formulation of entropy. Pressure is accounted by means of the mechanical entropy related to volume. This is based on the relationship existing between entropy and exergy, either for heat interaction (thermal exergy based on thermal entropy) or for work interaction (mechanical exergy based on mechanical entropy).

In order to simplify this analysis, the following assumptions have been posited: the system is considered as 'simple' according to the terminology reported in literature [1]; the

Correspondence: ppalazzo@technip.com
Technip, 68 Viale Castello della Magliana, Rome 00148, Italy

system consists in a perfect and single-phase homogeneous gas; there are no phase changes or chemical and nuclear reaction mechanisms inside the system; the existence of an external reference system that behaves as a reservoir is assumed [1]; the analysis is focused on stable equilibrium states; the kinetic and potential energy of the system as a whole are neglected. On the basis of these assumptions, chemical exergy will be out of the domain of present study and reference will be made to physical exergy only.

Methods

The methods adopted in the present analysis are based on the dualism deduced by the existence of the thermal and mechanical aspects characterizing the concept of energy and by the correlations existing between them; this dualism is developed in order to identify the two components of entropy and exergy properties correlated to heat interaction and work interaction exchanged by the system with a thermo-mechanical reservoir. In addition, the symmetry of the implications of temperature with respect to thermal energy and of pressure with respect to mechanical energy constitutes the paradigm for achieving the definition of the extended physical exergy that depends on both temperature and pressure.

Thermal and mechanical contribution of entropy property

As reported in literature, entropy property is related to the second law and the thermal aspect of thermodynamic systems behavior and properties. An investigation on the mechanical aspect can also be accomplished with an analysis of thermodynamic processes considered as ideal without thermal and mechanical irreversibilities. In particular, the isothermal reversible process is evaluated here in which the internal energy of a system with no chemical and nuclear reactions and constant amount of constituents is formulated by the classical expression of thermodynamic potential known as Euler relation:

$$U = TS - PV + \mu n \tag{1}$$

which implies that the variation of internal energy alongside a reversible isothermal process is null as follows for a homogeneous single-phase system without chemical reactions:

$$dU = d(TS) - d(PV) = \delta Q + \delta W = 0 \tag{2}$$

where δQ and δW both represent positive energy received into the system by convention for reference transfer direction. The mechanical term $d(PV)$ is equal to zero, in fact:

$$d(PV) = PdV + VdP = 0 \tag{3}$$

being $PdV = -VdP$ alongside an isothermal process. The mechanical term $d(PV)$ is equal to zero also because, having assumed the state equation $PV = \bar{R}T$ as valid, the term PV is constant, being T as constant by definition alongside the isothermal process. The following result is obtained:

$$dU = T \cdot dS = 0 \quad \text{that implies} \quad dS = 0 \tag{4}$$

that is valid if and only if entropy remains constant but, on the other side, it should change due to heat interaction occurring alongside the isothermal process. Thus, entropy should be constant and should change at the same time. This is an apparent inconsistency that can be resolved if S is intended to be the total entropy. In fact, it is here posited that total entropy is the result of contribution of entropy due to heat interaction, related to temperature, and entropy due to work interaction, related to pressure.

Another consideration addresses to the adiabatic reversible process, namely isoentropic, that is accomplished at constant thermal entropy, while temperature and pressure change according to the following formula:

$$S(T,P) - S_0 = C_P \ln\frac{T}{T_0} - \bar{R}\ln\frac{P}{P_0} \tag{5a}$$

or

$$S(T,P) - S_0 = C_V \ln\frac{T}{T_0} + \bar{R}\ln\frac{V}{V_0}. \tag{5b}$$

Thus, thermal entropy variation related to heat interaction is null depending on a compensation effect due to the combination of either temperature and pressure variation.

The above considerations in the special case of isothermal reversible process suggest that both internal energy and entropy remain constant while a heat-to-work conversion occurs isothermally due to equal quantities of heat and work interactions between the system and the reservoir. Nevertheless, the heat interaction implies a transfer of entropy; therefore, a transfer of entropy under 'thermal' form requires an entropy conversion into 'mechanical' form in order to render the balance of entropy (total) equal to zero as required by the expression (Equation 4). Therefore, it can be posited that this mechanical form of entropy is associated to work interaction by means of pressure and volume variations. Mechanical entropy is complementary to and does not contradict the classical expression of entropy calculated alongside an isothermal reversible process that corresponds to the thermal entropy associated to heat interaction only.

This analysis suggests that the entropy (total) can be considered as constituted by two components, namely 'thermal entropy' that is constant in adiabatic reversible (isoentropic) processes where work interaction only occurs, and a 'mechanical entropy' that is constant in isovolumic reversible processes where heat interaction only occurs. In addition, it can also be posited that entropy, appearing in the expression of internal energy, represents the thermal component of entropy (or thermal entropy) that, thus, results consistent with the properties requested to comply with the above behavior. The relationship between entropy and exergy represents the basis for assuming and proving that thermal and mechanical components set forth for entropy can be retrieved in exergy property as well.

Thermal aspect of entropy-exergy relationship

The definition of exergy here, considered for the scope of the present analysis, is the one based on heat and work interactions; in particular, the exergy formulated as the maximum net useful work obtained from the available energy on the basis of the thermodynamic efficiency of the Carnot direct cycle operating between the variable temperature T of system A and the constant temperature T_R of reservoir R, considered as the external reference system, is discussed in this section:

$$dEX^T = \delta W_{REV}^{NET} = \delta W_{REV}^{CONVER} + \delta W_{REV}^{TRANSF} \quad (6)$$

where $\delta W_{REV}{}^{CONVER}$ is the net amount of work resulting as the balance of a direct cycle that converts the available heat at a temperature T into work by means of a cyclic machinery in combination with the thermal reservoir at constant temperature T_R; $\delta W_{REV}{}^{TRANSF}$ is the net amount of available energy transferred by means of a work interaction in a cyclic machinery, resulting from system volume variation (cyclic machinery is an ideal device that returns to its initial state at the end of whatsoever process it accomplishes). For sake of generality, heat and work interactions are considered occurring either successively or simultaneously, and both result from generalized available energy of a simple system as defined in the study of Gyftopoulos and Beretta [1]. Therefore, in differential terms:

$$\begin{aligned} dEX^T &= \delta W_{REV}^{NET} \\ &= \eta_{id}^{CARNOT \cdot DIR} \cdot \delta Q^{HT} + \delta W_{REV}^{TRANSF} \\ &= \frac{\delta W}{\delta Q_{ISOTHERMAL}^{HT}} \cdot \delta Q^{HT} - P \cdot dV + P_R \cdot dV \\ &= \left(1 - \frac{T_R}{T}\right) \cdot \delta Q^{HT} + \left(1 - \frac{P_R}{P}\right) \cdot \delta W^{HP} \end{aligned} \quad (7)$$

in which δQ^{HT} represents the infinitesimal heat interaction along the process at temperature T different from the temperature T_R of the reservoir; δW^{HP} is the infinitesimal work interaction at high pressure P, alongside the process, different with respect to the reservoir pressure P_R. The formula (Equation 7) corresponds to the already known classical definition of physical exergy [2,3]. This expression is used to define the exergy that is identified by the superscript T (that stands for 'thermal') according to the definition reported in the literature [2-4] as above pointed out. In finite terms, considering that, $\delta W^{HP} = - PdV$:

$$\begin{aligned} EX^T &= W_{10} \\ &= \int_0^1 \left(1 - \frac{T_R}{T}\right) \cdot \delta Q^{HT} + \int_0^1 \left(1 - \frac{P_R}{P}\right) \cdot \delta W^{HP} \\ &= Q_{10}^{HT} - T_R \int_0^1 \frac{\delta Q^{HT}}{T} + W_{10}^{HP} + P_R \cdot (V_1 - V_0) \end{aligned} \quad (8)$$

where W_{10} is the maximum net useful work output from the generalized available energy [1] as the result of interaction between the system and the reservoir; Q_{10}^{HT} is the heat interaction from higher isothermal curve at T to lower isothermal curve at T_R (as a particular case, heat interaction can occur alongside an isovolumic process); and W_{10}^{HP} is the work interaction from higher isothermal curve at T to lower isothermal curve at T_R. The sum of Q_{10}^{HT} and W_{10}^{HP} can also be expressed as:

$$Q_{10}^{HT} + W_{10}^{HP} = U_1 - U_0 = C_V \cdot (T_1 - T_0) \quad (9)$$

that represents the equivalence with the amount of heat interaction only in the isovolumic process between two different temperatures. Thus, thermal exergy can also be associated to a sequence of isovolumic-isothermal process connecting the generic state 1 with a different generic state 0 of the system A. The integration function in the expression of thermal exergy (Equation 8) represents, in infinitesimal terms, the definition of entropy as per Clausius formulation or, as here proposed, the thermal component of entropy property identified by the superscript thermal, and therefore:

$$\begin{aligned} EX^T = W_{10} = (U_1 - U_0) - T_R \cdot (S_1^T - S_0^T) \\ + P_R \cdot (V_1 - V_0) \end{aligned} \quad (10)$$

as the result of combination of the system and the reservoir at constant temperature T_R and constant pressure P_R. This formulation is also consistent with the formulation presented by Gyftopoulos and Beretta [1], deduced from the definition of generalized available energy with respect to an external reference system at constant temperature T_R and constant pressure P_R that behaves as a reservoir.

Mechanical aspect of entropy-exergy relationship

The correlation between thermal entropy and thermal exergy clarified so far suggests to assume the mechanical aspect of entropy-exergy relationship to attempt the evaluation of the concept of exergy related to work and pressure, based on the existence of the mechanical component of entropy previously assumed. This assumption derives from the fact that equality of pressure between the system and the reservoir is a further condition for stable equilibrium state of the composite system-reservoir other than the equality of temperature.

Now that we have defined exergy formulated by the direct cycle as thermal exergy which highlights the role of temperature in heat-to-work conversion, we may now search a definition of mechanical exergy that is expressed by the inverse cycle with the intent of evaluating the role of pressure in the opposite process, that is, in a work-to-heat conversion. Also for the mechanical aspect, the general formulation of exergy, in infinitesimal terms, derives from the relationship built around the Carnot cycle and its expression of thermodynamic efficiency.

The Carnot cycle used in the formulation of thermal exergy is defined in literature as a symmetric cycle consisting of four processes, each pair of which is of the same type (isodiabatic) as represented in Figure 1.

If the working system is a perfect gas as assumed, then the alternating polytropic (isoentropic) processes provide the following property to the cycle 0-1-1 C-0 C in Figure 1:

$$\frac{V_1}{V_0} = \frac{V_{1C}}{V_{0C}}; \frac{P_1}{P_0} = \frac{P_{1C}}{P_{0C}}; \frac{T_1}{T_0} = \frac{T_{1C}}{T_{0C}}. \tag{11}$$

From which, it may be inferred that, since the ends of the processes are proportional, the work interaction between the system and the external environment (reservoir) is the same both along the isoentropic compression from 0 to 1 and in isoentropic expansion from 1 C to 0 C. The amount of work interaction

$$W = \frac{1}{K-1} P_0 V_0 \left[\left(\frac{P_1}{P_0} \right)^{\frac{K-1}{K}} - 1 \right] \tag{12}$$

depends on $\frac{P_1}{P_0} = \frac{P_{1C}}{P_{0C}}$, and therefore, W is equal for the two different adiabatic reversible (isoentropic) processes.

As a result, the balance of the work interaction between the system and the reservoir along the isoentropic processes alone is null, and the balance contribution over the whole cycle is due exclusively to the isothermal processes where heat and work is exchanged simultaneously in directly proportional and equal amounts. This property enables to express the thermodynamic efficiency of the Carnot cycle of a closed nonbulk-flow system both in terms of heat and work interactions. The efficiency can be expressed either in terms of heat only or in terms of work only due to the equality of heat and work interactions alongside the isothermal process:

$$\eta_{id}^{CARNOT \cdot DIR} = \frac{W}{Q^{HT}} = \frac{W^{HP} - W^{LP}}{Q^{HT}} = \frac{W^{HP} - W^{LP}}{W^{HP}} = \frac{Q^{HT} - Q^{LT}}{Q^{HT}} \tag{13}$$

With regard to the inverse cycle, if the roles of used heat Q^{HT} and utilized total work W are replaced by used work W^{HP} and utilized total heat Q, the following expression applies (that does not correspond to the coefficient of performance reported in literature):

$$\eta_{id}^{CARNOT \cdot INV} = \frac{Q}{W^{HP}} = \frac{Q^{HT} - Q^{LT}}{W^{HP}} = \frac{Q^{HT} - Q^{LT}}{Q^{HT}} = \frac{W^{HP} - W^{LP}}{W^{HP}} = \eta_{id}^{CARNOT \cdot DIR} \tag{14}$$

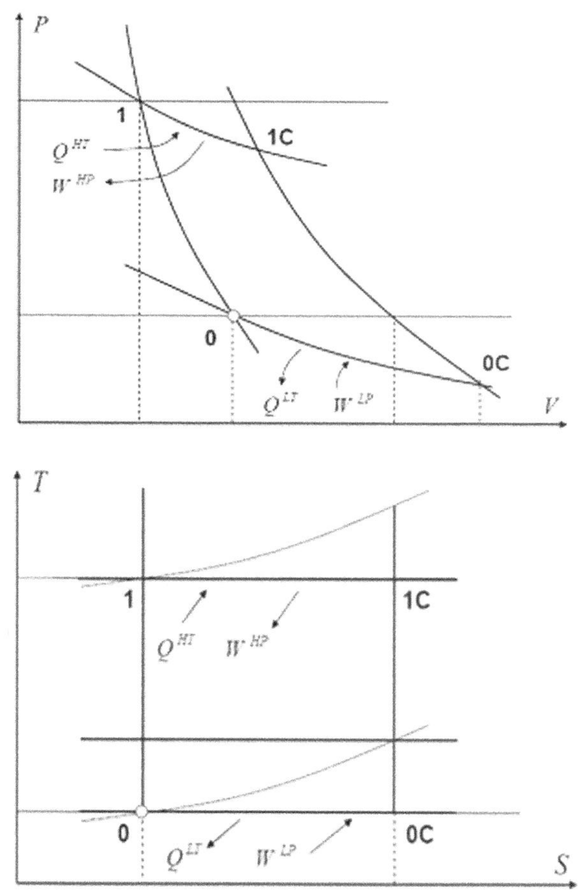

Figure 1 Symmetry property of Carnot cycle. P, pressure; Q^{HT}, high temperature heat; W^{HP}, high pressure work ; Q^{LT}, low temperature heat; W^{LP}, low pressure work; V, volume; T, absolute temperature; S, entropy; 0 C, point 0 of Carnot cycle; 1 C, point 1 of Carnot cycle.

As may be noted, since there is a linear proportionality between the two expressions due to the identity of two of the terms, it may be inferred that the meaning of used heat Q^{HT} in the direct cycle corresponds to the meaning of used work W^{HP} in the inverse cycle. The meaning of the utilized total work W in the direct cycle corresponds to the meaning of the utilized total heat Q in the inverse cycle as well. The conclusion is that the efficiency of Carnot cycle, that depends on isothermal process only, does not change if inverse cycle is compared to direct cycle. In fact:

$$\eta_{id}^{CARNOT \cdot INV} = \frac{Q}{W_{ISOTHERMAL}^{HP}} = \frac{W}{Q_{ISOTHERMAL}^{HT}} = 1 - \frac{T_0}{T} = \eta_{id}^{CARNOT \cdot DIR} \quad (15)$$

Now that we have defined exergy formulated by the direct cycle as thermal exergy which highlights the role of temperature in heat-to-work conversion, we may now define as mechanical exergy the exergy that may be expressed by means of the inverse cycle. In this case, the role of pressure in the opposite conversion, that is from work into heat, is due to the pressure level of work interaction alongside the higher temperature (and higher pressure) isothermal process of inverse Carnot cycle. As a consequence of the concept of interconvertibility (Gaggioli et al. [5-7]): 'useful work is not better than useful heat, and the available energy results in maximum net useful heat or, equivalently, maximum net useful work or the combination of both'. Thus, the definition of mechanical exergy that represents, in this case, the maximum net useful heat obtained from the available energy in infinitesimal terms may be expressed by the following relationship:

$$dEX^M = \delta Q_{REV}^{NET} = \delta Q_{REV}^{CONVER} + \delta Q_{REV}^{TRANSF} \quad (16)$$

where δQ_{REV}^{CONVER} is the net amount of heat resulting as the balance of an inverse cycle that converts available work at a pressure P into heat by means of a cyclic machinery in combination with the mechanical reservoir at constant pressure P_R; δQ_{REV}^{TRANSF} is the net amount of available energy transferred by means of heat interaction in a cyclic machinery resulting from system thermal entropy variation (cyclic machinery is an ideal device that returns to its initial state at the end of whatsoever process it accomplishes). For sake of generality, work and heat interactions are considered occurring either successively or simultaneously, and both result from generalized available energy of a simple system as defined by Gyftopoulos and Beretta [1]. Therefore, in differential terms:

$$\begin{aligned} dEX^M &= \delta Q_{REV}^{NET} \\ &= \eta_{id}^{CARNOT \cdot INV} \cdot \delta W^{HP} + \delta Q_{REV}^{TRANSF} \\ &= \frac{\delta Q}{\delta W_{ISOTHERMAL}^{HP}} \cdot \delta W^{HP} + TdS - T_R dS \\ &= \left(1 - \frac{T_R}{T}\right) \cdot \delta W^{HP} + \left(1 - \frac{T_R}{T}\right) \cdot \delta Q^{HT} \end{aligned} \quad (17)$$

in which δW^{HP} represents the infinitesimal work interaction along the process at the pressure P different from the pressure P_R of reservoir. It is noteworthy that the role of pressure here corresponds to the role of temperature with respect to heat in thermal exergy.

The formulation of thermal exergy is now reversed to define the mechanical exergy, identified by the superscript M (that stands for mechanical) that is not referred to potential and kinetic energy; in fact, in this case, 'potential exergy' and 'kinetic exergy' terms and definitions have already been adopted as components of exergy, namely kinetic exergy and potential exergy [2-4]. After replacing work with heat, mechanical exergy is formulated as follows:

$$\begin{aligned} dEX^M &= \delta Q_{REV}^{NET} \\ &= \left(1 - \frac{T_R}{T}\right) \cdot \delta W^{HP} + (T - T_R) \cdot dS^T. \end{aligned} \quad (18)$$

The expression in finite terms is:

$$\begin{aligned} EX^M &= Q_{10} \\ &= \int_0^1 \left(1 - \frac{T_R}{T}\right) \cdot \delta W^{HP} + \int_0^1 (T - T_R) \cdot dS^T \\ &= W_{10}^{HP} - T_R \int_0^1 \frac{\delta W^{HP}}{T} + Q_{10}^{HT} - T_R \cdot \left(S_1^T - S_0^T\right) \end{aligned} \quad (19)$$

where Q_{10} is the maximum net useful heat output from the generalized available energy [1] as the result of interaction between the system and the reservoir; W_{10}^{HP} is the work interaction from higher isothermal curve at T, and in correspondence of the point at P, to lower isothermal curve at T_R (as a particular case, work interaction can occur alongside an adiabatic reversible process); the sum of Q_{10}^{HT} and W_{10}^{HP} can also be expressed as:

$$W_{10}^{HP} + Q_{10}^{HT} = U_1 - U_0 = C_V \cdot (T_1 - T_0) \quad (20)$$

that represents the equivalence with the amount of work interaction only in the isoentropic process between two different temperatures. Thus, mechanical exergy can also be associated to a sequence of isoentropic-isothermal process connecting the generic state 1 with stable equilibrium state 0 of the composite of system and reservoir.

If the state equation applicable to ideal gas $PV = \vec{R}T$ is used with the proper substitutions, the mechanical exergy can be written as follows:

$$EX^M = Q_{10} = (U_1 - U_0) - \bar{R}T_R \int_0^1 \frac{\delta W^{HP}}{PV} - T_R \cdot (S_1^T - S_0^T) \tag{21}$$

The integrand function $\frac{\delta W^{HP}}{PV}$ depending on the integration operator symbol $\int$ is formally similar to the integrand function $\frac{\delta Q^{HT}}{T}$ that represents the definition of entropy property as per Clausius formulation and, in particular, thermal entropy according to the definition proposed here. On the basis of this formal analogy applied to work interaction, it is now possible to define a 'mechanical entropy' by means of the expression:

$$dS^M = \frac{\delta W^{HP}}{PV} \tag{22}$$

where the factor $\frac{1}{PV}$ would represent the integrating factor of the infinitesimal work δW^{HP} that makes the integration function an exact differential function. In fact, going back to the expression of mechanical exergy (Equation 21) set forth and considering that $\delta W^{HP} = - PdV$ it is now possible to write:

$$\begin{aligned} EX^M &= Q_{10} \\ &= (U_1 - U_0) - \bar{R}T_R \int_0^1 \frac{\delta W^{HP}}{PV} - T_R \cdot (S_1^T - S_0^T) \\ &= (U_1 - U_0) + T_R \int_0^1 \bar{R} \frac{dV}{V} - T_R \cdot (S_1^T - S_0^T) \\ &= (U_1 - U_0) + T_R \cdot (\bar{R} \ln V_1 - \bar{R} \ln V_0) \\ &\quad - T_R \cdot (S_1^T - S_0^T) \end{aligned} \tag{23}$$

that relates to the work interaction with environmental system (mechanical reservoir); therefore:

$$\begin{aligned} EX^M = (U_1 - U_0) + P_R V_R \cdot (\ln V_1 - \ln V_0) \\ - T_R \cdot (S_1^T - S_0^T) \end{aligned} \tag{24}$$

where the formal analogy with the equation of thermal exergy can be noted. In order to complete the proposed analogy, if reference is made to the integrating function of Equation 23, the thermal entropy according to Clausius formulation is as follows:

$$dS^T = \frac{\delta Q^{HT}}{T} \tag{25}$$

whereas the mechanical entropy is formulated as:

$$dS^M = \frac{\bar{R} \cdot \delta W^{HP}}{PV} = \frac{\bar{R} \cdot dV}{V} \geq S^M = \bar{R} \ln V + C \tag{26}$$

Thus, being dS^M an exact differential function, then S^M is a state property and can be adopted as the formal definition of mechanical entropy. It depends on the volume that is a state property, and therefore, S^M behaves as a state property as well. Furthermore, since volume is additive, it can be proved that mechanical entropy is additive as well. With regard to the dimensional analysis, since logarithmic function is dimensionless, then the dimension of mechanical entropy is related to $\bar{R}$ and is $(J \cdot \mathrm{kg}^{-1} \cdot K^{-1})$ that is identical to thermal entropy dimension.

The relationship between mechanical exergy and volume (and the pressure as a consequence) is the reason why the equality of pressure can be assumed as a further condition of mutual stable equilibrium of the composite system-reservoir that can be considered in the definition of entropy property. The physical meaning of mechanical exergy can be ascribed to the combination of pressure level that characterizes the internal energy of the system and the pressure level of work interaction at that pressure level of the system itself.

Physical exergy related to thermo-mechanical reservoir

The definition of thermal entropy and mechanical entropy, deduced and formulated from thermal exergy and mechanical exergy respectively, is adopted here to extend the concept of physical exergy to the work interaction other than heat interaction between the system and the reservoir as defined by Gyftopoulos and Beretta [1]. By virtue of the concepts of equivalence and interconvertibility conceived by Gaggioli et al. [5-7], the available energy of a system results in two outcomes:

- Available work or maximum net useful work that can be evaluated as thermal exergy
- Available heat or maximum net useful heat that can be evaluated as mechanical exergy.

The extension of physical exergy proposed here is, therefore, implicated with thermal exergy underpinned by the efficiency of Carnot direct cycle and the high temperature heat together with the mechanical exergy underpinned by the efficiency of Carnot inverse cycle and the high pressure work, whereas the environment is adopted as reservoir. A more specific definition of reference external system that would be a thermo-mechanical reservoir should behave as a thermal reservoir at constant temperature and variable pressure and as mechanical reservoir at constant pressure and variable temperature. This extension includes the effect of pressure in work interaction that results in different heat amounts depending on different pressure level of the same work amount. In other terms, the concept is that the same amount of available work can be used at different pressure of the system (at different thermodynamic potential at constant temperature and variable pressure), with respect to the constant pressure of the mechanical reservoir, to be converted into heat at different temperatures. Therefore, the useful work resulting from the available energy is further evaluated in terms of second law process to calculate the amount of heat that it is capable to produce. Thus, heat is converted into work and work is converted into heat or thermal exergy is converted into mechanical exergy, and vice versa, mechanical exergy is converted into thermal exergy. With different wording, 'extended physical exergy' can be translated into 'exergy of exergy' that makes becoming work interaction equivalent to and interconvertible with heat interaction and vice versa.

Before arriving to the formulation of the extended physical exergy, the internal energy as expressed by the Gibbs relation:

$$dU = T{\cdot}dS - P{\cdot}dV = \delta Q + \delta W \tag{27}$$

can be reformulated in different terms introducing the concepts of thermal entropy and mechanical entropy:

$$dU = T\cdot(dS^T) - \frac{PV}{\bar{R}}\cdot(dS^M) = \delta Q + \delta W \tag{28}$$

that can be transformed using the state equation in the following form:

$$\begin{aligned} dU &= T\cdot(dS^T) - T\cdot(dS^M) \\ &= T\cdot(dS^T - dS^M) = \delta Q + \delta W \end{aligned} \tag{29}$$

where $dS^{TOTAL} = dS^T - dS^M$, and in finite terms

$$S^{TOTAL} = S^T - S^M \tag{30}$$

that, associated to the temperature, takes into account either heat or work interactions that contribute to the variations of internal energy. The expression (Equation 28) can be also written as:

$$dU = T\cdot(dS^{TOTAL}) = \delta Q + \delta W \tag{31}$$

and in finite terms, considering that U is a state property that depends on two independent variables:

$$U = U(S, V) = T\cdot(\Delta S^{TOTAL}) = Q + W. \tag{32}$$

If the term ΔS^{TOTAL} is expressed by means of Equation 30 also using Equation 5, the total entropy, resulting from the addition of thermal and mechanical components of entropy, resolves the apparent inconsistency that thermal entropy would be constant in an isothermal reversible process that requires heat interaction by means of thermal entropy exchange as pointed out in relation to Equation 4.

In fact, a consequence deduced from the definition of total entropy is that Euler equation now results in finite terms:

$$\Delta U = Q + W = T\cdot\Delta S^{TOTAL} = 0 \tag{33}$$

that is verified being Equation 30 valid; in fact, in the particular case of a perfect single-phase homogeneous gas:

$$\Delta S^T_{ISOTHERMAL} = C_V \ln\frac{T}{T_0} + \bar{R}\ln\frac{V}{V_0} \tag{34}$$

and

$$\Delta S^M_{ISOTHERMAL} = \bar{R}\ln\frac{V}{V_0} \tag{35}$$

$$\begin{aligned} \Delta S^{TOTAL} &= \Delta S^T - \Delta S^M \\ &= C_V\ln\frac{T}{T_0} + \bar{R}\ln\frac{V}{V_0} - \bar{R}\ln\frac{V}{V_0} \\ &= C_V\ln\frac{T}{T_0} \end{aligned} \tag{36}$$

The typical thermodynamic processes are analyzed more in detail to prove that the above expression of total entropy is valid in general for all processes.

Isothermal process, referring to Equation 36:

$$\begin{aligned} \Delta S^{TOTAL}_{ISOTHERMAL} &= \bar{R}\ln\frac{V}{V_0} - \bar{R}\ln\frac{V}{V_0} \\ &= C_V\ln\frac{T}{T_0} = 0 \end{aligned} \tag{37}$$

and confirms that $\Delta U = 0$ for an isothermal reversible process since $\Delta S^{TOTAL} = 0$ as required to resolve the inconsistency of condition (Equation 4) discussed in 'Thermal and mechanical contribution of entropy property' section.

Isovolumic process, following the same procedure:

$$\Delta S^T_{ISOVOLUMIC} = C_V\ln\frac{T}{T_0} \tag{38}$$

$$\Delta S^M_{ISOVOLUMIC} = 0 \tag{39}$$

using Equation 36:

$$\Delta S^{TOTAL}_{ISOVOLUMIC} = C_V\ln\frac{T}{T_0}. \tag{40}$$

In this case, the total entropy coincides with the thermal entropy.

Isobaric process, using the expression (Equation 5a) for thermal entropy:

$$\Delta S^T_{ISOBARIC} = C_P\ln\frac{T}{T_0} - \bar{R}\ln\frac{P}{P_0} \tag{41}$$

$$\Delta S^{M}_{ISOBARIC} = \bar{R}\ln\frac{V}{V_0} \tag{42}$$

considering again Equation 36:

$$\Delta S^{TOTAL}_{ISOBARIC} = C_P\ln\frac{T}{T_0} - \bar{R}\ln\frac{V}{V_0} = C_P\ln\frac{V}{V_0} - \bar{R}\ln\frac{V}{V_0} = C_V\ln\frac{T}{T_0} \tag{43}$$

Adiabatic reversible (isoentropic) process, considering the validity of Equation 36, the following applies:

$$\Delta S^{T}_{ADIABATIC} = 0 \tag{44}$$

$$\Delta S^{M}_{ADIABATIC} = \bar{R}\ln\frac{V}{V_0} \tag{45}$$

$$\Delta S^{TOTAL}_{ADIABATIC} = -\bar{R}\ln\frac{V}{V_0} = C_V\ln\frac{T}{T_0} \tag{46}$$

that demonstrate the existence of the relationship between pressure that changes with volume and total entropy also, in case the thermal entropy, is null due to the absence of heat interaction. Thus, the procedure adopted so far to explicit the mechanical entropy leads to clarify the apparent inconsistency highlighted in Equation 4 of 'Thermal and mechanical contribution of entropy property' section and underlines the role of pressure in the formulation of mechanical entropy and mechanical exergy.

Another conclusion that arises from the above procedure is that the total entropy is, in all above cases, reduced to an expression that depends on temperature only. This result complies with the fact that internal energy, under the assumption of system constituted by a perfect single-phase homogeneous gas, depends on and is characterized by the temperature only.

On the basis of the relationship between total entropy and internal energy (Equation 32), if the reservoir behaves as a mechanical reservoir at constant pressure other than as a thermal reservoir at constant temperature, the internal energy balance of system and reservoir, adopting the symbology in [1], is:

$$\begin{aligned} EX^{PHYSICAL} &= -\left(W^{AR\rightarrow}\right) - \left(Q^{AR\rightarrow}\right) \\ &= \Delta U^{SYSTEM} + \Delta U^{RESERVOIR} \\ &= \Delta U^{SYSTEM}_{W} + \Delta U^{R,W} \\ &\quad + \Delta U^{SYSTEM}_{Q} + \Delta U^{R,Q}, \end{aligned} \tag{47}$$

where $\Delta U^{SYSTEM}_{W} + \Delta U^{R,W}$ = mechanical exergy converted into thermal exergy and $\Delta U^{SYSTEM}_{Q} + \Delta U^{R,Q}$ = thermal exergy converted into mechanical exergy

$$\begin{aligned} EX^{PHYSICAL} &= -\left(W^{AR\rightarrow}\right) - \left(Q^{AR\rightarrow}\right) \\ &= \Delta U^{SYSTEM}_{W,Q} + \Delta U^{R,W} + \Delta U^{R,Q} \end{aligned} \tag{48}$$

or, in other terms,

$$EX^{PHYSICAL} = (U - U_0) - Q_R - W_R, \tag{49}$$

where Q_R is the minimum heat interaction with the thermal reservoir and W_R is the minimum work interaction with the mechanical reservoir. The symbol $EX^{PHYSICAL}$ (or $X^{PHYSICAL}$), instead of W and Q, is proposed for identifying physical exergy to generalize its meaning that here addresses either to useful work or useful heat.

Going back to the Figure 1, it is to be noted that $Q^{AR} = W^{AR}$ alongside the isothermal process; however, $T_R = const$ and $P_R \neq const$. Therefore, W^{AR} at decreasing pressure constitutes an amount of exergy that is lost since it is released isothermally to the reservoir while exchanging heat acquired to lead the system A in the condition of stable equilibrium with the reservoir. This isothermal process, as recognized in 'Thermal and mechanical contribution of entropy property' section, makes energy conversion and entropy conversion accounted in the expression of extended physical exergy:

$$\begin{aligned} EX^{PHYSICAL} &= \Delta U^{SYSTEM} + \Delta U^{RESERVOIR} \\ &= (U - U_0) - T_R \cdot \Delta S^{TOTAL,R} \\ &\quad + P_R \cdot \Delta V^{R} - T_R \cdot \Delta S^{T,R}, \end{aligned} \tag{50}$$

where the term $-T_R \cdot \Delta S^{TOTAL,R}$ represents the contribution to entropy conversion (symbol 'CONVER') only occurring inside the reservoir and the terms $+P_R \cdot \Delta V^{R} - T_R \cdot \Delta S^{T,R}$ represent the contribution transferred (symbol 'TRANSF') from the system to the reservoir. It is noteworthy that entropy conversion is inherent to energy conversion, as demonstrated in 'Thermal and mechanical contribution of entropy property' section, and that entropy conversion requires the additional term that contributes to exergy balance expressed in the above extended formulation (Equation 50) that, therefore, considers the effect of both energy and entropy conversion processes.

Therefore, it can be deduced, substituting the expressions (Equations 10 and 24), that:

$$\begin{aligned} EX^{PHYSICAL} &= (U - U_0) - T_R\left(S^{T} - S^{T}_{0}\right)^{CONVER} \\ &\quad + T_R\left(S^{M} - S^{M}_{0}\right) \\ &\quad + P_R(V - V_0) \\ &\quad - T_R\left(S^{T} - S^{T}_{0}\right)^{TRANSF} \end{aligned} \tag{51}$$

using the expression of mechanical entropy:

$$\begin{aligned} EX^{PHYSICAL} &= (U - U_0) - T_R\left(S^{T} - S^{T}_{0}\right)^{CONVER} \\ &\quad + \bar{R}\,T_R(\ln V - \ln V_0) \\ &\quad + P_R(V - V_0) \\ &\quad - T_R\left(S^{T} - S^{T}_{0}\right)^{TRANSF} \end{aligned} \tag{52}$$

that can be considered as the formulation of the extended physical exergy expressed by means of the contribution of thermal entropy and mechanical entropy. Using the state equation $P_R V_R = \bar{R}\,T_R$, the above can be also written as:

$$\begin{aligned} EX^{PHYSICAL} &= (U - U_0) - T_R\left(S^T - S_0^T\right)^{CONVER} \\ &+ P_R V_R \ln\frac{V}{V_0} \\ &+ P_R(V - V_0) \\ &- T_R\left(S^T - S_0^T\right)^{TRANSF} \end{aligned} \quad (53)$$

The term $\bar{R}\,T_R(\ln V - \ln V_0)$ or $P_R V_R \ln\frac{V}{V_0}$ of the above expressions can be defined as an 'entropic-mechanical' term that takes into account the entropy conversion that occurs alongside the isothermal reversible process as pointed out in the previous sections. The entropy conversion balance takes into account the mechanical entropy other than thermal entropy that is included in the above entropical mechanical term of the extended formulation.

Being an additive property, it can also be written as:

$$EX^{PHYSICAL} = W_{10} + Q_{10} \quad (54a)$$

$$= \left[\left(U - U_0\right) - T_R\cdot\left(S^T - S_0^T\right) + P_R\cdot\left(V - V_0\right)\right]^V$$

$$+\left[(U - U_0) + \bar{R}\,T_R\cdot(S^M - S_0^M) - T_R\left(S^T - S_0^T\right)\right]^S \quad (54b)$$

where Equation 54a is the component related to the variation of internal energy due to heat interaction only (in particular, alongside an isovolumic process), namely the thermal exergy; Equation 54b is the component related to the variation of internal energy due to work interaction only (in particular, alongside an isoentropic process), namely the mechanical exergy.

In general, being the internal energy additive, it follows that both contributions to internal energy due to heat and work interaction can occur successively or simultaneously along any process. The term (Equation 54a) of extended physical exergy represents the thermal exergy calculated alongside an isovolumic-isothermal process; on the other side, the term (Equation 54b) represents the mechanical exergy calculated alongside an isoentropic-isothermal process.

Results and discussion

For a more clear understanding of the implications deriving from the extended formulation, reference can be made to an adiabatic reversible process as a particular case. This process is characterized by null variation of thermal entropy (due to absence of heat interaction) and a non-null variation of mechanical entropy (due to work interaction occurring). As a consequence of the extended formulation, if the adiabatic reversible process is evaluated in terms of exergy, the available energy (in the form of pressure mechanical energy withdrawn from the system) is accounted in terms of its capability to be converted (not transferred) into maximum net useful heat, and therefore, the exergy calculation implies a lower amount if compared with the classical concept that identifies exergy exclusively with work interaction as it is. From a different perspective, the entropic-mechanical addendum of the extended formulation behaves as a reduction term that takes into account the amount of work interaction undergoing the (reversible) entropy conversion occurring alongside the isothermal process related to the thermodynamic conditions of the reservoir that renders this work not more useful to be converted back into heat interaction. In a broader view, the application of the extended exergy can be suitable in exergy method and thermoeconomic analysis of multi-conversion complex systems realized to perform direct and inverse cycle processes with work and heat flows productions for either civil and industrial engineering applications.

Conclusions

If reference is made to the Equations 54a and 54b, the physical exergy can be expressed by the sum of the two components namely thermal exergy and mechanical exergy:

$$\begin{aligned} EX^{PHYSICAL} &= EX^T + EX^M \\ &= W_{REV}^{CONVER} + W_{REV}^{TRANSF} \\ &+ Q_{REV}^{CONVER} + Q_{REV}^{TRANSF} \end{aligned} \quad (55)$$

that do not depend on a particular process adopted for its definition; thus, it can be considered as a general formulation valid for whatever process, reversible or irreversible, connecting two different thermodynamic states.

One of the outcomes of the present proposal is that the extended physical exergy takes into account the equality of pressure, other than equality of temperature, as a further condition of mutual stable equilibrium between system and reservoir. Looking forward to the implications of this additional condition and the generalization to 'all systems (large and small) in all states (in particular, nonequilibrium)' in the sense formalized and worded by Gyftopoulos and Beretta, the concept of extended physical exergy would require the reference to a mechanical reservoir in addition to the thermal reservoir [1,8,9].

To generalize the results of the present proposal, the formulation of entropy reported [1,8,9]:

$$(S_1 - S_0)^T = \frac{1}{T_R}\left[(E_1 - E_0) - \left(\Omega_1^R - \Omega_0^R\right)\right]^T \quad (56)$$

should remain valid if the concept of mechanical reservoir is introduced, and the equality of pressure between the system and the mechanical reservoir is considered as a further condition for mutual stable equilibrium between system and reservoir. The additive property of entropy would lead to assume that:

$$(S_1 - S_0)^M = \frac{\bar{R}}{P_R V_R}\left[(E_1 - E_0) - \left(\Omega_1^R - \Omega_0^R\right)\right]^M \quad (57)$$

where the mechanical component of entropy would be defined with reference to a mechanical reservoir at constant pressure.

Finally, the additivity of entropy components

$$\left(S_1 - S_0\right)^{TOTAL} = \left(S_1 - S_0\right)^T - \left(S_1 - S_0\right)^M \quad (58)$$

should be proved to complete the formulation of total entropy, taking into account the general definitions proposed for thermal entropy and mechanical entropy.

Looking forward to the possible future research studies, the logical further extension would be the assumption of equality of total potentials as an additional condition of mutual stable equilibrium of the composite system-reservoir. Thus, the set of all conditions of mutual stable equilibrium would lead to a more complete formulation of exergy with the contribution of chemical and nuclear exergy related to a 'thermo-chemical-nuclear-mechanical' reservoir that implies the definition of chemical and nuclear entropy related to the chemical and nuclear architecture of any system in any state, in particular, nonequilibrium.

Abbreviations

A, System;
C_p, Specific heat under constant pressure ($J \cdot kg^{-1} \cdot K^{-1}$);
C_V, Specific heat under constant volume ($J \cdot kg^{-1} \cdot K^{-1}$);
E, Energy (J);
EX^M, Mechanical exergy (J);
EX^T, Thermal exergy (J);
HP, High pressure;
HT, High temperature;
LP, Low pressure;
LT, Low temperature;
P, Pressure (MPa);
Q, Total heat interaction over the whole Carnot cycle (J);
Q^{HT}, High temperature heat (J);
Q^{LT}, Low temperature heat (J);
R, Reservoir;
Universal gas constant ($J \cdot kgmol^{-1} \cdot K^{-1}$);
S^M, Mechanical entropy ($J \cdot kgmol^{-1} \cdot K^{-1}$);
S^T, Thermal entropy ($J \cdot kgmol^{-1} \cdot K^{-1}$);
T, Absolute temperature (K);
U, Internal energy (J);
V, Volume (m^3);
W, Total work interaction over the whole Carnot cycle (J);
W^{HP}, High pressure work (J);
W^{LP}, Low pressure work (J);
$\eta_{id}^{CARNOT\text{-}DIR}$, Carnot direct cycle efficiency;
$\eta_{id}^{CARNOT\text{-}INV}$, Carnot inverse cycle efficiency;
ε, $\frac{\gamma-1}{\gamma}$
γ, $\frac{C_P}{C_V}$
Ω, Available energy

Competing interests

The author declares that he has no competing interests.

Author's information

Pierfrancesco Palazzo graduated with a masters degree in Mechanical Engineering from the Department of Mechanical and Aeronautical Engineering, University of Roma 1 'La Sapienza' in Italy. He has been studying about basic thermodynamics, energy processes, exergy method and thermoeconomic analysis and the related applications, and participated to international congresses. At present, he is an engineering manager in Technip, a world leader in project management, engineering and construction for the energy industry operating in the onshore, offshore and subsea segments.

References

1. Gyftopoulos, EP, Beretta, GP: Thermodynamics: Foundations and Applications. Dover Publications, Mineola, New York (2005)
2. Moran, MJ, Sciubba, E: Exergy analysis: principles and practice. Journal of Engineering for Gas Turbine and Power **116**, 285–290 (1994)
3. Kotas, TJ: The Exergy Method of Thermal Plant Analysis. Krieger Publishing Company, FL, USA (1995)
4. Dunbar, WR, Lior, N, Gaggioli, RA: The component equations of energy and exergy. Journal of Energy Resources Technology **114**, 75 (1992)
5. Gaggioli, RA: Available energy and exergy. International Journal of Applied Thermodynamics **1**(1–4), 1–8 (1998)
6. Gaggioli, RA, Richardson, DH, Bowman, AJ: Available energy – Part I: Gibbs revisited. Journal of Energy Resources Technology **124**, 105–115 (2002)
7. Gaggioli, RA, Paulus Jr, DM: Available Energy – Part II: Gibbs extended. Transaction of the ASME. Journal of Energy Resources Technology **124**, 110–115 (2002)
8. Beretta, GP: Axiomatic definition of entropy for nonequilibrium states. International Journal of Thermodynamics **11**, 39–48 (2008)
9. Zanchini, E, Beretta, GP: Removing heat and conceptual loops from the definition of entropy. International Journal of Thermodynamics **13**, 67–76 (2010)

11

Modelling and estimation of photosynthetically active incident radiation based on global irradiance in Indian latitudes

Kumarasamy Sudhakar[1*], Tulika Srivastava[1], Guddy Satpathy[1] and Manicam Premalatha[2]

Abstract

Photosynthetically active radiation (PAR) as a component of solar radiation plays a major role in different applications dealing with plant canopies, biomass production and microalgae growth. The amount of PAR energy depends upon location, time of the year and atmospheric conditions. Understanding the PAR and its availability is very essential for modelling biological growth system. The objective of this study is to estimate PAR for latitudes ranging from 9° to 34° and for the entire year based on hourly and monthly average of the daily global radiation (Hg). Based on the estimated data, a power regression model showing the relationship between PAR and Hg is presented for six Indian latitudes. The ratios of hourly sum of PAR to Hg and monthly average of hourly global radiations (Ig) vary smoothly with significant seasonal variations and are influenced by several other local climatic conditions. The power regression equation between PAR and Hg obtained are as follows: (a) For 9° latitude, PAR = 0.040 $(Hg)^{0.924}$, R^2 = 0.970, (b) 14° latitude, PAR = 0.029 $(Hg)^{0.952}$, R^2 = 0.894, (c) 19° latitude, PAR = 0.256 $(Hg)^{0.748}$, R^2 = 0.721, (d) 24° latitude, PAR = 0.159 $(Hg)^{0.775}$, R^2 = 0.896, (e) 29° latitude, PAR = 0.052 $(Hg)^{0.886}$, R^2 = 0.830 and (f) 34° latitude, PAR = 0.016 $(Hg)^{1.013}$, R^2 = 0.768.

Keywords: Photosynthetically active radiation, Solar radiation, Latitude, PAR modelling, India

Background

The amount of solar radiation reaching the top of the Earth's atmosphere is around 1,367 W m^{-2}, also called as solar constant [1]. On a bright sunny day, roughly around 800 to 1,200 W m^{-2} of total radiation reaches the Earth's surface owing to atmospheric scattering and absorption. The sun radiates energy in the spectral range from 280 to 4,000 nm (Figure 1), with a maximum in the blue-green (480 nm). The three major regions of the solar spectrum are (a) ultraviolet region (wavelength less than 400 nm): constitutes 9% of the irradiance, (b) visible region (wavelength ranging from 400 to 700 nm): constitutes 45% of the irradiance and (c) infrared region (wavelength greater than 700 nm): constitutes 46% of the irradiance.

The photosynthetically active radiation (PAR) in visible spectrum ranging from 400 to 700 nm is absorbed by plants to carry out photosynthesis. The PAR provides energy for photosynthesis and regulates plant growth and development. Within the PAR, the solar radiation peaks at 590 nm. The amount of PAR highly depends on environmental conditions. The measurement of the PAR is very important for environmental systems, agriculture, forest, architecture, solar greenhouse and many other applications. The PAR estimation based on simple methodologies would be highly useful for ecological modelling, algal photosynthesis, plant yield and land use efficiency. Therefore, a simple regression model that accurately predicts the PAR from global irradiance is necessary to design the growth system of phototropic microorganism.

The energy generation from solar radiation and performance correlations has been investigated by many authors [2–6]. The estimation of PAR based on solar global radiation has been studied by different authors for various specific locations [7–12]. The theoretical background and applications of PAR are described by [13]. Relatively, very little information regarding the availability of PAR in Indian context has been published in literature. Considering that the PAR measurement is very scarce in India,

* Correspondence: sudha_k@sify.com
[1]Energy Centre, Maulana Azad National Institute of Technology Bhopal, Madhya Pradesh 462 051, India
Full list of author information is available at the end of the article

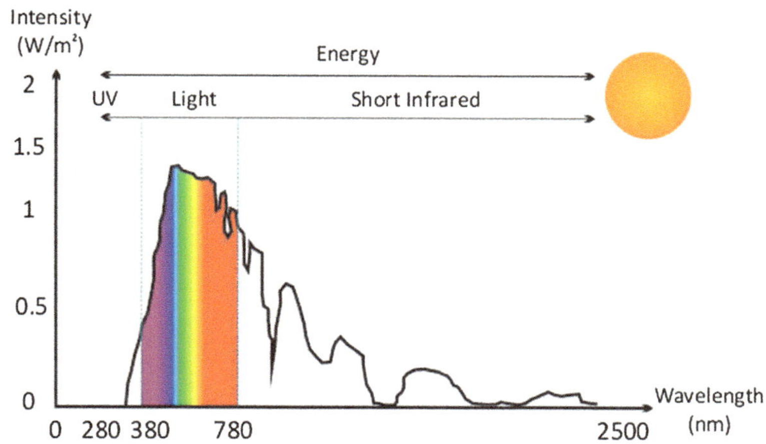

Figure 1 Solar spectrum.

it becomes more necessary to estimate PAR based on global solar radiation as a function of local climatic condition. The most probable idealised conversion coefficient (PAR/ Total global solar radiation, μmol J^{-1}) of 1.814 has been reported [13].

The objective of this study is to estimate PAR based on hourly average and daily average of monthly solar radiation for the Indian latitudes ranging from 9° to 34°. This study is initiated to develop a simpler approach for estimating and validating the PAR and provide monthly estimates of solar radiation that can be used as input for biological growth model.

The specific activities of this study are as follows: to predict the monthly and daily average of global solar radiation, to estimate the daily average PAR at various Indian latitudes, and to evaluate the correlation between estimated monthly average of hourly global radiation and the PAR radiation at the Earth's surface based on regression model.

Methods

PAR modelling

The characteristics of studied location

India lies to the north of the equator between 6°44' and 37°30' north latitude and 68°7' and 97°25' east longitude. The Indian subcontinent undergoes four major climatic seasons when every season is about 2 months. These are winter (January to February), summer (March to May), a monsoon (rainy) season (June to September) and a post-monsoon season (October to December). Various climatic factors are responsible for the seasonal changes in India.

Estimation of daily average of monthly horizontal global radiation

The extraterrestrial solar radiation incident on a horizontal surface can be determined for any location and day of the year according to previous researchers [14,15]. Solar declination angle can be calculated for each day based on the Cooper equation. It varies throughout the year and is the function of the day number of the year (N) as shown in Table 1. N ranges from 1 to 365.

$$\text{del} = 23.45 \sin\left(\frac{360\,(284+N)}{365}\right) \tag{1}$$

The solar hour angle for a location on Earth is zero when the sun is directly overhead, negative before noon and positive in the afternoon. The solar hour varies from 1 to 24.

$$\omega = 15\,(Lst-12) \tag{2}$$

The sunrise hour angle for a location is a function of solar declination angle and the latitude.

$$\omega_s = \cos^{-1}(-\tan(\text{del})\tan(\text{lat})) \tag{3}$$

Table 1 Days on which extraterrestrial radiation is equal to monthly mean value [16]

Month	Day	Day of the year (*N*)
January	17	17
February	16	47
March	16	75
April	15	105
May	15	135
June	11	162
July	17	198
August	16	228
September	15	258
October	15	288
November	14	318
December	10	344

The total daily average extraterrestrial radiation incident on Earth varies seasonally due to the atmospheric transmissivity associated with cloud cover.

$$\mathrm{Hg} = \left(\frac{24\, I_{sc}}{\pi}\right)\left(1 + 1.033\cos\left(\frac{360\, N}{365}\right)\right) \times ((\cos(\mathrm{lat})\cos(\mathrm{del})\sin\omega_s - (\omega_s \sin(\mathrm{lat})\sin(\mathrm{del}))) \quad (4)$$

I_{sc} is called as solar constant which is roughly equal to 1,367 W m^{-2}.

Estimation of hourly average of monthly horizontal global radiation

The hourly extraterrestrial radiation on a horizontal surface is obtained from the following expression:

$$\mathrm{I_g} = \left(\frac{3.142\, \mathrm{H_g}}{24}\right)(\mathrm{a} + \mathrm{b}\cos\omega_s)\left|\frac{\cos\omega_s - \cos\omega}{\sin\omega_s - \omega_s\cos\omega_s}\right| \quad (5)$$

The empirical constants by regression parameters for a particular location were obtained by the fitting data:

$$a = 0.409 - (0.5016\sin(\omega_s - 60)) \quad (6)$$

$$b = 0.6609 - (0.4767\sin(\omega_s - 60)) \quad (7)$$

Estimation of photosynthetically active global radiation

The PAR (μmol m^{-2} s^{-1}) is a function of various solar angles and the total global solar radiation. The total PAR over a horizontal surface can be determined from the monthly average hourly radiation directly by utilising the following expression [17]:

$$\mathrm{PAR} = \mathrm{Ig} \times 1.83 \quad (8)$$

Table 2 The estimation of Hg, Ig and PAR on a horizontal surface estimated for 9° latitude in India

Month	del	Hg kJ m^{-2} day^{-1}	Ig kJ m^{-2} h^{-1}	PAR μmol m^{-2} s^{-1}
Jan	−20.9098	40,914	486.7981	890.84
Feb	−12.9402	51,905	518.3203	948.5262
Mar	−2.399	60,682	578.3328	1,058.3
Apr	9.4335	38,612	366.2531	670.243
May	18.805	16,411	153.1378	280.2423
Jun	23.09	55,774	435.5227	797.008
Jul	21.1729	38,410	330.5119	603.8368
Aug	13.433	6,387	64.789	118.563
Sep	2.1886	9,122.4	101.2441	185.2767
Oct	−9.6267	28,694	309.6061	566.5791
Nov	−18.9306	55,080	604.9113	1,107
Dec	−23.0557	9,738.6	129.2328	236.496

Table 3 The estimation of Hg, Ig and PAR on a horizontal surface estimated for 14° latitude in India

Month	del	Hg kJ m^{-2} day^{-1}	Ig kJ m^{-2} h^{-1}	PAR μmol m^{-2} s^{-1}
Jan	−20.9098	40,579	523.984	958.891
Feb	12.9402	66,806	631.9202	1,156.4
Mar	−2.399	17,258	191.0942	349.7023
Apr	9.4335	12,625	124.5059	227.8458
May	18.805	56,542	403.64	738.6611
Jun	23.09	13,978	114.9241	210.3111
Jul	21.1729	42,435	323.444	591.9024
Aug	13.433	19,069	173.2918	317.1239
Sep	2.1886	46,186	449.6715	882.8988
Oct	−9.6267	51,407	508.4048	930.3808
Nov	−18.9306	15,685	212.6837	389.5406
Dec	−23.0557	48,933	614.4773	1,124.5

Equations 1, 2, 3, 4, 5, 6, 7 and 8 were solved in MATLAB (MathWorks, Inc., Natick, MA, USA) to find the PAR at any given latitude and day. MATLAB program calculates the values of the PAR by substituting values of N ranging from 17 to 344 in Equation 1.

Results

Variation of daily average of monthly extraterrestrial global solar radiations

The methodology proposed in Equations 1, 2, 3, 4, 5, 6, 7 and 8 has been used to estimate the hourly and daily average of monthly global solar radiation on a horizontal surface as a function of position and location. By knowing the hourly average of monthly global solar radiation

Table 4 The estimation of Hg, Ig and PAR on a horizontal surface estimated for 19° latitude in India

Month	del	Hg kJ m^{-2} day^{-1}	Ig kJ m^{-2} h^{-1}	PAR μmol m^{-2} s^{-1}
Jan	−20.9098	21,924	329.5871	603.1444
Feb	−12.9402	27,172	325.7783	596.1742
Mar	−2.399	5,802	66.9735	122.5614
Apr	9.4335	3,128.3	30.3794	55.5943
May	18.805	63,917	389.353	712.5163
Jun	23.09	22,221	157.0416	287.3862
Jul	21.1729	58,963	370.9429	678.8255
Aug	13.433	6,549	447.2846	818.5308
Sep	2.1886	67,468	591.2964	1,082.1
Oct	−9.6267	61,206	575.3947	1053
Nov	−18.9306	2,790.7	42.9229	78.549
Dec	−23.0557	45,290	629.6278	1,152.2

Table 5 The estimation of Hg, Ig and PAR on a horizontal surface estimated for 24° latitude in India

Month	del	Hg	Ig	PAR
		kJ m^{-2} day^{-1}	kJ m^{-2} h^{-1}	μmol m^{-2} s^{-1}
Jan	−20.9098	3,125	55.4156	101.4106
Feb	−12.9402	10,325	140.9995	258.0291
Mar	−2.399	46,509	451.1772	825.6543
Apr	9.4335	2,321.6	21.5012	39.3471
May	18.805	61,260	330.4533	604.7295
Jun	23.09	40,167	227.2903	415.9413
Jul	21.1729	42,745	255.0006	466.6511
Aug	13.433	13,560	108.7657	199.0412
Sep	2.1886	39,728	361.5716	661.676
Oct	−9.6267	60,757	569.1226	1,040
Nov	−18.9306	2,736.2	45.9098	84.015
Dec	−23.0557	16,963	295.5098	540.7829

data on the horizontal surface, the corresponding average PAR can be determined. Tables 2, 3, 4, 5, 6 and 7 show the estimated declination angle, monthly average of the daily values of global radiation, monthly average of the hourly values of global radiation and the PAR for Indian latitudes. The values obtained for 34° latitude are the lowest and ranged from 16,938 to 4,129.9 kJ m^{-2} day^{-1} as shown in Table 7. The total annual average global solar radiation is highest for 14° latitude (35,958 kJ m^{-2} day^{-1}), and 34° latitude has the lowest value of (18,738 kJ m^{-2} day^{-1}) as shown in Table 8. The results obtained for the studied locations show that irradiance profile is quite similar. The daily global variations for all the latitude are also analogous. It is observed that the estimated daily global radiation results are in close agreement with the other literature data [16].

Table 6 The estimation of Hg, Ig and PAR on a horizontal surface estimated for 29° latitude in India

Month	del	Hg	Ig	PAR
		kJ m^{-2} day^{-1}	kJ m^{-2} h^{-1}	μmol m^{-2} s^{-1}
Jan	−20.9098	56,579	871.1424	1,590
Feb	−12.9402	57,924	595.6629	1,090.1
Mar	−2.399	59,268	523.9165	958.7672
Apr	9.4335	3,097.5	27.0581	49.5164
May	18.805	49,033	246.9362	451.8933
Jun	23.09	53,528	237.00	433.3325
Jul	21.1729	13,957	84.1	153.7891
Aug	13.433	50,563	302.5928	553.7448
Sep	2.1886	2,856.5	30.4309	55.6882
Oct	−9.6267	50,511	505.964	925.9142
Nov	−18.9306	22,507	378.2182	692.1392
Dec	−23.0557	7,722.9	158.0091	289.1566

Table 7 The estimation of Hg, Ig and PAR on a horizontal surface estimated for 34° latitude in India

Month	del	Hg	Ig	PAR
		kJ m^{-2} day^{-1}	kJ m^{-2} h^{-1}	μmol m^{-2} s^{-1}
Jan	−20.9098	16,938	354.4352	648.6163
Feb	−12.9402	15,198	223.4562	408.9248
Mar	−2.399	11,756	132.4627	242.4067
Apr	9.4335	19,738	148.7651	272.2402
May	18.805	2,601.4	15.3687	28.1247
Jun	23.09	52,384	191.0395	349.6022
Jul	21.1729	6,846.5	37.1997	68.0754
Aug	13.433	2,432.7	17.4003	31.8426
Sep	2.1886	27,305	255.5551	467.6658
Oct	−9.6267	18,757	240.8388	440.735
Nov	−18.9306	46,777	742.5132	1,358.8
Dec	−23.0557	4,129.9	98.8037	180.8108

Variation of the PAR

The observed values of the PAR across the six locations have means of 621.9093, 656.5131, 604.1243, 436.4398, 557.8364 and 377.8203 μmol m^{-2} s^{-1}, respectively. The annual average values of PAR, Hg, Ig and irradiance are shown in Table 8. It is observed that 34° latitude has the lowest value of irradiance and PAR. The seasonal and location-specific variation of the PAR is clearly evident from the data presented in Tables 2, 3, 4, 5, 6 and 7. These data show the availability of PAR at a specific latitude that is highly useful for locating an open pond algae cultivation system that transmits sufficient PAR to produce a higher biomass yield. However, low disperse irradiance level results in higher photosynthetic efficiency of plants and microorganisms when their cells are growing under irradiance level far from saturation. The analysis of solar radiation performed in this paper suggests that 34° latitude is the best option for large-scale microalgal cultivation because of the fact that microalgae requires

Table 8 Annual average of PAR and irradiance

Latitude	H_g	I_g	PAR	Irradiance
	kJ m^{-2} day^{-1}	kJ m^{-2} h^{-1}	μmol m^{-2} s^{-1}	W m^{-2}
9	34,310	627.4	621.9093	397.996
14	35,958	356	656.5131	417.113
19	32,202	329.7	604.1243	373.54
24	28,349	440.41	436.4398	328.85
29	35,628	609.37	557.8364	413.28
34	18,738	204.8	377.8203	217.36

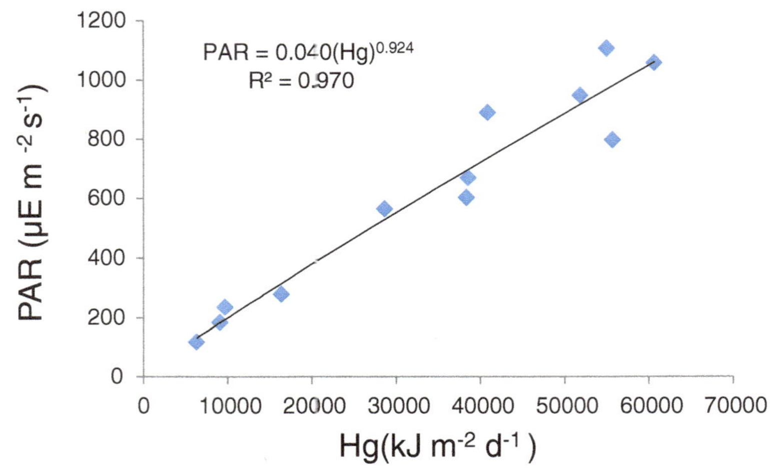

Figure 2 Estimated PAR vs Hg (Φ = 9°).

light intensity below 400 μmol m^{-2} s^{-1} for optimum growth [17].

Regression model

Regression analyses between the PAR and the monthly average of daily global solar radiation are carried out for each latitude. A power regression in the form $y = C(x)^n$ was chosen. The graphs of the PAR against daily average global irradiance are plotted for various locations (Figures 2 3, 4, 5, 6 and 7). The statistic performance of power regression equation to calculate the PAR for each latitude is shown in Table 9.

The power regression model provides estimates with some degree of correlation between the PAR and Hg. The estimated results are correlated for various latitudes in the following manner: 9° (R^2 = 0.931), 14° (R^2 = 0.829), 19° (R^2 = 0.721), 24° (R^2 = 0.766), 29° (R^2 = 0.830) and 34° (R^2 = 0.768). The power regression analysis between the estimated PAR and the Hg values shows a strong correlation for 9° latitude. A weaker correlation is observed for the rest of the latitudes. On the other hand, the spreading of these differences is higher at higher latitudes. The results for the PAR model suggest that the PAR could be estimated with an average level of confidence using global irradiance as a parameter .This suggests that the PAR values obtained from the analysis present a greater local dependency of climatic conditions and show marked differences. The proposed estimation scheme tries to highlight the necessity of local and seasonal calibration of the PAR/Hg ratio. Nevertheless, considering the differences in climatic conditions, it is possible that some differences exist in the correlation applicable to these different latitudes. This proves that there are marked tendencies to overestimate or underestimate the range of the PAR for other latitudes.

Discussion

PAR must be determined for applications related to microalgae growth, biomass yield, natural lighting and

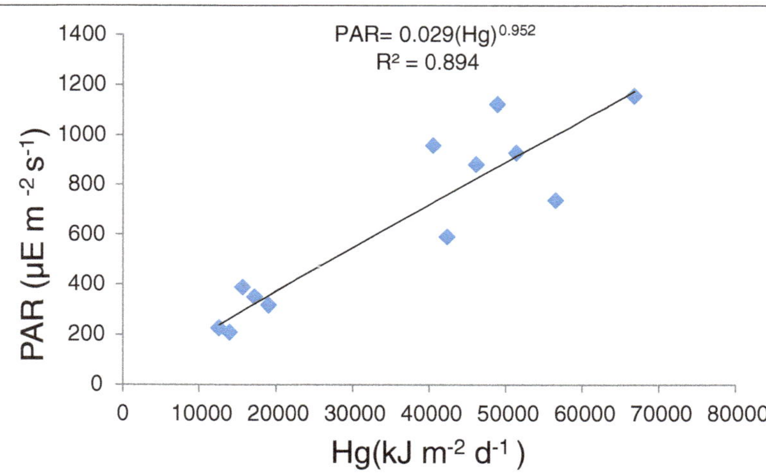

Figure 3 Estimated PAR vs Hg (Φ = 14°).

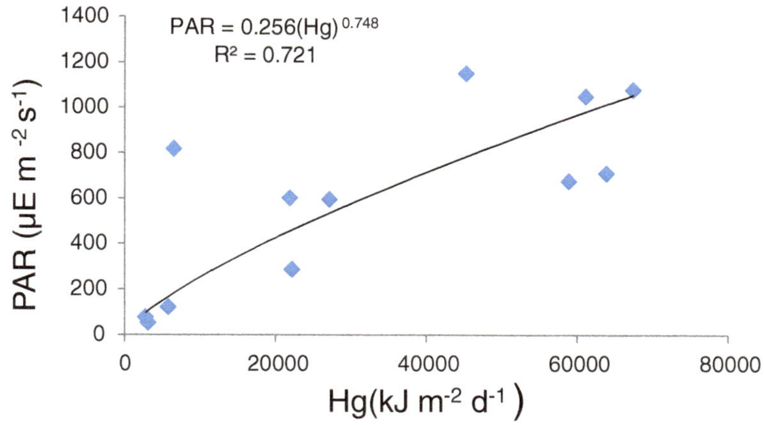

Figure 4 Estimated PAR vs Hg (Φ = 19°).

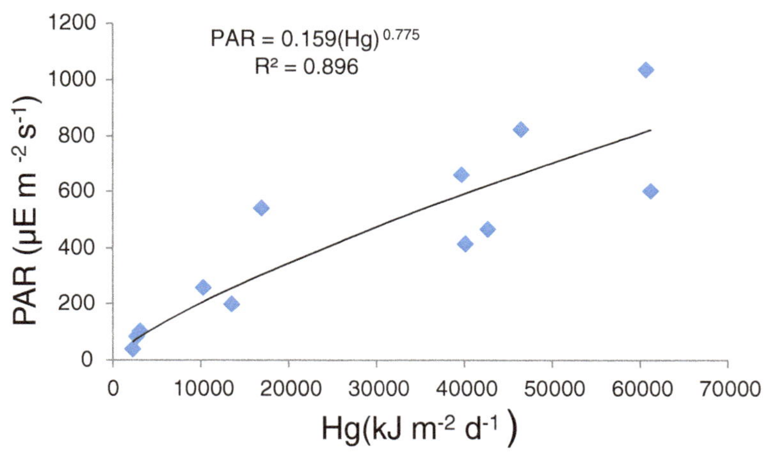

Figure 5 Estimated PAR vs Hg (Φ = 24°).

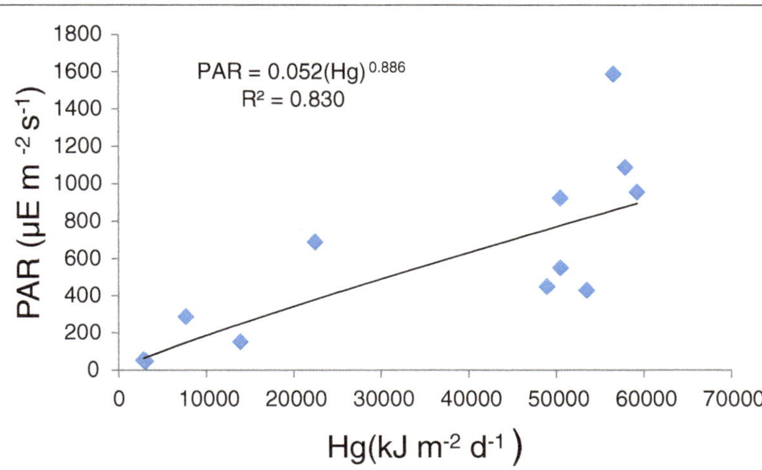

Figure 6 Estimated PAR vs Hg (Φ = 29°).

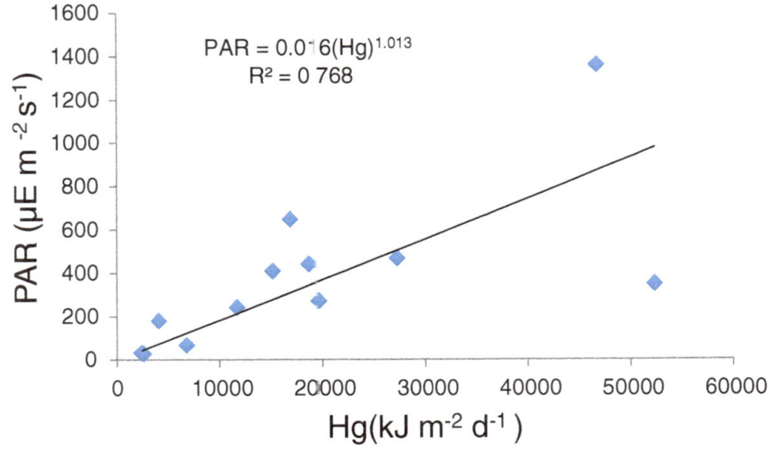

Figure 7 Estimated PAR vs Hg (Φ = 34°).

greenhouse design. The feasibility of microalgal cultivation system at a given location is greatly dependent on the availability of photosynthetically active component of natural sunlight. The models developed for this study were able to estimate PAR for a given day, latitude, surface and atmospheric conditions using MATLAB. This study is done to develop a simpler approach for estimating and validating the PAR . This can be used to provide rough estimates of hourly, daily, monthly and yearly global solar radiation. It is also evident that the annual average PAR can lie between 621 and 377 μmol m^{-2} s^{-1} which is close to established literature values [7–12]. With the relationships among irradiance, the PAR was examined using regression model. The results are used to determine if regression relationships differed significantly. For the 9° latitude data set, it is found that the power regression models provide better estimates. The estimates of PAR obtained through this method could be improved through actual ground measurements. The results of this model can be easily connected to biological process models for plant growth modelling and feasible design of algae cultivation system.

Conclusions

This study is a step towards modelling and estimation of PAR for a geographic location and day of the year. It takes into account the various solar angles, cloudiness, weather conditions and relates them to the global irradiance. The variation in the PAR during the whole year for various latitudes ranging from 9° to 34° geographical locations is quantified. The paper is the first attempt to quantify the PAR for various Indian latitudes; thus, the results can be used to obtain the important PAR values for similar conditions where Hg data are available. The performance of this method to estimate monthly average of hourly and daily global solar radiation is verified and validated with the other literature models. This approach represents a new opportunity to estimate the PAR for any location inside the study area. This could be particularly useful for designing algae cultivation system for optimising the PAR energy utilisation for maximum yield.

Table 9 Power regression equations obtained to estimate the PAR

Latitude (°)	R^2	Regression model
9	0.970	PAR = 0.040 $(Hg)^{0.924}$
14	0.894	PAR = 0.029 $(Hg)^{0.952}$
19	0.721	PAR = 0.256 $(Hg)^{0.748}$
24	0.896	PAR = 0.159 $(Hg)^{0.775}$
29	0.830	PAR = 0.052 $(Hg)^{0.886}$
34	0.768	PAR = 0.016 $(Hg)^{1.013}$

Abbreviations

a,b: empirical constants by regression parameters for a particular location, obtained by fitting data; del: declination angle (degrees); Hg: monthly average of daily global radiations on a horizontal surface (kJ m^{-2} day^{-1}); Ig: monthly average of hourly global radiations on a horizontal surface (kJ m^{-2} h^{-1}); I_{SC}: solar constant (W m^{-2}); lat: latitude (degrees); L_{ST}: local solar time (minutes); *N*: day number of the year; PAR: photosynthetically active radiation (μmol m^{-2} s^{-1}); ω_S: solar angle (degrees); ω: solar hour angle (degrees).

Competing interests

The authors declare that they have no competing interests.

Authors' contributions

MP gave guidance to KS who carried out the PAR modelling and drafted the manuscript. TS compiled them into a journal format. GS went through the text of the paper and critically checked the correlation for correctness. All authors read and approved the final manuscript.

Authors' information

KS is an assistant professor in the Department of Energy at Maulana National Institute of Technology, Bhopal. His research interests are algae photosynthesis, biofuels, solar thermal and PV, solar energy modelling, climate change and carbon sequestration. MP is an associate professor in Centre for Energy and Environmental Science and Technology at the National Institute of Technology, Tiruchirappalli. Her research interests are energy conservation, carbon sequestration treatment, energy modelling and simulation. TS and GS are M. Tech. scholars at the Energy Centre, Maulana National Institute of Technology, Bhopal.

Acknowledgements
The authors are grateful to the Director of the Maulana National Institute of Technology Bhopal, MP and to the director of the National Institute of Technology Tiruchirappalli, Tamilnadu.

Author details
[1]Energy Centre, Maulana Azad National Institute of Technology Bhopal, Madhya Pradesh 462 051, India. [2]CEESAT, National Institute of Technology Tiruchirappalli, Tamil Nadu 642051, India.

References

1. Iqbal, M: An Introduction to Solar Radiation. Academic Press, Toronto (1983)
2. Siddiqui, R, Bajpai, U: Correlation between thicknesses of dust collected on photovoltaic module and difference in efficiencies in composite climate. Int. J. Energy Env. Eng. **3**(26), 1–7 (2013)
3. Sharma, P, Harinarayana, T: Enhancement of energy generation from two layer solar panels. Int. J. Energy Env. Eng. **3**(12), 1–9 (2012)
4. Kandasamy, S, Vellingiri, M, Sengottain, S, Balasundaram, J: Performance correlation for single-basin double-slope solar still. Int. J. Energy Env. Eng. **4**(4), 1–6 (2013)
5. Panchal, H, Shah, PK: Investigation on solar stills having floating plates. Int. J. Energy Env. Eng. **3**(8), 1–5 (2012)
6. Oyedepo, S: Efficient energy utilization as a tool for sustainable development in Nigeria. Int. J. Energy Env. Eng. **3**, 11–22 (2012)
7. Papaioannou, G, Papanikolaou, N, Retails, D: Relationships of photosynthetically active radiation and shortwave irradiance. Theor. Appl. Climatol. **48**, 23–27 (1993)
8. Alados, I, Foyo-Moreno, I, Alados-Arboledas, L: Photosynthetically active radiation: measurements and modelling. Agric. For. Meteorol. **78**, 121–131 (1996)
9. Zhou, Y, Xiang, Y, Luan, L: Climatological estimation of photosynthetically active quantum flux. Acta meteorologica sinica **54**, 447–454 (1996)
10. Nojd, P, Hari, P: Instantaneous PAR estimated using long records of daily temperature and rainfall. Agric. For. Meteorol. **109**, 47–59 (2001). doi:10.1016/S0168-1923(01)00258-1
11. Tsubo, M, Walker, S: A model of radiation interception and use by a maize-bean intercrop canopy. Agric. For. Meteorol. **110**, 203–215 (2002). doi:10.1016/S0168-1923(01)00287-8
12. Jacovides, CP, Timvios, FS, Papaioannou, G, Asimakopoulos, DN, Theofilou, CM: Ratio of PAR to broadband solar radiation measured in Cyprus. Agric. For. Meteorol. **121**, 135–140 (2004). doi:10.1016/j.agrformet.2003.10.001
13. Ritchie, RJ: Modelling photosynthetic photon flux density and maximum potential gross photosynthesis. Photosynthetica **48**, 596–609 (2010)
14. Duffie, JA, Beckman, WA: Solar engineering of thermal processes. Wiley, New York (1980)
15. Liu, BYH, Jordan, RC: The interrelationship and characteristic distribution of direct, diffuse and total solar radiation. Sol. Energy **7**, 53–65 (1968)
16. Klein, SA: Calculation of monthly average insolation on titled surfaces. Sol. Energy **19**, 325 (1997)
17. Molina Grima, E, Fernandez, FGA, Camacho, FG, Chisti, Y: Photobioreactors: light regime, mass transfer, and scale up. J. Biotechnol. **70**, 231–247 (1999)

Hydrothermal synthesis of carbon-coated $LiCoPO_4$ cathode material from various Co sources

Masashi Kotobuki

Abstract

$LiCoPO_4$ has been recognized as a promising cathode material for lithium batteries due to its high stability and high operation voltage. However, poor electronic conductivity of $LiCoPO_4$ prohibits its practical use. A carbon coating can improve electronic conductivity of $LiCoPO_4$. A hydrothermal synthesis is a very convenient method because it allows us easy preparation of small particles of carbon-coated $LiCoPO_4$; however, an effect of precursor for $LiCoPO_4$ preparation on the performance of the synthesized $LiCoPO_4$ has yet to be cleared. In this paper, the effect of Co source for carbon-coated $LiCoPO_4$ ($LiCoPO_4$/C) preparation on performance as a cathode material for Li-ion battery is investigated. The Co source strongly affects the pH value in the starting solution and final products. The single-phase $LiCoPO_4$ is obtained only when $CoSO_4$ or $CoCl_2$ are used as the Co sources. A quality of carbon layer on the $LiCoPO_4$ is also affected by the Co source. The carbon layer on the $LiCoPO_4$ synthesized from $CoSO_4$ contains graphite carbon with high concentration which provides high electronic conductivity compared with that from $CoCl_2$. Accordingly, the $LiCoPO_4$/C synthesized from $CoSO_4$ shows a superior performance than that from $CoCl_2$ due to high-quality carbon layer.

Keywords: Hydrothermal synthesis, Co sources, Lithium-ion battery, Carbon-coated $LiCoPO_4$

Background

In order to use natural resources such as solar power and wind power, numerous research are done [1-3]. However, a shortcoming on the usage of the natural resources is its difficulty for stable electricity supply. Therefore, energy storage devices must be developed. Lithium rechargeable batteries have high energy and power densities and have been recognized as the most suitable battery for this purpose. The lithium batteries are composed of mainly cathode, anode, and electrolyte. $LiCoO_2$ and graphite have been used as cathode and anode, respectively, and organic solvent-resolving Li salt has been used for the electrolyte thus far. Both cathode and anode can intercalate Li ion, i.e., Li ion moves from anode to cathode during the discharge process. In the charge process, the Li ion goes back from cathode to anode.

One of the issues of $LiCoO_2$ is instability of its crystal structure especially under overcharge state. During the charge process, $LiCoO_2$ releases Li ion, and $Li_{1-x}CoO_2$ is formed. The safety circuits in the devices stop charging at $x = 0.5$ (3.9 V). In $x > 0.5$, the structure of $LiCoO_2$ collapses, and oxygen is released into the electrolyte. This oxygen causes the decomposition of electrolyte by oxidation and sometimes makes serious problems such as fire hazard and battery explosion.

Great attention has been paid to the olivine-type $LiMPO_4$ (M = Fe, Mn, Ni, Co) cathode materials for lithium-ion batteries in recent years [4-7]. The olivine materials contain PO_4 group in its structure and this strong P-O bond avoids the release of oxygen. Therefore, it is expected that the olivine materials improve the safety of the lithium batteries. Among the olivine-type materials, $LiCoPO_4$ has been given much attention owing to its high operation voltage close to 4.8 V vs. Li/Li^+ [8], which allows us to fabricate the lithium batteries with high operation voltage and high energy density because the energy density of the battery is estimated by multiplying the current density by the operation voltage. The electrochemical reaction of $LiCoPO_4$ in the cathode is shown as follows:

$$LiCoPO_4 \rightleftarrows Li^+ + CoPO_4 + e^-.$$

Correspondence: kotobuki@hakodate-ct.ac.jp
Department of Material and Environmental Engineering, Hakodate National College of Technology, 14-1 Tokura-cho, Hakodate, Hokkaido 042-8501, Japan

During the charge process, $LiCoPO_4$ releases a Li^+ into the electrolyte, and an electron (e^-) into a current collector (usually Al foil). Co^{2+} is oxidized to Co^{3+} at the same time. In the discharge process, reverse reaction occurs. The redox voltage of Co is high compared with other olivine materials, and it contributes to the high operation voltage of $LiCoPO_4$. According to the equation, immigration of the Li^+ and the electron in the cathode is needed. Therefore, the cathode materials require high Li^+ and electronic conductivity. However, the most serious problem of these olivine-type compounds is low intrinsic Li-ion and electronic conductivities [9,10]. To improve the Li-ion conductivity, preparation of small particles of the $LiMPO_4$ which derives a short Li-ion diffusion path is very effective [11,12]. Drezen et al. claimed that the particle size of $LiMnPO_4$ prepared by a sol–gel method strongly affected their performance [13]. On the other hand, to overcome the low electronic conductivity, electronic conductive carbon-coating on the surface of the particles has been applied in many groups [11,14-18]. Zhou et al. reported that the discharge capacity of $LiFePO_4$ prepared by the coprecipitation method had an increase of 110 to 147 mAh g^{-1} by carbon coating [19].

By hydrothermal synthesis, the carbon-coated small particles can be prepared easily with only an addition of a carbon source to starting materials. In fact, Mizuno et al. have succeeded in the preparation of carbon-coated small particles of $LiFePO_4$ [20] and $LiMnPO_4$ [7], and they have exhibited a good performance as the cathode material. In the hydrothermal synthesis, starting materials strongly affect the morphology and performance of the final products. Zhao et al. synthesized $LiCoPO_4$ from various phosphate sources such as H_3PO_4 and $(NH_4)_2HPO_4$ through the hydrothermal route and observed a hexagonal prism-like crystal in the range of 10 to approximately 15 μm synthesized from H_3PO_4 and irregular agglomeration particles with the size of 10 to approximately 40 μm synthesized from $(NH_4)_2HPO_4$ [21]. However, studies on different start materials, Li and Co sources, have yet to appear. In order to find a better preparation procedure for $LiCoPO_4$, an effect of Li and Co sources on properties of produced $LiCoPO_4$ should be tested. Therefore, the purpose of this paper is to investigate the effect of the Co source for $LiCoPO_4$ prepared by the hydrothermal route on its performance as the cathode of lithium-ion battery.

Methods

$LiCoPO_4$ was synthesized by a hydrothermal process according to previous reports for $LiFePO_4$ and $LiMnPO_4$ [22-24]. Li_3PO_4 was dissolved into purified water with various Co sources, i.e., $CoSO_4 \cdot 7H_2O$, $CoCl_2$, Co $(CH_3COO)_2 \cdot 4H_2O$, $Co(NO_3)_2 \cdot 6H_2O$, and $Co(OH)_2$. To prevent oxidation of Co^{2+} to Co^{3+}, this process was performed under N_2 atmosphere. A molar ratio of Li/Co/P was 3:1:1. Carboxy methyl cellulose was used as a carbon source and was put into the precursor solution. A glass-lined Parr reactor with the precursor solution and N_2 gas was sealed in a stainless steel autoclave. After that, the hydrothermal synthesis was performed at 200°C for 3 h with stirring at 680 rpm.

Precipitation was separated centrifugally and then freeze-dried at −50°C for 12 h. Yield was always higher than 95%. Synthesized powder was annealed at 700°C for 1 h under 3% H_2/Ar flow to convert carbon species to graphite with high electronic conductivity.

To clarify the crystalline phase of the synthesized powder, X-ray diffraction (XRD, RINT-Ultima, Rigaku Co., Shibuya-ku, Japan) with Cu Kα radiation was used. Raman measurements (NRS-1000, JASCO Ltd., Tokyo, Japan) were carried out at room temperature by a laser radiation of 532 nm. Morphologies of the synthesized powder were observed by means of a scanning electron microscope (SEM, JEOL Ltd., Akishima-shi, Japan).

Performance of a synthesized $LiCoPO_4$ as a positive electrode was evaluated by a charge–discharge test. A composite $LiCoPO_4$ electrode was prepared by mixing $LiCoPO_4$, Ketjen black, and polyvinylidene difluoride in the weight ratio of 75:15:10 in 1.2 ml of N-methyl pyrrolidone. The $LiCoPO_4$ electrode was painted onto a thin Al sheet and dried overnight at 85°C under reduced pressure. The $LiCoPO_4$ composite electrode (14 mm in diameter) was set in a coin cell 2032 with a mixture of ethylene carbonate and ethyl methyl carbonate (volume ratio = 3:7) containing 1 mol·dm^{-3} $LiPF_6$ as an electrolyte and with lithium metal as a negative electrode. The galvanostatic charge/discharge tests of the coin cell with $LiCoPO_4$ positive electrode were performed using the HJ1001SM8A (Hokuto Denko Ltd., Tokyo, Japan). In the test, the charge process was performed at constant current (CC)-constant voltage (CV) mode, charging at a constant current of 0.1 C until 5.1 V vs. Li/Li^+, then at a constant voltage of 5.1 V until the current dropped to 0.01 C. A discharge process proceeded at a CC mode of 0.1 C. Cut-off voltages of charge and discharge processes were 5.1 and 3.0 V vs. Li/Li^+, respectively.

Results and discussion

Results

In the hydrothermal synthesis, it has been well known that pH value in the starting solution influences the phase components and shapes and sizes of particles of the final products [25,26]. Thus, the pH value in the starting solution was checked at first. The pH values were widely changed depending on the Co sources, i.e., 3.7, 3.7, 4.4, 5.5, and 7.8 in the starting solution containing $Co(NO_3)_2$, $CoCl_2$, $CoSO_4$, $Co(CH_3COO)_2$, and $Co(OH)_2$, respectively.

Figure 1 shows XRD patterns of samples synthesized under hydrothermal condition using various Co sources. Obviously, the Co sources strongly influenced the final products. The diffraction peaks were completely different in various Co sources. $CoSO_4$ and $CoCl_2$ could produce a single-phase $LiCoPO_4$, i.e., the diffraction peaks were consistent with the standard peaks of $LiCoPO_4$. On the contrary, only diffraction peaks of starting materials (Li_3PO_4 and $Co(OH)_2$) were detected from $Co(OH)_2$, suggesting no reaction under this condition. The product from Co $(CH_3COO)_2$ also contained Li_3PO_4 with some unknown peaks. In the case of $Co(NO_3)_2$, the diffraction peaks of $LiCoPO_4$ was confirmed. The diffraction peaks of $Co(OH)_2$ and $Co(NO_3)_2$ were also observed.

Pure $LiCoPO_4/C$ was obtained only from $CoSO_4$ and $CoCl_2$. Thus, $LiCoO_4/C$ synthesized $LiCoPO_4/C$ from $CoSO_4$ and $CoCl_2$ (denoted as $LiCoPO_4/C(SO_4)$ and $LiCoPO_4/C(Cl_2)$, respectively) were supplied for further analysis. SEM images of synthesized $LiCoPO_4/C$ are displayed in Figure 2. In both cases (Figure 2a,c), the $LiCoPO_4/C$ particles aggregated each other and formed secondary particles whose sizes were below 20 μm. Interestingly, primary particles had different shapes depending on the Co source. Square-like particles whose sizes were 0.5 to approximately 2 μm were observed in $LiCoPO_4/C(Cl_2)$, while the shapes of the primary particles of $LiCoPO_4/C(SO_4)$ were irregular, and the sizes were 200 to approximately 500 nm. In both cases, the sizes of primary particles were below 2 μm and small enough as the cathode.

Raman measurement (Figure 3) was performed to detect small amounts of impurities on the surface of $LiCoPO_4/C$, which could not be confirmed in the XRD patterns. Nakano et al. have succeeded in the detection of even small amounts of impurities on the $LiFePO_4$ synthesized by the hydrothermal process [20]. In both the Raman spectra, shown in Figure 3, a small peak at 948 cm^{-1} was observed. This was attributed to the symmetric vibration of the PO_4 group [27]. Therefore, it is said that the synthesized powder contained a PO_4 group. As mentioned above, clear diffraction peaks of the $LiCoPO_4$ were observed in the XRD patterns. The peak at 948 cm^{-1} was consistent with the result of the XRD. It is concluded that $LiCoPO_4$ has been synthesized successfully from $CoSO_4$ and $CoCl_2$. A peak at 1,610 cm^{-1} was also observed in both spectra and was assigned to the graphite carbon (G band) [28]. This G band is generated by a presence of SP^2 carbon, which has high electronic conductive nature. However, the peak was very weak in both spectra. The carbon amounts of $LiCoPO_4/C(SO_4)$ and $LiCoPO_4/C(Cl_2)$ determined by thermogravimetry measurement were 0.9 and 3.6 wt.%, respectively. It is sure that carbon species existed but not very sensitive for the Raman measurement.

In a previous paper [20], Nakano et al. reported that electronic conductivity of hydrothermally synthesized $LiFePO_4/C$ was improved by annealing at 700°C in reductive or inert atmosphere. Consequently, the synthesized $LiCoPO_4/C$ was annealed at 700°C for 1 h under 3% H_2/Ar flow.

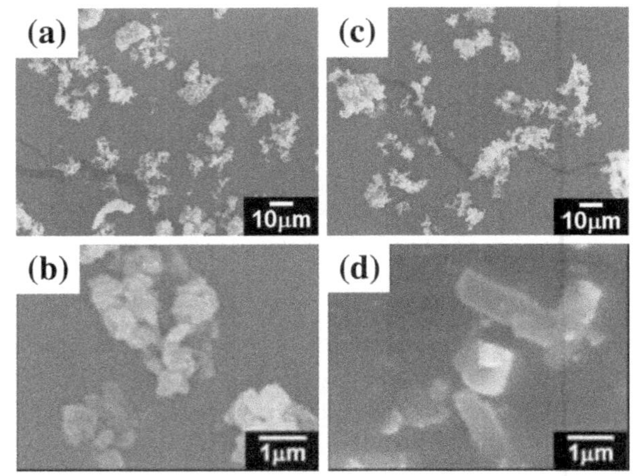

Figure 2 SEM images of hydrothermally synthesized $LiCoPO_4$ from various Co sources. **(a)**, **(b)** $CoSO_4$ and **(c)**, **(d)** $CoCl_2$.

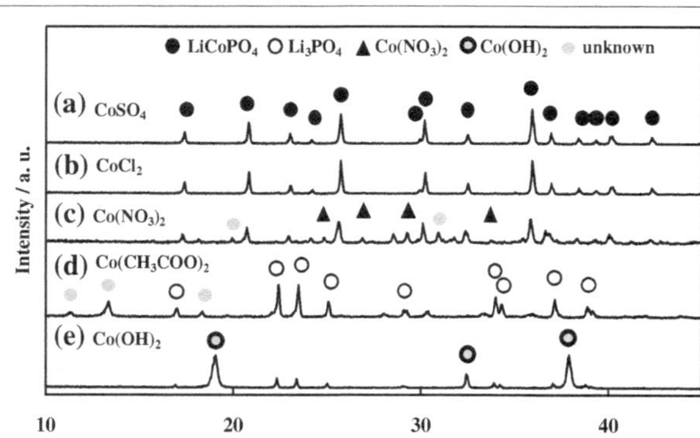

Figure 1 XRD patterns of hydrothermally synthesized $LiCoPO_4$ from various Co sources. **(a)** $CoSO_4$, **(b)** $CoCl_2$, **(c)** $Co(NO_3)_2$, **(d)** Co $(CH_3COO)_2$, and **(e)** $Co(OH)_2$.

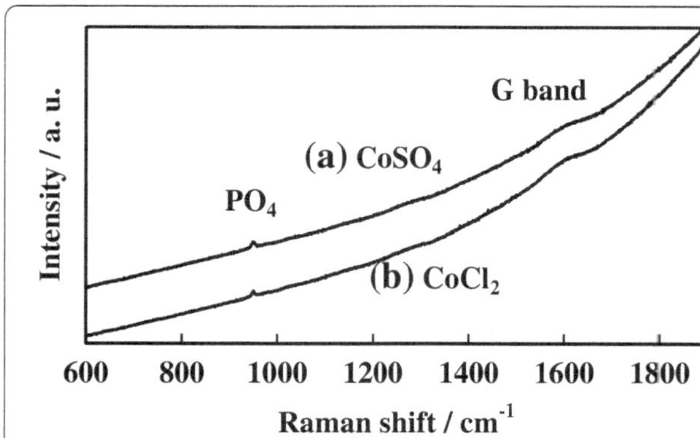

Figure 3 Raman spectra of hydrothermally synthesized $LiCoPO_4$ from various Co sources. **(a)** $CoSO_4$ and **(b)** $CoCl_2$.

Figure 4 shows SEM images of $LiCoPO_4/C$ after annealing. $LiCoPO_4$ particles aggregated each other and formed secondary particles like those before annealing. In the case of $LiCoPO_4/C(Cl_2)$, the sizes of the secondary particles were <20 μm, and the size of the primary particle was estimated as 0.5 to approximately 2 μm. The size of primary particles were the same as those before annealing, indicating that the annealing did not affect the particle size of $LiCoPO_4/C(Cl_2)$. On the contrary, large secondary particles (<60 μm) were obtained in $LiCoPO_4/C(SO_4)$. Additionally, the size of the primary particle increased as 1 to approximately 2 μm, and the particles became round shaped by annealing.

XRD patterns of $LiCoPO_4$ after annealing are depicted in Figure 5. All diffraction peaks were attributed to $LiCoPO_4$, indicating that the annealing did not produce any impurity phase. In order to estimate lattice parameters of $LiCoPO_4/C$, Si internal standard was added in the sample (A peak at $2\theta = 28.4°$ is assigned as Si internal standard). Lattice parameters of the $LiCoPO_4$ estimated using these XRD patterns were revealed in Table 1. $LiCoPO_4/C(Cl_2)$ possessed shorter lattice constant in the *b*-axis direction (5.929 Å in $LiCoPO_4/C(Cl_2)$ and 5.935 Å in $LiCoPO_4/C(SO_4)$), but larger ones in the *a*- and *c*-axis directions compared with $LiCoPO_4/C(SO_4)$. As a result, a lattice volume of $LiCoPO_4/C(Cl_2)$ was 0.40% larger than $LiCoPO_4/C(SO_4)$.

Figure 6 displays the Raman spectra of the $LiCoPO_4$ after annealing. By annealing, the Raman spectra changed dramatically, especially the Raman shift of 1,200 to approximately 1,700 cm^{-1} region. A new peak appeared at 1,345 cm^{-1} and was assigned as the disordered carbon (D band) [28], which has poor electronic conductivity, and a distinct G band could also be confirmed. The peak area ratios of G to D bands of $LiCoPO_4/C(SO_4)$ and $LiCoPO_4/C(Cl_2)$ were 1.4 and 0.7, respectively. The peak at 948 cm^{-1} was attributed to the symmetric vibration of the PO_4 group as described above. Therefore, it is concluded that $LiCoPO_4$ structure did not change by annealing.

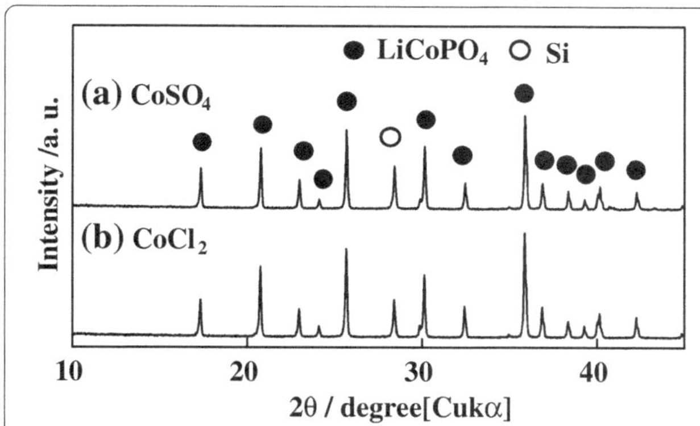

Figure 5 XRD patterns of hydrothermally synthesized $LiCoPO_4$ from various Co sources after annealing. **(a)** $CoSO_4$ and **(b)** $CoCl_2$.

Figure 7 revealed charge and discharge curves of $LiCoPO_4/C(SO_4)$ and $LiCoPO_4/C(Cl_2)$. Three coin cells were constructed in each sample to check the experimental error. In both cases, the charge capacity was much larger than the theoretical capacity (167 mAh g^{-1}) [6]. Many authors have reported a decomposition of electrolyte at high voltage like 5.1 V vs. Li/Li^+ [29,30]. Therefore, the reason for the large charge capacity is attributed to delithiation of the $LiCoPO_4$ coincident with the decomposition of the electrolyte. A discharge curve of $LiCoPO_4/C(SO_4)$ showed a much longer plateau at approximately 4.6 V vs. Li/Li^+ compared with $LiCoPO_4/C(Cl_2)$. Discharge capacities of $LiCoPO_4/C(SO_4)$ and $LiCoPO_4/C(Cl_2)$ were 99 ± 5 and 32 ± 3 mAh g^{-1}, respectively. The experimental error was very small. It is obvious that $LiCoPO_4/C(SO_4)$ showed better performance as the cathode.

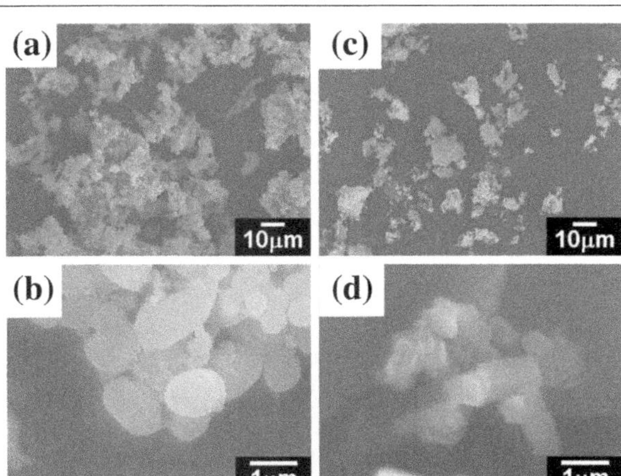

Figure 4 SEM images of hydrothermally synthesized $LiCoPO_4$ from various Co sources after annealing. **(a)** and **(b)** $CoSO_4$ and **(c)** and **(d)** $CoCl_2$.

Discussion

An effect of Co source for hydrothermally synthesized $LiCoPO_4$ on performance as cathode for the lithium-ion battery was studied.

Single-phase $LiCoPO_4$ particle was obtained only from $CoSO_4$ and $CoCl_2$. The pH value in the starting solution affected the formation of the final products. Huang et al. reported that the formation of pure $LiCoPO_4$ phase

Table 1 Lattice parameters of hydrothermally synthesized $LiCoPO_4$ after annealing estimated from Figure 5

Co source	*a*-Axis (Å)	*b*-Axis (Å)	*c*-Axis (Å)	Volume (Å^3)	ΔV/V
$CoSo_4$	10.23(2)	5.935(5)	4.708(6)	285.84(6)	-
$CoCl_2$	10.24(5)	5.929(2)	4.727(5)	286.99(0)	+0.40%

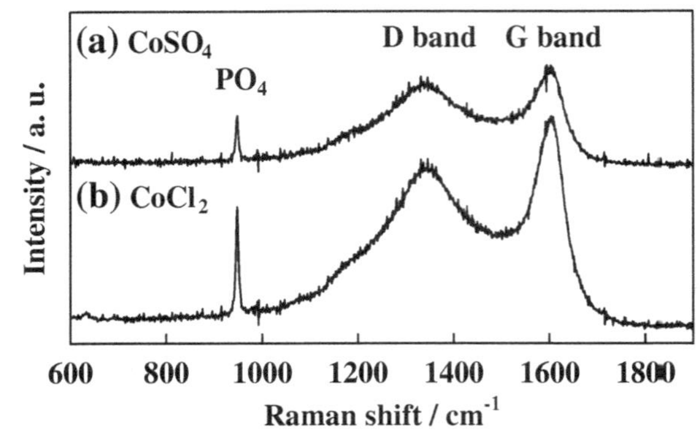

Figure 6 **Raman spectra of hydrothermally synthesized LiCoPO$_4$ from various Co sources after annealing. (a)** CoSO$_4$ and **(b)** CoCl$_2$.

single-phase LiCoPO$_4$ was obtained only under acidic condition. This discrepancy could be attributed to PO$_4$ and Li sources. They used LiOH and H$_3$PO$_4$ as Li and PO$_4$ sources, respectively. On the other hand, I used Li$_3$PO$_4$ as both Li and PO$_4$ sources. It is thought that solubility of Li$_3$PO$_4$ in the range of weak acidic to alkaline conditions would be low. In fact, diffraction peaks of Li$_3$PO$_4$ (Figure 1d,e) were observed only in the case of Co(OH)$_2$ and Co(CH$_3$COO)$_2$, whose pH values were 7.8 and 5.5, respectively. In these conditions, Li$_3$PO$_4$ could not dissolve in water even under hydrothermal condition. Therefore, Li$_3$PO$_4$ was precipitated on the bottom and could not participate in the reaction. This would be a reason that non-reacted Li$_3$PO$_4$ still remain in the final products. As for Co(NO$_3$)$_2$, a formation of LiCoPO$_4$ was confirmed, but some were unknown, and Co(NO$_3$)$_2$ peaks were also observed. The pH value of the starting solution containing Co(NO$_3$)$_2$ was 3.7, same as that of CoCl$_2$. Consequently, the counter anion may also play an important role for the formation of LiCoPO$_4$, not only the pH value of precursor solution.

After annealing, morphology change of LiCoPO$_4$/C (SO$_4$) was observed, while no large difference of LiCoPO$_4$/C(Cl$_2$) particle shape was confirmed. Thermal stability of produced LiCoPO$_4$/C particles would also be affected by the Co source.

Discharge capacity of LiCoPO$_4$/C(SO$_4$) and LiCoPO$_4$/C(Cl$_2$) were 99 ± 5 and 32 ± 3 mAh g^{-1}, respectively. It has been known that Li-ion transport in LiMPO$_4$ (M = Fe, Mn, Co, Ni) olivine materials occurs along with the *b*-axis [31]. The lattice constant in the *b*-axis direction of LiCoPO$_4$/C(Cl$_2$) was shorter than that of LiCoPO$_4$/C (SO$_4$) as shown in Table 1. Additionally, primary particle size of LiCoPO$_4$/C(Cl$_2$) was smaller than that of LiCoPO$_4$/C(SO$_4$). From the view of the Li-ion transport in the particles, LiCoPO$_4$/C(Cl$_2$) has an advantage against LiCoPO$_4$/C(SO$_4$) due to a short distance of Li-ion transport. Thus, LiCoPO$_4$/C(Cl$_2$) was expected to demonstrate better electrochemical performance; however, much larger discharge capacity was obtained in LiCoPO$_4$/C(SO$_4$). This should relate to the quality of carbon layer on the LiCoPO$_4$ particle, which can provide high electronic conductivity. As shown in the Raman spectra, there are two kinds of carbon, graphite, and disorder carbons, on the LiCoPO$_4$. The graphite carbon possesses high electronic conductivity, and the disorder carbon does not. The ratios of the graphite to the disorder carbon of LiCoPO$_4$/C(SO$_4$) and LiCoPO$_4$/C(Cl$_2$) were 1.4 and 0.7, respectively. The disorder carbon could not give a high electronic conductivity to the carbon-coating layer on the LiCoPO$_4$ whereas the graphite carbon could. Therefore, LiCoPO$_4$/C(SO$_4$) has superior electronic conductivity. This would be a reason why the LiCoPO$_4$/C(SO$_4$) showed better performance than LiCoPO$_4$/C(Cl$_2$). It is found that the Co sources affect not only the crystal structure and morphology of LiCoPO$_4$ but also the nature of the carbon layer, resulting in different cathodic performance among the products. Much attention should be paid for Co source selection to prepare the carbon-coated LiCoPO$_4$.

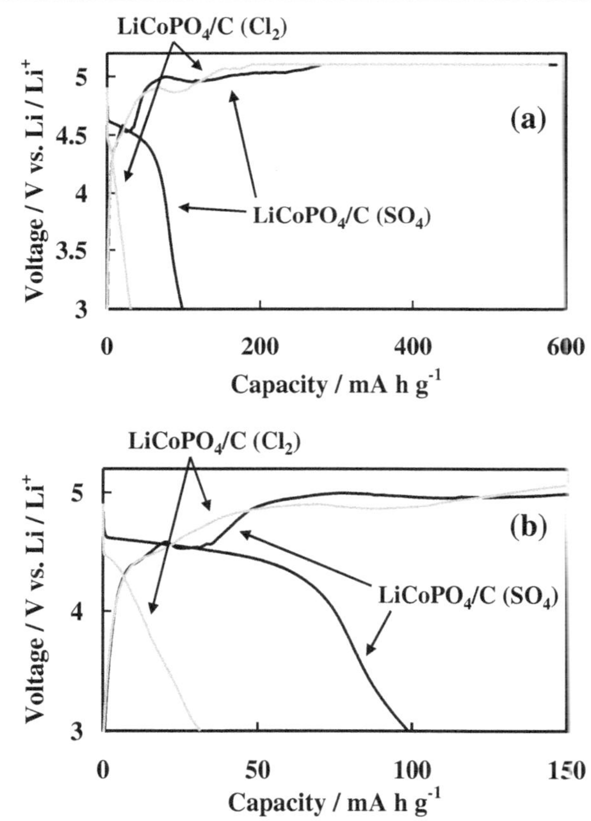

Figure 7 **Charge and discharge curves of hydrothermally synthesized LiCoPO$_4$ from various Co sources after annealing.** A charge process was performed at a constant current of 0.1 C until 5.1 V vs. Li/Li$^+$ then at constant voltage of 5.1 V until current dropped to 0.01 C. A discharge process proceeded at a constant current mode at 0.1 C. **(a)** Whole charge and discharge curves; **(b)** detailed charge and discharge curves until capacity of 150 mAh g^{-1}.

It was reported that the ratio of the graphite to disorder carbon on the $LiMnPO_4$ depended on the carbon sources [7]. Therefore, further improvement of the performance of $LiCoPO_4/C$ would be possible if suitable carbon source for $LiCoPO_4/C(Cl_2)$ was found because $LiCoPO_4/C(Cl_2)$ is thought to possess a favorable structure with respect to Li-ion transport. This attempt is now underway, and it will be reported.

Conclusions

The $LiCoPO_4$ cathode material for lithium batteries were synthesized by hydrothermal method using various Co sources. Crystal structure and morphology of hydrothermally synthesized $LiCoPO_4$ were strongly affected by the Co source. Single-phase $LiCoPO_4$ could not be obtained from $Co(NO_3)_2$, $Co(OH)_2$, and $Co(CH_3COO)_2$. Only $CoCl_2$ and $CoSO_4$ could produce single-phase $LiCoPO_4$. Before annealing, $LiCoPO_4$ was not very sensitive against Raman measurement. After annealing, the carbon species on the $LiCoPO_4$ particles became active for the Raman measurement and clear G and D bands could be observed. $LiCoPO_4$ synthesized from $CoCl_2$ possessed shorter lattice constant along with the *b*-axis than that from $CoSO_4$, suggesting favorable Li-ion transport. Additionally, Co source also affected a nature of carbon layer on the $LiCoPO_4$ after annealing. Ratios of G band to D band in the Raman spectra were 1.4 and 0.7 for $LiCoPO_4/C(SO_4)$ and $LiCoPO_4/C(Cl_2)$, respectively. The carbon layer on $LiCoPO_4$ synthesized from $CoSO_4$ was converted to graphite carbon easily, resulting in the investment of high electronic conductivity to the $LiCoPO_4$ particle. As a result, $LiCoPO_4$ synthesized from $CoSO_4$ revealed better performance compared with that from $CoCl_2$ although its particle seems to be inferior in terms of Li-ion transport. Further improvement of performance of $LiCoPO_4$ from $CoCl_2$ could be possible if high-quality carbon layer could be prepared because particles of $LiCoPO_4$ from $CoCl_2$ possess a favorable structure with respect to Li-ion transport.

Competing interests

The author has no competing interests.

Authors' information

Dr. MK graduated in 1996 and received his M.S. in 1998 from Toyohashi University of Technology. After working at Hino Motors as an engine designer from 1998 to 2002, he received a Ph.D. degree in Engineering from the University of Yamanashi in 2006 under the direction of Prof. Watanabe. He studied on hydrogen production and hydrogen purification catalysts for fuel cell application. After receiving the Ph.D., he worked at the University Ulm, Germany, as an Alexander von Humboldt fellow. He researched on the CO oxidation mechanism on Au catalyst mainly using Temporal Analysis of Product (TAP) reactor under Prof. R. J. Behm. Then, he came back to Japan as an associate professor in Tokyo Metropolitan University in 2009 and researched on lithium battery. Now, he is an associate professor at the Department of Material and Environmental Engineering in Hakodate National College of Technology, Japan. He specializes in lithium battery, solid electrolyte, electrochemistry, and heterogeneous catalysts. More than 40 papers with respect to fuel cell, lithium battery, and heterogeneous catalysts have been published under his name.

Acknowledgments

The authors appreciate Rigaku Co, JASCO Ltd, JEOL Ltd, and Hokuto Denko Ltd for their technical support.

References

1. Oyedepo, O, Adaramola, M, Paul, S: Analysis of wind speed data and wind energy potential in three selected locations in south-east Nigeria. Int. J. Energy. and Environ. Eng. **3**, 7 (2012)
2. Islam, S: Increasing wind energy penetration level using pumped hydro storage in island micro-grid system. Int. J. Energy. and Environ. Eng. **3**, 9 (2012)
3. Pippo, W, Luengo, C: Sugarcane energy use: accounting of feedstock energy considering current agro-industrial trends and their feasibility. Int. J. Energy and Environ. Eng. **4**, 10 (2013)
4. Kang, B, Ceder, G: Battery materials for ultrafast charging and discharging. Nature **458**, 190–193 (2009)
5. Zhou, F, Cococcioni, M, Kang, K, Ceder, G: The Li ion insertion potential of $LiMPO_4$ and $LiSiO_4$ olivines with M = Fe, Mn, Co. Ni. Electrochem. Comm. **6**, 1144–1148 (2004)
6. Kotobuki, M, Mizuno, Y, Munakata, H, Kanamura, K: Electrochemical properties of hydrothermally synthesised $LiCoPO_4$ as a high voltage cathode material for lithium secondary battery. Phosporous Res. Bull **24**, 12–15 (2010)
7. Mizuno, Y, Kotobuki, M, Munakata, H, Kanamura, K: Effect of carbon source on electrochemical performance of carbon coated $LiMnPO_4$ cathode. J. Ceramic Soc. Jpn. **117**, 1225–1228 (2009)
8. Amine, K, Yasuda, H, Yamachi, M: Olivine $LiCoPO_4$ as 4.8V Electrode Material for Lithium Batteires. Electrochem. and Solid-State Lett **3**(4), 178 (2000)
9. Delacourt, C, Laffont, L, Bouchet, R, Wurm, C, Leriche, JB, Morcrette, M, Tarascon, JM, Masquelier, C: Toward understanding of electrical limitations (electronic, ionic) in $LiMPO_4$ (M = Fe, Mn) electrode materials. J. Electrochemical Soc. **152**, A913 (2005)
10. Yonemura, M, Yamada, A, Takei, Y, Sonoyama, N, Kanno, R: Synthesis, characterization, and electrochemical properties of nanocrystalline silver thin films obtained by spray pyrolysis. J. Electrochemical Soc. **151**(1), A1352 (2004)
11. Yamada, A, Chung, SC, Hinokuma, K: Optimised $LiFePO_4$ for lithium battery cathodes. J. Electrochemical Soc. **148**(2), A224 (2001)
12. Prosini, PP, Carewska, M, Scaccia, S, Wisniewski, P, Pasquali, M: Long-term cyclability of nanostructured $LiFePO_4$. Electrochim. Acta **48**, 4205 (2003)
13. Drezen, T, Kwon, NH, Bowen, P, Teerlinck, I, Isono, M, Exnar, I: Effect of particle size on $LiMnPO_4$ cathodes. J. Power Sources **174**, 949–953 (2007)
14. Chen, ZH, Dahn, JR: Reducing carbon in $LiFePO_4/C$ composite electrodes to maximise specific energy, volumeric energy and tap density. J. Electrochemical Soc. **149**(9), A1184 (2002)
15. Huang, H, Yin, SC, Nazar, LF: Approaching theoretical capacity of $LiFePO_4$ at room temperature at high rates. Electrochem. Solid-State Lett. **4**, A170 (2001)
16. Kim, HS, Cho, BW, Cho, WI: Cycling performance of $LiFePO_4$ cathode material for lithium secondary batteries. J. Power Sources **132**, 235–239 (2004)
17. Dominko, R, Goupil, JM, Bele, M, Gaberseek, M, Remskar, M, Hanzel, D, Jamnik, J: Impact $LiFePO_4/C$ composites porosity on their electrochemical performance. J Electrochem Soc. **152**(5), A858 (2005)
18. Fey, GTK, Liu, TL: Morphological characterization of LiFePO4/C composite cathode materials synthesized via a carboxylic acid route. J Power Sources **178**, 807–814 (2008)
19. Zhou, W, He, W, Li, Z, Zhao, H, Yan, S: Biosynthesis and electrochemical characteristics of $LiFePO_4/C$ by microwave processing. J. Solid State Electrochem. **13**, 1819 (2009)
20. Nakano, H, Dokko, K, Koizumi, S, Tannai, H, Kanamura, K: Hydrothermal synthesis of carbon-coated $LiFePO_4$ and its application to lithium polymer battery. J. Electrochemical Soc. **155**(12), A909 (2008)
21. Zhao, Y, Wang, S, Zhao, C, Xia, D: Synthesis and electrochemical performance of $LiCoPO_4$ micros-rods by dispersant-aided hydrothermal method for lithium ion battery. Rare Metals **28**, 117 (2009)

22. Dokko, K, Shiraishi, K, Kanamura, K: Identification of surface impurities of on $LiFePO_4$ particles prepared by a hydrothermal process. J. Electrochemical Soc. **152**(11), A2199 (2005)
23. Dokko, K, Koizumi, S, Koizumi, K: Electrochemical reactivity of $LiFePO_4$ prepared by hydrothermal method. Chem. Lett. **35**, 338 (2006)
24. Dokko, K, Koizumi, S, Shiraishi, K, Kanamura, K: Electrochemical properties of $LiFePO_4$ prepared via hydrothermal route. J. Power Sources **165**, 656–659 (2007)
25. Huang, X, Ma, J, Wu, P, Hu, Y, Dai, J, Zhu, Z, Chen, H, Wang, H: Hydrothermal synthesis of $LiCoPO_4$ cathode materials for rechargeable lithium ion battery. Material Lett. **59**, 578 (2005)
26. Yang, S, Zavalij, PY, Whittingham, MS: Hydrothermal synthesis of lithium iron phosephate cathodes. Electrochem. Commum. **3**, 505–508 (2001)
27. Salah, AA, Mauger, A, Zaghib, K, Goodenough, JB, Ravet, N, Gauthier, M, Gendron, F, Julien, CM: Reduction Fe^{3+} of impurities in $LiFePO_4$ from pyrolysis organic precursor used for carbon deposition. J. Electrochem. Soc. **153**(9), A1692 (2006)
28. Kotobuki, M, Mizuno, Y, Munakata, H, Kanamura, K: Improved performance of hydrothermally synthesized $LiMnPO_4$ by Mg doping. Electrochemistry **79**, 467 (2011)
29. Aurbach, D, Markovsky, B, Levi, MD, Levi, E, Schechter, A, Moshkovich, M, Cohen, Y: New insights into the interactions between electrode materials and electrolyte solutions for advanced nonaqueous batteries. J. Power. Sources **81–82**, 95–111 (1999)
30. Talyosef, Y, Markovsky, B, Salitra, G, Aurbach, D, Kim, HJ, Choi, S: The study of $LiNi_{0.5}Mn_{1.5}O_4$ 5-V cathodes for Li-ion batteries. J. Power Sources **146** 664–669 (2005)
31. Morgan, D, Van der Ven, A, Ceder, G: Li conductivity in Li_xMPO_4 (M = Fe, Mn, Co, Ni) olivine materials. Electrochem. Solid-State Lett. **7**(2), A30 (2004)

13

Determination of optimum tilt angles for solar collectors in low-latitude tropical region

Oloketuyi S Idowu*, Oyewola M Olarenwaju and Odesola I Ifedayo

Abstract

This work optimises the collection of solar energy within the period of its availability in order to increase its utilisation and also to enhance performance of heating systems that depend on it through appropriate determination of optimum solar collector tilt angles. Buoyancy-induced flow equation in solar collector pipe was established by continuity, energy and Navier–Stokes equations in cylindrical coordinates. Fundamental solar radiation equations were programmed to determine optimum tilt angles in locations within latitudes 1° and 14°. A set of data recorded from a pryanometer located on latitude 6.45° north of the equator was used to generate average monthly radiation over the latitudes. Graphs obtained from latitude 6° and 13° data were analysed to investigate solar radiation on some tilt angles. The optimum tilt angles for solar heating for periodic tracking of the sun in the region within latitudes 1° and 14° were predicted as ∅ + 25° for November, December and January; ∅ + 15° for February, September and October; ∅ – 15° for August; ∅ – 25° for May, June and July; and ∅ for March and April. The results of this work confirmed that solar radiation on tilted surface increases with latitude.

Keywords: Solar energy; Optimization; Buoyancy; Pryanometer; Tilt angle; Tracking; Latitude

Background

Renewable energy obtained from the sun is very important because of the fact that it is free and environment-friendly [1]. The importance of detailed knowledge of solar radiation received from the sun at a site in the design and selection of solar devices cannot be overstated [2]. In order to optimise solar isolation on solar collectors, appropriate method to determine solar tilt angles at any given time is essential to increase the efficiencies of the collectors and that of the devices connected to them [3,4]. The position of the earth relative to the sun changes with time; the change must be monitored adequately in order to increase the amount of energy being received by solar devices [5].

Solar radiation in low-latitude tropical region is not as intensified as that in high latitude tropical region due to cloudiness and high humidity in the former region [2]. For example, Nigeria is located between longitude 3° and 14° east of Greenwich and latitude 4° and 14° north of equator. Ojosu's [6] study on the iso-radiation map of Nigeria showed that the annual average solar radiation along the coastal areas is 3.7 kW/m^2/day, while that in the semi-arid zone is 7.0 kW/m^2/day. Interestingly, Abubakar [7] in an independent study showed that the mean annual average of total solar radiation varies from 3.5 kW/m^2/day in the coastal latitudes to about 7.0 kW/m^2/day along the semi-arid areas in the far north. The emerging conclusion from the past studies was that solar energy must be managed effectively in the southern region of Nigeria.

The magnitude of solar radiation received by a collector is a function of many factors such as location latitude, the declination angle (the angular position of the sun at solar noon with respect to the plane of the equator), tilt angle, the sunrise hour angle and the azimuth angle [8]. According to Benghanem [9], both the orientation and tilt angles have significant effects on the magnitude of the solar radiation reaching the surface of a collector. Many investigations have been carried out to determine the optimum tilt angle for solar collector. Honsberg [10] suggested the following tilt angles: ∅ + 15° for optimum, ∅ – 15° for maximum and ∅ for uniform power production. Some other research works suggested two values for optimum tilt angle, one for summer (rainy season in tropical region) and the other for winter (dry season in tropical region): ∅ ± 8° [11,12] and ∅ ± 5°

* Correspondence: oketuus@yahoo.com
Department of Mechanical Engineering, University of Ibadan, Ibadan, Nigeria

[11,13] ('+' is for winter or dry season, while '–' is for summer or rainy season).

In a solar system where thermal fluid is to be heated, mathematical simulation shows that for maximum flow rate, the solar collector tilt angle is 90° [14]. This specification is not suitable for solar collector, either in low or high latitude region for maximum solar energy collection. Therefore, the required optimum tilt angle of the solar collector should be less than 90°.

A lot of awareness is ongoing worldwide on solar energy utilisation. This research study determines appropriate periodic optimum tilt angles for solar collectors in different locations at different days of the year using Nigeria as a case study.

Fundamentals of solar radiation on earth surface

When solar rays hit a flat surface at an oblique angle, the rays are more spread out [5]. This means, the power of the beam is being spread out over a larger area. Figure 1 shows the relationship between radiation on horizontal surface and that on tilted surface. The ratio of normal radiation on tilted surface (H_T) to that on horizontal surface (H) is given in terms of θ_z (the angle between the beam from the sun and vertical), θ_T and H_n as shown below [15]:

$$R_b = \frac{H_T}{H} = \frac{H_n \cos\theta_T}{H_n \cos\theta_Z} \tag{1}$$

$$\frac{\cos\theta_T}{\cos\theta_Z} = \frac{\cos(\phi - s)\cos(\delta)\cos(\omega) + \sin(\phi - s)\sin(\delta)}{\sin(\delta)\sin(\phi) + \cos(\delta)\cos(\phi)\cos(\omega)} \tag{2}$$

$$\delta = 23.45 \sin\left(\frac{360}{365}(284 + d)\right). \tag{3}$$

Maximum power into the solar collector occurs when the surface is normal to the incident solar radiation; however, it is not always possible with fixed solar collector since the relative position of the earth to the sun varies [10]. Outside the earth's atmosphere, at any given point in space, the energy given off by the sun (insolation) is nearly constant. On earth, however, that situation changes as a result of earth rotation about its axis, change in earth position in space as indicated in Figure 2 and the earth atmosphere which includes gases, clouds and dust [5].

Fundamentals of induced thermo-siphoned flow through circular pipes

For simplicity without compromising the usefulness of the results, steady flow analysis can be adopted to determine the fluid flow regimes [20]. The three fundamental steady flow equations that govern the flow of thermal fluid in solar collector pipes are continuity, momentum and energy equations.

The continuity equation is

$$\frac{\partial(\rho u)}{\partial x} + \frac{\partial(\rho v)}{\partial y} = 0. \tag{4}$$

Navier–Stokes equation (momentum equation) for steady flow is

$$\rho \frac{D}{Dt} v = -\nabla p + \mu \nabla^2 + \rho g. \tag{5}$$

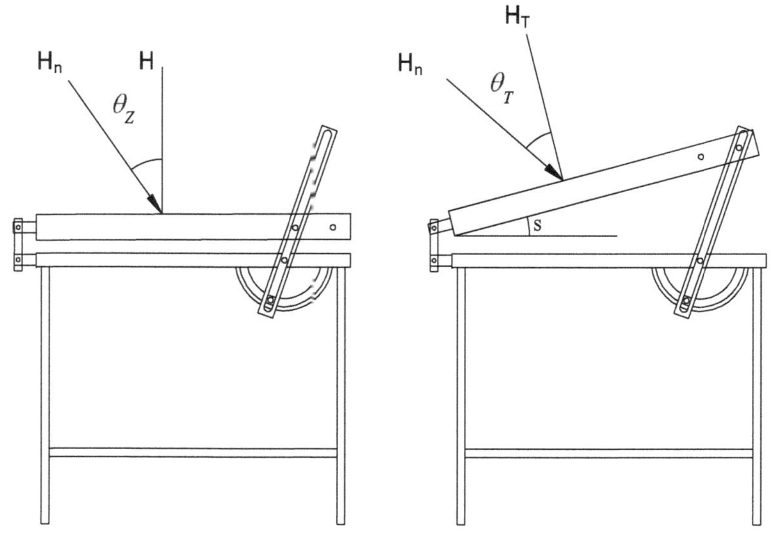

Figure 1 Radiation on horizontal and tilted surfaces.

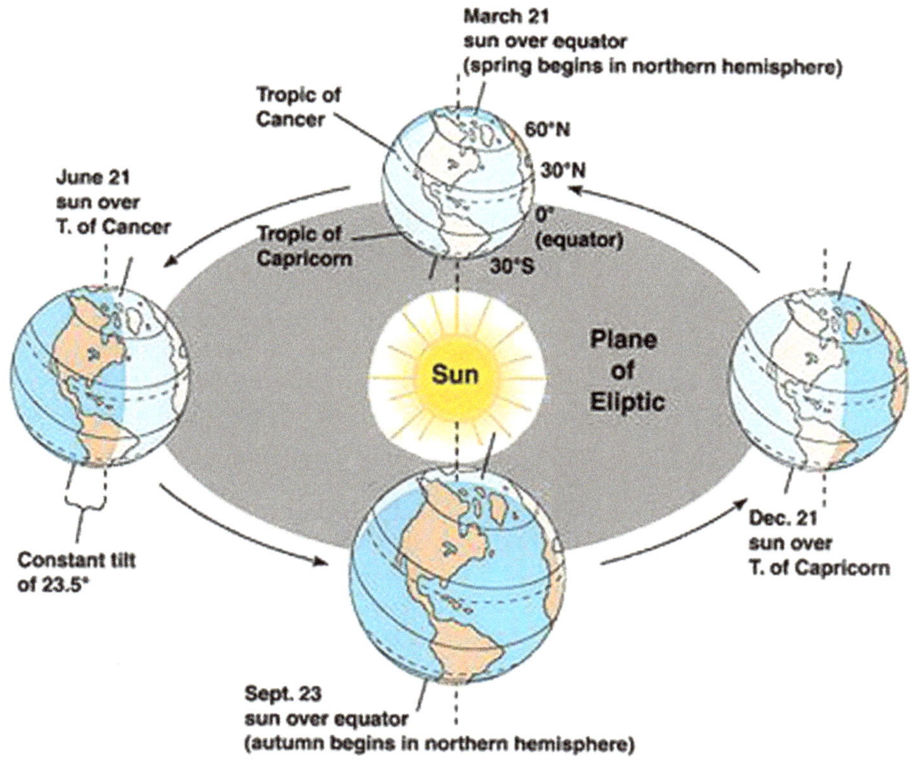

Figure 2 The positions of the earth at different seasons of the year [16].

Below is the modified Navier–Stokes equation in cylindrical coordinate:

$$\left(u\frac{\partial u}{\partial x}+v\frac{\partial u}{\partial v}\right)\rho=-\frac{\partial p}{\partial x}+\frac{1}{r}\frac{\partial}{\partial r}\left(\mu r\frac{\partial u}{\partial r}\right)-\rho g. \quad (6)$$

Energy equation for temperature distribution analysis is presented below:

$$u\frac{\partial T}{\partial x}+v\frac{\partial T}{\partial r}=\frac{k}{\rho c_p}\left(\frac{\partial^2 T}{\partial r^2}+\left(\frac{1}{r}\frac{\partial T}{\partial r}\right)+\left(\frac{\partial^2 T}{\partial x^2}\right)\right). \quad (7)$$

The following assumptions prevailed [14]:

1. Fluid properties are independent of temperature, i.e. μ, c_p, β and k are constants.
2. The Boussinesq approximation characterizes the buoyancy effects.
3. The flow is fully developed, $\frac{\partial u}{\partial x}=0$.

The momentum equation model for inclined pipe flow is shown in Figure 3.

$$\left(u\frac{\partial u}{\partial x}+v\frac{\partial u}{\partial v}\right)\rho=-\frac{\partial p}{\partial x}+\frac{1}{r}\frac{\partial}{\partial r}\left(\mu r\frac{\partial u}{\partial r}\right)-\rho g\sin\theta. \quad (8)$$

Equation (8) can be simplified using Boussinesq approximation for free convection. By putting the assumption (3) above into consideration and by assuming that radial flow equals zero (v = 0)

$$-\frac{\partial p}{\partial x}+\frac{1}{r}\frac{\partial}{\partial r}\left(\mu r\frac{\partial u}{\partial r}\right)-\rho g\sin\theta=0. \quad (9)$$

Volumetric coefficient of thermal expansion is expressed through this equation: $\beta=-\frac{1}{\rho}\left(\frac{\partial p}{\partial T}\right)_p$

$$\rho_\infty-\rho=\rho\beta(T-T_\infty) \quad (10)$$

For incompressible liquids, the term $\rho\beta$ is nearly constant. This approximation is good for gravity-induced flow over vertical surfaces with negligible temperature variation of T_∞ with x. It should be noted that Equation

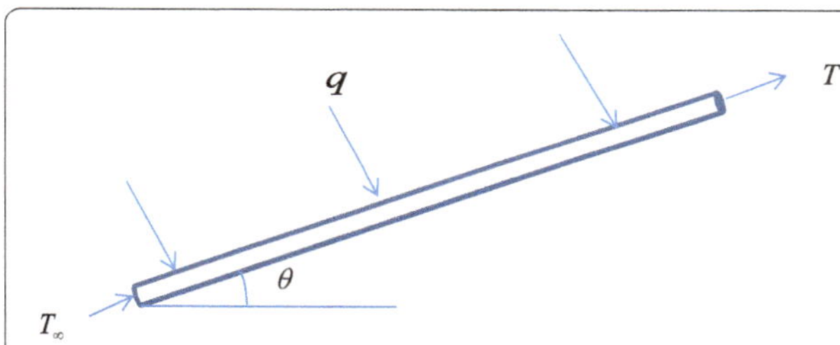

Figure 3 Inclined pipe showing thermal fluid and energy flows.

(10) should not be applied if the dependence between density and temperature is nonlinear in the temperature range of interest [17].

Momentum (9) was integrated to generate

$$u(r) = -\frac{R^2}{4\mu}\left(\frac{dp}{dx} + \rho g \sin\theta\right)\left(1 - \frac{r^2}{R^2}\right) \tag{11}$$

$$\bar{u} = \frac{2}{R}\int_0^R u(r) r dr \tag{12}$$

$$\bar{u} = \frac{(\Delta p - \rho g L \sin\theta) D^2}{32\mu L}. \tag{13}$$

By finding the average velocity of the flow in the pipe based on Equation (12), hence, Equation (13) is produced. The pressure-gradient term in the momentum equation is due to the hydrostatic pressure field outside the boundary layer [17], with

$$\frac{\Delta p}{L} = \rho_\infty g \sin\theta \tag{14}$$

By substituting, Equation (14) in (13),

$$\bar{u} = \frac{(\rho_\infty - \rho) g \sin\theta D^2}{32\mu}. \tag{15}$$

Furthermore, by substituting Equation (10) in Equation (15),

$$\bar{u} = \frac{(T - T_\infty)\beta\rho g \sin\theta D^2}{32\mu}. \tag{16}$$

Simplifying the energy balance equation for the pipe line under steady state condition yields

$$\frac{q\pi DL}{\dot{m}} = c_p(T - T_\infty) \tag{17}$$

$$\Pr = \frac{\mu c_p}{k}. \tag{18}$$

By eliminating $T - T_\infty$ in Equation (16) using Equation (17) and by rational mathematical manipulation using Equation (18),

$$\bar{u} = \frac{q\pi\beta\rho g \sin\theta D^3}{32\,\dot{m}\cdot \Pr k}, \tag{19}$$

given that modified Grashof number is $\mathrm{Gr}^* = \frac{q\beta g D^4}{k v^2}$ [14] and $\dot{m} = \frac{\pi}{4}\rho\bar{u}D^2$.

Equations (20) and (21) are produced from Equation (19) by substituting the relevant dimensionless parameters:

$$\mathrm{Re} = 0.354\left(\frac{\mathrm{Gr}^* H_p}{\Pr D}\right)^{0.5}, \tag{20}$$

where $H_p = L\sin\theta$

$$\dot{m} = 0.28\left(\frac{\mu^2 \mathrm{Gr}^* D H_p}{\Pr}\right)^{0.5}. \tag{21}$$

From Equation (21), it can be deduced that the fluid mass flow rate increases as the collector tilt angle increases; the maximum mass flow rate occurs at $\theta = 90°$ for any given length of the pipe. It is obvious from Equation (21) that Prandlt number $\left(\Pr = \frac{\mu c_p}{k}\right)$ plays an important role in the determination of magnitude of thermal fluid flow rate induced by buoyancy. The following deductions about thermal fluid used in solar collector are valid:

1. Highly viscous thermal fluid will reduce heat transfer in the solar collector.
2. Thermal fluid with low specific heat capacity is required for effective heat transfer from the solar collector.
3. Thermal fluid with relatively high thermal conductivity is required for effective heat transfer from the solar collector.

Methods

A computer program was written to investigate suitability of water or methanol as thermal fluid for heat transfer in solar collector using Equations (20) and (21) above. The pipe length used for the simulation was 700 mm, while the pipes nominal diameters used were 12 and 20 mm, respectively. The simulation was carried out with solar radiation ranging between 0 and 1,000 W/m^2, while the tilt angle of the collector was 30°. Methanol and water at average temperature (70°C) were used as thermal fluids in succession. The graphs in Figure 3 obtained from the simulation showed that the induced

Table 1 Measured solar radiation intensity for the simulation of collector tilt angle at different latitudes

Time	Solar radiation intensity (H/W/m^2)
7:00 am	62
8:00 am	105
9:00 am	200
10:00 am	450
11:00 am	330
12:00 noon	600
1:00 pm	800
2:00pm	960
3:00 pm	575
4:00 pm	422
5:00 pm	205

Table 2 Simulated solar radiation intensity with collector tilt angle equals to location latitude plus 25° (∅ + 25)

Latitude	January	February	March	April	May	June	July	August	September	October	November	December	Tilt angle
1	478.5873	443.5266	398.442	349.9231	311.0312	291.8618	301.7392	336.0522	382.0572	430.7677	469.9777	489.2786	26
2	482.7909	446.0119	398.9691	348.8276	308.9313	289.3829	299.4431	334.559	381.9966	432.6307	473.7505	494.0414	27
3	487.2193	448.664	399.6209	347.8456	306.9529	287.3026	297.2718	333.1809	382.0538	434.6443	477.7315	499.0516	28
4	491.8814	451.4872	400.3985	346.9755	305.0926	284.8063	295.2213	331.9158	382.2286	436.8119	481.929	504.3197	29
5	496.787	454.4868	401.303	346.2159	303.3468	282.6998	293.2879	330.7615	382.5215	439.1371	486.3511	509.8574	30
6	501.9465	457.668	402.336	345.5657	301.7129	280.7093	291.4682	329.716	382.9329	441.6239	491.0069	515.677	31
7	507.3711	461.0368	403.499	345.0237	300.1877	278.8309	289.7589	328.7779	383.4633	444.2767	495.9061	521.792	32
8	513.0729	464.5995	404.7939	344.5892	298.7689	277.0615	288.157	327.9455	384.1138	447.10002	501.0592	528.2167	33
9	519.065	468.3627	406.2229	344.2615	297.454	275.3977	286.6599	327.2175	384.8851	450.0995	506.4773	534.9667	34
10	525.3613	472.334	407.7882	344.0401	296.2411	273.8369	285.2649	326.5929	385.7788	453.28	512.1727	542.0591	35
11	531.9771	476.5211	409.4925	343.9245	295.1278	272.3762	283.9698	326.0705	386.7959	456.6475	518.1583	549.5117	36
12	538.9288	480.9323	411.3385	343.9149	294.1125	271.0132	282.7724	325.6497	387.9384	460.2086	524.4482	557.3445	37
13	546.2339	485.577	413.3294	344.0109	293.1935	269.7456	281.6707	325.3297	389.2078	463.9699	531.0576	565.5787	38
14	553.9119	490.4648	415.4686	344.2127	292.3695	268.5713	280.6629	325.1101	390.6064	467.939	538.003	574.2377	39

Table 3 Simulated solar radiation intensity with collector tilt angle equals to location latitude minus 25° (Ø − 25)

Latitude	January	February	March	April	May	June	July	August	September	October	November	December	Tilt angle
1	307.0781	337.8821	381.4402	425.6233	462.5434	479.495	471.165	439.328	395.808	351.447	314.2318	297.9407	−24
2	309.7679	339.7551	381.9106	424.2598	459.4031	475.4218	467.5694	437.3463	395.7123	352.9392	316.743	300.8405	−23
3	312.6017	341.7543	382.5003	423.0346	456.444	471.5599	464.469	435.5155	395.7385	354.5535	319.3932	303.8908	−22
4	315.5853	343.8835	383.2101	421.9459	453.6607	467.9017	460.9576	433.8328	395.886	356.2929	322.1875	307.0983	−21
5	318.7247	346.1465	384.041	420.9919	451.0485	464.4404	457.9293	432.2953	396.1569	358.1604	325.1318	310.4699	−20
6	322.0267	348.5473	384.9947	420.171	448.6026	461.1694	455.0787	430.9006	396.5497	360.1591	328.2318	314.0131	−19
7	325.4985	351.0902	386.0723	419.4821	446.319	458.0829	452.4006	429.6464	397.0658	362.2924	331.494	317.7362	−18
8	329.1476	353.7802	387.2758	418.924	444.1937	455.1753	449.8906	428.5307	397.7059	364.5642	334.9254	321.6478	−17
9	332.9827	356.6221	388.607	418.4958	442.2233	452.4415	447.5442	427.5517	398.4708	366.9785	338.5334	325.7575	−16
10	337.0125	359.6217	390.0681	418.197	440.4045	449.8766	445.3576	426.708	399.3621	369.5397	342.3261	330.0756	−15
11	341.2467	362.7847	391.6615	418.0269	438.7343	447.4763	443.3268	425.9981	400.381	372.2525	346.3122	334.6131	−14
12	345.696	366.1176	393.3899	417.9854	437.2099	445.2366	441.4489	425.421	401.5292	375.122	350.5009	339.382	−13
13	350.3714	369.627	395.2563	418.0725	435.8289	443.1535	439.7205	424.9758	402.8083	378.1538	354.9024	344.3953	−12
14	355.2853	373.3206	397.2636	418.2881	434.5893	441.2239	438.1389	424.6617	404.2208	381.3537	359.5277	349.6671	−11

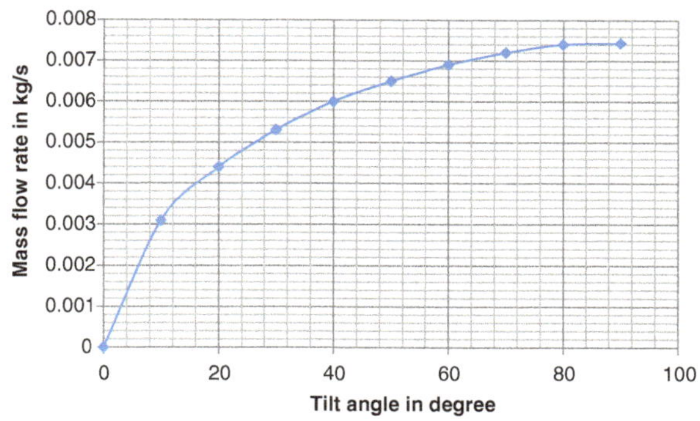

Figure 4 Measured solar radiation intensity in W/m^2 against hour of the day.

flow rates in the pipes increase with solar radiation intensity.

A pyranometer was used to measure horizontal radiation intensity at latitude 6.45° on a clear and sunny day at an hour interval from 7:00 am to 5:00 pm. The measurements are as shown in Table 1, and a bar chart plotted from them is shown in Figure 4. The assumption made in this study is that horizontal solar radiation received by similar solar collectors located at different latitudes within 1.0° and 14.0° is the same throughout the year. In other words, the measured data was assumed hypothetically for the different locations within latitude 1.0° and 14.0° throughout the year. The fundamental objective of this assumption is to investigate the resultant effect of the latitudes at the various locations, days of the year and tilt angles on the quantity of heat energy collected by the solar collectors. However, it should be noted that solar radiation measured at different latitudes are affected by humidity and other environmental conditions. The effects of these conditions are not considered in this paper to reduce avoidable complexities. The results from this analysis are expected to provide good platforms to predict the influence of the unconsidered factors. The appropriate tilt angles at different latitudes for different sessions of the year based on the above assumption were determined from a program developed from Equations (2) and (3) using C sharp software in Microsoft.net studio. The data simulated between latitudes 1° and 14° north of the equator are as shown in Tables 2 and 3.

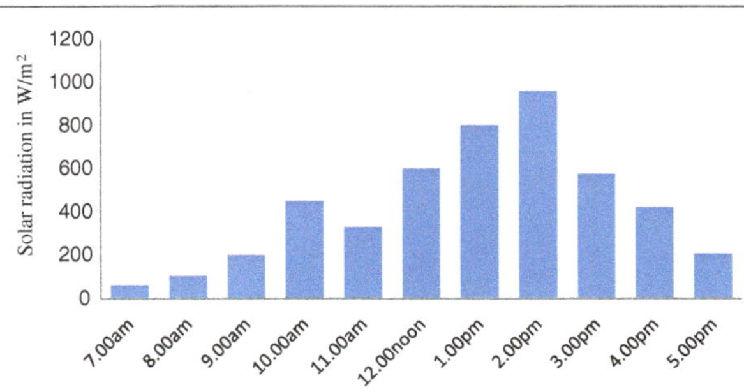

Figure 5 Flow regime in 30° tilted pipeline in solar collector.

Results and discussions

The graphs in Figure 5 show that methanol is better than water as thermal fluid for heat transfer in solar collector due to its low specific heat capacity and low viscosity. However, water is preferred to methanol where heat energy storage is required due to its high specific heat capacity. It is also observed that the size of the pipe, radiation intensity, tilt angle and properties of the thermal fluid determine the flow regime. The graph in Figure 6 shows the effect of solar collector tilting on flow rates of methanol in a 20-mm nominal diameter pipe under average solar intensity of 500 W/m^2. It can be observed that tilting increases the flow rate of the thermal fluid in the solar collector. In the flow rate Equation (21) above, it was observed that maximum flow rate is produced when the tilt angle is 90°; this angle is not appropriate for maximum solar radiation collection.

Several tilt angles for solar collectors as recommended from past and recent literatures were tested with the program developed from Equations (2) and (3). Tables 2 and 3 contain the data simulated for locations within latitudes 1° and 14° . The solar collector tilt angle in each location in Table 2 is equal to the location latitude ($\varnothing$) plus 25°, while that of each location in Table 3 is equal to the location latitude ($\varnothing$) minus 25°. It is important to note that a positive tilt angle in the northern hemisphere means that the collector is tilted facing south, while negative tilt angle in the northern hemisphere means that the collector is tilted facing north [1,18,19]. Figures 7 and 8 displayed the graphs of average monthly solar radiation against different months of the year for regions

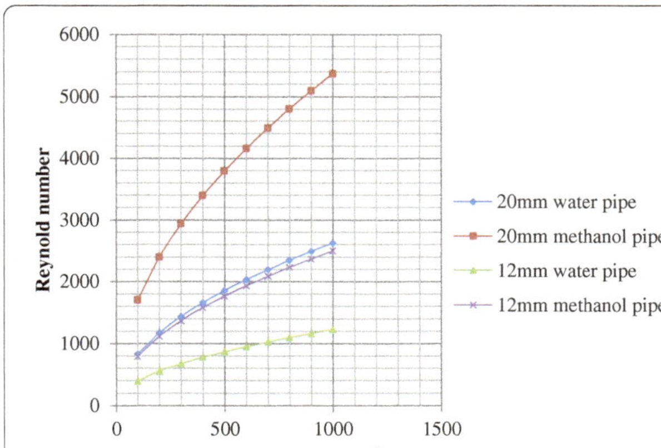

Figure 6 Methanol flow rate variation with tilt angles under 500 W/m^2 average solar intensity.

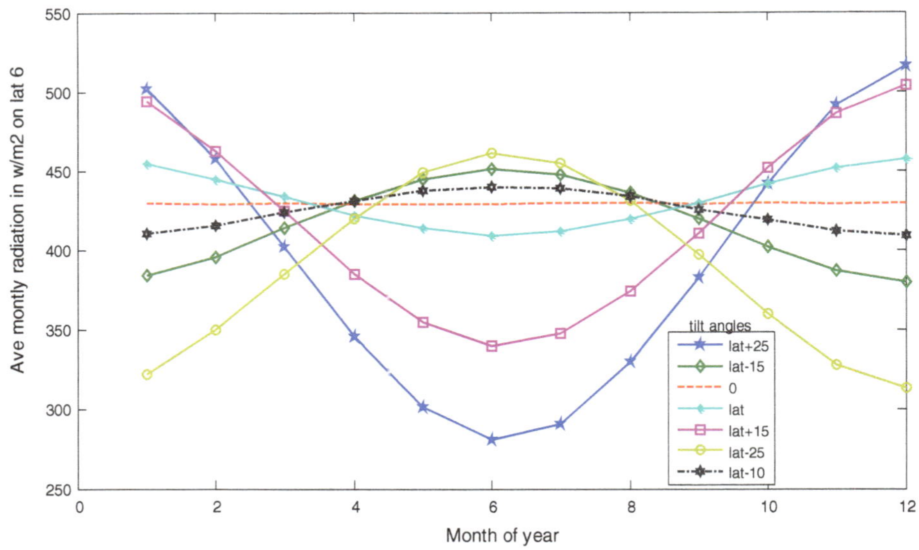

Figure 7 **Average monthly solar intensity on tilted solar collector on location latitude 6°.**

located on latitude 6° and 13°. The graphs show the different tilt angles investigated.

The average monthly solar radiation on horizontal solar collector surface for this simulation is 430 W/m^2. It is indicated by dotted line in Figure 7 and Figure 8. Many publications on optimum tilt angle for solar collector favoured ∅ + 15° and ∅ + 10°. The inclusion of Tables 2 and 3 was deliberate in order to show data distributions on latitudes from 1° to 14° for tilt angles ∅ + 25° and ∅ – 25°. From Figures 7 and 8, it can be inferred approximately that from November to mid of January, the tilt angle obtained from ∅ + 25° produces more solar radiation intensity in the solar collector than tilt angles ∅ + 15° and ∅ + 10°. Close observation further revealed that tilt angle ∅ + 15° has better performance from middle of January to March and from middle of September to middle of October. The figures also show that appropriate tilt angle for the first 2 weeks of April and the last 2 weeks of July is ∅ – 15°. After the first 2 weeks of April to the last 2 weeks of July, appropriate tilt angle for maximum solar radiation in the solar collector is ∅ – 25°. About the last week of March and middle of August, none of the tilt angles investigated produced radiation above the average radiation on horizontal solar collector. However, to sustain

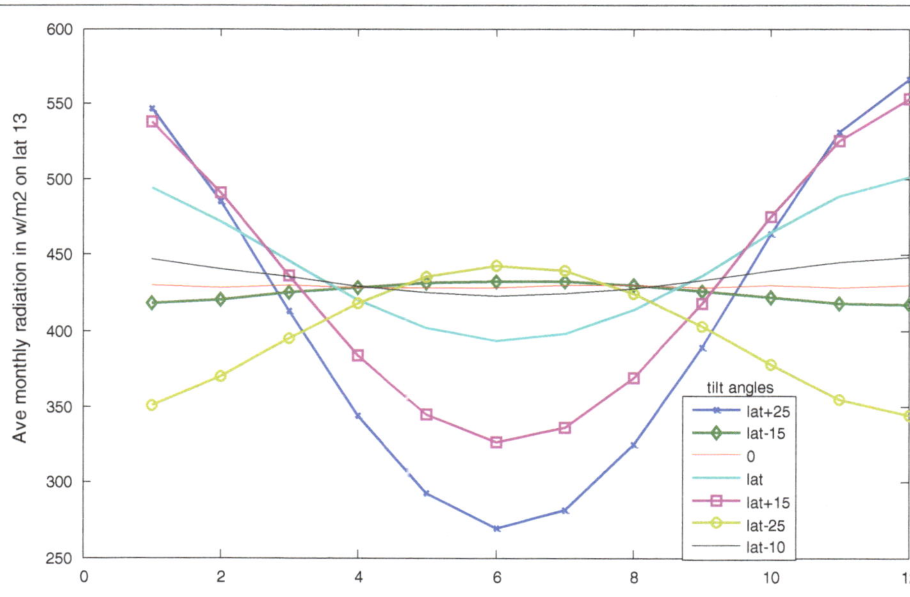

Figure 8 **Average monthly solar intensity on tilted solar collector on location latitude 13°.**

Table 4 Recommended tilt angles for optimum radiation collection for different months of the year

January	February	March	April	May	June	July	August	September	October	November	December
∅ + 25	∅ + 15	∅	∅	∅ − 25	∅ − 25	∅ − 25	∅ − 15	∅ + 15	∅ + 15	∅ + 25	∅ + 25

thermal fluid flow in the collector, the tilt angle recommended is the solar collector location latitude.

The above recommendations on tilting are ideal for system, the efficiency of which depends on instantaneous thermal energy received rather than average thermal energy received over a period of time. Solar collector with large surface area may be difficult to tilt as recommended; in that respect, such collector may be tilted permanently to its location latitude. An important observation from Figures 7 and 8 above is that when solar collector is tilted to a permanent angle without periodical tilting to track the sun position at certain months of the year, the received solar radiation on tilted surface is less than that on horizontal surface. Under this condition, solar collector tilting is counter-productive.

The thermo-siphon mass flow rate equation in solar collector for fluid heating based on periodic tilting at different latitudes is given by

$$\dot{m} = 0.28\left(\frac{\mu^2 Gr^* DL \sin(\phi + c)}{\text{Pr}}\right)^{0.5}. \qquad (22)$$

Observations from the Figures 7 and 8 are summarised in Table 4.

The chart shown in Figure 9 above was produced from Figures 7 and 8. It can be used to estimate appropriate tilt angles of solar collectors at locations having different latitudes.

Conclusion

The induced flow generated from tilted solar collector has been established for effective monitoring of solar collector performance. This work has successfully investigated recommendations from past research activities on solar collector tilting for enhanced performances with simplified approach, as far as data collection for analyses is concerned. The results obtained can be summarised as follows: for periodic tracking of solar collector throughout the year, the following are the recommended tilt angles: ∅ + 25° for November, December and January; ∅ + 15° for February, September and October; ∅ − 15° for August; ∅ − 25^0 for May, June and July; and ∅ for March and April. These recommendations are appropriate for solar collectors with tilting devices. The tilt angles ∅ + 25° and ∅ − 25° are not common in the literature; however, the results showed that they are relevant for periodic tracking of solar collector even though they are for a short period. It is further showed that solar radiation on tilted surface increases with latitude. This method of analysis demonstrated efficient management of time, energy and capital investment required in data collection to analyse appropriate tilt angles of solar collectors in comparison with past works done in this area. Finally, this paper has been able to provide formula for estimating

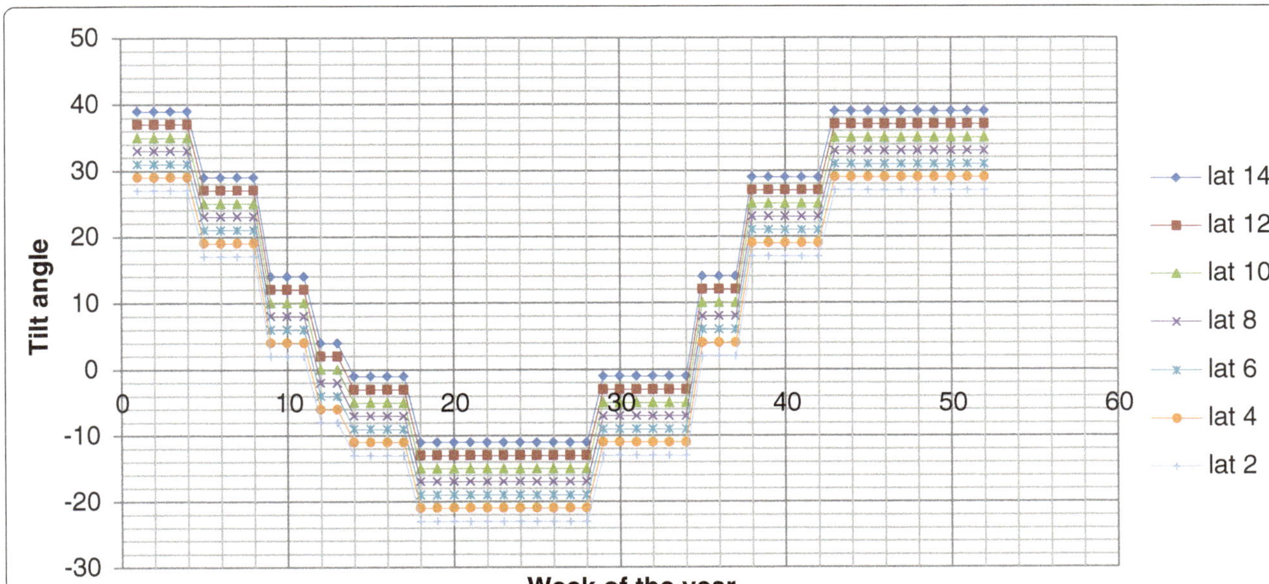

Figure 9 Chart for estimating weekly tilting for optimum solar radiation collection.

thermo-siphon-induced flow rate in solar collector, and it has established the need to incorporate tilting mechanisms in solar collectors for better thermal efficiency.

Abbreviations

B: Volumetric coefficient of thermal expansion; *δ*: Declination; *Ø*: Latitude (north positive); *M*: Dynamic viscosity of thermal fluid; *Ω*: Hour angle, solar noon being zero, and each hour equals to 150 of longitude with morning positive and afternoon negative; *P*: Fluid density in kg/m^3; *θ*: Thermal fluid tilt angle in radian; θ_T: Angle between the beam from the sun and the vertical on the tilted plane; θ_Z: Zenith angle; *c*: Constant (values 0 or ±15 or ±25, etc.); c_p: Specific heat capacity of thermal fluid at constant pressure KJ/kgk; *d*: Number of days starting from 1 January to the given date; *g*: Acceleration due to gravity; Gr*: Modified Grashof number; *H*: Solar radiation on horizontal surface; H_p: Vertical distance between inlet and exit of the collector pipe; H_T: Solar radiation on tilt surface; *k*: Thermal conductivity of thermal fluid (w/mk); *L*: Pipe length; Lat: Latitude; *m*: Mass flow rate in kg/s; *p*: Fluid pressure in N/m^2; Pr: Prandt number; *q*: Heat flux per unit length (W/m^2); *r/R*: Pipe radius in m; R_b: Ratio of radiation on tilted surface to that of horizontal surfaceS, Angle between the solar collector and the horizontal plane (angle of tilt); *T*: Temperature in Kelvin; T_∞: Temperature at infinity in K; *u/v*: Fluid velocities in m/s; *x/y*: Coordinates in m.

Competing interests

The authors declare that they have no competing interests.

Authors' contributions

SIO carried out the preliminary studies, programming and prepared the draft of the manuscript. MOO and IFO both provided technical guide and review of the manuscript. All authors read and approved the final manuscript.

Authors' information

SIO is a chief lecturer in the Department of Industrial Maintenance Engineering, Yaba College of Technology, Lagos, Nigeria. He received his B. Eng. in Mechanical Engineering from the University of Benin, Nigeria in 1987 and M.Sc. in Chemical Engineering from the University of Lagos, Nigeria in 1996. He is a corporate member of the Nigerian Society of Engineers as well as a registered mechanical engineer of the Council for the Regulation of Engineering in Nigeria (COREN). His present area of research is optimisation of heat and mass transfer phenomena in solar adsorption refrigeration system using hybrid solar collector at University of Ibadan, Nigeria. He has published several journal papers on virtual prototyping of systems and renewable energy.

MOO obtained B.Sc. (First Class Honours) in Mechanical Engineering from the Obafemi Awolowo University (OAU), Ile-Ife, Nigeria in 1997 and Ph.D. in Mechanical Engineering from the University of Newcastle, NSW, Australia in 2004. He is an editorial board member of several international engineering journals. He has been a member of the American Society of Engineers (ASME) since 2003. MOO is a reader in the Mechanical Engineering Department, University of Ibadan, Nigeria. He specialises in energy analyses and thermo fluids. He has scores of publications in learned international journals and conference proceedings.

IFO obtained B.Eng. and M.Sc. in Mechanical Engineering from the University of Ibadan in 1986 and 1990 respectively. He also obtained his Ph.D. in Mechanical Engineering from the same University in 2009. He is a corporate member of the Nigerian Society of Engineers as well as a registered mechanical engineer of the Council for the Regulation of Engineering in Nigeria (COREN). He is also a corporate member of Institute of Environmental Engineers in Nigeria. IFO is a senoir lecturer in the Mechanical Engineering Department, University of Ibadan, Nigeria. He has several publications in peer-reviewed international and local journals.

Acknowledgements

The authors wish to acknowledge the World Bank for initiating STEP-B project in higher institutions of learning in Nigeria. The project provided the pyranometer and computer used for this work. We appreciate people that contributed in one way or the other towards successful completion of this publication.

References

1. Chandrakumar, BP, Jiwanlal, LB: Development and performance evaluation of mixed-mode solar dryer with forced convection. IJEEE **4**, 23 (2013)
2. Fagbenle, RL: Optimum collector tilt angles and annual global radiation for Nigerian locations. Nigerian J. Renew. Energy **2**(17), 1–8 (1991)
3. Oko, COC, Nnamchi, SN: Optimum collector tilt angles for low latitudes the open renewable. Energy Journal **5**, 7–14 (2012)
4. Gunerhan, H, Hepbasli, A: Determination of the optimum tilt angle of solar collectors for building applications. Build. Env. **42**, 779–783 (2007)
5. Gavin, DJN: Solar Energy Projects for the evil Genius, 1st edn. McGraw-Hill, Chicago (2007)
6. Ojosu, JO: The iso-radiation map of Nigeria. Solar and Wind Technology **7**(5), 563–575 (1990)
7. Abubakar, SS: Strategic developments in renewable energy, International Association for Economics, Sept-Dec, Nigeria (2009). http://www.iaee.org/en/publications/newsletterdl.aspx?id=75. Accessed July 2013
8. Kumarasamy, S, Tulika, S, Guddy, S, Manicam, P: Modelling and estimation of photosynthetically active incident radiation based on global irradiance in Indian latitudes. IJEEE **4**, 21 (2013)
9. Benghanem, M: Optimization of tilt angle for solar panel: case study for Madinah. Saudi Arabia Applied Energy **88**, 1427–1433 (2011)
10. Honsberg, C: Solar radiation. University of Delaware. http://www.eecis.udel.edu/~honsberg/Eleg620/02_Solar_radiation.pdf (2008). Accessed 2 September 2012
11. Pavlović, T, Pavlović, ZL, Pantić, L, Kostić, J: Determining optimum tilt angles and orientations of photovoltaic panels in Nis Serbia. Contemporary Materials. **2**, 151–156 (2010)
12. Lewis, G: Optimum tilt of solar collectors. Solar and Wind Technology. **4**, 407 (1978)
13. Garp, HP, Gupta, GL: Flat plate collectors: experimental studies and design data for India. Paper presented at the international solar energy congress, New Delhi (16–21 January 1978)
14. Prayagi, SV, Thombre, SB: Parametric studies on buoyancy induced flow through circular pipes in solar water heating system. Journal of Engineering Science and Technology **3**(1), 616–627 (2011)
15. Duffie, JA, Beckman, WA: Solar engineering of thermal processes, 1st edn. Wiley, New York (1974)
16. St. Rosemary Education Institution. The effect of the Earth's rotation and revolution. St. Rosemary Education Institution's Schoolworkhelper. http://schoolworkhelper.net/wp-content/uploads/2010/08/01m.jpg (2013). Accessed 9 December 2012
17. Kays, W, Crawford, M, Weigan, B: Convective heat and mass transfer, 4th edn. McGraw-Hill, New York (2005)
18. Landau, CR: Optimum tilt of solar panels. MACS Lab, Inc.. www.macslab.com/optsolar.html (2012). Accessed 15 December 2012
19. Pragya, S: Solar energy generation potential along national high ways. IJEEE. **4**, 16 (2013)
20. Bruce, RM, Donald, FY, Theodore, HO: Fundamentals of Fluid Mechanics, 1st edn. Wiley, New York (1990)

14

Study of combustion in residential oil burning equipment of animal by-products and derived products not intended for human consumption

Julio F San José Alonso*, I Gobernado Arribas and S Alonso Miñambre

Abstract

This article studies the use of animal by-products and derived products not intended for humans as possible fuels in residential oil burning facilities. We first offer a chemical and physical description of the various types of animal by-products and derived products not intended for humans with a view to their possible use as fuels. Animal by-products and derived products not intended for humans have an extremely high viscosity for the pressure pulverisation burners used in residential oil burning equipment. We therefore mixed diesel with animal by-products and other derived products not intended for humans in different percentages so as to obtain suitable viscosity. To achieve this, we carried out a study of the miscibility of animal by-products in diesel. We subsequently performed a series of combustion experiments for mixtures of diesel and animal by-products, in varying (a) percentage of by-products in diesel, (b) injection pressure and (c) excess air in combustion. We analysed the experimental combustion results based on (a) energy efficiency of combustion, (b) CO and NO_x emissions and (c) fossil-based greenhouse effect gases. Finally, we present the conclusions that combustion of mixed diesel with animal by-products and other derived products not intended for humans for use as boiler fuel low power does not require specific technology when using a conventional fuel oil burner for proper combustion. One only needs to adjust the burner factors: pressure and air flow. In the study of the combustion of mixtures, once the burner factor is adjusted, it appears that the combustion efficiency and greenhouse gas and emission gases are acceptable. The costs of removing fat through incineration or landfill range from 34 to 59 €/ton. The solution proposed in this work not only avoided the cost but also mentioned the value of residue use as fuel.

Keywords: Animal fat; Mixtures; Oil; Cocombustion; Boiler

Background

In 2007, the Council of Europe set out a series of ambitious goals for 2020, given the name of '20-20-20', which seek to cut greenhouse gas emissions by 20%, cut primary energy consumption by 20% through enhanced energy efficiency and promote up to 20% use of renewable final energy consumption [1].

Through the Directive 2009/28/CE, the European Parliament set a target for 2020 that 20% of final energy should be from renewable sources. The directive promotes energy generated from biofuels and bioliquids. For the latter to be considered sustainable, they must contribute to cutting greenhouse gas emissions by at least 35%. As of 1 January 2017, their contribution to emission reduction must reach 50% [2].

The directive stipulates that in order to be granted financial support, biofuels and bioliquids must be classified as 'sustainable.' This requires biofuels and bioliquids to be generated using raw materials from both outside as well as inside the various regions of UE. In addition, they must not originate from land that has high biodiversity value or large carbon reserves [3].

Bioliquids may be transformed for use in engines or used directly as fuels in burners. When used as fuels in burners, certain advantages are evident such as (a) requiring no specific transformation processes, thereby enabling them to be obtained relatively cheaply; (b) less rigorous specifications when used in burners than when used in engines; and (c) a wider range of technologies linked to burners than to heat engines. In addition, burners have a wider range of regulation vis-à-vis fuel than heat engines [4].

* Correspondence: julsan@eii.uva.es
Departamento Ingeniería Energética y Fluidomecánica, E. Ingenierías Industriales, Universidad de Valladolid, Paseo del Cauce 59, 47014 Valladolid, Spain

Some authors have explored the use of bioliquids. Batey [5] conducted studies into combustion of biodiesel and mixtures of oil and diesel in residential combustion facilities; Raheman et al. [6] evaluated a 10.3-kW single-cylinder water-cooled direct-injection diesel engine using blends of biodiesel (B10 and B20) obtained from a mixture of mahua and simarouba oils (50:50) with high-speed diesel (HSD) in terms of brake-specific fuel consumption; Vanlaningham et al. [7] analysed combustion of oil and soya seed for heating; San José et al. [8] studied combustion of mixtures of soya oil, rapeseed and sunflower oil with diesel in various proportions over a range of combustion parameters in a pressure pulverization burner and also studied combustion of biodiesel that does not meet biofuel specifications [9].

The fat used for heating purposes has been analysed at the University of Budapest by Lezsovits and Könczöl [10] who researched the use of animal fats in industrial steam generators that conventionally ran on diesel or natural gas. The use of this by-product was intended not only to remove it as a waste product but also to provide an alternative non-fossil fuel. In Germany, F. Pfab [11] treated animal fat from processing plants and subsequently used it in combustion to generate steam in a tubular boiler. Before feeding the fat into the burner, the burner must be cleaned with a decanter and separator. Our study breaks new ground in current technology by performing combustion of animal fat at an injection pressure below 10 bars, compared to conventional 30 bar pressure. In Spain, Dr. J. San José [12] has conducted a number of studies of the combustion of lard in a commercial burner and into the miscibility of lard in diesel. These studies posit an important field of study in the description of animal fats and their use as fuels in conventional diesel facilities.

Animal by-products and other derived products not intended for human consumption, which are recoverable in energy terms, are in abundant supply as waste and are found in the animal-based food production chain in (a) intensive livestock holdings, hen manure and other similar products, and in (b) slaughterhouses and retailers. The former are usually processed in the actual livestock holdings themselves, whereas the latter, which provide the case study for the present article, are dealt with by officially accredited firms.

The amount generated in slaughterhouses and by retailers in kilogram per animal slaughtered depends on the animal species. Studies carried out by the Ministry of the Environment, Rural and Marine Affairs in Spain put the estimates for mean production of waste per species at (a) bovines, 166.25 kg/animal; (b) dairy cows, 260.45 kg/animal; (c) swine, 10.79 kg/animal and (d) ovine, 8.05 kg/animal [13].

Handling these animal by-products as waste, together with the use of transformed animal fat as a biomass resource in combustion processes, is controlled under European Union Regulation 1774/2002. The said regulation lays down strict rules concerning animal welfare and public health to be applied when collecting, transporting, storing, handling, processing and using or eliminating this kind of by-product. These rules have been applicable throughout the whole of the EU since 1 May 2003 [14].

Processing any animal by-products and derived products not intended for human consumption generated in slaughterhouses and livestock resource processing companies firstly involves grinding the raw material followed by high-pressure heat treatment to remove excess moisture and kill microorganisms. One fraction of the subsequent by-product is in the form of flour and the other in the form of fat. In processing plants, the fraction of flour obtained accounts for between 25% and 40% wt. f.m. of

Table 1 Animal fat not intended for human consumption linked to animal census of EU in tons

	Bovine	Dairy cows	Sheep	Swine	Fat (total)
Belgium	2,509.5	517.7	0	6,176.3	36,105.5
Bulgaria	531.1	291.6	1,368.0	664	11,060.9
Czech Republic	1,319.4	375.4	0	1,846.0	19,919.0
Denmark	1,630.0	573	0	12,293.0	31,151.0
Germany	12,706.2	4,181.7	1,799.7	26,900.8	206,250.0
Estonia	236.3	96.5	0	371.7	4,044.6
Ireland	5,917.7	1,027.0	3,122.0	1,500.4	77,823.9
Greece	679	144	8,980.0	1,087.0	15,362.0
Spain	6,075.1	845.3	18,551.6	25,704.0	98,234.0
France	18,992.0	3,641.0	6,904.0	14,063.0	257,605.2
Italy	5,832.5	1,746.1	7,009.0	9,321.1	94,505.2
Cyprus	54.7	23.4	328.9	463.7	1,348.2
Latvia	379.5	164.1	0	389.7	6,538.1
Lithuania	748	359.8	58.5	929.4	13,570.7
Luxembourg	194	46	7.5	89.4	2,702.0
Hungary	681	239	1,181.0	3,168.0	12,822.6
Malta	15	6.4	12.4	69.3	291.1
Netherlands	3,960.0	1,518.0	1,211.0	12,206.0	69,902.3
Austria	2,013.3	532.7	358.4	3,134.2	30,153.0
Poland	5,561.7	2,529.40	213.7	14,775.7	102,279.9
Portugal	1,374.8	275.1	2,512.0	2,144.7	20,674.8
Rumania	2,008.8	1,190.5	8,264.2	5,358.8	46,580.3
Slovenia	470.1	109.5	0	395.6	6,591.5
Slovakia	467.1	159.3	394.2	687.3	7,735.9
Finland	908.9	284.3	0	1,339.9	14,156.2
Sweden	1,474.5	348.6	274.3	1,520.1	21,067.5
United Kingdom	9,896.0	1,847.0	21,295.0	4,385.0	143,480.0
EU 27	86,636.2	23,072.4	83,845.4	150,984.1	1,351,955.3

EU livestock census (1,000 animals) in December 2010 and animal fat not intended for human consumption linked to animal census in tons.

Table 2 Characteristics of animal fats and diesel

Property	Test method	Type of fat			
		FF	NF1	NF2	Diesel
Density at 15°C (kg/m^3)	ASTM D-4052	911	916	910	848
Density at 35°C (kg/m^3)	ASTM D-4052	899	904	898	635
Kinematic viscosity at 40°C (mm^2/s)	ASTM D341-09	40.87	51.97	41.64	2.7
Kinematic viscosity at 100°C (mm^2/s)	ASTM D341-09	8.46	9.06	8.64	1.2
Percentage of S by mass	ASTM 1552	0.03	0.04	0.03	0.07
Percentage of C by mass	ASTM 5291	76.8	75.1	75.1	86.6
Percentage of H by mass	ASTM 5291	11.6	11.6	11.7	12.3
Percentage of O by mass	ASTM 5291	11.3	12.9	13.0	1.03
Percentage of N by mass	ASTM 5291	0.2	0.16	0.05	-
Ashes	EN 6245	0.002	0.164	0.002	-
HHV (MJ/kg)	ASTM 240	39.4	38.8	39.0	44.9
LCV (MJ/kg)	ASTM 240	36.9	36.3	36.5	42.3

HHV higher heating value, *LCV* lower heating value.

the total processed by-product, whereas for the fat fraction this percentage ranges from 6% to 15% wt f.m. of the total by-product [15].

The amount of fat obtained in Europe is shown in Table 1 [16]. This waste has a high-energy quality which ranges between 38.7 and 42.1 MJ/kg and which may be used as raw material to obtain biodiesel [17-19]. However, the economic crisis, the strict specifications imposed by the European Union on biodiesel and the investment required by such processing plants have made animal fat not intended for human consumption a waste product, the fat from which must be burnt in application of the current legislation.

The present article presents co-combustion of animal fats not intended for human consumption as an alternative fuel for conventional diesel facilities due to the following advantages:

- It is a waste product which must be destroyed, as a result of which its calorific potential may be used and the waste thereby being removed.
- It is biomass not intended for human consumption such that it does not cause price distortions in food products.
- It allows for the use of conventional technology such that no major investments are required in equipment.
- As it is biomass, it complies with all policies concerning reduction of greenhouse gas emissions and use of fossil fuel alternatives.

Methods

Animal fats are glycerine esters (triglycerides) and fatty acids. Fatty acids found in animal fats may be saturated CnH_2nO_2 and unsaturated [20]. One characteristic of saturated fatty acids is that at room temperature they are solid. Animal fats tend to be rich in this kind of acid. These include the following:

- Formic or methanoic acid, with a single carbon and formula CH_2O_2, which in the developed form is H-COOH

Figure 1 Type FN1. **(a)** Room temperature. **(b)** Heated to a temperature of 40°C.

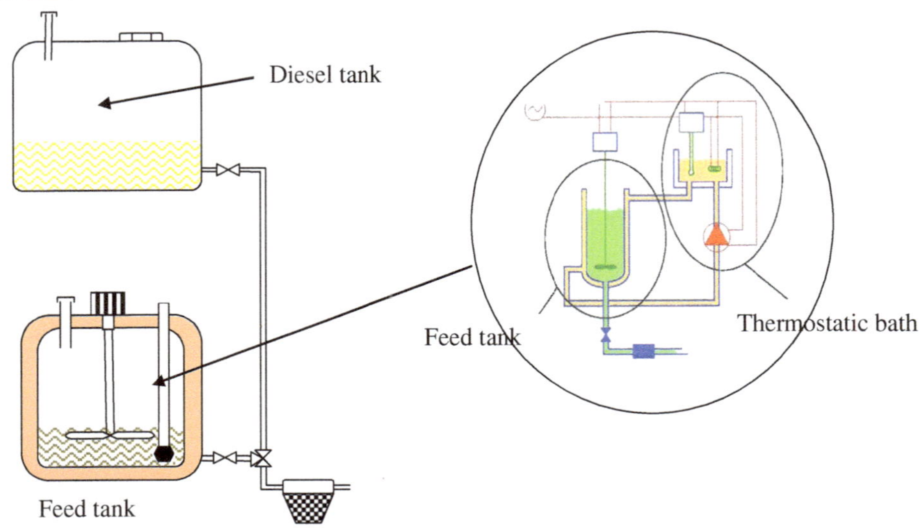

Figure 2 System for feeding the fuel to the burner.

- Acetic acid or ethanoic acid with two carbons. It is an intermediate product of animal metabolism
- Butyric acid or butanoic acid, $C_4H_8O_2$ or in the developed form $CH_3\text{-}(CH_2)_2\text{-}COOH$
- Palmitic acid, $C_{16}H_3O_2$, the developed formula which is $CH_3\text{-}(CH_2)_{14}\text{-}COOH$
- Stearic acid, $C_{18}H_{36}O_2$, the developed formula which is $CH_3\text{-}(CH)_{16}\text{-}COOH$

Unsaturated fatty acids are generally liquid at room temperature. Due to their chemical structure, they can be mono- or poly-unsaturated. The following can be found in animal fat:

- Oleic acid, $C_{18}H_{34}O_2$, with the formula $CH_3\text{-}(CH_2)_7\text{-}CH{=}CH\text{-}(CH_2)_7\text{-}COOH$
- Linoleic acid with the formula $CH_3\text{-}(CH_2)_4\text{-}CH\text{-}CH_2\text{-}CH{=}CH\text{-}(CH_2)_7\text{-}COOH$

Table 3 Characteristics of the mixtures of animal fats and diesel

Name	Percentage of C	Percentage of H	Percentage of O	Percentage of S	Density (15°C) (kg/m^3)	Viscosity (100°C) (mm^2/s)	LHV (MJ/kg)
FF-10	85.62	12.23	2.06	0.06	854.3	1.36	41.7
FF-20	84.64	12.16	3.08	0.06	860.6	1.62	41.2
FF-30	83.66	12.09	4.11	0.06	866.9	2.07	40.7
FF-40	82.68	12.02	5.14	0.05	873.2	2.45	40.1
FF-50	81.70	11.95	6.17	0.05	879.5	3.12	39.6
FF-100	76.80	11.6	11.3	0.03	911.0	8.61	36.9
FN1-10	85.45	12.24	2.23	0.07	854.8	1.36	41.7
FN1-20	84.30	12.16	3.40	0.06	861.6	1.64	41.1
FN1-30	83.15	12.09	4.59	0.06	868.4	1.95	40.5
FN1-40	82.00	12.02	5.78	0.05	875.2	2.44	39.9
FN1-50	80.85	11.95	6.965	0.06	882.0	3.13	39.3
FN1-100	75.1	11.6	12.9	0.04	916.0	9.06	36.3
FN2-10	85.45	12.24	2.23	0.06	860.3	1.38	41.7
FN2-20	84.30	12.18	3.42	0.06	860.4	1.60	41.1
FN2-30	83.15	12.12	4.62	0.06	866.6	1.89	40.5
FN2-40	82.00	12.06	5.82	0.05	872.8	2.48	40.0
FN2-50	80.85	12.00	7.02	0.05	879.0	3.10	39.4
FN2-100	75.1	11.7	13.0	0.03	910.0	8.64	36.5

Table 4 Boiler and burner characteristics

	Description/value
Boiler model - AR/25GT (ROCA)	
Boiler type	Cast iron
Heating capacity	26.7 kW
Water capacity	26 l
Approximate weight	210 kg
Maximum working temperature	110°C
Burner model KADET-TRONIC (ROCA)	
Nozzle	0.5 GPH
Angle	60°
Pump pressure	8 to 16 bars
Flow fuel	1.86 kg/h
Air regulation	0 to 10 m^3/h

The chemical nature (composition) of fuels is insufficient when determining their behaviour or efficiency during combustion, as it is the physical-chemical characteristics which provide such knowledge. Table 2 shows the main physical-chemical characteristics of three fats: FF, food fat, NF1, non-food fat type 1 and NF2, non-food fat type 2.

The features of the three fats in Table 2 show that the three types of fat evidence very similar characteristics. The main differences with diesel are viscosity and percentage of oxygen, which means that a specific burner must be used for combustion of fats due to their high viscosity and that combustion must take place with lower excess air due to the oxygen content of the fat.

Since the present study seeks to achieve combustion of fats using conventional diesel combustion facilities which are normally mechanical pressure pulverization burners, we conducted a study of the mixture of fats with diesel in order to reduce the viscosity of the fats.

Analysis of the miscibility of the fats in diesel

Above their melting point, oils and fats are totally, and in all proportions, miscible with organic solvents [21,22].

At room temperature, fats are doughy, preventing them from being handled and mixed for use as fuels. We carried out a melting analysis based on the principle of turbidity. Figure 1 shows a sample at room temperature and a melted sample.

After numerous tests, we concluded that not all fats undergo phase change at the same temperature, and we set 30°C as the minimum melting point temperature for all of them.

The miscibility and/or solubility of the diesel samples is tested by mass and because of the apolar nature of both fluids, dissolution is not a problem. Since there are no functional groups and the molecular structure is similar, there are no chemical reactions between the two products. In certain cases, some problems of stratification and turbidity of the fats could be seen at low temperatures. The conclusion is that for combustion assays using mixtures of animal fats and diesel, it is important to heat and stir the (animal fat-diesel) mixture before it is fed into the burner for subsequent use as fuel. During heating, the mixture must reach a temperature of above

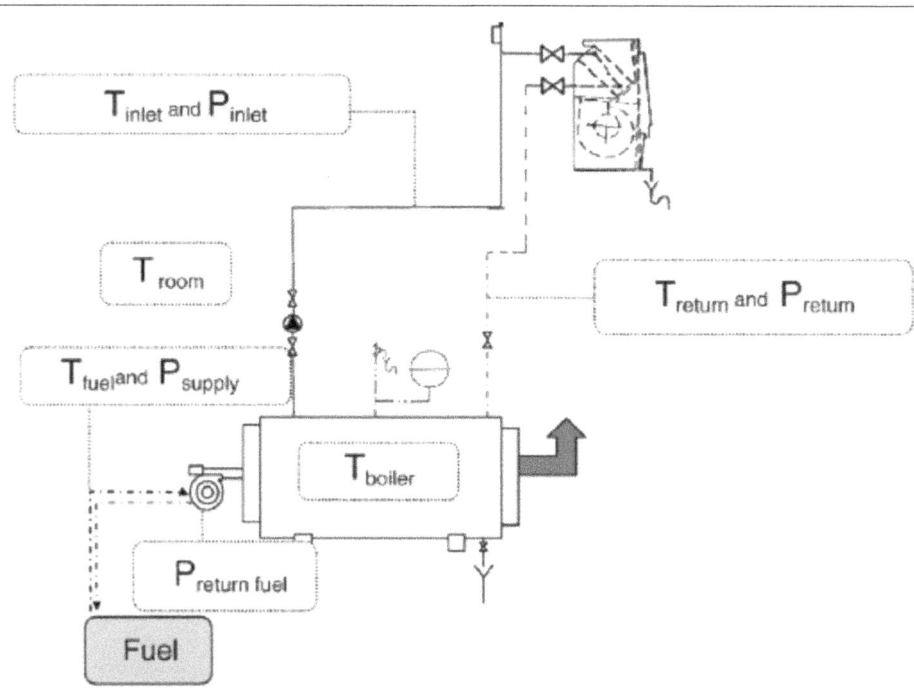

Figure 3 Diagram of the experimental facility.

Table 5 Characteristics of the generating group used in the combustion assays

Parameter	Unit	Operating value	Error	Description
T_{inlet}	°C	90	±1	Represents the water temperature in the distributor located in the inlet circuit
T_{return}	°C	80	±1	Indicates the water temperature in the return collector, through which al the water flows prior to returning to the boiler
P_{supply}	Bar	2.5	±0.2	Input water pressure in the main circuit, immediately before the distributor
P_{return}	Bar	2	±0.2	Water pressure at the water pump exhaust located before entry into the distributor
$P_{supply\ fuel}$	Bar	10 to 30	±0.1	Input pressure at the fuel pump
$P_{return\ fuel}$	Bar	5	±0.1	Exhaust pressure at the fuel pump
T_{fuel}	°C	20	±0.5	Fuel temperature located in the tank
T_{boiler}	°C	85	±1	Water temperature when passing through the boiler
T_{room}	°C	20	±1	Air temperature

30°C and the stirring must be constant to ensure that the mixture is homogeneously balanced.

A tank has been designed to ensure the balance of the mixture. This consists of a jacketed container which, together with the water pumped in by a thermostatic bath and coupled with an electrical stirrer, keeps the mixture warm, thereby ensuring that the animal fat-diesel mixture is kept uniform. Figure 2 shows a diagram of the tank in which the samples were kept.

Characteristics of the mixtures

The burner used has a limitation on fuel viscosity; this forced us to use mixtures of diesel and up to 40% lard. Our study of the mixture allows us to conclude that to have an even mixture of animal fat and diesel, the mixture cannot contain over 40% animal fat in mass. Each mixture is labelled with an identification code consisting of letters and two numbers: FC-XX, FN1-XX and FN2-XX.

The meanings of the letters are the following:

- FF refers to edible fat.
- NF1 refers to type 1 non-edible fat.
- NF2 refers to type 2 non-edible fat.

Table 6 TESTO measuring equipment, model 342-3

	Parameter
Fume analyzer TESTO model 342-3	
Electrochemical O_2 sonde	Range 0% to 21%
CO sonde (with H compensation)	Range 0 to 4,000 ppm
NO sonde with electrolyte cell	Range 0 to 3,000 ppm
Temperature sonde (five sondes)	
Termopar type	K
Measuring interval	−40° to 1,000°C ± 0.5°C
Pressure sonde (four sondes)	
Manometers	Spiral
Impulse pressure interval	1 to 40 bars; 1% accuracy
Aspiration pressure interval	0.2 to 1 bar; 1% accuracy

The numbers represent the percentage (%) of fat used in the mixture. Four mixtures in varying percentages (10, 20, 30 and 40) were prepared for each type of fat.

Describing each of the mixtures performed, based on the knowledge of the properties of the pure components that make up the (theoretical) mixtures, we use the additivity rule with regard to a physical property:

$$P_{mt}(T) = x_a \times P_a(T) + (1-x_a) \times P_d(T),$$

where $P_{mt}(T)$ is the physical property of the theoretical mixture at temperature T, x_a is the molar fraction of the fat in the mixture and $P_a(T)$ and $P_d(T)$ are the physical properties of the fat and diesel. A significant deviation can be seen in the viscosity calculated based on the values of the original compounds and their value percentage measured in the laboratory. As a result, the viscosity data are measured in the laboratory (empirical measurements) since the mixtures of fats and diesel cannot be considered as Newtonian liquid [23].

Table 3 shows the results of the mixtures of the various fats and diesel in different proportions.

The experimental facilities are divided into four parts: (a) the boiler-burner unit, (b) the fuel feed system, (c) the heat dissipation system and (d) the control and measuring system.

Boiler-burners. Combustion for heating is performed using burners prepared for the controlled and optimised combustion of a specific fuel. The basic conditions to be met by the burner are the following:

1. It should adapt to the combustion chamber where it is to be used.
2. It should be equipped with a suitable margin of regulation.
3. It should be possible to control the shape and dimension of the flame.
4. It should function in a stable manner.

Table 7 Reference values analysis of diesel fumes

Fuel	*P* (bar)	Excess air (n)	CO (ppm)	O_2 (vol.%)	T_H (°C)	NO_x (ppm)	Combustion performance (%)
	10	1.2	88.3	13.17	171.6	46	86.0
	10	1.3	251	14.30	202.8	37	84.4
	10	1.4	1530.7	15.37	212.5	14.7	83.3
	12	1.2	96.3	13.83	172.9	41.0	84.1
Diesel	12	1.3	150	14.3	214.4	39.3	82.9
	12	1.4	412.7	15.7	219.4	24.3	81.7
	14	1.2	-	-	-	-	-
	14	1.3	70.3	14.8	243.3	54.3	79.2
	14	1.4	385.8	14.8	247.9	37.0	77.3

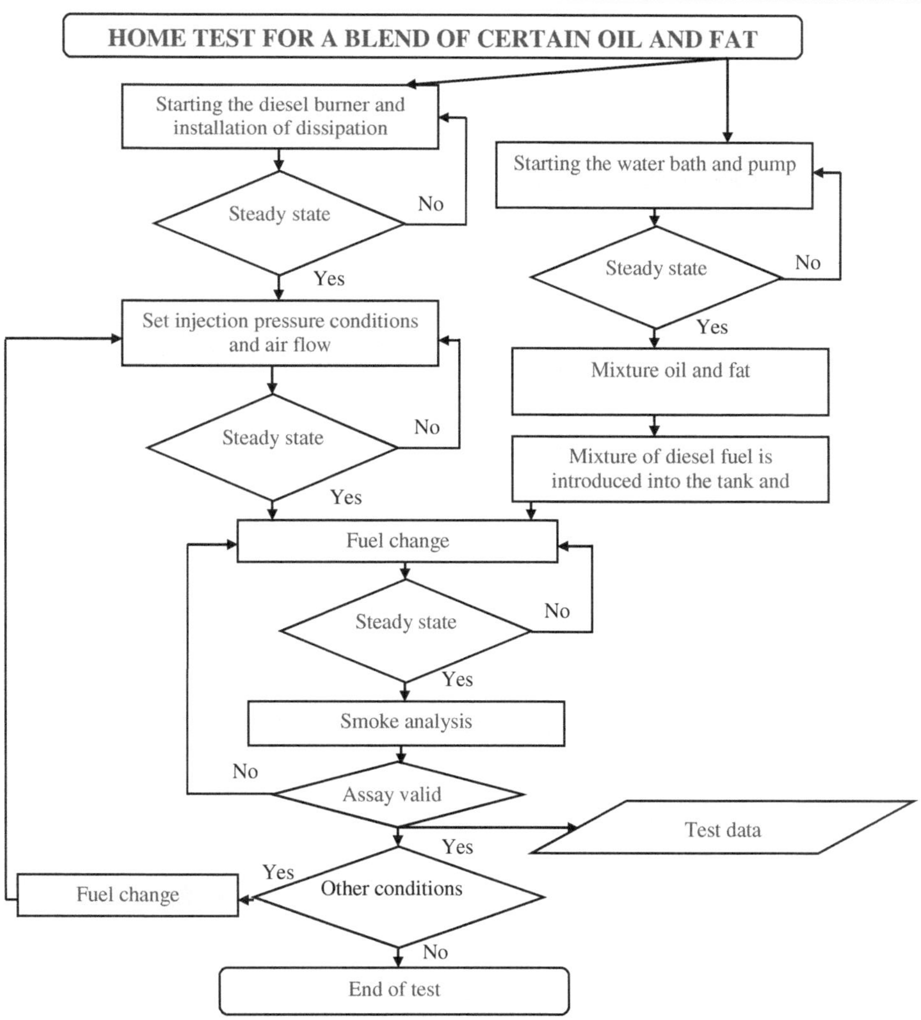

Figure 4 Diagram of the measuring process.

Table 8 Combustion analysis of the diesel and food fat

	P (bar)	Excess air (n)	CO (ppm)	O_2 (vol.%)	T_H (°C)	T_A (°C)	NOx (ppm)
FF-10	10	1.2	103	5.4	211.7	23	30
		1.3	129	5.6	217.1	23	28
		1.4	189	6	217.8	23	27
	12	1.2	76	5.2	220.6	23	29
		1.3	81	5.6	222.3	23	27
		1.4	130	5.9	223.9	23	25
	14	1.2	47	5.1	227.5	23	28
		1.3	54	5.5	229.8	23	26
		1.4	95	5.9	231.9	23	25
FF-20	10	1.2	120	5.1	214.6	23	32
		1.3	131	5.6	217.2	23	30
		1.4	196	6	218	23	29
	12	1.2	61	5.2	222.4	23	33
		1.3	65	5.6	224.1	23	31
		1.4	94	5.8	227.9	23	29
	14	1.2	36	5	233.7	23	33
		1.3	37	5.3	238.3	23	32
		1.4	58	5.6	241.3	23	30
FF-30	10	1.2	122	5.3	212.5	23	32
		1.3	185	5.7	216.1	23	28
		1.4	300	6	221.3	23	26
	12	1.2	74	5	225.3	23	34
		1.3	84	5.4	227.7	23	33
		1.4	117	5.8	229.6	23	31
	14	1.2	46	5.1	233.1	23	35
		1.3	41	5.3	239.1	23	34
		1.4	69	5.6	243	23	33
FF-40	10	1.2	150	5.1	222.7	23	28
		1.3	198	5.4	225	23	27
		1.4	198	5.4	225	23	26
	12	1.2	40	4.8	229.1	23	35
		1.3	36	5.1	232.1	23	34
		1.4	68	5.4	240.9	23	32
	14	1.2	21	4.9	241.3	23	38
		1.3	25	5.2	246.9	23	36
		1.4	41	5.4	250.3	23	35

5. It should be equipped with safety and regulation systems to ensure automatic functioning

In the case of burners for liquids, the following features are also required:

1. Capacity to achieve correct combustion intensity (amount of fuel per specific volume)

Table 9 Combustion analysis of the diesel and non-edible type 1 fat mixtures

	P (bar)	Excess air (n)	CO (ppm)	O_2 (vol.%)	T_H (°C)	T_A (°C)	NOx (ppm)
FN1-10	10	1.2	145	5	213.6	23	46
		1.3	167	5.3	224.3	23	45
		1.4	213	5.6	233.1	23	46
	12	1.2	112	4.8	237.3	23	50
		1.3	94	5	245.2	23	51
		1.4	124	5.4	252.1	23	48
	14	1.2	68	4.8	255.9	23	53
		1.3	66	5.1	261.8	23	52
		1.4	89	5.3	269.4	23	52
FN1-20	10	1.2	215	4.8	212.1	23	56
		1.3	329	5.2	220.7	23	61
		1.4	425	5.5	230.6	23	64
	12	1.2	277	4.9	236.5	23	70
		1.3	252	5.4	239.5	23	68
		1.4	316	5.6	245.5	23	67
	14	1.2	186	4.7	254.6	23	78
		1.3	161	5.1	260.2	23	76
		1.4	211	5.3	266.6	23	74
FN1-30	10	1.2	624	4.4	235.2	23	78
		1.3	558	5.1	239.7	23	81
		1.4	638	5.5	246.6	23	81
	12	1.2	603	4.7	248	23	88
		1.3	521	5	253.2	23	89
		1.4	506	5.4	258	23	90
	14	1.2	300	4.4	266.8	23	94
		1.3	226	4.9	270.8	23	98
		1.4	315	5.2	276.5	23	96
FN1-40	10	1.2	872	4.3	229.6	23	96
		1.3	634	5.1	237	23	98
		1.4	910	4.4	249	23	104
	12	1.2	728	4.5	252.9	23	105
		1.3	636	4.9	259.2	23	102
		1.4	737	5.2	261.5	23	105
	14	1.2	637	4.6	266.2	23	107
		1.3	534	4.7	271.1	23	109
		1.4	626	5	276	23	110

2. Capacity to operate with a minimum amount of excess air (maximum flame temperature)

Burners for liquid fuels are based on two technologies: fuel vaporisation and fuel pulverisation in small drops. This study uses a pressure pulverisation burner, which can burn fuels with a viscosity between 1 and 5°E at 20°C,

Table 10 Combustion analysis of the diesel and non-edible type 2 fat mixtures

	P (bar)	Excess air (n)	CO (ppm)	O_2 (vol.%)	T_H (°C)	T_A (°C)	NOx (ppm)
FN2-10	10	1.2	68	5.1	217.7	23	29
		1.3	74	5.3	221.2	23	27
		1.4	88	5.5	229.1	23	26
	12	1.2	39	4.8	231.4	23	28
		1.3	42	5.2	236.5	23	27
		1.4	53	5.4	243.1	23	26
	14	1.2	24	4.8	244.5	23	31
		1.3	26	4.9	249.7	23	30
		1.4	36	5.2	255.3	23	28
FN2-20	10	1.2	69	4.9	219.1	23	29
		1.3	72	5.2	224.9	23	28
		1.4	94	5.4	227.3	23	27
	12	1.2	43	4.7	231.9	23	31
		1.3	47	5.1	235.7	23	30
		1.4	54	5.3	243.6	23	28
	14	1.2	27	4.7	246.5	23	31
		1.3	27	4.9	253.6	23	30
		1.4	33	5.3	258.9	23	29
FN2-30	10	1.2	60	4.8	222.9	23	29
		1.3	63	5.3	224.4	23	27
		1.4	87	5.5	230	23	26
	12	1.2	41	4.8	234.2	23	32
		1.3	43	5.1	239.9	23	30
		1.4	49	5.2	246	23	29
	14	1.2	28	4.4	249.9	23	33
		1.3	30	4.8	255.9	23	31
		1.4	36	5.1	260.5	23	30
FN2-40	10	1.2	52	4.6	224.4	23	29
		1.3	62	4.9	228.2	23	28
		1.4	70	5.2	236.9	23	27
	12	1.2	36	4.7	239.4	23	31
		1.3	36	4.9	246.7	23	30
		1.4	50	5.1	255.9	23	29
	14	1.2	28	4.5	260.1	23	33
		1.3	28	4.7	265.4	23	33
		1.4	32	5	272.1	23	32

enabling regulation of the pressure between 10 and 20 atm and the air flow through which combustion is performed. If we add the mixture percentages to these two control parameters, we have the three parameters to be controlled and optimised in combustion for generating heat using mixtures of fat and diesel. The characteristics of the boiler and the burner are shown in Table 4. The burner and boiler are widely used commercial equipment in Spain. *The system for feeding the fuel* to the burner comprises two tanks, one containing the diesel used for the experiment to be considered stationary and the other containing the fuel mixtures. Figure 2 shows the system for feeding the fuel to burner.

Heat dissipation is through a two-tube radiator facility, with direct return. Figure 3 shows the layout of the facility.

The facility control parameters are shown in Table 5. These values should be kept within a specific interval for the test to be considered stationary.

The control and measuring system. The fume analysing equipment and the parameters obtained during each test are those used to describe the combustion and are shown in Table 6.

The results from the combustion of the mixtures were analysed based on three aspects:

- Emissions of pollutant gases
- Combustion efficiency
- Greenhouse gases

Before analyzing the combustion of a mixture, it is found that the set boiler-burner works properly by an analysis of diesel fumes, which is considered patron fuel; Table 7 presents the reference values analysis of diesel fumes.

Experimental designs

Excess air, compared to the stoichiometric flow rate, is modified among values 1.2, 1.3 and 1.4 so as to establish the optimal combustion conditions for these mixtures in the burner used.

Having determined the variables, the test matrix to be conducted is posited to ensure that all the levels of all variables will be tested with the levels of the remaining variables in the same number of times. In other words, the design will maintain orthogonality. The order of the assays will therefore be carried out thus

1. Three types of animal fat are studied (three levels).
2. Four mixtures with different proportions of each animal fat with diesel are carried out (four levels).
3. The injection pressure is varied using three values for each mixture (three levels).
4. Excess air is varied using three values for a single sample of one type of fat at a specific percentage (three levels).

The number of tests to be conducted in order to maintain orthogonality coincides with the product of the number of levels of all the variables:

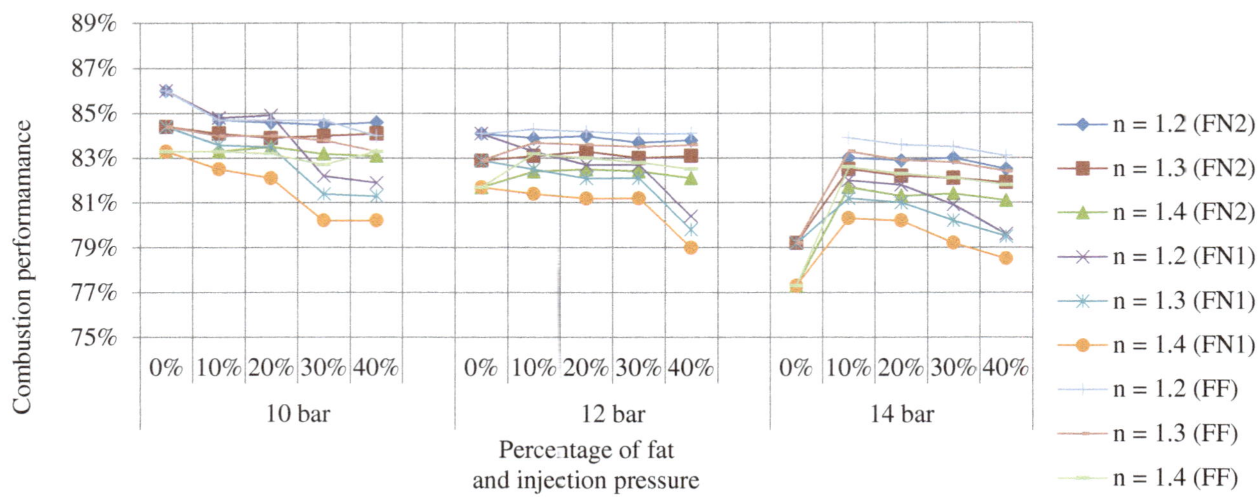

Figure 5 Combustion performance for FF, FN1 and FN2 in terms of the three factors.

$$3 \times 4 \times 3 \times 3 = 108 \text{ tests.}$$

Finally, we present the procedure followed in the experimental tests and the requirements the facility must meet if results are to be accepted as valid (Figure 4).

Results and discussion

A total of 108 combustion tests were carried out in which the following conditions were met: (a) stable working conditions for the boiler, (b) air flow and fuel injection pressure adapted to the test, (c) uninterrupted running of the facility for each mixture and (d) the values for each test within the error limits for the measuring equipment.

Tables 8, 9 and 10 show the parts per million in CO fumes, the percentage of oxygen in fumes, fume temperature, air temperature or reference temperature and the concentration of nitrogen oxides in fumes obtained from the combustion.

With regard to the emissions of pollutant gases, we analysed emissions of carbon monoxide and nitrogen oxides. Combustions were considered acceptable when the concentration of fumes is less than 500 ppm of CO and/or is less than 300 ppm of NO_x, the limits that are imposed under current legislation.

Combustion efficiency is defined by the following expression:

$$\eta_{\text{comb}} = \frac{\text{LHV} - P_s - P_i}{\text{LHV}},$$

where P_s is sensible heat loss (MJ/kg_{comb}), P_i, is heat loss from incomplete combustion (MJ/kg_{comb}) and LHV is lower heating value (MJ/kg_{comb}). Sensible heat loss (P_s) corresponds to the energy the combustion gases have when extracted compared to the reference conditions of the combustion process. When these losses are higher, the internal energy of the flue gas (flue gas temperature) is greater. They also increase with flue gas flow,

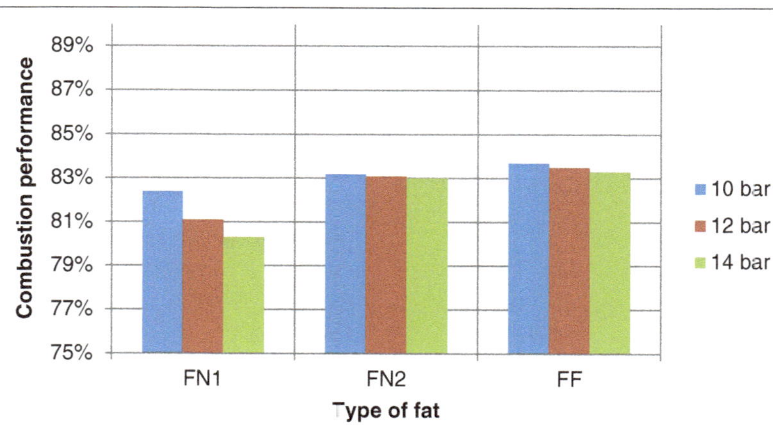

Figure 6 Combustion performance for FF, FN1 and FN2 in terms of three injection pressures.

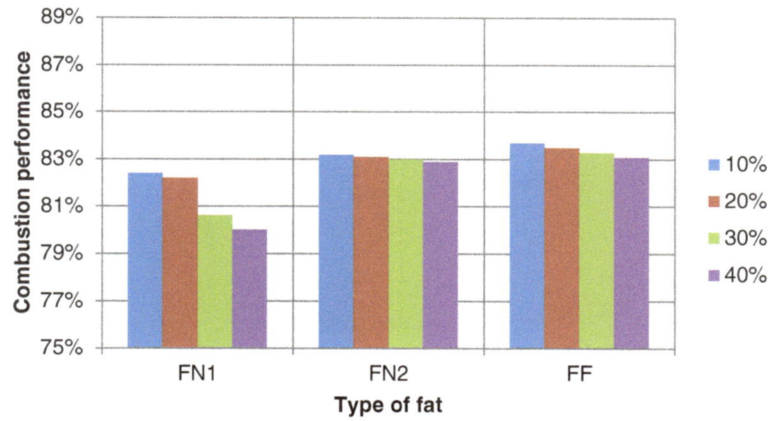

Figure 7 Combustion performance for FF, FN1 and FN2 in terms of percentage of fat.

which increases with excess air, and this also increases the flow of flue gas, transporting a greater amount of heat.

Sensible heat loss is defined by the following percentage expression:

$$P_s = \frac{M^1_{Hs} \times Cp_A \times (t_H - t_{ref})}{LHV},$$

where M^1_{Hs} is mass of combustion of flue gas in (kg/kg$_{comb}$); Cp_A, heat capacity at constant pressure of flue gas heat (MJ/kg °C$_{comb}$); LHV, lower heating value of the fuel in (MJ/kg); t_H, temperature of flue gas at output in (°C); and t_{ref}, the ambient temperature in degrees Celsius (°C).

Heat loss from incomplete combustion (P_i) corresponds to the energy of the combustion products which have not been totally oxidised. This occurs in cases of incomplete combustion, the loss being greater with the higher amount of unburnt gas. When total carbon oxidation does not occur, the combustion product will be CO rather than CO_2:

$$P_i = \frac{V^1_{Hs} \times LHV_{CO} \times \rho_{co} \times (CO)}{LHV},$$

where V^1_{Hs} is the volume of dry of flue gas in (nm^3/kg$_{com}$), LHV_{CO} is the lower heating value of CO in (MJ/kg$_{CO}$), ρ_{co} is the density of CO in (kg/m^3) (CO), percent by volume in the flue gas (nm$^3_{CO}$/nm^3) and LHV is the lower heating value of the fuel in megajoules per kilogram of combustion (MJ/kg$_{comb}$).

The combustion efficiency represents the amount of energy transferred to the heat system of the energy released by the fuel. The combustion performance values are shown in Figure 5 for FF, food fat, NF1, non-food fat type 1 and NF2, non-food fat type 2 mixtures in terms of the injection pressure, the mixture percentage and excess air during combustion.

As seen in Figure 5, the combustion efficiency is reduced when increasing the percentage of non-fat foot type 1, being very significant when it exceeds 30% in the mixture

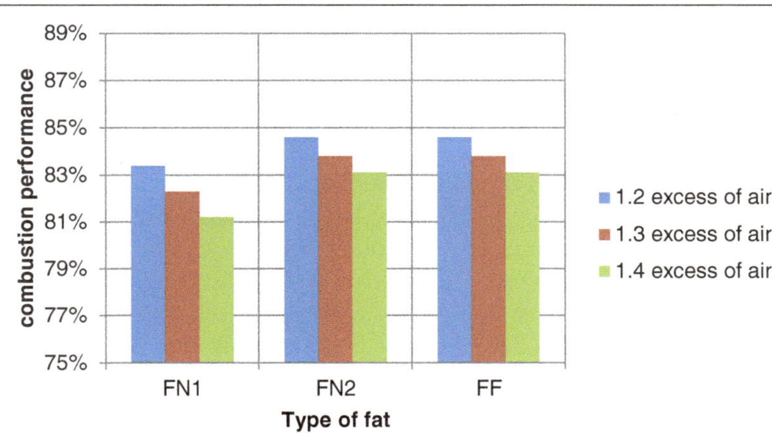

Figure 8 Combustion performance for FF, FN1 and FN2 in terms of excess air.

Table 11 Greenhouse gas emissions of the combustion of fat and diesel mixtures

	Total CO_2 (kg/kg fuel)	Non-fossil CO_2 (kg/kg fuel)
FF-10	3.14	0.28
FF-20	3.10	0.56
FF-30	3.07	0.84
FF-40	3.03	1.13
FN1-10	3.13	0.28
FN1-20	3.09	0.55
FN1-30	3.05	0.83
FN1-40	3.01	1.10
FN2-10	3.13	0.28
FN2-20	3.09	0.55
FN2-30	3.05	0.83
FN2-40	3.01	1.10

and also seen in Table 9, which increases the unburned (CO); this is due to be a fat containing many solid particles impinging on the phenomenon of the fuel spray.

The results are similar to those obtained in diesel combustion. It is worth noting that fats have a lower LHV than diesel and that the energy released per unit of mass in the combustion process is therefore lower. However, fats have a higher percentage of oxygen in their composition and thus require less airflow, which reduces loss due to fumes. These two phenomena together mean that the performance values of the mixtures and diesel are similar.

The average combustion performance values are shown for each fat in terms of injection pressure. For FN1, performance decreases as injection pressure increases. As a result, increasing injection pressure does not enhance combustion, as can be seen in Figure 6.

The average combustion performance values are shown for each fat in terms of the percentage of fat in the mixture (Figure 7). Combustion efficiency decreases as the percentage of fat in the mixture increases. This is because the fats have a lower calorific value and contain oxygen, which produces higher excess air during combustion.

The average combustion performance values are shown for each fat in terms of excess air (Figure 8). The performance follows a clearly linear trend, diminishing as the airflow involved in the combustion increases.

By increasing the percentage of fat in the fuel mix, an ascent is observed in the evolution of NO_x. This is because one side has more oxygenated fuel component as we increase the percentage of fat, which increases the flame temperature and results to higher nitrogen content in the fat as we increase the percentage of NO_x.

The greenhouse gases considered in this study were emissions of CO_2 associated with diesel combustion, as it is a fossil fuel, since the CO_2 generated by combustion of fats is renewable. The evaluation was carried out in terms of the basic composition of the diesel and fats and the mixture percentage. This value is shown in Table 11, which also shows the kilogram of CO_2 emitted during full combustion per kilogram of fuel, distinguishing between the kilogram originating from fossil diesel and those from non-fossil fat.

Conclusions

Animal fat as an animal by-product not intended for human consumption is an abundant resource. Each year, Europe generates 1.35 million tons. Taking an average LHV of 42.1 MJ/kg, the energy potential per year is 56.8 million GJ/year which would mean a reduction in CO_2 emissions of 4.43 million tons/year compared to diesel.

Providing a basic description of fats has enabled us to determine that their composition is similar to the composition of diesel, with a lower carbon and hydrogen content and higher oxygen content. This is due to the type of fatty acids contained in animal fats compared to the hydrocarbons that make up diesel. We also see that the LCV of fats is around 10% lower than that of diesel in all cases and that the density of fats is just under 10% higher than that of diesel. As a result, the energy flow by volume is similar in fats and diesel. The main difference between diesel and fats is in the viscosity. At 40°C, the viscosity of fats is 20 times higher than that of diesel. This property has a major impact on pulverization and proper combustion of liquid fuels and it should be remembered that at room temperature animal fats are mostly solid.

When fats are melted at a temperature above 38°C they are totally soluble in diesel, the mixture being fully homogeneous, and no dissolution limit having been found. Studies performed into the stability of the mixture and its homogeneity indicate that the mixture should be kept at 38°C and constantly stirred to ensure it does not lose its homogeneity.

We observed that combustion performance diminishes when the percentage of fat in the mixture is increased, which concurs with the lower LCV of fats compared to that of diesel.

We observed that combustion performance diminishes when pressure is increased, concurring with the theoretical studies carried out which indicate that above a certain injection pressure, no improvements in the pulverization process are seen, although it does contribute towards a detachment of the flame.

We also observed that performance diminishes when the combustion air flow is increased, which is accounted for by the amount of oxygen contained in the fats, meaning that they require less excess air for combustion.

Authors' contributions
JFSNA coordinator working, IGA responsible for the installation and testing, SAM processing and analysis of experimental results. All authors read and approved the final manuscript.

References

1. Commission of the European Communities: GREEN PAPER, A European strategy for sustainable, competitive and secure energy; COM(2006) 105 final, Brussels, 8.3.2006
2. Decision No 406/2009/EC, On the effort of Member States to reduce their greenhouse gas emissions to meet the Community's greenhouse gas emission reduction commitments up to 2020. Off. J. Eur. Union 5.6.2009
3. Directive 2009/28/EC, on the promotion of the use of energy from renewable sources and amending and subsequently repealing, the European Parliament and of the Council of 23.4.2009
4. Ndayishimiye, P, Masimalai, S, Tazerout, M: Preparation and properties analysis of methyl esters of palm oil and waste cooking oil mixture to use as fuel in diesel engines, 15th European biomass conference & exhibition. , Hamburg Germany. 7–11.5. 2009
5. Batey, J: Combustion testing of a bio-diesel fuel oil blend in residential oil burning equipment, Tech. rep., Energy Research Centre, prepared for: Massachusetts Oil hest Council & National Oil heat Research Alliance. (2003)
6. Raheman, H, Jena, PC, Jadav, SS: Performance of a diesel engine with blends of biodiesel (from a mixture of oils) and high-speed diesel. Int. J. Energy Environ. Eng., 4–6 (2013)
7. Vanlaningham, N, Gibson, H, Kaufman, B: Evaluation of soybean heating oil blends for use in residential applications, no. 046082. ASAE Paper, annual meeting held, 23–24. 8. 2004
8. San José Alonso, JF, López Sastre, JA, Romero-Avila, C: A note on the combustion of blends of diesel and soya, sunflower and rapeseed vegetable oils in a light boiler. Biomass Bioenergy **32**, 880–886 (2008)
9. San José, J, Al-Kassir, A, López Sastre, JA, Gañán, J: Analysis of biodiesel combustion in a boiler with a pressure operated mechanical pulverization burner. Fuel Process. Tech. **92**, 271–277 (2011)
10. Lezsovits, F, Könczöl, S: Animal-fat investigation and combustion test. J. Therm. Anal. Alorimetry **107**, 271–278 (2012)
11. Pfab, F: Processing and combustion of animal fat from animal waste rendering plants for steam generation in a smoke tube boiler. 2nd World Conference on Biomass for Energy, Industry and Climate protection, Rome (2004)
12. San José, J, Alonso, S, Gobernado, I: Study of thermal energy production by combustion of lard and diesel mixtures. J. Combustion **2012**
13. Ministerio de Medio Ambiente y Medio Rural y Marino: Anuario de estadística agroalimentaria y pesca 2007., Madrid (2008)
14. Decision No 1774/2002/EC, laying down health rules concerning animal by-products not intended for human consumption. Official Journal of the European Communities 3.10.2002
15. Consejería de Sanidad: Agencia de Protección de la Salud y Seguridad Alimentaría, Datos estadísticos 2010. Junta de Castilla y León (2010)
16. Issue number 6/2009, Main stages in the meat food chain in Europe. Euro stat Publish. 11.4 .2009
17. Ghassan, M, Tashtoush, Mohamad, I, Al-Widyan, Aiman, M: Types of biofuels. J. Environ. Manage. **84**, 401–411 (2007)
18. Calais, P, Clark, T: Waste vegetable oil as a diesel replacement fuel. Environmental Science, Murdoch University, Perth, Australia (2006)
19. Senthil Kumar, M, Kerihuel, A, Bellettre, J, Tazerout, M: Experimental investigations on the use of preheated animal fat as fuel in a compression ignition engine. Renew. Energy **30**, 1443–1456 (2005)
20. Madrid, A: Aprovechamiento de los subproductos cárnicos. Capítulo 1. Los subproductos cárnicos: definición, composición y características. AMV Ediciones, 17–22 (1999). Madrid
21. Bills, CE, with the cooperation of McDonald F.G: Antiricketic substances: iii. The catalytic formation of an antiricketic cholesterol derivative. J. Biol. Chem. **67**, 753–758 (1926)
22. Formo, MW, Jungermmann, E, Norris, FA, Sonntag, NO: In: Swern, D (ed.) Bailey's industrial oil and fat products, vol. 1, 4th edn. John Wiley and Sons, New York, USA (1997)
23. Flores Luque, V, Gómez Herrera, C, Tabernero de la Linde, P: Estudio de volumen molar y refracción molar de miscelas de triglicéridos (triacetina, tributirina o tricaprilina) y alcoholes (etanol, 1-butanol o 1-hexanol). Grasas y aceites **4**, 216–220 (1992)

Effect of domestic water use on air pollutant emissions in Abu Dhabi, United Arab Emirates

Nicholas B DeFelice and Jacqueline MacDonald Gibson*

Abstract

The members of the Cooperation Council for the Arab States of the Gulf have typically addressed water scarcity problems by building energy-intensive desalination plants. Few efforts have addressed water scarcity through metering, pricing, and other efficiency measures to reduce demand. This paper examines how decreased leakage in the water distribution system and decreased residential water use in Abu Dhabi, United Arab Emirates, could decrease air pollutant and greenhouse gas emissions from desalination plants. We developed a probabilistic model to predict the effects of water use reductions on pollutant emissions from Abu Dhabi's major independent water and power plants, which use a combination of multi-stage flash distillation and multi-effect distillation to produce fresh water from seawater drawn from the Arabian Gulf. We examine three categories of scenarios for reducing water use: increasing the price signal to residential users, instituting demand management programs among residential users, and reducing water loss in the distribution system. Our analysis suggests that water conservation price incentives could reduce air pollutant and greenhouse gas emissions by 1% to 5%, depending on assumptions about how households respond to the incentives. Demand-side management programs curbing per capita water use to levels typical of the Singapore or the UK would curb emissions by 10% or 11%, respectively. Reducing water loss during distribution from the current high level of 35% to 15% (similar to loss rates in other developed nations) could cut emissions by more than 3%. Overall, our analysis suggests that high per capita water use contributes to ambient air pollution and greenhouse gas emissions in Abu Dhabi.

Keywords: Water-energy nexus; Desalination; Air pollution; Domestic water consumption; Water scarcity

Background

The countries that make up the Cooperation Council for the Arab States of the Gulf (GCC) have the scarcest per capita renewable freshwater resources in the world, and their demand for water continues to increase [1]. Within the GCC, the United Arab Emirates (UAE) has one of the highest per capita consumption levels, at 525 to 600 liters per capita per day (l/c/day) in 2007 [2,3]. The UAE is able to maintain this high consumption level by producing fresh water through desalination plants. These plants burn fossil fuels that contribute to air pollution and greenhouse gas emissions [4]. The objective of this paper is to examine how decreases in residential water use and water loss during distribution in Abu Dhabi City (the capital of the UAE) could decrease air pollutant and greenhouse gas emissions from desalination plants. We quantify the total annual mass reductions in nitrogen oxides (NO_x), SO_2, particulate matter (PM), carbon monoxide (CO), and carbon dioxide (CO_2) emissions expected to occur as a result of plausible interventions to reduce water use.

Air quality is an increasing concern in Abu Dhabi. A recent analysis indicated that more than 545 deaths per year may be attributable to ambient air pollution in the UAE, with 208 of those in Abu Dhabi emirate [5,6]. A recent project to prioritize environmental risks in the UAE ranked air quality as the highest concern [7]. The government of Abu Dhabi is actively exploring additional options to improve air quality. As Abu Dhabi considers its options, an important question is the extent to which reducing emissions from specific pollution sources could improve air quality. The analysis presented in this paper can help to inform decisions about air quality management in Abu Dhabi City. Since the UAE has high per

* Correspondence: jackie.macdonald@unc.edu
Department of Environmental Sciences and Engineering, Gillings School of Global Public Health, University of North Carolina-Chapel Hill, Campus Box 7431, Chapel Hill 27599-7431, NC, USA

capita emissions of greenhouse gases, we also assess the extent to which water use contributes to Abu Dhabi's releases of CO_2.

Water in Abu Dhabi City is heavily subsidized. Citizens (UAE nationals) receive water for free, whereas non-nationals pay 2.2 Dhs/m^3 (US $0.60) - about 25% to 30% of the total cost to produce and supply water to consumer taps. The total annual cost of producing residential water in 2008 was US $572 million, of which the government subsidized US $423 million [8,9]. Partly as a result of this large subsidy, water consumers have little incentive to reduce their consumption. Residential water demand and related water subsidy costs are expected to increase as Abu Dhabi City's population continues its rapid growth (5.2%/year) [10].

Abu Dhabi's major independent water and power plants (IWPPs) simultaneously generate energy for electricity and produce desalinated water, using some of the waste heat from energy production to power the desalination process. However, waste heat alone is insufficient to meet all the power needs for desalination; a considerable amount of fossil fuel is burned for water production, contributing to the release of air pollutants and greenhouse gases [2]. Siddiqi and Anadon recently estimated that desalination uses 12% to 22% of the total electricity produced in the UAE [11].

We developed a probabilistic model to examine the trade-offs between domestic water consumption and emissions of air pollutants and greenhouse gases (PM_{10}, NO_x, SO_2, CO, CO_2) generated from the major IWPPs located along the coast of Abu Dhabi emirate (at Taweelah, Mirfa, Umm Al Nar, and Shuweihat). In this paper, we use the model to explore how different options for reducing residential water consumption in Abu Dhabi City could affect water demand and atmospheric pollutant emissions. Raluy et al. showed that more than 90% of the environmental loads (energy use, raw materials, emissions, and waste products) of desalination plants are associated with the operating stage of the plants, with less than 10% of the load attributed to construction and end-of-life emissions [12]. Therefore, this paper focuses on the energy requirements of the water production stage since this stage causes majority of the environmental impacts.

Desalination effects on atmospheric pollution: previous studies

Several previous life cycle assessments have examined the environmental effects (including air quality effects) of different desalination technologies. A study by Raluy et al. found that multi-stage flash distillation (MSF) technology, followed by multi-effect desalination (MED), generated the most atmospheric pollution. However, when residual heat is used to help power the thermal processes of MSF and MED, emissions can be reduced to an order comparable to that of a reverse osmosis (RO) plant [13,14].

Stokes and Horvath conducted a full life cycle assessment of water management options in California, concluding that the energy and air emission footprint in California would be 1.6 to 3.4 times larger if the state used RO desalinated seawater than if water were imported or recycled [15]. Using the California energy mix and life cycle emissions, the study determined that desalination would consume 2.3 to 2.5 times more energy and emit 2.3 to 2.4 times more carbon dioxide, 1.8 to 3.4 times more nitrogen oxides, 1.6 to 1.9 times more particulate matter, and 2.4 times more sulfur oxides than if water were imported or recycled.

A study by Alameddine and El-Fadel [16] used the Industrial Source Complex-3 model to simulate plume dispersion of SO_2 from a typical IWPP located in the Arabian Gulf. Simulation results indicated that the impact of SO_2 stack emissions in several instances can lead to concentrations exceeding international health standards for SO_2, depending on the sulfur concentration of the fuel source [16].

Rubio-Maya et al. [17] conducted a feasibility analysis of a combined cooling-heating power plant with a desalination unit to supply a tourist resort on the Mediterranean coast of Spain. They concluded that although RO consumes less energy than MED when considered in isolation from power production and heating/cooling operations, overall, the use of MED offered greater potential for reductions in greenhouse gas emissions due to the ability to link the production schedule for desalinated water to the availability of waste heat from the power generation and cooling units [17].

Although studies have shown that atmospheric emissions are a major component of the environmental impact caused by desalination plants, to our knowledge, no studies have been conducted to determine the fraction of total air pollutant and greenhouse gas emissions arising from desalination plants in energy-rich and water-scarce regions of the world. This is the first study to examine how steps to reduce water consumption through demand management and reduced water losses can benefit air quality in Abu Dhabi.

Current water tariff

According to the World Bank, nations in the Arabian Gulf region have typically dealt with water scarcity from the supply side by building desalination plants to generate fresh water to meet the growing demand. Few, if any, efforts are in place to improve demand management of urban water supplies through metering, pricing, or other efficiency-improving measures [1], despite ample evidence that people change consumption habits based on price or alternative demand-side management policies (information campaigns, retrofit subsidies, water rationing, or use restrictions) [18].

According to information from the Abu Dhabi Distribution Company (ADDC), water tariffs are 2.2 Dhs/m^3 (US $0.60) for non-nationals and free for nationals [9]. Furthermore, 35% of water entering the distribution system goes unaccounted for: 16% due to physical losses from leaky pipes and the remainder due to mismanagement (such as unbilled connections) [2]. Ironically, nationals are charged 0.05 Dhs/kW h (US $0.014) for electricity, yet pay nothing for water even though production of 1 m^3 of desalinated water uses an average of 22.4 kW h (which would cost Dhs 1.12 or US $0.31 if bought at current rates) [19].

A number of different options are available for governments to employ to try to decrease residential water demand. One option is to increase the price of water. A previous research study by Abu Qdais and Al Nassay provides an indication of how nationals and non-nationals might respond to changes in domestic water prices in Abu Dhabi in the short term [8]. Based on a short-term study of 90 households, they determined that in Abu Dhabi City, the price elasticity of water (in other words, the percent change in water quantity demanded in response to the percent change in water price) is −0.072 (standard deviation (SD) = 0.0089). This elasticity is low, which is consistent with previous research showing that residential water demand is comparatively inelastic, for three reasons: (1) there is no close substitute for water in most of its uses, (2) the amount spent on water is a relatively small fraction of the typical household budget, and (3) water is often jointly demanded with some other good or service [18]. Nonetheless, as this study and others have demonstrated, financial incentives can reduce water demand. Furthermore, over the longer term, one would expect to see a higher absolute value of elasticity - and thus further reductions in water use - in response to a financial incentive.

Water tariffs are a management tool that the UAE could examine more closely in order to create an efficient system for the residential water sector. Yet, pricing water is a complex and controversial issue for several reasons. First, policymakers often disagree about the objectives of a water pricing system; some view the primary objective as a provision of affordable water, while others view it as a sustainable management of scarce resources. Second, water managers may disagree about the effects of implementing a particular water tariff because empirical studies on customers' reactions to long-term price changes are often lacking. Third, no market test exists for different tariff structures; many tariff structures are tried simply because they are feasible and can accomplish a portion of the objective [20]. Finally, adding to these complexities in the Arabian Gulf region, water subsidies are viewed as an essential service that governments provide to citizens, and hence, reducing these subsidies is likely to be highly controversial.

Policy tools other than raising water price have also shown considerable promise in reducing water consumption. Demand-side management programs established in California and Las Vegas have been effective in conserving water. Las Vegas introduced a turf removal program, replacing turf with more water-efficient landscapes. Replacing grass with a drought-tolerant natural landscape for the desert ecosystem reduced water use by an average of 76%, saving 2,260 l/m^2 of irrigated land each year [21]. In a study in California, Renwick and Green concluded that public information campaigns and retrofit subsidies reduced household water consumption by 8% and 9%, while water rationing and water restriction dropped consumption by 19% and 29%, respectively [18].

With the population growth that Abu Dhabi is experiencing and the high costs of producing desalinated water, current water pricing policies, as well as a range of options for curtailing demand, should be examined more closely. This paper examines how such incentives could reduce air pollutant and greenhouse gas emissions from major desalination plants serving Abu Dhabi City.

Methods

The model created for this study incorporates probabilistic inputs and site-specific data to run Monte Carlo simulations (using Analytica 4.2 software, Lumina Decision Systems, Los Gatos, CA, USA). The model estimates total reductions in PM, NO_2, SO_2, CO, and greenhouse gas emissions from Abu Dhabi's IWPPs under different water consumption scenarios. We use 2008 as the base year for our estimates since it was the most recent year for which water production figures were available at the time our analysis was conducted.

The amount of water produced, along with the type of technology employed, determines the amount of energy needed to desalinate the water. This energy use, in turn, determines how much fuel is consumed in producing water. The type and amount of fuel combusted then determine how much pollution is emitted. In estimating atmospheric emissions, our model uses local water demand data along with data about specific desalination technologies and fuels employed at each IWPP. The following sections provide details on the model's components.

Current residential water demand

Residential water production in Abu Dhabi for 2008 was calculated using reports from Abu Dhabi Water and Electricity Company and ADDC. The ADDC receives 70% of desalinated water produced, and 44% of this travels to the residential sector. Of the water that ADDC distributes, 35% is lost, 16% due to physical leakage and the rest from technical and administrative losses [2].

Our model divides residential water users into two main groups, nationals and non-nationals, and then subdivides these groups by housing type. Based on information from the Regulation and Supervision Bureau (responsible for overseeing potable water quality), non-nationals living in flats consume 160 to 220 l/c/day, whereas those in villas use 270 to 730 l/c/day. This trend is also seen among UAE nationals: families in flats use around 165 l/c/day, villa households use 460 to 1,760 l/c/day, and families in shabias (a form of housing traditionally provided by the government for citizens) use 610 to 1,010 l/c/day [3]. The variation in consumption between users suggests that many different water policies will be needed to reach targeted goals. On average across all these groups, residential water demand was 544 l/c/day during the study year [9]. These current water use rates constitute the baseline scenario.

Scenarios

We examine the potential reduction in air pollutant and greenhouse gas emissions under several scenarios of decreased water use. First, we assess the potential effects of price changes. Even though increasing water prices may not be politically feasible in the UAE, quantifying the effects of price changes serves as a proxy for the potential effects of other, politically feasible demand-side management programs, such as rebates for xeriscaping or water-efficient fixtures. Our model examines how residential consumption could change as water price (or equivalent rebates offered to households that reduce water use) fluctuates near its average cost, 7.4 Dhs/m^3 (US $2.00). Then we consider air emission reductions if per capita residential water consumption levels in Abu Dhabi were reduced to those of several other developed countries (Australia, Spain, Singapore, and the UK) - all of which have lower per capita residential water use rates than Abu Dhabi - through demand-side management. We also examine how air emissions would decrease if water use were curtailed to per capita rates in the UAE in 2000. Last, we consider the effects of decreasing the volume of water lost during distribution below the current rate of 35%.

Elasticity

The scenarios in which water prices are increased (or rebates are offered) require an estimate of the price elasticity of water. As noted above, Abu Qdais and Nassay, in a study of 90 households, determined that the short-term price elasticity of water is approximately −0.07 [8]. However, for two reasons, this elasticity may not adequately represent how water demand in Abu Dhabi might respond to financial incentives. First, since Abu Dhabi's water consumers to date have paid only a very small fraction of the cost of production, Abu Qdais and Nassay's observed elasticity may not apply to the large financial incentive (whether cash rebate for decreased water use or charge for use) needed to reflect the full cost of production. It is possible that elasticity would increase (i.e., become increasingly negative) as the price increases and water consumes a larger fraction of the household budget (or offers a larger potential rebate). Second, Abu Qdais and Nassay's study estimated the short-run elasticity of water, but the long-run elasticity of demand should be used to predict long-term changes in consumption. Based on previous studies, the long-run elasticity tends to be approximately 0.2 to 0.3 points lower than the short-run elasticity of water [22]. Using this estimate, Abu Dhabi City's long-run price elasticity should be around −0.37 to −0.27. This elasticity range is closer to the average elasticity of water that Dalhuisen and Florax found in their meta-analysis of price elasticity of water demand, averaging −0.38 (range, −1.3 to +0.1) in 282 observations [23]. An Espey et al. meta-analysis showed an average elasticity of −0.51, while Nauges and Whittington showed that price elasticity for water from private connections ranges from −0.3 to −0.6 in developing countries [24,25]. Based on this information, our model examines three different scenarios of demand elasticity: (1) −0.072 (SD = 0.0089) from Abu Qdais and Al Nassay [8], (2) −0.27, and (3) −0.37.

We use the following function to estimate how water demand changes with price:

$$w = \alpha p^{\varepsilon} \quad (1)$$

where w = demand, p = price, ε = elasticity, and α = a constant determined by dividing current demand by current price raised to a power equal to the elasticity. The function is based on the assumption that elasticity is constant. The parameter alpha is estimated from current demand and price information. Elasticity does not always stay constant as price changes. Nonetheless, the assumption of constant elasticity represents a conservative estimate of water reduction due to financial incentives. In theory, water becomes more elastic as its price rises, resulting in a greater subsequent reduction because the water bill makes up a greater portion of a typical household income or because the value of the jointly demanded good or service is less desirable at the higher water price.

Energy use

The amount of energy used to produce water depends on the desalination technology that the plant employs and the plant's overall operational efficiency (considering both the fraction of fossil fuel converted to useable energy and the fraction of waste heat captured to support desalination). Three different technologies are used in the GCC to process seawater into potable water: MSF, MED, and RO. Table 1 shows information on the energy requirements of each of these technologies, as estimated from a meta-analysis of previous studies. As shown, MSF is the most energy-intensive of the three. Despite this drawback, the majority (85.3%) of the desalination units at Abu Dhabi's IWPPs are MSF systems, with 8.9% using MED and 5.8% using RO [2]. All of the major desalination facilities located within the emirate of Abu Dhabi in 2008 used MSF and MED; the use of RO occurred only in a facility in the emirate of Fujairah, which exports water to Abu Dhabi, and in small facilities serving remote communities or desalinating brackish groundwater [2]. None of the major installations included in this analysis use RO. The high reliance on MSF systems among these major installations is driven by the high salinity levels in the Arabian Gulf due to the performance limitations of the other technologies (especially RO) under high-salinity conditions [26].

Each major desalination facility in this analysis employs a different mix of MSF and MED technologies and therefore uses a different amount of energy to produce a unit volume of water. We employed data available from the Abu Dhabi Water and Electricity Company and Environment Agency-Abu Dhabi to determine the percentage of total annual produced water volume treated by each technology in each desalination unit [2]. Then for each plant, we estimated the energy needed to produce 1 m^3 of water as a weighted average of the energy requirements of each technology:

$$\text{Energy}_{\text{water}} = E_{\text{MSF}} \times W_{\text{MSF}} + E_{\text{MED}} \times W_{\text{MED}} \tag{2}$$

where $\text{Energy}_{\text{water}}$ is the amount of energy (kW h) required to produce 1 m^3 of water from the given desalination unit, E_{MSF} and E_{MED} are the energy requirements per 1 m^3 of water produced by MSF and MED systems, respectively, and W_{MSF} and W_{MED} are the percentages of

Table 1 Energy (kW h) required per cubic meter of water produced by MSF, MED, and RO desalination plants

Technology	Number of studies[a]	Minimum[b] (kW h/m^3)	Mode[b] (kW h/m^3)	Maximum[b] (kW h/m^3)
MSF	5	10	22	33
MED	4	7.5	15	40
RO	6	2.5	4.5	7

[a]Refers to the number of previous research studies used to estimate the statistical parameters shown. The mode was estimated from central values reported in relevant previous studies [12,14,15,19,26-32]. [b]The minimum and maximum values were determined as the lowest and highest reported energy use for the particular technology in previous studies.

water processed by the two technologies at the given desalination unit. We represent E_{MSF} and E_{MED} as triangular probability distributions with the parameters (determined from the review of previous studies) shown in Table 1.

To calculate the total amount of energy actually used to produce 1 m^3 of water at each plant, we divided the amount of energy required to produce water ($\text{Energy}_{\text{water}}$) by an efficiency factor. The efficiency factor accounts for energy lost when converting fossil fuel to useable energy and energy gained by reusing waste heat. The efficiency factor is computed from the following equation:

$$\text{Eff} = \frac{E_{\text{Gross}}(\text{kW h}) + E_{\text{wasteheat}}(\text{kW h})}{\text{totalfuel}(\text{MMBtu}) \times 293.1\left(\frac{\text{kW h}}{\text{MMBtu}}\right)} \tag{3}$$

where E_{Gross} is the gross amount of energy produced by the plant, $E_{\text{wasteheat}}$ represents the energy content of waste heat used in water production, and totalfuel is the energy content of the fuel consumed by the plant. Therefore, the total amount of energy actually used to produce 1 m^3 of water (denoted $\text{Energy}_{\text{water,total}}$) is given by

$$\text{Energy}_{\text{water,total}} = \frac{\text{Energy}_{\text{water}}}{\text{Eff}} \tag{4}$$

Atmospheric emissions

The fuel mix at each desalter determines the mass of pollutants emitted per unit volume of fuel consumed. The calculation of emissions per unit of fuel is based on the U.S. Environmental Protection Agency (U.S. EPA) estimates for the different boiler types in use at each desalting unit [33]. Based on U.S. EPA emission factors for each boiler-fuel combination, we calculate statistical emission factors (g/MMBtu) for each plant. More than 99% of the energy mix burned at these plants is natural gas, with the remainder provided by fuel oil, crude oil, and gas oil [34]. To estimate the total emissions for each plant attributable to residential water production under each scenario, we multiply the emission rate (g/kW h) for each plant by the total energy ($\text{Energy}_{\text{water,total}}$, in kW h/$m^3$) used for water production and by the total amount of water produced to meet residential demand in Abu Dhabi City in 2008.

Results and discussion

We present results for three scenarios of decreased water use:

- financial incentives alone,
- a decrease in per capita consumption (via a package of demand-side management incentives) to levels observed in other developed countries or to those in Abu Dhabi in 2000, and

- a decrease in water loss in the distribution system.

Although we express financial incentives as a price increase, a more politically feasible option might be to offer cash rebates to households that decrease their volumetric water use from prior years or use less than some predetermined threshold volume. Table 2 summarizes atmospheric pollutant emissions from the IWPPs under the baseline scenario as well as under these alternative scenarios.

Scenario 1: financial incentives

As noted above, the actual elasticity of water in Abu Dhabi in response to a long-term price signal is unknown because of a lack of previous studies in Abu Dhabi. Thus, we consider three elasticity scenarios. First, under the most conservative assumption (i.e., assuming the smallest decrease in water use in response to a financial incentive), using projections from Abu Qdais and Al Nassay [8], per capita consumption decreases by a relatively small amount: from 544 to 500 l/c/day. These figures suggest an approximately 1% decrease in emissions of all atmospheric pollutants from IWPPs. Assuming that water is slightly less inelastic at –0.27, per capita consumption would decline to 390 l/c/day, resulting in a reduction of approximately 4% of IWPP emissions for all pollutants. When the price elasticity is –0.37, water consumption drops to 350 l/c/day, cutting emissions by approximately 5%.

Scenario 2: package of incentives

If Abu Dhabi City could establish incentives (whether price signals, information campaigns, rebate programs, water reuse programs, or other initiatives) sufficient to curtail per capita consumption to levels observed in the year 2000 (323 l/c/day), results suggest an approximately 6% reduction of atmospheric pollutant emissions. Because Australia consumes large volumes of water and has regions that suffer from physical water scarcity, it can serve as another benchmark for comparison. If Abu Dhabi City's per capita consumption levels could be decreased to match those in Australia (490 l/c/day on average), atmospheric emissions would decrease by approximately 1%. If Abu Dhabi City instead consumed water at a level matching that in Spain (320 l/c/day), where the northern region is approaching water scarcity, emissions would drop by almost 6%. Singapore also relies heavily on reclaiming wastewater and desalination due to its limited water resources. Singapore has a per capita consumption rate of 160 l/c/day, which if matched would reduce emissions from Abu Dhabi IWPPs by about 10%. If Abu Dhabi City were able to decrease consumption to levels observed in the UK (a water-rich country using only minimal water, 150 l/c/day), emissions would decrease by approximately 11%.

Scenario 3: decreased water loss

If water consumption is assumed constant but less water is lost during distribution, emissions will also decrease. Typically, in high-income countries, about 10% to 15% of water produced at a treatment plant is lost somewhere between the plant and the consumer, mostly through leaks in the water distribution system [35]. This figure increases to above 50% in low-income countries [36]. During 2008 in Abu Dhabi, 16% of desalinated water was lost through physical leakage and another 19% from mismanagement (such as illegal and unmonitored connections) [2]. If 5% more water is used (decreasing the loss rate to 30%), approximately 1% fewer atmospheric emissions from IWPPs will follow. If losses are reduced to 25%, emissions would be 2% lower. A loss rate of 20% corresponds to an approximately 3% reduction in emissions. The model shows that every 5% decrease in water loss results in a 1% emission reduction.

Comparison of options

All scenarios of reduced water consumption lead to reductions in atmospheric emissions (Table 2). Each scenario changes just one parameter (water financial signals, per capita water use, or water loss) while holding all other parameters constant. Based on these results, the effectiveness of different water management options in reducing atmospheric emissions can be compared. Also, our model can be used to estimate emission reductions achievable under combinations of scenarios.

At a long-term price elasticity of –0.07, implementing financial incentives (such as rebates or water charges) that reflect the full price of producing and distributing residential water decreases emissions by amounts comparable to emission reductions achieved by decreasing the amount of unaccounted-for water to 30% (from its current level of 35%). These expected reductions are similar to the reductions predicted to occur if water consumption is equivalent to that in Australia. If the elasticity is actually –0.37 (the lowest value assumed in this analysis), indicating a stronger response to financial signals, then financial incentives would lead to consumption levels similar to those of Spain or of the UAE in 2000.

If Abu Dhabi desires to achieve water consumption levels consistent with those in the most water-efficient developed countries, then a combination of demand-side management practices will be required. Financial incentives and the patching of leaks in the distribution system do not result in water use reductions substantial enough

Table 2 Annual pollutant emissions from IWPPs (tons/year) under the baseline and alternative residential water use scenarios

	NO_x	SO_2	CO_2	PM_{10}	CO
Baseline (current residential water use)	59,300 (26,700-94,800)	500 (310–700)	23,481,700 (23,481,600-23,481,700)	1,500 (680–2,200)	10,500 (5,500-15,500)
Scenario 1 [a]					
−0.07 [a]	750 (320–1,300)	6.1 (3.3-9.6)	295,000 (212,000-389,000)	18 (8.1-32)	130 (58–230)
	1.2% (0.9%-1.6%)	1.2% (0.8%-1.6%)	1.2% (0.9%-1.6%)	1.2% (0.9%-1.6%)	1.2% (0.9%-1.6%)
−0.27 [a]	2,500. (1,100-4,100)	20 (11–30)	974,009.9 (769,569.6-1,163,634)	61.62 (27.26-99.09)	436.4 (197.6-717.2)
	4.1% (3.2%-4.9%)	3.9% (2.8%-5.1%)	4.1% (3.2%-4.9%)	4.1% (3.2%-4.9%)	4.1% (3.2%-4.9%)
−0.37 [a]	3,100 (1,400-5,300)	26 (15–39)	1,257,000 (993,000-1,494,000)	80 (35–130)	560 (250–920)
	5.3% (4.2%-6.3%)	5.1% (3.6%-6.6%)	5.3% (4.2%-6.3%)	5.3% (4.2%-6.3%)	5.3% (4.2%-6.3%)
Scenario 2					
UAE 2000 (323 l/c/day)	3,500 (1,500-5,900)	29 (17–43)	1,407,000 (1,109,000-1,658,000)	89 (40–140)	630 (290–1,000)
	5.9% (4.7%-7.0%)	5.7% (4.0%-7.4%)	5.9% (4.7%-7.0%)	5.9% (4.7%-7.0%)	5.9% (4.7%-7.0%)
Australia (490 l/c/day)	870 (380–1,400)	7.1 (4.1-11)	344,000 (271,000-406,000)	22 (9.8-35)	150 (70–250)
	1.4% (1.1%-1.7%)	1.3% (0.9%-1.8%)	1.4% (1.1%-1.7%)	1.4% (1.1%-1.7%)	1.4% (1.1%-1.7%)
Spain (320 l/c/day)	3,600 (1,500-6,000)	29 (17–44)	1,428,000 (1,125,000-1,682,000)	90 (40–150)	640 (290–1,000)
	6.0% (4.7%-7.1%)	5.8% (4.0%-7.5%)	6.0% (4.7%-7.1%)	6.0% (4.7%-7.1%)	6.0% (4.7%-7.1%)
Singapore (160 l/c/day)	6,200 (2,700-10,000)	50 (29–75)	2,448,000 (1,929,000-2,884,000)	150 (70–250)	1,100 (500–1,800)
	10% (8.2%-12%)	10% (6.9%-13%)	10% (8.2%-12%)	10% (8.2%-12%)	10% (8.2%-12%)
UK (150 l/c/day)	6,300 (2,800-11,000)	51 (30–77)	2,512,000 (1,979,000-2,959,000)	160 (72–260)	1,100 (510–1,800)
	11% (8.4%-12%)	10% (7.1%-13%)	11% (8.4%-12%)	11% (8.4%-12%)	11% (8.4%-12%)
Scenario 3					
30% [b]	630 (280–1,000)	5.1 (2.963-7.655)	248,000 (197,000-292,000)	16 (6.9-25)	110 (51–180)
	1.1% (0.8%-1.2%)	1.0% (0.7%-1.3%)	1.1% (0.8%-1.2%)	1.1% (0.8%-1.2%)	1.1% (0.8%-1.2%)
25% [b]	1,200 (520–2,000)	9.5 (5.5-14)	462,000 (367,000-546,000)	29 (13–47)	210 (94–340)
	2% (1.6%-2.3%)	1.9% (1.3%-2.4%)	2% (1.6%-2.3%)	2% (1.6%-2.3%)	2% (1.6%-2.3%)
20% [b]	1,600 (740–2,800)	13 (7.8-20)	650,000 (516,000-767,000)	41 (18 66)	290 (130–480)
	2.8% (2.2%-3.3%)	2.7% (1.9%-3.4%)	2.8% (2.2%-3.3%)	2.8% (2.2%-3.3%)	2.8% (2.2%-3.3%)
16% [b]	2,000 (890–3,300)	16 (9.4-24)	784,000 (622,000-926,000)	50 (22–80)	350 (160–570)
	3.3% (2.7%-3.9%)	3.2% (2.3%-4.1%)	3.3% (2.7%-3.9%)	3.3% (2.7%-3.9%)	3.3% (2.7%-3.9%)
15% [b]	2,100 (920–3,400)	16 (9.8-25)	815,000 (647,000-963,000)	52 (23–83)	370 (170–600)

Table 2 Annual pollutant emissions from IWPPs (tons/year) under the baseline and alternative residential water use scenarios *(Continued)*

	3.5% (2.8%-4.1%)	3.3% (2.4%-4.3%)	3.5% (2.8%-4.1%)	3.5% (2.8%-4.1%)	3.5% (2.8%-4.1%)
10%	2,400 (1,100-4,100)	20 (12–30)	963,000 (764,000-1,140,000.)	61 (27–97)	430 (200–700)
	4.1% (3.3%-4.8%)	3.9% (2.8%-5.1%)	4.1% (3.3%-4.8%)	4.1% (3.3%-4.8%)	4.1% (3.3%-4.8%)

Numbers at the top line of each cell show annual reduction in tons of emissions (tons/year); numbers in parentheses represent 95% confidence intervals of these estimates. Numbers at the bottom line of each cell show the percentage by which total emissions from all IWPPs would be reduced under the scenario (with 95% confidence intervals in parentheses). Scenario 1, implement financial incentives equal to cost of water production (US \$2.00/m^3); Scenario 2, implement package of demand management programs to match per capita consumption in other countries; Scenario 3, reduce water loss in the distribution system (through leak reduction and improved management). [a] Assumed price elasticity of water. [b] Assumed percentage of water lost during distribution.

to decrease UAE's per capita consumption to the conservatively low levels observed in Singapore or the UK.

Limitations

Emission reductions depend on three major inputs in this model. Emissions per unit of fuel estimates are based on U.S. EPA data, which provide the most extensive estimates available on emissions associated with boiler technologies used in the UAE. In addition, our analyses may have overstated the efficiency of the IWPPs due to overweighting the value of waste heat, causing an underestimate of emissions from water production. Furthermore, our estimate of the energy used to produce water is based on a nine-study meta-analysis using triangular probability distributions to represent the possible range of energy consumption needed to produce potable water for different kinds of desalination plants. A further limitation of our analysis is that it focuses on the production of potable water and does not perform a full water-energy analysis; it does not include distribution, end use (heating water for washing, cooking, or cleaning), or wastewater treatment due to the lack of data. It also includes residential water use only, which accounts for only 44% of total desalinated water use in Abu Dhabi [2]. Including all the other water users, plus conducting a full life cycle analysis of the energy consumed through water use, would result in much greater reductions in emissions from efforts to reduce water use.

Conclusions

Despite the scarcity of water resources in the GCC, the traditional approach to address increased water demand has been to increase water supply by boosting production from desalination plants. The UAE has averaged an 11.7% annual increase in desalinated water production over the past decade [37]. This approach results in unintended consequences: increases in both amounts of energy consumed and amounts of air pollution emitted.

This analysis demonstrates that establishing incentives to reduce residential water demand can result in substantial reductions in water use and air emissions. Charging UAE households full price for water (rather than subsidizing water consumption) could reduce total emissions of particulate matter, NO_2, SO_2, CO, and greenhouse gases from Abu Dhabi's water and power plants by 1% to 5%, depending on assumptions about the price elasticity of water. Reducing water loss in the distribution system from the current high level of 35% to 15% (similar to loss rates in other developed nations) could cut emissions by more than 3%. Demand-side management programs curbing per capita water use to levels typical of Singapore or the UK would curb emissions by 10% (95% CI 8.2% to 12%) or 11% (95% CI 8.4% to 12%), respectively.

Notably, the estimates presented here consider residential water use only, accounting for only 44% of use of desalinated water produced in Abu Dhabi. Of the remaining desalinated water, 33% is used for government facilities and schools, 11% is used for commercial facilities, and 11% is used for agriculture [2]. Considering options to reduce demand by other users could yield substantial additional air quality benefits.

In order to build on this analysis, the UAE could extend the calculations to nonresidential water users. An additional step would employ air quality and risk assessment models to characterize the health benefits gained by the estimated reductions in NO_x, PM_{10}, CO, and SO_2 emissions that occur when water demand is curbed. Additional research in this area could enhance understanding on how the reduction of water consumption results in health benefits due to improved ambient air quality. Ambient air quality was recently ranked as the highest environmental health risk in the country [7], which in itself may encourage new research to understand how reducing emissions from specific sources could benefit air quality. Other important topics for future research include studies on the effects of financial incentives (rebates, retrofit subsidies, water reuse programs, public information campaigns, water rationing, and water restriction policies) on water demand.

The reduction in water demand will play a key role in the development of Abu Dhabi City over the next 20 years as it grows to over 3 million residents. Using excess resources today to produce water to meet extremely high consumption rates comes at a cost to ambient air quality. The value of improved air quality should be accounted for when setting policies associated with water consumption, especially when large portions of energy-intensive potable water are lost during distribution.

Abbreviations

ADDC: Abu Dhabi Distribution Company; CO: Carbon Monoxide; CO_2: Carbon Dioxide; GCC: Cooperation Council for the Arab States of the Gulf; IWPP: Independent water and power plant; MED: Multi-effect distillation; MSF: Multi-stage flash distillation; NO_x: Nitrogen Oxides; PM: Particulate Matter; RO: Reverse Osmosis; UAE: United Arab Emirates; U.S. EPA: U.S. Environmental Protection Agency.

Competing interests

The authors declare that they have no competing interests.

Authors' contributions

NBD developed the probabilistic model. NBD and JMG drafted the manuscript. Both authors read and approved the final manuscript.

Authors' information

NBD received his master of science in environmental engineering from the University of North Carolina at Chapel Hill in 2010 and started his doctorate in the same year. JMG received a dual Ph.D. degree in Civil and Environmental Engineering and Engineering and Public Policy from Carnegie Mellon University; an M.S. in Environmental Science in Civil Engineering from the University of Illinois; and a B.A. in Mathematics from Bryn Mawr College.

Acknowledgements

Thanks are due to the Environment Agency-Abu Dhabi for their support of this research. In particular, we wish to thank H. E. Majid Al Mansouri, H. E. Razan Al Mubarak, Dr. Ahmed Bashir, and Dr. Frederic Launay. Financial support for this research was provided by a Progress Energy Fellowship from the University of North Carolina at Chapel Hill.

References

1. World Bank: A water sector assessment report on the countries of the cooperation council of the Arab States of the gulf. Report no.32539-MNA World Bank, Washington, D.C. (2005)
2. Environment Agency-Abu Dhabi: Water policy and master plan, Emirate of Abu Dhabi. Environment Agency-Abu Dhabi, Abu Dhabi (2008)
3. Regulation and Supervision Bureau: Water and electricity consumption by residential customers. Regulation and Supervision Bureau, Abu Dhabi (2008)
4. Mohamed, H, Hmaeid, B: Influence of turbulent mixing intensity on the MILD combustion and the pollutant formation. Int. J. Energy Environ. Eng. **3**, 22 (2012). doi:10.1186/2251-6832-3-22
5. Li, Y, Gibson, JM, Jat, P, Puggioni, G, Hasan, M, West, JJ, Vizuete, W, Sexton, K, Serre, M: Burden of disease attributed to anthropogenic air pollution in the United Arab Emirates: estimates based on observed air quality data. Sci. Total Environ. **408**, 5784–5793 (2010). doi:10.1016/j.scitotenv.2010.08.017
6. MacDonald Gibson, J, Brammer, A, Davidson, C, Folley, T, Launay, FJP, Thomsen, JTW: Environmental Burden of Disease Assessment: A Case Study in the United Arab Emirates. Springer, Dordrecht (2013)
7. Willis, HH, Gibson, JM, Shih, R, Geschwind, S, Olmstead, S, Hu, J, Curtright, AE, Cecchine, G, Moore, M: Prioritizing environmental health risks in the UAE. Risk Anal **30**, 1842–1856 (2010). doi:10.1111/j.1539-6924.2010.01463.x
8. Abu Qdais, HA, Al Nassay, HI: Effect of pricing policy on water conservation: a case study. Water Policy **3**, 207–214 (2001)
9. Abu Dhabi Distribution Company Financial Reports, Energy and Water, 2004 to 2008. Abu Dhabi Distribution Company, Abu Dhabi (2008)
10. Abu Dhabi Urban Planning Council: Plan Abu Dhabi 2030: urban structure framework plan. Abu Dhabi Urban Planning Council, Abu Dhabi (2007)
11. Siddiqi, A, Anadon, LD: The water-energy nexus in the Middle East and North Africa. Energy Policy **39**, 4529–4540 (2011)
12. Raluy, RG, Serra, L, Uche, J, Valero, A: Life-cycle assessment of desalination technologies integrated with energy production systems. Desalination **167**, 445–458 (2004)
13. Palazzo, P: Thermal and mechanical aspect of entropy-exergy relationship. Int. J. Energy Environ. Eng. **3**, 4 (2012)
14. Raluy, R, Serra, L, Uche, J: Life-cycle assessment of water production technologies–part 1: life-cycle assessment of different commercial desalination technologies (MSF, MED, RO). Int. J. Life Cycle Assess. **10**, 285–293 (2005)
15. Stokes, JR, Horvath, A: Energy and air emission effects of water supply. Environ. Sci. Technol. **43**, 2680–2687 (2009)
16. Alameddine, I, El-Fadel, M: Stack emissions from desalination plants: a parametric sensitivity analysis for exposure assessment. Desalination **177**, 15–29 (2005)
17. Rubio-Maya, C, Uche, J, Pacheco-Ibarra, JJ: Feasibility analysis of a combined cooling-heatingpower and desalted water plant for a non-residential building. Int. J. Energy Environ. Eng. **2**, 33–41 (2011)
18. Renwick, ME, Green, RD: Do residential water demand-side management policies measure up? An analysis of eight California water agencies. J. Environ. Econ. Manag. **40**, 37–55 (2000)
19. Darwish, M: Desalting fuel energy cost in Kuwait in view of $75/barrel oil price. Desalination **208**, 306–320 (2007)
20. Whittington, D, Boland, J, Foster, V: Water tariffs and subsidies in South Asia: understanding the basics. The World Bank, Washington, D.C. (2003)
21. Southern Nevada Water Authority (SNWA): SNWA water conservation plan 2004–2009. SNWA, Las Vegas (2004)
22. Griffin, R: Water Resource Economics: The Analysis of Scarcity, Policies, and Projects. The MIT Press, Cambridge (2006)
23. Dalhuisen, J, Florax, R: Price and income elasticities of residential water demand: a meta-analysis. Land Econ. **79**(2), 292–308 (2003)
24. Espey, M, Espey, J, Shaw, W: Price elasticity of residential demand for water: a meta-analysis. Water Resour. Res. **33**(6), 1369–1374 (1997)
25. Nauges, C, Whittington, D.: Estimation of water demand in developing countries: an overview. World Bank Res. Observer **25**(2), 263–294 (2010)
26. Hamed, O: Evolutionary developments of thermal desalination plants in the Arab Gulf region. In: 2004 Beirut Conference. (2004)
27. Darwish, M, Yousef, F, Al-Najem, N: Energy consumption and costs with a multi-stage flashing (MSF) desalting system. Desalination **109**(3), 285–302 (1997)
28. Darwish, M, Al Asfour, F, Al-Najem, N: Energy consumption in equivalent work by different desalting methods: case study for Kuwait. Desalination **152**, 83–92 (2003)
29. Zander, A, Elimelech, M, Furukawa, D, Gleick, P: Desalination: A National Perspective. National Research Council, Washington, D.C. (2008)
30. Buros, O: The ABCs of desalting. International Desalination Association, Topsfield (2000)
31. Lattemann, S, Höpner, T: Environmental impact and impact assessment of seawater desalination. Desalination **220**, 1–5 (2008)
32. Al-Shammiri, M, Safar, M: Multi-effect distillation plants: state of the art. Desalination **126**, 45–59 (1999)
33. U.S. Environmental Protection Agency: AP-42: compilation of air pollutant emissions factors, stationary point and area sources, vol. 1, 5th edn. U.S. EPA, Washington, D.C (1995). With updates in 2009
34. Environment Agency-Abu Dhabi: Waste and pollution sources in Abu Dhabi Emirate. Environment Agency-Abu Dhabi, Abu Dhabi (2006)
35. Yepes, G, Dianderas, A: Water & Wastewater Utilities: Indicators. World Bank, Washington, D.C. (1996)
36. Worthington, A, Hoffman, M: An empirical survey of residential water demand modeling. J. Econ. Surv. **22**(5), 842–871 (2008)
37. Abu Dhabi Water and Electricity Company: Statistical reports water and energy 2008. Abu Dhabi Water and Electricity Company, Abu Dhabi (2008)

Laboratory evaluation of surface amendments for controlling greenhouse gas emissions from beef cattle feedlots

Orlando A Aguilar[1*], Ronaldo Maghirang[2], Steven L Trabue[3], Charles W Rice[4] and Larry E Erickson[5]

Abstract

Pen surface amendments for mitigating emissions of greenhouse gases (GHGs), such as nitrous oxide (N_2O), methane (CH_4), and carbon dioxide (CO_2), from beef cattle feedlots, were evaluated under controlled laboratory conditions. Amendments were organic residues (i.e., sorghum straw, prairie grass, woodchip), biochar from those organic residues and from beef cattle manure, and activated carbon. Manure samples were collected from several randomly selected pens from two beef cattle feedlots in Kansas and used in the experiments, either as dry (0.10 $g \cdot g^{-1}$ wet basis water content) or moist (0.35 $g \cdot g^{-1}$ wet basis). For each amendment, four different treatment levels (i.e., amounts of material) were placed on top of manure samples in glass containers and analyzed for GHG emissions using a photo-acoustic infrared multi-gas analyzer. From measured concentrations, emission rates were determined. For the dry manure conditions, all amendment materials showed significant reduction of N_2O and CO_2 emissions compared to the control (i.e., no amendment). For the moist manure conditions, none of the amendments showed significant effects on GHG emissions during the first 8 days; at days 10 and 15 after application, however, the biochar materials performed significantly better than the control (i.e., no surface amendment) in reducing N_2O and CH_4 emissions. No significant difference was observed in GHG emissions when the amendments were placed on top or mixed within the top surface layer of the manure.

Keywords: Biochar; Feedlot emission; Greenhouse gas emission; Greenhouse gas control; Organic residue; Surface amendment

Background

Animal feeding operations (AFOs), such as open beef cattle feedlots, emit a variety of air pollutants, including particulate matter (PM), ammonia (NH_3), odor, and volatile organic compounds (VOCs) that have the potential to cause health problems to workers and neighbors. In addition, they emit greenhouse gases (GHGs), including carbon dioxide (CO_2), methane (CH_4), and nitrous oxide (N_2O) [1], and their contribution to climate change is a growing environmental concern [2]. Global increases in anthropogenic CO_2 concentrations are largely due to fossil fuel use and industrial processes. For CH_4, increases have come through both industrial and agricultural activities, whereas increases in N_2O are primarily from agricultural activities with soil management as its main source [3,4]. Ruminant livestock operations are considered significant contributors to global CH_4 concentration [3,4].

Numerous factors can influence the emission of GHGs from beef cattle feedlots. Woodbury et al. [5] reported that pen surfaces in cattle feedlots were aerated and highly organic and favorable for both mineralizing and nitrification, while the sub-surfaces were compacted with anaerobic zones making them susceptible to denitrification. This co-existence of both nitrification and denitrification processes in tandem have also been reported to occur with manure composting [6,7]. However, the process of denitrification and subsequent N_2O emission is highly variable with surface water content controlling surface emission flux rates [8]. This observation is also supported by Woodbury et al. [5], who found that denitrifying enzyme activity was highly variable both

* Correspondence: orlando.aguilar@utp.ac.pa
[1]Department of Mechanical Engineering, Technological University of Panama, Panama, Panama
Full list of author information is available at the end of the article

seasonally and spatially. Aguilar [8] reported that moist/muddy surface conditions (0.52 $cm^3 \cdot cm^{-3}$) had the largest median emission flux of 2.03 $mg \cdot m^{-2} \cdot h^{-1}$ compared to either dry or flooded conditions with median fluxes ranging from 0.10 to 0.16 $mg \cdot m^{-2} \cdot h^{-1}$. The highly variable nature of emissions from feedlot surfaces has also been shown to occur for other compounds, including NH_3, VOCs, and CO_2 [9].

Considering that cattle feedlots could potentially contribute significantly to global N_2O levels, some mitigation strategies must be undertaken to reduce emissions. Nitrification inhibitors have been suggested as a possible mitigation strategy and research has shown the effectiveness of these compounds [10-14]. Nitrification inhibitors are chemicals used to reduce the rate ammonium converts to nitrate and they have been shown to be effective in reducing N_2O emissions from soils. However, costs and animal safety issues make the use of nitrification inhibitors less attractive for AFOs. Alternatively, Adams et al. [15] reported that the manipulation of manure carbon/nitrogen (C:N) ratio by direct application of organic matter to the pen surfaces, might be an effective mitigation strategy used to decrease N losses. Consequently, soil amendments that can change the C/N ratio might be a cost-effective means of reducing GHG emissions from cattle feedlots.

This study was conducted to evaluate the effectiveness of surface amendments (i.e., organic residues and biochar) in reducing emission of GHGs from feedlot manure. The effects of manure water content (i.e., dry vs. moist) and the means of application of the amendment (i.e., topical vs. mixed) were also considered.

Methods

Experiments

A series of laboratory experiments was conducted to evaluate the effectiveness of pen surface amendments in reducing emissions of N_2O, CH_4, and CO_2 from feedlot manure. Amendments were organic residues (i.e., sorghum straw, prairie grass, woodchip), biochar from those organic residues and from beef cattle manure, and activated carbon. Table 1 summarizes the experimental parameters for the different experiments. Approximately 130 kg of manure was collected from several randomly selected pens in two beef cattle feedlots in Kansas. The collected manure was completely mixed and air dried for several days until the average gravimetric water content (wet mass basis) reached approximately 0.10 $g \cdot g^{-1}$. Large clods were removed manually from the dry manure and from the amendment materials. The dry manure and amendment materials were sieved using an ASTM E-II No 4 (4.75 mm) standard sieve. For each amendment, four different amounts of material were applied on top of manure samples within glass containers and analyzed for GHG emissions using a photo-acoustic infrared multi gas analyzer.

The elemental composition of each organic residue and biochar is summarized in Table 2. The elemental composition was measured by grinding a sample (30 g) of each material to 0.5 mm using a sample mill (Model 3010–018, Udy Corp., Fort Collins, CO, USA) and then analyzing each ground sample (2 to 3 mg ± 0.001 mg) in an Elemental Analyzer (Model 2400, Series II Perkin Elmer, Norwalk, CT, USA).

Experiment 1 - topical application of organic residues and biochar on moist manure

Experiment 1 considered the effects of topical application of organic residues and biochar on moist manure. Samples were prepared by mixing 238 g of dry manure (0.10 $g \cdot g^{-1}$ water content wet basis) and 92 g of water at 22°C in 1-L wide-mouth glass containers, which were used as static flux chambers (SFCs). The water content of the moist manure was 0.35 $g \cdot g^{-1}$ wet basis, which is similar to the average manure water content for pen surfaces observed in the field [8]. The moist manure was then compacted at 1.1 $g \cdot cm^{-3}$. Containers were kept uncapped in an enclosed space at approximately constant humidity and temperature for stabilization purposes during a period of 12 h before treatment application (Figure 1b). In this experiment, amendments were woodchip, sorghum straw, prairie grass, woodchip biochar, sorghum straw biochar, and prairie grass biochar (Figure 1a). Biochars were obtained from gasification of the organic residues in a laboratory updraft reactor.

The prepared glass containers were randomly selected and fixed amounts (treatments) of the amendment were applied on top of the compacted manure within the containers, as indicated in Table 1. The containers with manure but without any amendment served as the control. The amendment materials showed large differences in wet bulk density (Table 1). Due to the small volume of the containers (1 L) used in the experiment, treatments were designed such that each would result in the same headspace volume for all amendment materials. Therefore, for each amendment material, treatments consisted of different thicknesses of the material applied on top of the manure within the containers. There were four treatments (i.e., 0 mm or control, 1 mm, 3 mm, and 5 mm), with three replicates each. The amount of amendment corresponding to each treatment was computed based on the actual wet bulk density of each amendment. As a consequence of the different wet bulk densities, the same treatment for different amendments required different masses per unit surface area (Table 1).

The initial gas sampling of each container was performed 45 min after treatment application. Before

Table 1 Experimental parameters

Experiment	Manure conditions (wet basis)		Amendment			
	Water content ($g \cdot g^{-1}$)	Bulk density ($g \cdot cm^{-3}$)	Material	Wet bulk density ($g \cdot cm^{-3}$)	Treatment	
					Amount (mm or g)	Mass/Surface area ($kg \cdot m^{-2}$)
1 Topical application of organic residues and biochar on moist manure[a]	0.35	1.1	WC	0.36	0, 1, 3, 5 mm	0, 0.36, 1.08, 1.80
			SS	0.14	0, 1, 3, 5 mm	0, 0.14, 0.42, 0.71
			PG	0.12	0, 1, 3, 5 mm	0, 0.12, 0.35, 0.58
			WCB	0.41	0, 1, 3, 5 mm	0, 0.41, 1.23, 2.05
			SSB	0.16	0, 1, 3, 5 mm	0, 0.16, 0.48, 0.79
			PGB	0.14	0, 1, 3, 5 mm	0, 0.14, 0.43, 0.71
2 Topical application of biochar and activated carbon on moist manure[a]	0.35	1.1	WCB	0.41	0, 1, 3, 5 mm	0, 0.41, 1.23, 2.05
			SSB	0.16	0, 1, 3, 5 mm	0, 0.16, 0.48, 0.79
			PGB	0.14	0, 1, 3, 5 mm	0, 0.14, 0.43, 0.71
			PMB	1.02	0, 1, 3, 5 mm	0, 1.02, 3.05, 5.08
			LMB	0.53	0, 1, 3, 5 mm	0, 0.53, 1.59, 2.65
			EAC	0.62	0, 1, 3, 5 mm	0, 0.62, 1.86, 3.09
			PAC	0.38	0, 1, 3, 5 mm	0, 0.38, 1.14, 1.89
3 Topical application of organic residues and biochar on dry manure[b]	0.10	0.48	WC	0.36	0, 1, 3, 5 mm	0, 0.36, 1.08, 1.80
			SS	0.14	0, 1, 3, 5 mm	0, 0.14, 0.42, 0.71
			PG	0.12	0, 1, 3, 5 mm	0, 0.12, 0.35, 0.58
			WCB	0.41	0, 1, 3, 5 mm	0, 0.41, 1.23, 2.05
			SSB	0.16	0, 1, 3, 5 mm	0, 0.16, 0.48, 0.79
			PGB	0.14	0, 1, 3, 5 mm	0, 0.14, 0.43, 0.71
4 GHG emission from organic residues and biochars[c]	-	-	WC	0.36	0, 10 g	0, 1.5
			SS	0.14	0, 10 g	0, 1.5
			PG	0.12	0, 10 g	0, 1.5
			WCB	0.41	0, 10 g	0, 1.5
			SSB	0.16	0, 10 g	0, 1.5
			PGB	0.14	0, 10 g	0, 1.5
			PMB	1.02	0, 10 g	0, 1.5
			LMB	0.53	0, 10 g	0, 1.5
			EAC	0.62	0, 10 g	0, 1.5
			PAC	0.38	0, 10 g	0, 1.5
5 Mixing of organic residues and biochars with manure[a]	0.35	0.60	LMB	0.53	0, 20 g	0, 3.0
			EAC	0.62	0, 20 g	0, 3.0
6 Adsorption as mechanism of GHG mitigation[d]	-	-	WCB	0.41	0, 25 g	0, 3.8
			LMB	0.53	0, 60 g	0, 9.0
			EAC	0.62	0, 50 g	0, 7.5

WC, woodchip; SS, sorghum straw; PG, prairie grass; WCB, woodchip biochar; SSB, sorghum straw biochar; PGB, prairie grass biochar; PMB, pellet manure biochar; LMB, loose manure biochar; EAC, extruded (pellet)-activated carbon; PAC, powder-activated carbon. [a]Control was moist manure with no amendment (0.35 $g \cdot g^{-1}$ gravimetric water content wet basis and 1.1 $g \cdot cm^{-3}$ wet bulk density). [b]Control was dry manure with no amendment (0.10 $g \cdot g^{-1}$ gravimetric water content wet basis and 0.5 $g \cdot cm^{-3}$ wet bulk density). [c]Two controls and one treatment per material. Control 1 was indoor air. Control 2 was moist manure with no amendment (0.35 $g \cdot g^{-1}$ gravimetric water content wet basis and 1.1 $g \cdot cm^{-3}$ wet bulk density). [d]Control was indoor air.

sampling, the headspace of each glass container was flushed with ambient air to ensure that GHG concentrations were at ambient levels [16]. Then, the container being sampled was capped and immediately, air from its headspace was circulated through a photo-acoustic infrared multi-gas analyzer or PIMA (Model 1312, AirTech Instruments, Ballerup, Denmark) [17,18] equipped with optical filters for measuring N_2O, CH_4, and CO_2,

Table 2 Elemental composition of the materials used as surface amendments

Material	Composition (%)				C/N
	Carbon	Hydrogen	Nitrogen	Sulfur	
Woodchip (WC)	46.85	6.13	0.53	1.03	89:1
Sorghum straw (SS)	43.11	5.93	0.92	1.04	47:1
Prairie grass (PG)	44.19	6.07	0.90	1.10	49:1
Woodchip biochar (WCB)	59.82	2.42	0.94	0.40	64:1
Sorghum straw biochar (SSB)	58.38	1.63	1.13	0.28	52:1
Prairie grass biochar (PGB)	63.32	2.51	1.65	0.43	38:1
Pellet manure biochar (PMB)	10.39	0.46	1.05	0.31	10:1
Loose manure biochar (LMB)	14.13	0.51	0.92	0.25	15:1
Pellet-activated carbon (EAC)	83.44	0.43	0.83	0.41	101:1
Powder-activated carbon (PAC)	86.69	0.53	0.54	0.06	162:1

and water vapor. The PIMA was connected by two 0.5-m-long Teflon tubes as inflow and outflow to the glass container, as shown in Figure 1c. Readings of the head-space concentrations were taken every 50 s from 0 to 10 min. Gas emissions were determined for each container. Sampling was conducted once a day for 3 days within a 5-day period (i.e., days 1, 3, and 5, with the day of treatment application serving as day 1). During this period, containers were kept uncapped in an enclosed space at approximately constant humidity and temperature conditions (Figure 1b). For each sampling day, the laboratory air temperature and pressure were recorded with liquid-in glass thermometer and barometer, respectively. Manure gravimetric water content and temperature from each container were also measured during the sampling period.

Experiment 2 - topical application of biochar and activated carbon on moist manure

Based on results from Experiment 1, Experiment 2 was conducted to further evaluate the effectiveness of different biochars in mitigating GHG emission from moist manure. Sample preparation, treatments, and the experimental setup were similar to those for Experiment 1. However, sampling was conducted once a day for 6 days within a 15-day period (i.e., days 1, 3, 5, 8, 10, and 15). In addition to the biochars in Experiment 1 (i.e., woodchip, sorghum straw, and prairie grass biochars), Experiment 2 also included manure biochar and activated carbon as amendments.

Experiment 3 - topical application of organic residues and biochar on dry manure

Experiment 3 was conducted to evaluate the effectiveness of several amendment materials in mitigating GHG emissions from dry manure. In this experiment, 238 g of dry manure (0.10 $g \cdot g^{-1}$ water content wet basis and 0.55 $g \cdot cm^{-3}$ wet bulk density) were placed into 1-L glass containers; no water was added. Amendments and treatments were the same as those for Experiment 1. Control was dry manure without any amendment. Gas sampling was performed in the same fashion as for Experiments 1 and 2; however, in this case, sampling was conducted once a day for 3 days within a 5-day period (i.e., days 1, 3, 5).

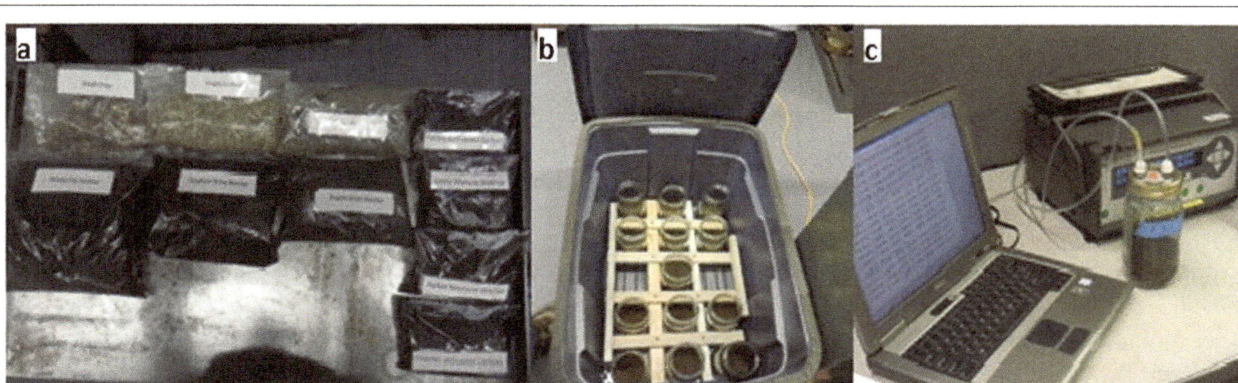

Figure 1 Photographs of the experiment. (a) Amendment materials, **(b)** glass containers with the compacted moist manure, within a plastic container with water at the bottom to maintain constant manure water content, and **(c)** measurement set up.

Experiment 4 - GHG emission from the organic residues and biochars

To assess the potential contribution of the amendments to the GHG emission, 10 g of each amendment material were placed in 1-L glass containers. Treatments were the organic residues and biochars. Two controls were considered: (1) empty containers and (2) containers containing moist manure with gravimetric water content of 0.35 $g \cdot g^{-1}$ wet basis and bulk density of 1.1 $g \cdot cm^{-3}$. Each treatment had two replications. All amendment materials used in the previous experiments were assessed (Figure 1a). Gas samples were collected once a day for 3 days within a 5-day period (i.e., days 1, 3, and 5).

Experiment 5 - mixing of biochars with manure

Experiment 5 was conducted to evaluate the effectiveness of biochars in mitigating GHG emissions when mixed within the top manure layer. Amendment materials to mitigate GHG emissions from pen surfaces in beef cattle feedlots are meant to be placed on the pen surfaces; however, with animal activity, some of the amendments are expected to be mixed with the top surface layer of the moist and loose areas of the pen, while others will remain on the top of the harder and drier pen surfaces. Fixed amounts of manure and water, as described in experiment 1, were mixed in the 1-L glass containers within 2 min (0.35 $g \cdot g^{-1}$ wet gravimetric water content and 0.66 $g \cdot cm^{-3}$ wet bulk density). As soon as each manure sample was prepared, the treatment was mixed within the first 5-cm top layer in the container. Treatments included 20 g of manure biochar and 20 g of activated carbon. Control was moist manure without any amendment. There were two replications for each treatment. Gas samples were collected once a day for 4 days within a 10-day period (i.e., days 1, 3, 5, and 10).

Experiment 6 - mechanism of GHG emission reduction

Experiment 6 was conducted to determine if gas adsorption is a possible mechanism in mitigating GHG emission from pen surfaces. In this experiment, 500-cc glass containers were used as sealed chambers, in which, 150 cc of standard N_2O gas (3.5 ppm) were injected into the container. Amendment materials (treatments) were woodchip biochar, loose manure biochar, and activated carbon. They were first oven-dried at 125°C for 12 h to desorb any trace gases. During sampling, 25, 60, and 50 g of woodchip biochar, manure biochar, and activated carbon, respectively, were placed into the containers and capped with a lid prepared for syringe sampling. The control treatment was an empty container with indoor air. There were two replications for each treatment. The amendments' masses were computed to allow a headspace volume of 400 cc. Within 2 min after treatment preparation, 3-cc air samples were collected from the containers and analyzed for N_2O concentration using a GC (Model GC-14B, Shimadzu Scientific Instrument, Columbia, MD, USA). The GC had a Porapak-Q (80/100 mesh) stainless steel column (3.175×10^{-3} m diameter by 1 m length), electron-capture detector, and UHP/zero nitrogen carrier gas. The oven, injector, and detector temperatures were 60°C, 100°C, and 300°C, respectively, as described by Bremer [19]. The first sample was considered as the base line N_2O concentration for each treatment. As soon as the first sample was collected, 100 cc of air were extracted from the containers and then, 150 cc of N_2O 3.5 ppm standard gas were injected into each container. In this manner, a low positive pressure was always present in the containers even after the final sampling event. Ten minutes after the N_2O standard gas injection, a second 3-cc headspace air sample was collected from each container and analyzed in the GC for N_2O concentration. Fifty minutes later, gas sampling was repeated. After the first day, headspace gas sampling was then repeated once a day for 4 more days within a 6-day period (i.e., days 2, 3, 4, and 6).

As soon as the last gas samples were taken, the containers were placed into an oven (Model OV-500B-1, Blue M Electric Co., Blue Island, IL) and heated to reach different temperatures (35°C, 40°C, and 75°C). Each temperature setting was kept for 2 h and then, 3-cc gas samples were taken from the containers and analyzed in the GC for N_2O gas concentration. Finally, once the final temperature was reached and gas samples collected, the oven was turned off, letting the containers to cool down to room temperature (23°C) for 24 h. A final gas sample was collected from each container and analyzed in the GC.

Data analysis

The emission flux from each container was calculated from mass balance, that is

$$F = k(V/A)(\Delta C/\Delta t), \tag{1}$$

where k is a conversion constant, V is headspace volume, A is surface area, and $\Delta C/\Delta t$ is the change in gas concentration with time within the enclosed space. In general, the relationship between gas concentration and time within the enclosed space was linear; as such, the slope (S) of the linear regression between gas concentration and time was used to represent $\Delta C/\Delta t$ ($ppm \cdot min^{-1}$). For experiments involving moist manure (experiments 1, 2, and 5), the average R^2 values ± standard deviation for the linear relationship between gas concentration and time were 0.99 ± 0.01, 0.99 ± 0.01, and 0.90 ± 0.20 for N_2O, CO_2, and CH_4, respectively.

In general, the gas emissions depended on their previous day's flux. To account for the correlation between emission flux values, the Autoregressive One, AR(1), structure was used on the residuals. Data were analyzed using Proc Glimmix of SAS [20] using a 5% level of significance. *P* values were adjusted by Tukey [21]. When the Type III test of fixed effects indicated no significant (treatment) × (time) interaction, the treatment effects were analyzed and compared to the control. When (treatment) × (time) interaction was significant, analysis was done based on sampling days.

Results and discussion

Experiment 1 - topical application of organic residues and biochar on moist manure

For each treatment and control, emissions of N_2O and CH_4 increased with sampling day; the increase was generally much higher from sampling days 3 to 5 than from samplings day 1 to 3. The emissions of CO_2, on the other hand, did not change much with sampling day. Figure 2 shows results of GHG emissions from moist manure amended with woodchip and woodchip biochar. Statistical analysis showed significant (treatment) × (time) interactions, as such, comparison of treatments with control was based on sampling days. In general, topical application of the organic residues on the manure sample showed some reduction, although not significant, in emissions. However, the biochars at 3- and 5-mm levels significantly reduced GHG emissions at sampling day 5; CO_2 emissions were also significantly reduced on day 5 even with a 1-mm amendment of woodchip biochar (Figure 2).

Experiment 2 - topical application of biochar and activated carbon on moist manure

Similar to experiment 1, there were significant (treatment) × (time) interactions. As such, treatment effects were analyzed based on sampling days. Topical application of 3 mm and 5 mm of loose manure biochar (Figure 3b) and pellet manure biochar (Figure 3c) showed similar effects as pellet-activated carbon (Figure 3a) in reducing N_2O emissions after day 10. For all sampling days, N_2O emission flux from the control was larger than those from the 3-mm and 5-mm treatments of both manure biochars, as also occurred with the three treatments of pellet-activated carbon (Figure 3a). However, in the case of manure biochars, those differences were significant only on day 15, while activated carbon significantly reduced N_2O emissions starting at day 10. Powder-activated carbon, as expected, showed the same behavior in GHG reduction as pellet-activated carbon. The 3-mm treatment of loose manure biochar and pellet manure biochar, compared to the control, reduced N_2O emissions by 63% and 57%, respectively, on day 15. The reduction was slightly lower than that from the 3-mm treatment of activated carbon, which had a reduction of 73%. Reductions of N_2O emissions by activated carbon and manure biochar increased with sampling day (Figure 3a, b, c). Moreover, with the exception of activated carbon, the 1-mm treatment did not result in any significant reduction in N_2O emission flux possible because of poor surface covering.

The 3-mm and 5-mm treatments of woodchip biochar also resulted in significant reduction of N_2O starting at day 10 (Figure 3d). Moreover, 5-mm treatment of prairie grass biochar also reduced N_2O emissions on day 15 (Figure 3f). No treatment of sorghum straw biochar showed any significant effect on N_2O emissions (Figure 3e). Among biochar materials, the 3-mm treatment of loose manure biochar was best in reducing N_2O emissions from the moist manure on day 15 (Figure 3b).

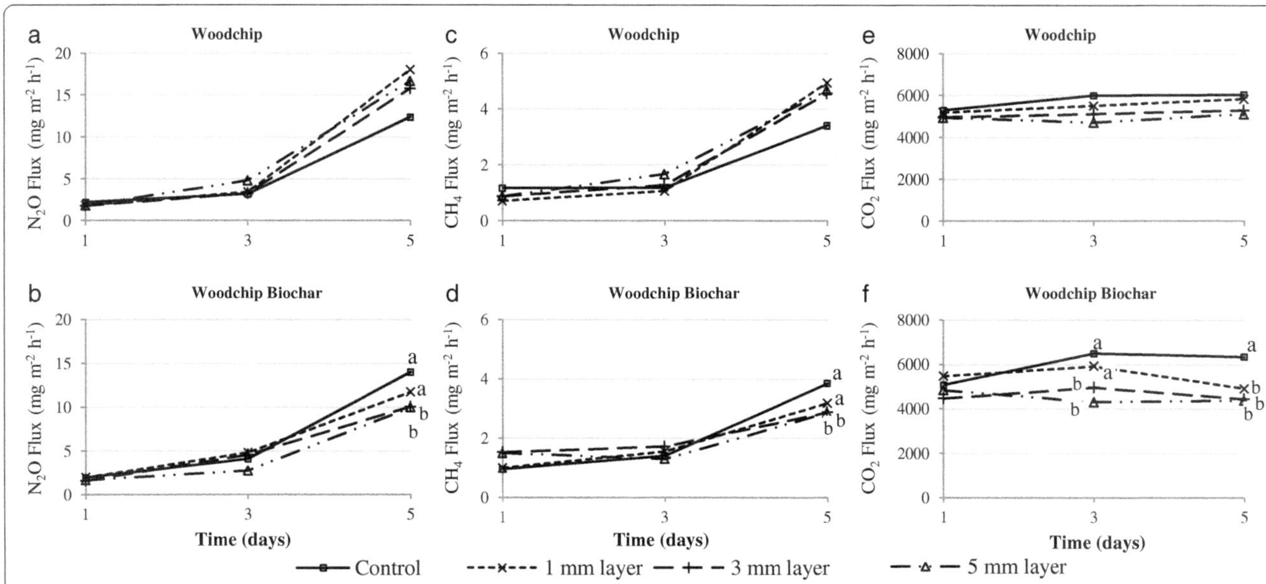

Figure 2 Effects of topical application of woodchips and woodchip biochar on greenhouse gas emissions from moist manure. **(a)** effect of WC in N_2O flux, **(b)** effect of WCB in N_2O flux, **(c)** effect of WC in CH_4 flux, **(d)** effect of WCB in CH_4 flux, **(e)** effect of WC in CO_2 flux, and **(f)** effect of WCB in CO_2 flux. Within the same day, treatments with the same letter or those with no letters are not significantly different at $\alpha = 5\%$.

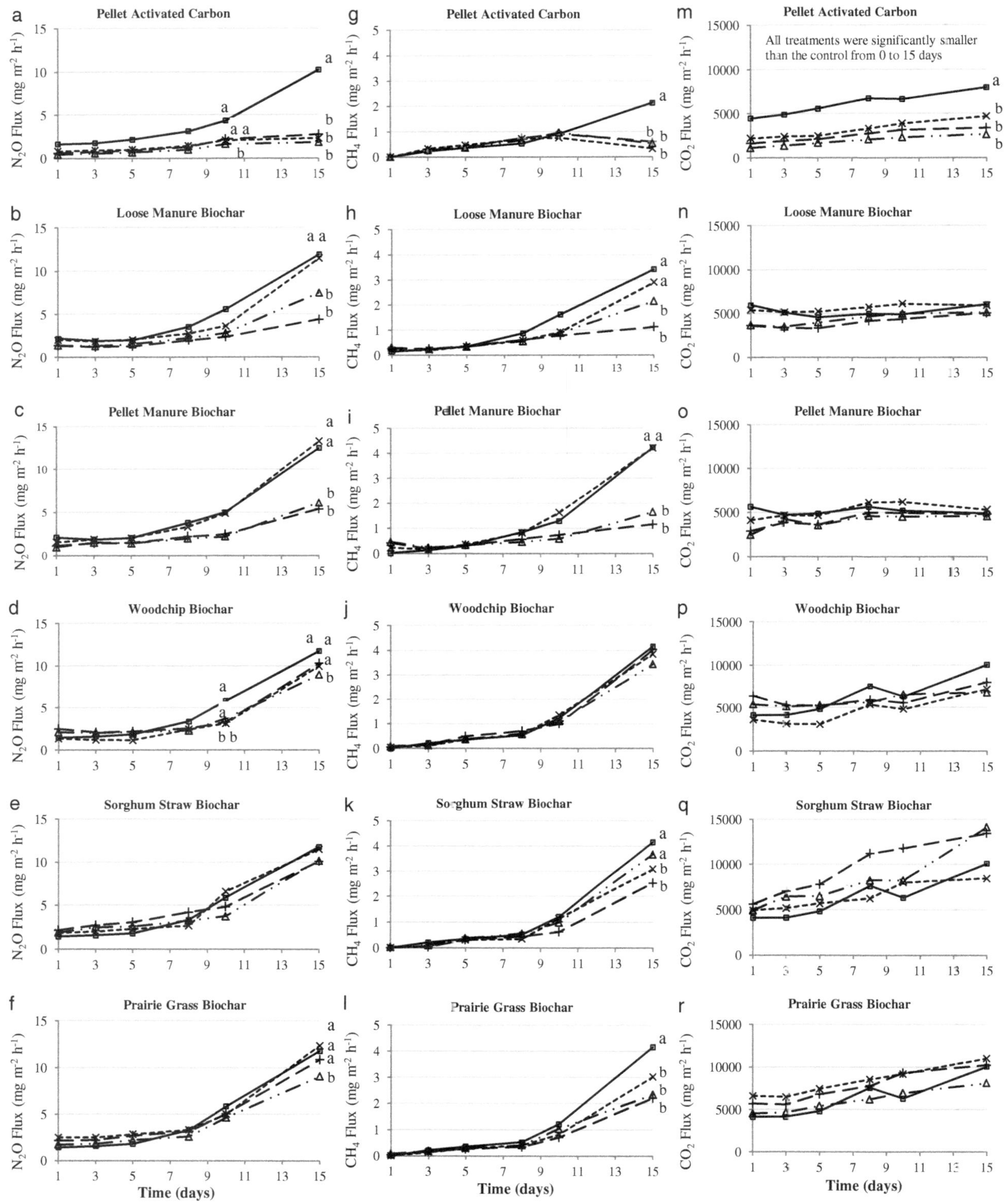

Figure 3 Effects of topical application of biochar and activated carbon on GHG emissions from moist manure. **(a)** effect of EAC in N_2O flux, **(b)** effect of LMB in N_2O flux, **(c)** effect of PMB in N_2O flux, **(d)** effect of WCB in N_2O flux, **(e)** effect of SSB in N_2O flux, **(f)** effect of PGB in N_2O flux, **(g)** effect of EAC in CH_4 flux, **(h)** effect of LMB in CH_4 flux, **(i)** effect of PMB in CH_4 flux, **(j)** effect of WCB in CH_4 flux, **(k)** effect of SSB in CH_4 flux, **(l)** effect of PGB in CH_4 flux, **(m)** effect of EAC in CO_2 flux, **(n)** effect of LMB in CO_2 flux, **(o)** effect of PMB in CO_2 flux, **(p)** effect of WCB in CO_2 flux, **(q)** effect of SSB in CO_2 flux, **(r)** effect of PGB in CO_2 flux. Within the same day, treatments with the same letter or those with no letters are not significantly different at $\alpha = 5\%$.

The reduction in N_2O emission with biochar-amended manure is not surprising since previous research on biochar have reported significant reductions in N_2O emissions from soils [22,23]. Taghizadeh-Toosi [24] reported reductions in N_2O fluxes by as much as 70% for pasture soils following the incorporation of 3 $kg \cdot m^{-2}$ of biochar into the soil. In addition, others have shown that emissions of N_2O decreased as soil was amended with increased amounts of biochar [25,26]. Even in rice paddy soils amended with biochar, there was a significant reduction (51%) in total N_2O emission, but higher levels of biochar amendments did not necessarily decrease N_2O emission rates [27,28].

Effects of application of biochar on CH_4 emissions were generally similar to those of N_2O emissions. Application of 3 mm and 5 mm of manure biochars showed significant reductions of CH_4 emissions compared to the control on day 15 (Figure 3h, i). All three treatments of activated carbon (Figure 3g) and prairie grass biochar (Figure 3l) resulted in significant reduction of CH_4 emission on day 15. The 1-mm and 3-mm treatments of sorghum straw biochar (Figure 3k) also showed significant reduction of CH_4 emission on day 15. The 3-mm treatment of activated carbon showed significant reduction of CH_4 emission on day 15 at 72% compared to the control treatment, while pellet manure biochar, loose manure biochar, sorghum straw biochar, and prairie grass biochar had significant reductions of 73%, 63%, 39%, and 47%, respectively, on day 15. There was no significant reduction in CH_4 emission treated with woodchip biochar. This is supported by Aguilar-Chávez et al. [22] who also did not find any significant effect on CH_4 emissions due to application of biochar to agricultural soils.

In the case of CO_2, the three treatments of activated carbon (Figure 3m) resulted in significant reduction of CO_2 emissions during the whole experimental period, with the larger reduction obtained from the largest amount of activated carbon placed on the moist manure surface, i.e., 5-mm treatment (3.09 $kg \cdot m^{-2}$). The 5-mm treatment of woodchip biochar (Figure 3p) also significantly reduced CO_2 emissions but only at day 15. No other material/treatment combination significantly influenced CO_2 emissions from moist manure.

Differences in CO_2 emissions from soil amended with several biochars have been reported as a result of the differences in the biochars used [22]. Cayuela et al. [23] reported that biochar, used as soil amendment, was the most stable residue with the lowest CO_2 loss with respect to the total C added. Rogovska et al. [26] reported that biochar sequestered large amounts of highly stable C, but either increased or decreased CO_2 emissions from the soils, depending on soil characteristics. Scheer et al. [29] reported no significant differences in net fluxes of GHGs between biochar-amended pastures and control plots. In that study, the biochar from cattle feedlot manure was applied at a rate of 1 $kg \cdot m^{-2}$ to a depth of 10 cm and the GHG emission was measured 28 months later.

Experiment 3 - topical application of organic residues and biochar on dry manure

Figure 4 summarizes the emissions from the dry manure samples as affected by application of biochars. As expected, emissions from the dry manure samples were considerably lower than those from the moist manure samples (Figures 2 and 3). Even though emissions were small, all amendment materials showed significant reduction in N_2O and CO_2 emissions. The three treatments of prairie grass and sorghum straw significantly reduced N_2O emissions. For the woodchip biochars, only the 3-mm and 5-mm treatments showed significant effect in reducing N_2O emissions. This might be a consequence of the poor surface area coverage by the 1-mm treatment of woodchip biochar.

In the case of CH_4, only the three treatments of prairie grass and sorghum straw biochars significantly affected CH_4 emissions but only during the first day of the experiment. This result was possibly a consequence of the low emission flux of CH_4 from the substrate due to its low water content. Note that for experiment 3, the manure samples were not compacted so that conditions were likely aerobic, which may significantly reduce denitrification or methanogenesis, the main mechanisms of N_2O and CH_4 formation in the soil, respectively [30,31].

In the case of CO_2, the three treatments of all amendments, with the exception of the 1-mm treatment of woodchip biochar, significantly reduced CO_2 emission flux (Figure 4g, h, i). Under aerobic conditions, most soil microorganism use O_2 as an electron acceptor, releasing CO_2 into the atmosphere [31].

Experiment 4 - GHG emission from the organic residues and biochars

Table 3 summarizes the GHG emissions from the amendment materials, tested without manure, and the control. Emissions from the organic residues and biochars were not significantly different from those for control 1 (indoor air). Also, emissions of all three GHGs from the amendment materials were significantly lower than those from control 2 (moist manure). As such, there was no significant contribution of GHGs from the organic materials and biochars when used as soil amendment.

Experiment 5 - mixing of biochars with manure

Table 4 summarizes the GHG emissions from the control and from the manure biochar and activated carbon treatments. Both manure biochar and activated carbon significantly reduced emissions of N_2O and CO_2

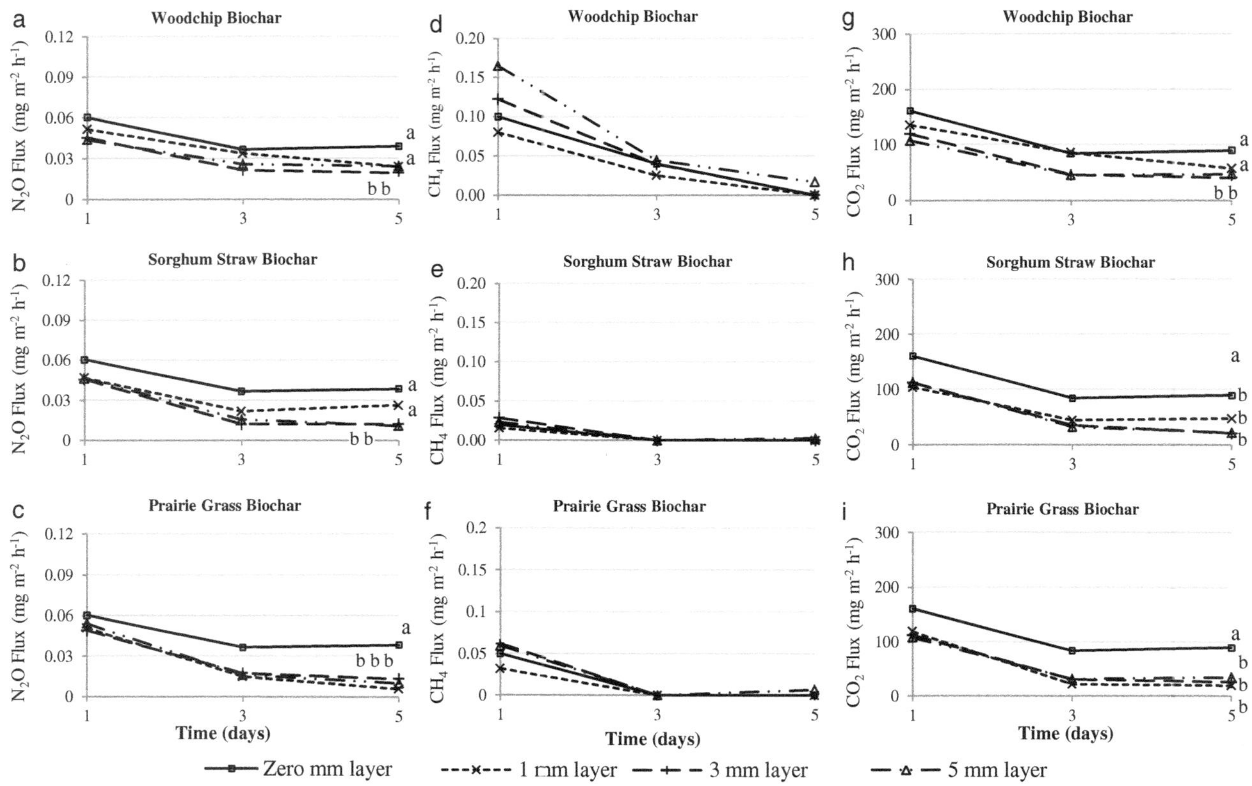

Figure 4 Effects of topical application of biochars on greenhouse gas emissions from dry manure. **(a)** effect of WCB in N_2O flux, **(b)** effect of SSB in N_2O flux, **(c)** effect of PGB in N_2O flux, **(d)** effect of WCB in CH_4 flux, **(e)** effect of SSB in CH_4 flux, **(f)** effect of PGB in CH_4 flux, **(g)** effect of WCB in CO_2 flux, **(h)** effect of SSB in CO_2 flux, **(i)** effect of PGB in CO_2 flux. Within the same day, treatments with the same letter and those with no letters are not significantly different at $\alpha = 5\%$.

compared with the control. There was no significant reduction of CH_4 emission flux; nevertheless, as in experiment 2, at day10, there was significant reduction in CH_4 emission flux for both moist manure treatments compared with the control. The manure biochar showed similar effect as pellet-activated carbon in reducing N_2O emissions (Table 4) and their mitigating effect increased with time. These results suggest that mixing the biochar with the top loose and moist surface layers in the pens would be at least as good as or better than the topical application of the amendments in controlling GHGs from pen surfaces. In this experiment, the manure samples were not compacted, which could help explain the larger fluxes compared with those from experiments 1 and 2.

Table 3 Mean emissions of greenhouse gases from the amendment materials themselves

Treatment	Flux ($mg \cdot m^{-2} \cdot h^{-1}$)		
	N_2O	CH_4	CO_2
Control-1 (indoor air)	0.01 a	0.00 a	0.7 a
Control-2 (moist manure)	6.00 b	2.01 b	5782 b
Loose manure biochar	0.01 a	0.00 a	11.7 a
Pellet-activated carbon	0.03 a	0.00 a	70.3 a
Pellet-manure biochar	0.01 a	0.00 a	2.8 a
Powder-activated carbon	0.03 a	0.00 a	46.8 a
Prairie grass	0.02 a	0.00 a	9.0 a
Prairie grass biochar	0.03 a	0.00 a	43.0 a
Sorghum straw	0.04 a	0.00 a	69.5 a
Sorghum straw biochar	0.04 a	0.00 a	82.0 a
Woodchip	0.02 a	0.00 a	13.8 a
Woodchip biochar	0.02 a	0.03 a	21.5 a

Values are mean fluxes for the 5-day experimental period. Mean values followed by the same letter within a specific GHG are not significantly different at $\alpha = 5\%$.

Table 4 Mean emissions of GHGs under mixed moist manure/amendment condition

Treatment	Flux ($mg \cdot m^{-2} \cdot h^{-1}$)		
	N_2O	CH_4	CO_2
Control (no amendment)	12.05 a	2.05 a	1,2051 a
Loose manure biochar mixed with moist manure	8.71 b	1.52 a	9,151 b
Activated carbon mixed with moist manure	8.11 b	1.58 a	6,735 c

Values are mean fluxes for the 10-day experimental period; column means followed by the same letter are not significantly different at $\alpha = 5\%$.

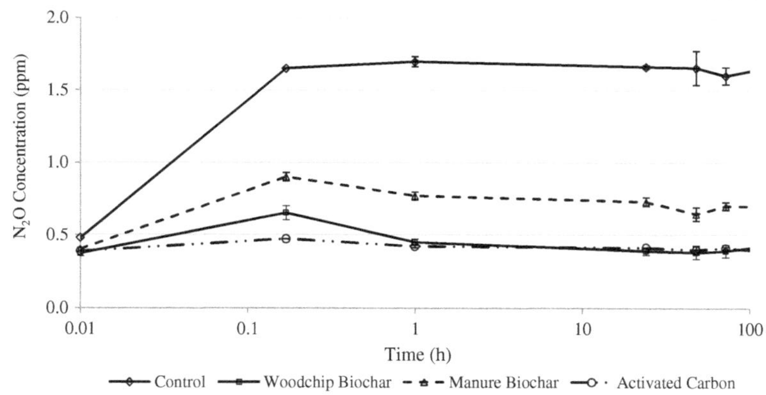

Figure 5 Nitrous oxide concentrations at headspace of containers without (control) and with amendments. A known amount of N_2O was injected into each container at 0.17 h.

Experiment 6 - mechanism of GHG emission reduction

Figure 5 plots the headspace concentration of N_2O in the different containers without (control) and with amendments. For all containers, a known amount of N_2O (i.e., 150 cc, 3.5 ppm) was injected into the containers at 0.17 h. As soon as the N_2O was injected, measured concentration of N_2O for the control (indoor air without any amendment) increased from 0.48 to 1.65 ppm. For containers with treatments (i.e., woodchip biochar, manure biochar, and activated carbon), the increase in concentrations after injection of the same standard N_2O gas was significantly lower. No further significant changes in N_2O concentrations were observed after the first hour of the experiment (Figure 5). The significant difference in N_2O concentration might be a consequence of several mechanisms, including adsorption. When biochar materials are added to the soil, they are able to adsorb organic molecules through several mechanisms [32]. Peng et al. [33] reported activated carbon with high pore volume as a good N_2O adsorbent.

To confirm that adsorption was a possible mechanism for N_2O concentration reduction within the containers, once the 120-h period of gas sampling at room temperature (23°C) was completed, each container still capped was heated to 35°C, 40°C, and 75°C. Then, gas samples were drawn from the container headspace and immediately analyzed for N_2O concentration in a GC. Results indicated that at temperatures higher than room temperature (23°C), the N_2O concentration within the containers increased but remained relatively constant for a given temperature (Figure 6). Apparently, the adsorbed N_2O at room temperature was released at higher temperature, but once that temperature stabilized, there was no any additional desorption.

Figure 7a shows the concentrations of N_2O inside the heated containers with biochars. For each material and control, there was no significant change in N_2O concentrations when samples were heated from 23°C to 40°C. At temperatures larger than 40°C, however, all materials showed significant N_2O desorption. Moreover, N_2O concentrations for the control did not significantly change with increasing temperatures. These results confirm that there was no chemical reaction between the biochar and the N_2O injected into the containers, suggesting that the main mechanism responsible for the increase in N_2O concentration within the containers when heated was desorption. Therefore, adsorption is a possible mechanism for the reduction of N_2O emission from the manure treated with biochar and/or activated carbon in the previous experiments. Once the containers were cooled to room

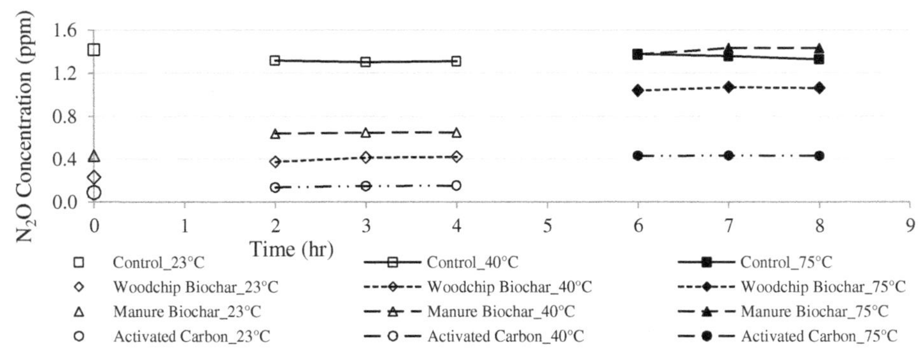

Figure 6 Effect of temperature on N_2O concentrations inside containers with various amendment materials.

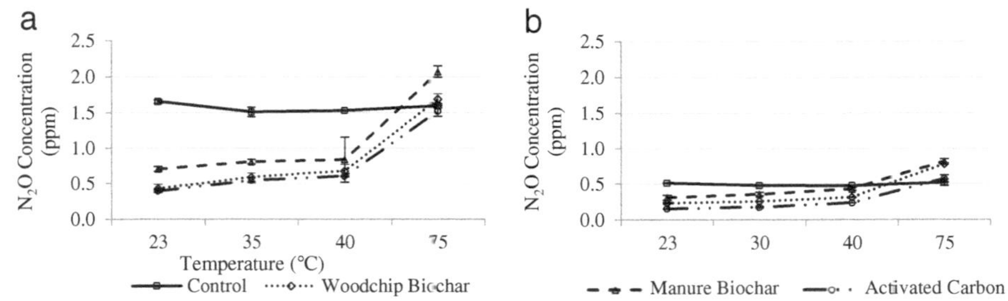

Figure 7 Nitrous oxide concentrations inside containers with amendment materials. **(a)** effect of temperature on desorption of N_2O after injection of 150 cc of 3.5 ppm N_2O standard gas and **(b)** effect of temperature on desorption of N_2O from the amendment materials without injection of N_2O standard gas.

temperature, the final N_2O concentrations decreased to levels comparable to those for the 23°C-to-40°C range.

The higher manure temperature in pen surfaces in a beef cattle feedlot in Kansas during 15 consecutively months was 40.5°C [8]. The lack of significant change in N_2O gas concentration within the containers when temperature rose from 23°C to 40°C is useful because this suggests that the GHGs adsorbed on the amendment materials in the feedlot surfaces would not be desorbed even during the higher summer temperatures. Moreover, the fact that there was a significant difference in N_2O gas concentration within the containers between the treatments (amendment materials) and the control (Figures 5 and 7a) supports the hypothesis that the biochar materials can be used as surface amendments to reduce GHG emissions from pen surfaces of beef cattle feedlots, even though a significant effect in GHG reduction was observed only from days 10 and 15 after biochar application (Figure 3).

As expected, significant N_2O desorption was observed when substrates were heated (Figure 7a); in the interval from 40°C to 75°C, all materials reached the N_2O concentration of the control. This suggests that at 75°C, both biochar and activated carbon have at least released 100% of the N_2O previously adsorbed. Figure 7b represents the N_2O gas desorption from the amendment materials themselves without any external addition of N_2O gas standard into the containers. At 23°C, the N_2O concentration in all containers with the three amendment materials was significantly lower than the one in the control at the same temperature. This finding suggests that the amendment materials adsorbed part of the N_2O concentration present in the indoor air within the containers. When containers were heated from 23°C to 40°C, there was no significant gas desorption from the amendment materials. Once the temperature exceeded 40°C, a larger amount of N_2O was released from the materials, reaching the N_2O concentration of the control. Therefore, because the materials did not react at temperatures between 23°C to 40°C, no gas desorption is expected from the biochars in the field since temperatures did not commonly reach 40°C [8].

The adsorption capacities of selected amendment materials were estimated from the results. From Figure 5, N_2O concentrations within the containers remained relatively constant over time, indicating that the adsorption capacity of the amendment materials might have been reached. Based on the average gas concentration, the adsorption capacity of each material was computed with respect to the control. Estimated adsorption capacities (<0.1 μg N_2O/g of material) were orders of magnitude lower than the reduction in N_2O emission observed from experiment 2 (Figure 3a, b, d). As such, adsorption did not appear to be the main N_2O mitigation mechanism.

From previous studies for soils, other possible N_2O mitigation mechanisms [34] include NH_4^+ immobilization, NO_3^- adsorption, and NH_4^+ adsorption (Figure 8). The NH_4^+ immobilization mechanism is related to the C/N ratios of biochars. Table 2 shows that with the exception of manure biochar, all other amendments have a C/N ratio greater than 20:1, which represents low N content [35]. When the biochars are mixed with the manure, the microorganisms' activity is expected to increase due to the extra C. Because the amendments have low N contents [35,36], the microorganisms will likely immobilize part of the available inorganic N (NH_4^+) in the manure surface. Once the NH_4^+ is used, nitrification and denitrification will decrease, with a net effect of reduced emissions of N_2O [36] from the manure. This is supported by Adams et al. [15], who reported lower N losses in feedlot pens under increased surface manure C/N ratio as result of the application of sawdust on pen surfaces. Other researchers have reported linear relationship between the organic matter content and the amount of N preserved in the manure [37]. Therefore, as manure C content increases, it is expected that less N will volatilize from the manure surfaces.

The N_2O mitigation mechanisms related to NO_3^- adsorption by the biochar (Figure 8) is supported by previous studies on soil, drinking water, and wastewater.

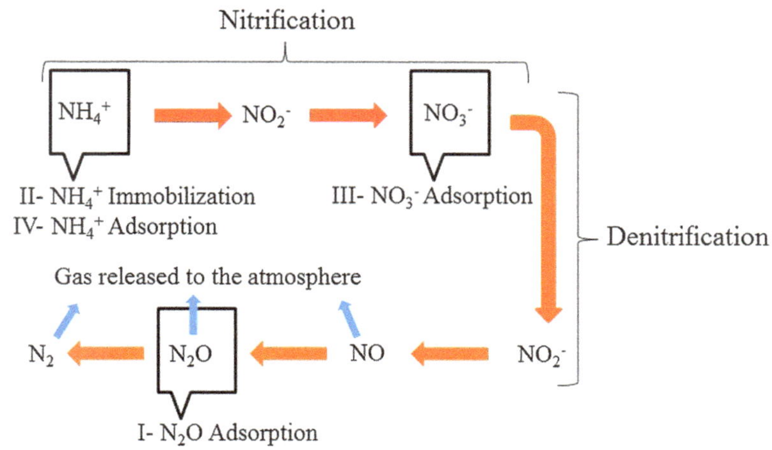

Figure 8 **Possible mechanisms for mitigation of N_2O emission from feedlot manure by biochar or activated carbon.**

Kameyama et al. [38], in a study on the effect of biochar on NO_3^- leaching in soil, reported significant NO_3^- adsorption on the biochar. Yao et al. [39] reported that several biochars significantly adsorbed NO_3^- from soils, with removal rates of up to 3.7%. Nunell et al. [40], in a study of NO_3^- removal from wastewater using activated carbon, reported high NO_3^- adsorption on wood saw dust activated with potassium hydroxide. They reported that a combined effect of carbon surface chemistry (high basic functional groups and low acidic groups) and carbon porous characteristics were responsible for the NO_3^- adsorption. In a study of NO_3^- removal from drinking water, Mizuta et al. [41] reported that the bamboo charcoal was 15% more effective in adsorbing NO_3^- than commercial activated carbon.

As previously indicated, another possible N_2O mitigation mechanism is nitrification inhibition through the adsorption of NH_4^+. Yao et al. [39] reported that nine biochars significantly adsorbed NH_4^+, with removal rates of up to 15.7%.

Based on results from those studies, biochar and activated carbon can adsorb N_2O, NO_3^-, and NH_4^+. If the adsorbed NH_4^+ from manure is not available for microbial activity, nitrification inhibition might result, with a reduction of NO_3^- generation. If NO_3^- is also directly adsorbed onto the biochar and not available for microbial activity, a net denitrification reduction is expected. The net result would be a reduction on N_2O emission rates.

Conclusions

Highlights

- Organic residues and biochar were evaluated in controlling GHGs from feedlot manure.
- Applying organic residues on feedlot manure had limited effects on GHG emissions.
- Applying biochar or activated carbon on feedlot manure reduced GHG emissions.
- Adsorption did not appear to be the primary mechanism for the N_2O emission reduction.

This research evaluated, under controlled laboratory conditions, the effectiveness of application of organic residues, biochar, and activated carbon in controlling emissions of N_2O, CH_4, and CO_2 from beef cattle feedlot manure. The following conclusions were drawn:

1. Topical application of organic residues (i.e., prairie grass, sorghum straw, and woodchip) and biochar on dry manure showed significant reduction of N_2O and CO_2 emissions but did not affect CH_4 emission. When applied on moist manure, the organic residues did not significantly affect GHG emission.
2. Topical application of biochar did not show significant reduction of GHG emissions for the first 8 days. From days 10 and 15, application of biochar materials significantly reduced N_2O and CH_4 emissions compared with the control (i.e., no amendment). Biochar from manure (loose or pelletized) appeared to be the most promising material for reducing GHG emissions from pen surfaces because their effect was similar to that of activated carbon.
3. The method of application of biochar (i.e., topical vs. mixed) did not significantly influence the effectiveness of the material in reducing GHG emissions.
4. Adsorption on biochar or activated carbon appeared to be a mechanism for reducing N_2O emission from feedlot manure; however, other mechanisms (e.g., NH_4^+ immobilization, NO_3^- adsorption, and NH_4^+ adsorption) might be important in N_2O mitigation.

Abbreviations

A: Surface area (m^2); AFOs: Animal feeding operations; ASTM: American Society for Testing and Materials; C/N: Carbon-to-nitrogen ratio; EAC: Extruded activated carbon; *F*: Emission flux ($mg \cdot m^{-2} \cdot h^{-1}$); GC: Gas chromatograph; GHGs: Greenhouse gases; LMB: Loose manure biochar; PAC: Powder-activated carbon; PG: Prairie grass; PGB: Prairie grass biochar; PIMA: Photo-acoustic infrared multi-gas analyzer; PMB: Pellet manure biochar; *S*: Slope of the least square regression line between concentration and time (ppm/min); SFC: Static flux chamber; SS: Sorghum straw; SSB: Sorghum straw biochar; *T*: Temperature (K); *V*: Volume of air within the static flux chamber (L); VOCs: Volatile organic compounds; WC: Woodchip; WCB: Woochip biochar; ΔC: Gas concentration difference (ppm); Δt: Sampling interval (min).

Competing interests

The authors declare that they have no competing interests.

Authors' contributions

OAA designed the experimental study, collected and analyzed the data, and drafted the manuscript. RM conceived the study, participated in its design, and made critical revisions to the manuscript. SLT participated in its design, provided measurement instrumentation, and contributed to the manuscript. CWR participated in the experimental design, and provided measurement instrumentation. LEE participated in the experimental design, and made critical revisions to the manuscript. All authors read and approved the final manuscript.

Acknowledgements

This study was supported in part by the government of the Republic of Panama through SENACYT/IFARHU/Universidad Tecnológica de Panama, Kansas Agricultural Experiment Station, and USDA-NIFA Special Research Grant 'Air Quality: Reducing Air Emissions from Cattle Feedlots and Dairies (TX and KS),' through the Texas AgriLife Research. Special thanks to Dr. Dale Bremer and Dr. Donghai Wang for allowing us to use their equipment and to Dr. Wenqiao Yuan and Arthur James for providing some of the amendment materials.

Author details

[1]Department of Mechanical Engineering, Technological University of Panama, Panama, Panama. [2]Department of Biological and Agricultural Engineering, Kansas State University, Manhattan, KS 66506, USA. [3]USDA Agricultural Research Service, National Laboratory of Agriculture and the Environment, Ames, IA 50011, USA. [4]Department of Agronomy, Kansas State University, Manhattan, KS 66506, USA. [5]Department of Chemical Engineering, Kansas State University, Manhattan, KS 66506, USA.

References

1. Mosier, A, Kroeze, C, Nevison, C, Oenema, O, Seitzinger, S, Van Cleemput, O: Closing the global N_2O budget: nitrous oxide emissions through the agricultural nitrogen cycle. Nutr. Cycl. Agroecosyst. **52**, 225–248 (1998)
2. Stackhouse, KR, Pan, Y, Zhao, Y, Mitloehner, FM: Greenhouse gas and alcohol emissions from feedlot steers and calves. J. Environ. Qual. **40**, 899–906 (2011)
3. Intergovernmental Panel on Climate Change, Fourth Assessment Report: Working group I: The Physical Science Basis. (2007). http://www.ipcc.ch/publications_and_data/ar4/wg1/en/spmsspm-human-and.html. Accessed 7 October 2010
4. Raupach, M, Fraser, P: Chapter 2: Climate and greenhouse gases. CSIRO Publishing (2011). http://dx.doi.org/10.1071/CSIRO_CC_Chapter 2. Accessed 29 June 2012
5. Woodbury, BL, Miller, DN, Nienaber, JA, Eigenberg, RA: Seasonal and spatial variations of denitrifying enzyme activity in feedlot soil. Transaction of the ASABE. **44**, 1635–1642 (2001)
6. Ma, WK, Farrell, RE, Siciliano, SD: Soil formate regulates the fungal nitrous oxide emission pathway. Appl. Environ. Microbiol. **74**, 6690–6696 (2008)
7. Maeda, K, Toyoda, S, Shimojima, R, Osada, T, Hanajima, D, Morioka, R, Yoshida, N: Source of nitrous oxide emissions during the cow manure composting process as revealed by isotopomer analysis of and amoA abundance in betaproteobacterial ammonia-oxidizing bacteria. Appl. Environ. Microbiol. **76**, 1555–1562 (2010)
8. Aguilar, OA: Measurement and control of greenhouse gas emissions from beef cattle feedlots. Dissertation, Kansas State University (2013)
9. Woodbury, BL, Miller, DN, Eigenberg, RA, Nienaber, JA: An inexpensive laboratory and field chamber for manure volatile gas flux analysis. Transaction of the ASABE. **49**, 767–772 (2006)
10. Malla, G, Bhatia, A, Pathak, H, Prasad, S, Jain, N, Singh, J: Mitigating nitrous oxide and methane emissions from soil in rice–wheat system of the Indo-Gangetic plain with nitrification and urease inhibitors. Chemosphere **58**, 141–147 (2005)
11. Menendez, S, Merino, P, Pinto, M, González-Murua, C, Estavillo, J: 3, 4-dimethylpyrazol phosphate effect on nitrous oxide, nitric oxide, ammonia, and carbon dioxide emissions from grasslands. J. Environ. Qual. **35**, 973–981 (2006)
12. Parkin, TB, Kaspar, TC: Nitrous oxide emissions from corn–soybean systems in the midwest. J. Environ. Qual. **35**, 1496–1506 (2006)
13. Weiske, A, Benckiser, G, Ottow, JCG: Effect of the new nitrification inhibitor DMPP in comparison to DCD on nitrous oxide (N_2O) emissions and methane (CH_4) oxidation during 3 years of repeated applications in field experiments. Nutr. Cycl. Agroecosyst. **60**, 57–64 (2001)
14. Weiske, A, Benckiser, G, Herbert, T, Ottow, JCG: Influence of the nitrification inhibitor 3, 4-dimethylpyrazole phosphate (DMPP) in comparison to dicyandiamide (DCD) on nitrous oxide emissions, carbon dioxide fluxes and methane oxidation during 3 years of repeated application in field experiments. Biol. Fertility Soils. **34**, 109–117 (2001)
15. Adams, JR, Farran, TB, Erickson, GE, Klopfenstein, TJ, Macken, CN, Wilson, CB: Effect of organic matter addition to the pen surface and pen cleaning frequency on nitrogen mass balance in open feedlots. J. Anim. Sci **82**, 2153–2163 (2004)
16. Chiyoka, WL, Hao, X, Zvomuya, F, Li, X: Nitrous oxide emissions from chernozemic soils amended with anaerobically digested beef cattle feedlot manure: a laboratory study. Anim. Feed Sci. Technol. **166–67**, 492–502 (2011)
17. Cayuela, ML, Velthof, GL, Mondini, C, Sinicco, T, van Groenigen, JW: Nitrous oxide and carbon dioxide emissions during initial decomposition of animal by-products applied as fertilisers to soils. Geoderma **157**, 235–242 (2010)
18. Predotova, M, Kretschmann, R, Gebauer, J, Buerkert, A: Effects of cuvette surface material on ammonia-, nitrous oxide-, carbon dioxide-, and methane-concentration measurements. J. Plant Nutr. Soil Sci. **174**, 347–349 (2011)
19. Bremer, DJ: Nitrous oxide fluxes in turfgrass. J. Environ. Qual. **35**, 1678–1685 (2006)
20. SAS Institute: SAS/STAT 9.2 User's guide, 9.2. , Cary (2008)
21. Milliken, GA, Johnson, DE: Analysis of Messy Data Vol. 1: Designed Experiments. Taylor & Francis, New York, NY (2009)
22. Aguilar-Chávez, A, Díaz-Rojas, M, Cárdenas-Aquino, M, Dendooven, L, Luna-Guido, M: Greenhouse gas emissions from a wastewater sludge-amended soil cultivated with wheat (triticum spp. L.) as affected by different application rates of charcoal. Soil Biol. Biochem. **52**, 90–95 (2012)
23. Cayuela, ML, Oenema, O, Kuikman, PJ, Bakker, RR, van Groenigen, JW: Bioenergy by-products as soil amendments? Implications for carbon sequestration and greenhouse gas emissions. Global. Change. Biol. Bioenergy. **2**, 201–213 (2010)
24. Taghizadeh-Toosi, A, Clough, TJ, Condron, LM, Sherlock, RR, Anderson, CR, Craigie, RA: Biochar incorporation into pasture soil suppresses in situ nitrous oxide emissions from ruminant urine patches. J. Environ. Qual. **40**, 468–476 (2011)
25. Bruun, EW, Müller-Stöver, D, Ambus, P, Hauggaard-Nielsen, H: Application of biochar to soil and N_2O emissions: potential effects of blending fast-pyrolysis biochar with anaerobically digested slurry. Eur. J. Soil Sci. **62**, 581–589 (2011)
26. Rogovska, N, Laird, D, Cruse, R, Fleming, P, Parkin, T, Meek, D: Impact of biochar on manure carbon stabilization and greenhouse gas emissions. Soil Sci. Soc. Am. J. **75**, 871–879 (2011)
27. Zhang, A, Bian, R, Pan, G, Cui, L, Hussain, Q, Li, L, Zheng, J, Zheng, J, Zhang, X, Han, X: Effects of biochar amendment on soil quality, crop yield and greenhouse gas emission in a Chinese rice paddy: a field study of 2 consecutive rice growing cycles. Field Crop Res. **127**, 153–160 (2012)
28. Zhang, A, Cui, L, Pan, G, Li, L, Hussain, Q, Zhang, X, Zheng, J, Crowley, D: Effect of biochar amendment on yield and methane and nitrous oxide emissions from a rice paddy from Tai Lake Plain, China. Agr Ecosyst Environ **139**, 469–475 (2010)

29. Scheer, C, Grace, PR, Rowlings, DW, Kimber, S, Van Zwieten, L: Effect of biochar amendment on the soil-atmosphere exchange of greenhouse gases from an intensive subtropical pasture in Northern New South Wales. Australia. Plant Soil. **345**, 47–58 (2011)
30. Hofstra, N, Bouwman, AF: Denitrification in agricultural soils: summarizing published data and estimating global annual rates. Nutr. Cycl. Agroecosyst. **72**, 267–278 (2005)
31. Li, C: Quantifying greenhouse gas emissions from soils: scientific basis and modeling approach. Soil Sc. Plant Nutr. **53**, 344–352 (2007)
32. Joseph, SD, Camps-Arbestain, M, Lin, Y, Munroe, P, Chia, CH, Hook, J, Van Zwieten, L, Kimber, S, Cowie, A, Singh, BP, Lehmann, J, Foidl, N, Smernik, RJ, Amonette, JE: An investigation into the reactions of biochar in soil. Soil Research. **48**, 501–515 (2010)
33. Peng, Y, Zhang, F, Xu, C, Xiao, Q, Zhong, Y, Zhu, W: Adsorption of nitrous oxide on activated carbons. J. Chem. Eng. Data **54**, 3079–3081 (2009)
34. Paul, EA (ed.): Soil Microbiology, Ecology, and Biochemistry. Academic Press, Burlington, MA (2007)
35. Barbarick, KA: Organic materials as nitrogen fertilizers. Fact sheet No. 0.546. (2006). http://www.ext.colostate.edu/pubs/crops/00546.pdf. Accessed 25 July 2012
36. Robertson, GP, Groffman, PM: Nitrogen transformation. In: Paul, EA (ed.) Soil Microbiology, Ecology, and Biochemistry, pp. 341–359. Academic Press, Burlington, MA (2007)
37. Erickson, GE, Klopfenstein, TJ: Managing N inputs and the effect on N losses following excretion in open-dirt feedlots in Nebraska. Scientif. World J. **1**, 830–835 (2001)
38. Kameyama, K, Miyamoto, T, Shiono, T, Shinogi, Y: Influence of sugarcane bagasse-derived biochar application on nitrate leaching in calcaric dark red soil. J. Environ. Qual. **41**, 1131–1137 (2012)
39. Yao, Y, Gao, B, Zhang, M, Inyang, M, Zimmerman, AR: Effect of biochar amendment on sorption and leaching of nitrate, ammonium, and phosphate in a sandy soil. Chemosphere **89**, 1467–1471 (2012)
40. Nunell, GV, Fernandez, ME, Bonelli, PR, Cukierman, AL: Conversion of biomass from an invasive species into activated carbons for removal of nitrate from wastewater. Biomass & Bioenergy. **44**, 87–95 (2012)
41. Mizuta, K, Matsumoto, T, Hatate, Y, Nishihara, K, Nakanishi, T: Removal of nitrate-nitrogen from drinking water using bamboo powder charcoal. Bioresour. Technol. **95**, 255–257 (2004)

Impact of biochar on the water holding capacity of loamy sand soil

Ok-Youn Yu[1*], Brian Raichle[1] and Sam Sink[2]

Abstract

With the rise in global population and increased agricultural requirements comes an increasing need for freshwater. Currently, it is estimated that approximately 75% of fresh water consumption is for the growth of agricultural crops, and only 10% to 30% of this water is actually made available to plants. It is widely accepted that farming practices which rely heavily on chemical fertilizers and unsustainable land management practices have led in many regions to infertile sandy soils with reduced water holding capacity and insufficient amounts of organic matter. Combined with increasing global population, the need to better manage fresh water use, particularly agricultural usage, is paramount. The use of biochar as a soil amendment has been suggested as a way to increase water holding capacity, but only limited quantitative studies exist in terms of the effectiveness of biochar in increasing a soil's water holding capacity. The main purpose of this study was to determine the effect of woody biochar amendment (yellow pine from pyrolysis at 400°C) on the water holding capacity of loamy sand soil with different mixture rates. Results show a doubling in water holding capacity by mass using a 9% mixture of biochar (equivalent to 195 metric ton/ha), which is an agriculturally relevant concentration. High percentage mixtures of biochar increase water holding capacity dramatically. These results suggest the use of biochar has potential to mitigate drought and increase crop yields in loamy sand soil.

Keywords: Biochar; Water holding capacity; Loamy sand soil; Pyrolysis

Background

Agriculture is the single largest consumer of fresh water and accounts for about 75% of anthropogenic fresh water use [1]. According to Wallace [1], on average 63% of fresh water applied to agricultural lands is lost to evaporation and runoff. Rainwater and irrigation water that is not absorbed by soils is capable of transporting fertilizers and pesticides into watersheds creating non-point pollution. Due to the increased use of chemical fertilizers and the highly inefficient use of water, there has been an increase in non-point pollution in agricultural areas around the world [2]. With the world's population set to increase by 65% (i.e., 3.7 billion people) by 2050, the additional food required to feed future generations will put further pressure on freshwater resources [1].

The United States Department of Agriculture (USDA) reported that 8.7 million tons of commercial nitrogen fertilizer and an additional 1.1 million tons of animal manure were added to agricultural lands in the USA [3]. Table 1 shows that over 500,000 tons of chemical nitrogen fertilizer and animal manure are annually distributed to agricultural lands in the southeastern USA, the highest areal rate in the country [3]. As seen in Table 2, nearly 50% of nitrogen fertilizer was lost through runoff and leaching [3]. Agricultural land in the southeastern USA is typically comprised of loamy sand soil, whose inability to hold water makes it more prone to fertilizer loss [3].

Biochar, or charcoal produced for agricultural usage, is produced from thermal decomposition of organic material under reduced oxygen conditions at temperatures above 700°C [4]. The microscopic structure of biochar is one of the primary determinants in its soil conditioning properties; the surface area of the pre-charred source material can be increased several thousand fold [5]. This increased surface area is the result of thermal decomposition of the organic material through which volatiles are driven off and the remaining structure is comprised of highly concentrated carbon chains. These chains can take on different organizational patterns based on the production

* Correspondence: yuo@appstate.edu
[1]Appalachian State University, Boone, NC 28608, USA
Full list of author information is available at the end of the article

Table 1 Sources of nitrogen inputs-by region and by crop [3]

	Acres		Commercial fertilizer		Manure		Atmospheric deposition		Bio-fixation		Sum of inputs	
	1,000 s	Percent	Tons	Percent	Tons	Percent	Tons	Percent	Tons	Percent	Tons	Percent
By region												
Northeast	13,642	4.6	388,655	4.5	146,867	13.6	48,523	5.8	1,081,687	10.4	1,665,742	7.9
Northeast Great Plains	72,397	24.3	1,815,130	20.9	154,986	14.3	122,474	14.6	907,910	8.7	3,000,645	14.3
South Central	45,350	15.2	1,290,546	14.8	85,795	7.9	197,007	23.4	1,993,185	19.2	3,566,628	17.0
Southeast	13,394	4.5	423,992	4.9	82,103	7.6	55,854	6.6	468,580	4.5	1,030,529	4.9
Southern Great Plains	32,096	10.8	952,920	11.0	74,517	6.9	61,552	7.3	106,041	1.0	1,195,054	5.7
Upper Midwest	112,581	37.7	3,504,461	40.3	466,355	43.1	344,878	41.0	5,579,239	53.6	9,894,962	47.1
West	9,018	3.0	318,839	3.7	71,619	6.6	10,313	1.2	263,443	2.5	664,220	3.2
All regions	298,487	100.0	8,694,553	100.0	1,082,242	100.0	840,601	100.0	10,400,085	100.0	21,017,780	100
By crop												
Barley	4,635	1.6	171,683	2.0	2,244	0.2	7,313	0.9	0	0.0	181,242	0.9
Bom	78,219	26.2	4,369,865	50.3	552,495	51.1	231,507	27.5	0	0.0	5,153,867	24.5
Corn silage	5,197	1.7	186,760	2.1	298,616	27.6	14,971	1.8	0	0.0	500,345	2.4
Cotton	16,858	5.6	560,237	6.4	12,369	1.1	51,108	6.1	0	0.0	623,746	3.0
Legume hay	24,776	8.3	444,358	5.1	18,299	1.7	65,268	7.8	4,512,759	43.4	5,040,690	24.0
Oats	3,772	1.3	43,934	0.5	852	0.1	9,303	1.1	0	0.0	54,096	0.3
Peanuts	1,843	0.6	17,372	0.2	1,629	0.2	7,207	0.9	63,490	0.6	89,699	0.4
Potatoes	987	0.3	72,952	0.8	1,315	0.1	2,268	0.3	0	0.0	76,535	0.4
Rice	3,637	1.2	190,001	2.2	7	<0.1	15,522	1.8	0	0.0	205,541	1.0
Spring wheat	20,503	6.9	423,081	4.9	4,206	0.4	32,545	3.9	0	0.0	459,961	2.2
Sorghum	10,897	3.7	444,695	5.1	16,465	1.5	27,978	3.3	0	0.0	489,144	2.3
Soybeans	67,543	22.6	143,954	1.7	59,224	5.5	221,487	26.3	5,823,836	56.0	6,248,524	29.7
Winter wheat	45,014	15.1	1,179,798	13.6	26,822	2.5	106,697	12.7	0	0.0	1,313,400	6.2
All crops	298,478	100.0	8,694,553	100.0	1,082,242	100.0	840,601	100.0	10,400,085	100.0	21,017,780	100.0

temperature, with increased temperatures leading to increased organization [4].

The most famous example of agricultural biochar usage is the pre-Columbian Terra Preta soils of the Amazon River basin. The ancient Amazonian people built up the Terra Preta by adding large quantities of charcoal to the soil to amend the nutrient poor soils of the rainforest [6]. Terra Preta is found near large, well-established Amazon Indian villages, suggesting the causal relationship between agricultural stability and population expansion [6].

The practice of adding charcoal to degraded soils was deemed obsolete after the industrialization of chemical fertilizers. However, due to the negative effects of chemical fertilizers on modern agricultural soils and the environment, and the concern for increasing atmospheric carbon, there has been a renewed interest in charcoal-based soil amendment [7]. Much of this interest is focused on biochar. The claims for biochar are many: larger crop yields, decreased fertilizer requirements, greater microbial activity, reductions in greenhouse gas emissions from fields, greater soil water holding capacity, drought mitigation, and increased soil organic carbon content (SOC), which can improve the physical properties of soil. Further, carbon sequestration benefits of biochar soil amendment have been heavily studied [8-11].

Sohi et al. [12] showed that soils with a high water holding capacity produce increased crop yields and a decreased need for irrigation. Singh et al. [13] suggested that the increased porosity of biochar increases water retention in soils, and the enhancement depends on biochar feedstock, soil type, and mixture rates. Nutrients dissolved in the water may also be retained in the soil so plants may be better able to access the nutrients [4]. To promote the use of biochar soil amendment, it is important to understand the mechanism of biochar-amended water retention, to characterize the effects of feedstock, biochar production, soil types, and mixtures, and to quantify these effects on plant growth.

The main objective of this research was to study how different biochar mixture rates affect the water holding capacity of biochar-amended loamy sand soil. This sandy

Table 2 Nitrogen loss estimates-by region and by crop [3]

	Acres	Volatilized		Dissolved in surface water runoff		Dissolved in leachate		Dissolved in lateral subsurface flow		Lost with waterbome sediment		Lost with windborne sediment		Sum of all loss pathways	
	Percent	Tons	Percent	Tons	Percent	Tons	Percent	Tons	Percent	Tons	Percent	Tons	Percent	Tons	Percent
By region															
Northeast	4.6	78,711	2.9	45,275	7.9	46,457	4.7	4,604	7.0	91,141	7.2	1,060	0.4	267,281	4.5
Northern Great Plains	24.3	596,583	21.6	64,928	11.4	36,852	3.7	11,222	17.0	164,394	12.9	131,371	49.4	1,005,360	17.0
South Central	15.2	398,622	14.5	174,590	30.6	304,219	30.5	11,747	17.8	247,942	19.5	9,657	3.6	1,146,779	19.3
Southeast	4.5	170,688	6.2	26,587	4.7	200,291	20.1	5,934	9.0	47,604	3.7	86	<0.1	451,191	7.6
Southern Great Plains	10.8	438,673	15.9	27,384	4.8	61,394	6.1	4,089	6.2	41,002	3.2	103,221	38.8	675,785	11.4
Upper Midwest	37.7	996,009	36.1	159,552	28.0	339,126	34.0	27,060	41.0	665,135	52.4	19,473	7.3	2,206,378	37.2
West	3.0	76,795	2.8	72,026	12.6	10,298	1.0	1,399	2.1	12,301	1.0	1,047	0.4	173,956	2.9
All regions	100.0	2,756,079	100.0	570,341	100.0	998,637	100.0	66,055	100.0	1,269,517	100.0	265,924	100.0	5,926,729	100.0
By crop															
Barley	1.6	52,993	1.9	12,562	2.2	1,765	0.2	1,235	1.9	11,585	0.9	6,659	2.5	86,798	1.5
Corn	26.2	956,074	34.7	124,161	21.8	389,473	39.0	25,796	39.1	562,179	44.3	92,247	34.7	2,149,929	36.3
Corn silage	1.7	46,467	1.7	15,244	2.7	20,805	2.1	2,448	.3.7	56,684	4.5	5,058	1.9	146,705	2.5
Cotton	5.6	101,326	3.7	55,777	9.8	114,922	11.5	2,574	3.9	51,654	4.1	48,920	18.4	375,172	6.3
Grass hay	4.9	59,044	2.1	66,011	11.6	4,287	0.4	1,660	2.5	13,986	1.1	34	<0.1	145,023	2.4
Legume hay	8.3	152,594	5.5	67,735	11.9	5,340	0.5	2,555	3.9	781	0.1	13	<0.1	229,193	3.9
Oats	1.3	18,264	0.7	3,289	0.6	4,435	0.45	512	0.8	14,784	1.2	2,515	0.9	43,797	0.7
Peanuts	0.6	16,915	0.6	3,124	0.5	40,268	4.0	1,637	2.5	3,957	0.3	2,455	0.9	68,355	1.2
Potatoes	0.3	20,253	0.7	8,181	1.4	21,245	2.1	785	1.2	4,006	0.3	0,710	0.6	56,178	0.9
Rice	1.2	11,869	0.4	60,612	10.6	39,659	4.0	177	0.3	13,268	1.0	60	<0.1	125,643	2.1
Spring wheat	6.9	129,671	4.7	21,068	3.7	1,248	0.1	1,666	2.5	48,548	3.8	29,990	11.3	232,189	3.9
Sorghum	3.7	147,017	5.3	10,851	1.9	30,979	3.1	2,495	3.8	36,176	2.8	34,269	12.9	261,785	4.4
Soybeans	22.6	581,091	21.1	90,757	15.9	282,995	28.3	17,786	26.9	352,233	27.7	24,865	9.4	1,349,726	22.8
Winter wheat	15.1	462,504	16.8	30,973	5.4	41,217	4.1	4,733	7.2	99,679	7.9	17,133	6.4	656,238	11.1
All crops	100.0	2,756,079	100.0	570,341	100.0	988,637	100.0	66,055	100.0	1,269,517	100.0	265,924	100.0	5,926,729	100.0

soil is the dominate soil type of the southeastern USA and is recognized as having poor water holding capacity and limited growth potential for crops [3]. More information pertaining to the water holding capacity of biochar will allow for a better understanding of the other attributes which biochar is stated to possess. The use of biochar as a soil amendment in areas prone to drought may increase by better understanding the water holding capacity characteristics of biochar.

Relevant previous work in biochar

The potential benefits of biochar soil amendment are well identified in the literature. These include carbon sequestration, improved crop yields, and enhanced water retention.

The choices of feedstock for biochar production are dependent upon the most readily available biomass as well as the handling mechanism [6,14]. Any organic material can be used to make biochar, including wood, grass, leaves, and manure [4]. It was also suggested by Karhu et al. [15] that it may be possible to develop specialized biochars to meet different needs of the end user. The most important determining factor in regards to the actual affect of these custom tailored biochars would be the feedstock and the temperature used to produce the biochar.

The conversion of biomass carbon to biochar leads to sequestration of about 50% of the initial carbon compared to 3% sequestration from burning and less than 20% from biological decomposition [16]. Biochar is resistant to decomposition and remains in the soil for centuries or millennia. In summary, pyrolysis can transfer 50% of the carbon stored in plant tissue from the active to an inactive carbon pool. The remaining 50% of plant carbon can be used to produce energy in the form of food and fuel. This enables carbon negative energy production if renewable resources are used. Pyrolysis would facilitate bio-energy production and carbon sequestration if the biochar is redistributed to agricultural fields. Lehmann et al. [10] proposed biochar from farm wood-waste as a promising method for integrating carbon sequestration and renewable energy generation with conventional agricultural production. It is clear that biomass conversion sequestration projects have the potential to contribute significantly to climate change mitigation, although they may not be economically attractive at current output production and carbon prices [17].

Biochar has been also shown to reduce the amount of methane (CH4) released from agricultural fields that utilize cover crops as a nutrient supply [15]. In addition, biochar has been shown to improve the environmental needs of Mycorrhizal bacteria in the soil; these microbes are fundamental aspects of a healthy soil bed [18]. It is currently speculated that the increased microbial activity is actually one of the largest determining factors in the positive effects seen in plants after the introduction of biochar.

Since the time of the Amazonian Terra Preta soils, increased crop yield has been a recognized benefit from biochar soil amendment [6]. Biochar is capable of increasing the levels of calcium, potassium, and phosphorus in loamy sand soil [19]. Biochar also proved an adequate medium for immobilization and retention of soluble cadmium and zinc as well as increased pH levels in acidic soils [5]. Biochar's ability to manipulate different nutrients at different rates, as well as raise pH at different intervals was also studied. The results showed that different feedstocks at different temperatures could have these effects and that it may be possible to create biochar to do a very specific task [13].

Chan et al. [20] investigated the effect of biochar produced from green waste on the yield of radish crops. They applied three rates of biochar (10, 50, and 100 t/ha) with and without supplemental nitrogen application of 100 kg, N/ha. Biochar alone did not increase radish yield, but in the presence of supplemental nitrogen, higher rates of biochar application resulted in higher radish yields up to 266%. Major et al. [21] reviewed existing work on the magnitude and dynamics of biochar's effect on nutrient leaching and discussed possible mechanism and processes by which this effect is observed. They observed that nutrient leaching is generally greatest under fertilized row crops such as corn or horticultural corps, and targeting these cropping systems may yield the best results for reducing leaching.

While many articles report on carbon sequestration potential and nutrient trapping, there have been only a few studies on the effect of biochar on water holding capacity. Novak et al. [14] reported an increase in the water holding capacity of a loamy sand soil with 2% mixtures of biochar made from various switchgrass feedstocks. They were interested in understanding the different effects of temperature and feedstock on the water holding capacity of biochar but all values were calculated at a 2% mixture rate only. By varying retort temperature from 250°C to 750°C, increases in water holding capacity ranging from 7% to 16% were observed. Another finding was an 11% increase in water holding capacity reported as an additional observation and was not validated through the use of control techniques [15]. The ability of biochar to increase water holding capacity could have profound effects on areas prone to drought [15]. Sohi et al. [12] summarized the current state of biochar knowledge and concluded that soil water holding capacity was an area of significance that was lacking in research.

Methods

Methodology

The design of this experiment consists of sampling different mixtures of biochar and loamy sand soil by mass to see the effects on the water holding capacity at different mixture rates. Due to the thousands of soil combinations and the large discrepancy from one soil to another, this study focuses on the most common agricultural soil in North Carolina and the southeastern USA as a whole which is loamy sand. The defined name of the soil to be used in this study is outlined by Novak et al. [14] as a loamy sand. This soil is characterized as less than 35% of clay (i.e., particle size is smaller than 0.002 mm) and less than 50% of fine sand (i.e., particle size is between 0.05 and 2 mm) [22]. Loamy sand has limited water holding capacity resulting in increased leaching of nutrients [22]. The use of poor farming practices and degradation from chemical fertilizers is a known factor in the transformation of traditionally fertile soils to the sandy textures commonly found in the southeastern USA [7].

The particle size of the biochar typically used in experiments ranges from 0.25 to 20 mm uniformed sieved size [4,19]. It was also suggested by Lehmann [4] that 2 mm of biochar be the most suitable for application to the agricultural lands as well as transportation. The actual percentages of biochar mixed with sample soils ranged from 0.5 to 91 metric ton/ha, which is approximately up to 4.5% of biochar amendment [14,23]. It was also reported by Jha et al. [23] that negative effects on plant development began to occur in mixtures, greater than 9% biochar to loamy sand by mass (e.g., 90 g/kg).

In this study, the samples were prepared using an unamended sample of loamy sand to establish a baseline measurement. The samples were first subjected to a characterization process where a sieve was used to insure a uniform size, less than 3 mm for the biochar and less than 2 mm for the loamy sand. Next, samples were mixed at different proportions between 0% and 100% of biochar to loamy sand by mass. In order to determine each sample weight, Howard's chart [24] was used. The use of Howard's chart will provide a relevant reference point for replication of the study and add validity to the experiment. Based on Howard's chart, a particle size, smaller than 2 mm, requires 20 g of saturated sample specimen [24].

The American Society for Testing and Materials (ASTM) standard D 2216-10, 'Laboratory determination of water (moisture) content of soil and rock by mass' states that soil samples be dried at (110°C) at relatively low humidity with a thermostatic control capable of maintaining 110°C ± 5°C. It also provides a standard drying time of 12 to 16 h or until percent moisture readings over 1-h periods are less than 0.1%. Drying times may be reduced to 4 h if a forced-draft oven is used [25]. In addition, saturation procedures utilized by Péron et al. [26] present that a saturation period of 1 day will provide homogenization of water content throughout the sample.

Each mixture used in this study was saturated with water by following the procedure found in Péron et al. [26] to establish sample's water holding capacity. Water was slowly applied to each mixture container, while gently agitating, until excess water was observed. The mixtures were then allowed to sit for 24 h to assure homogeneity of water content throughout the sample. After that, the mixtures were drained by gravity for another 24 h through a coffee filter. Three 90-mL stainless steel containers were then tared, filled to two third full, and massed using a 0.01-g digital balance to determine wet mass. The samples were then dried at 110°C for 24 h using a convection oven and remassed to determine the dry mass. The results yielded the amount of water being held by each mixture.

Sample preparation

The biochar used for this experiment was produced by the research using a propane-fueled retort (i.e., pyrolysis). The vessel had an internal temperature sensor and the observer recorded the temperature as the reaction took place. Uncombusted gasses were flared off and the temperature reached 400°C toward the end of the reaction. The feedstock was mostly untreated yellow pine scrap lumber. The biochar (pyrolysis in 400°C for 3 h, C%: 71.2, N%: 0.2, surface area: 0.19 m^2/g) was air dried, crushed by hand, and sieved to a 3-mm particle size. Representative feedstock and raw biochar are shown in Figure 1.

Figure 1 Representative raw yellow pine and resulting biochar from pyrolysis at 400°C.

The soil used in this study is loamy sand. A sample of loamy sand was acquired from the USDA-Agricultural Research Service (ARS) coastal plain soils, water, and plant research center in Florence, South Carolina. The USDA-ARS is a small research laboratory located on a historic tobacco farm and is designated as a creditable source for soil sampling [27]. The soil was air dried and sieved to a 2-mm particle size. Bulk mixtures with mass ratios between 0% and 100% biochar by mass were prepared. Representative samples are shown in Figure 2.

Results and discussion

Data collection and analysis

Collection of data occurred throughout the experiment. To determine the water holding capacity by mass, the following equation was used [25]:

$$\text{Water holding capacity}\,(\%) = \frac{(\text{mass}_{\text{wet}} - \text{mass}_{\text{dry}})}{\text{mass}_{\text{dry}}} \times 100\%$$

Statistical uncertainty (i.e., standard deviation) was determined in the way from the three replicate samples for each mixture. The average water holding capacities as a percentage of dry mass are shown in Table 3. It is noted that metric ton per hectare equivalent concentrations, calculated based on a soil density of 1,440 kg/m^3 and a 15-cm treatment depth, are listed for agricultural comparison. Standard deviation at each proportion is also calculated, and it shows very low variance until 20% of biochar mixture. However, it is noticed that standard deviation gradually increases (up to 9.5%) as the mixture rate increases. Percent increase in water holding capacity from unamended soil is reported, along with the percent increase per percent biochar amendment. Note that unamended sandy loam soil has a water holding capacity of 16%, while pure biochar can hold over 2.7 times (= 270%) its mass of water as shown in Figure 3.

Water holding capacities for biochar concentrations below 10% are shown in Figure 4. Notice that the error bar and the error ranges are also shown in the figure as the measurements repeated three times for each mixture. It was reported by Jha et al. [23] that mixture rates of less than 10% were safe for agricultural purposes to loamy sand soil (i.e., agriculturally meaningful biochar amendment). A clear positive correlation is shown in Figure 4 with doubled water holding capacity with around 9% of biochar amendment. An agriculturally relevant biochar amendment of 5% biochar (around 100 metric ton/ha) results in a water holding capacity of 24%, that is, a 50% increase over unamended soil. This is significant due to the tremendous amount of research, presenting that poor water holding capacity plays a major role in nutrient loss in the southeastern USA [12,13,19,21].

To quantify the *value* of biochar amendment, the increase in water holding capacity per unit of biochar amendment is shown in Figure 5. For biochar concentrations below 10%, higher biochar concentrations provide an increasingly greater enhancement of water holding capacity per unit amendment, suggesting a better value from higher concentrations. Above 10% biochar concentration, however, the increase in water holding capacity per unit of biochar amendment remains constant at around 12%, suggesting no further increase in value above 10% amendment. A possible interpretation is that 10% amendment maximizes the water holding value of biochar.

Conclusions

While previous research suggests that biochar is well suited for increasing the water holding capacity of soils, there have been few quantitative studies on the effect of

Figure 2 Representative prepared samples after sieving and mixing.

Table 3 Experiment results

% biochar by mass	Metric ton/ha equivalent	Average water holding capacity (%)	Standard deviation (%)	% increase	% increase/ % amendment
0	0	16	0.7	-	0
1	21.6	16.8	0.7	5.1	5.1
2	43.2	19	0.1	18.9	9.4
3	64.8	19.3	0.2	20.7	6.9
4	86.4	21.4	0.2	33.9	8.5
5	108	23.5	0.3	47.3	9.5
6	129.6	27.2	0.5	70.4	11.7
7	151.2	24.8	0.3	55.2	7.9
8	172.8	29.5	1.7	84.8	10.6
9	194.4	30.2	1.8	89.1	9.9
10	216	32.3	1.4	102.1	10.2
15	324	44.4	0.6	178	11.9
20	432	50.4	1.3	215.5	10.8
25	540	60.1	3	276.6	11.1
30	648	78.3	8.7	390.2	13
35	756	81	3.5	407.4	11.6
40	864	91.2	2	470.8	11.8
45	972	92.4	1.9	478.8	10.6
50	1,080	124.9	9.3	681.7	13.6
75	1,620	209.6	7.8	1,212.6	16.2
100	2,160	274.1	9.5	1,616.1	16.2

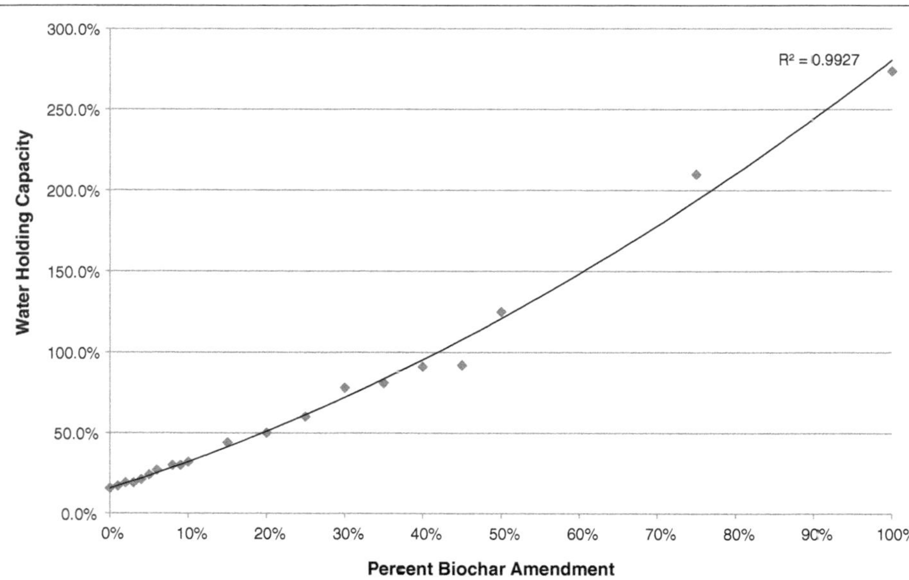

Figure 3 Water holding capacity at different mixture rates varied from 0% to 100%.

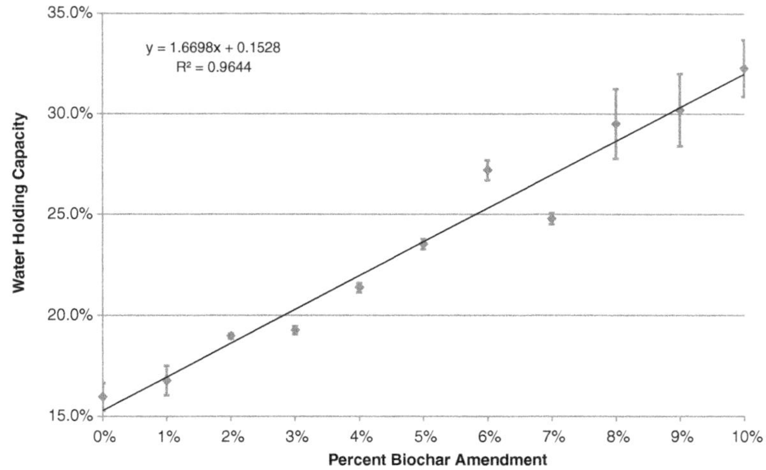

Figure 4 Water holding capacity at different mixture rates varied from 0% to 10%.

biochar on water holding capacity. This research found that the biochar used in this study increases water holding capacity of a loamy sand soil by around 1.7% by mass for each 1% of added biochar over the agriculturally relevant range up to 10% biochar concentrations. These lower end mixture values are representative of the allowable amounts of biochar that can be added without causing damage to plants. Water holding capacity of unamended sandy loam soil, 16%, is doubled by the addition of 9% by mass of biochar. This finding is important because it establishes biochar as an effective medium for increasing irrigation effectiveness, runoff mitigation, and reducing non-point source agricultural pollution.

Biochar's effects on water when mixed with soils are important to understand because it may be the most influential aspect of biochar in regards to microbial activity, plant growth, and nutrient usage. The current practices of irrigation are awaiting a revolution in order to more efficiently provide soils with water, but it will be important to understand the possible impact of preparing soils with amendments such as biochar. The significance of such information may allow for a shift in the conventional wisdoms associated with irrigation as well as fertilization practices.

This study only considered one type of biochar (yellow pine from pyrolysis at 400°C) mixed with one type of soil (loamy sand), and thus further research is clearly indicated. Additional research should address the effects of biochar feedstock and production methods, soil types, particle size, and saturation/drying cycles. The creation of a water release curve for biochar-amended soils would be a logical next step.

To promote the practice of agricultural biochar soil amendment, the full life cycle costs and benefits to biochar soil amendment must be estimated. The effect of water holding capacity on crop growth due to water holding capacity, nutrient retention, and microbial growth must be understood, in addition to the benefits of the likely reduced need for irrigation and fertilizer and pesticide usage. Inclusion of traditionally externalized costs associated with

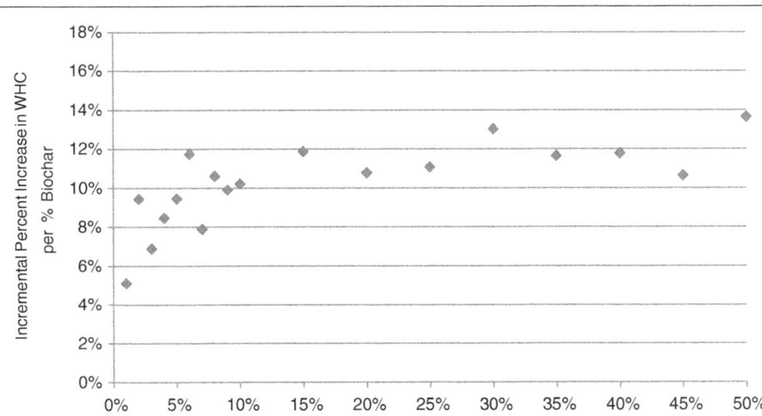

Figure 5 Incremental increase in percent water holding capacity normalized to percent biochar amendment.

carbon and environmental degradation, a side effect of current farming techniques, will further improve the cost/benefit analysis of agricultural biochar usage.

Competing interests
The authors declare that they have no competing interests.

Authors' contributions
OY supervised SS's Master of Science research regarding the biochar experiment and drafted the manuscript. BR participated in the design of the study and drafted the manuscript. SS conducted the biochar experiment for this research and helped to write the manuscript. All authors read and approved the final manuscript.

Acknowledgements
The authors would like to thank Drs. Jim Houser, Dave Domermuth, and Marie Hoepfl in the Department of Technology and Environmental Design at Appalachian State University for their involvement in this project. The authors also would like to acknowledge the US Environmental Protection Agency's P3 program for their partial funding support of this work.

Author details
[1]Appalachian State University, Boone, NC 28608, USA. [2]Strata Solar, Chapel Hill, NC 27517, USA.

References
1. Wallace, J: Increasing agricultural water use efficiency to meet future food production. Agr Ecosyst Environ **82**(1–3), 105–119 (2000)
2. EPA: Protecting Water Quality from Agricultural Runoff. U.S. Environmental Protection Agency, Washington, DC (2005)
3. USDA: Model Simulation of Soil Loss, Nutrient Loss, and Change in Soil Organic Carbon Associated with Crop Production. United States Department of Agriculture, Washington, DC (2006)
4. Lehmann, J, Joseph, S: Biochar for Environmental Management: Science and Technology. Earthscan/James & James, London (2009)
5. Beesley, L, Marmiroli, M: The immobilisation and retention of soluble arsenic, cadmium and zinc by biochar. Environ Pollut **159**(2), 474–480 (2011)
6. Taylor, P: The Biochar Revolution: Transforming Agriculture & Environment. Global Publishing Group Mt Evelyn, Victoria (2010)
7. Chia, CH, Munroe, P, Joseph, S, Lin, Y: Microscopic characterisation of synthetic Terra Preta. Soil Research **48**(7), 593–605 (2010)
8. Gaunt, JL, Lehmann, J: Energy balance and emissions associated with biochar sequestration and pyrolysis bioenergy production. Environ Sci Technol **42**(11), 4152–4158 (2008)
9. Lal, R: Carbon sequestration. Philosophical Transactions of the Royal Society B: Biological Sciences **363**(1492), 815–830 (2008)
10. Lehmann, J, Gaunt, J, Rondon, M: Bio-char sequestration in terrestrial ecosystems - a review. Mitigation and adaptation strategies for global change. **11**(2), 395–419 (2006)
11. McHenry, MP: Agricultural bio-char production, renewable energy generation and farm carbon sequestration in Western Australia: certainty, uncertainty and risk. Agr Ecosyst Environ **129**(1–3), 1–7 (2009)
12. Sohi, S, Lopez-Capel, E, Krull, E, Bol, R: Biochar, climate change and soil: a review to guide future research. CSIRO Land and Water Science Report **5**(09), 17–31 (2009)
13. Singh, B, Singh, BP, Cowie, AL: Characterisation and evaluation of biochars for their application as a soil amendment. Soil Research **48**(7), 516–525 (2010)
14. Novak, JM, Lima, I, Xing, B, Gaskin, JW, Steiner, C, Das, K, Ahmedna, M, Rehrah, D, Watts, DW, Busscher, WJ: Characterization of designer biochar produced at different temperatures and their effects on a loamy sand. Annals of Environmental Science **3**(1), 195–206 (2009)
15. Karhu, K, Mattila, T, Bergström, I, Regina, K: Biochar addition to agricultural soil increased CH4 uptake and water holding capacity - results from a short-term pilot field study. Agr Ecosyst Environ **140**(1), 309–313 (2011)
16. Steiner, C: Biochar Carbon Sequestration. University of Georgia, Biorefining and Carbon Cycling Program, Athens, GA (2008)
17. Cacho, OJ, Wise, RM, MacDicken, KG: Carbon monitoring costs and their effect on incentives to sequester carbon through forestry. Mitig Adapt Strat Glob Chang **9**(3), 273–293 (2004)
18. Warnock, DD, Lehmann, J, Kuyper, TW, Rillig, MC: Mycorrhizal responses to biochar in soil - concepts and mechanisms. Plant Soil **300**(1–2), 9–20 (2007)
19. Novak, JM, Busscher, WJ, Laird, DL, Ahmedna, M, Watts, DW, Niandou, MAS: Impact of biochar amendment on fertility of a southeastern Coastal Plain soil. Soil Sci **174**(2), 105–112 (2009)
20. Chan, K, Van Zwieten, L, Meszaros, I, Downie, A, Joseph, S: Agronomic values of greenwaste biochar as a soil amendment. Soil Research. **45**(8), 629–634 (2008)
21. Major, J, Steiner, C, Downie, A, Lehmann, J: Biochar effects on nutrient leaching. In: Biochar for environmental management: Science and technology, pp. 271–287. Earthscan, London (2009)
22. USDA: Keys to soil taxonomy. United States Department of Agriculture, Washington, DC (2006)
23. Jha, P, Biswas, A, Lakaria, B, Rao, AS: Biochar in agriculture - prospects and related implications. Curr Sci **99**, 1218–1225 (2010)
24. Howard, A: Minimum Test Specimen Mass for Moisture Content Determination. ASTM Geotechnical Testing Journal **12**(1), 39–44 (1989)
25. ASTM: Laboratory Determination of Water (Moisture) Content of Soil and Rock by Mass. In: D 2216 - 10ASTM International: West Conshohocken, PA. (2010)
26. Péron, H, Hueckel, T, Laloui, L: An improved volume measurement for determining soil water retention curves. Geotech Test J **30**(1), 1–8 (2007)
27. USDA-ARS: Coastal Plains Soil, Water, and Plant Research Center. (2013). http://www.ars.usda.gov/Main/docs.htm?docid=3536

An experimental analysis of a stand-alone standing-wave thermoacoustic refrigerator

Giulio Allesina

Abstract

In the last years, the thermoacoustic phenomena had intensified the scientific community's interest due to its implications in thermodynamics, heat transfer, and acoustic matters that make possible the conversion between sound and heat. This work is aimed at exploring the Russell-Weibull refrigerator, analyzing the behavior of the different parts in order to achieve a good level of thermal separation, together with the capability of working as stand-alone. Special attention was given to all those parts that need to be designed properly with explanation of the physics that ruled them, such as sound generator, woofer box, stack, and tube insulation. In the second part of the work, the results obtained are discussed, consistently showing that, if some expedients are adopted, this simple refrigerator can achieve a temperature difference of 24°C. The design choices concerning materials and geometries are shown to be of great influence on the device behavior.

Keywords: Thermoacoustic; Standing wave; Refrigerator

Background

Eleven years have elapsed since the work by Russell and Weibull on tabletop refrigerator had been published [1]. Within this time span, thermoacoustic devices have made much progress, improving technologies and efficiencies [2-5]. Despite this progress, the Russell-Weibull refrigerator still remains the simplest way to demonstrate the charm of thermoacoustic refrigeration and the physics phenomena on which it rests. In the last two decades, thermoacoustic technology has drawn the attention of many research groups that are looking for a different way to obtain cooling [6]. This attention resulted in the development of several refrigerators, from prototypes to industrial applications [7,8]. The fundamental advantages related to this technology are simplicity, compactness and reliability, absence of moving parts (i.e., neither wear nor lubrication), and inert and environmentally friendly working fluid. Moreover, refrigerating power is adjustable as a function of the sound amplitude (the compressor in common refrigerators works just in on-off mode) [9]. However, since the first Hofler's thermoacoustic cooling engine was built [10], these devices have dealt with some limitations, such as low thermal power, low efficiency, the need for an accurate tuning of the system, and noise pollution [5,11]. A great deal of effort was made in different directions to alleviate these problems. The proposed solutions resulted in the realization of traveling wave devices, complex geometry resonators, multiple speakers devices, etc. [3,4,11-14]. These innovations were responsible for the development of thermoacoustic devices ranging from microcooling devices up to 500-gal/day acoustic gas liquefiers [3,4,15,16]. Different solutions allowed to fix or restrain the weak points of this technology, but at this stage, more progress is needed to compete efficiently against common refrigeration plants.

While it is relatively easy to create simple working devices, it is however quite hard to produce efficient refrigerators. Although simple standing-wave refrigerators may seem obsolete compared with such a technically complex outlook, they remain a starting point for the development of more complex devices and test potential improvements.

The aim of this work is to explore the experimental feasibility of a stand-alone thermoacoustic refrigerator. In order to design a stand-alone device, the Russell-Weibull refrigerator was used as basis [1]. The thermoacoustic (T-A) refrigerator was then modified from the original design; furthermore, an oscillator circuit for sound generation was implemented. Furthermore, the woofer box

Correspondence: giulio.allesina@unimore.it
BEELab, Department of Engineering 'Enzo Ferrari', University of Modena and Reggio Emilia, Via Vignolese 905/b, Modena 41125, Italy

design was investigated with particular focus on the methods to increase speaker efficiency. Finally, the apparatus was tested with different stacks: one built as instructed by Russell and Weibull, and one built with parallel glass slides.

The adopted solutions have contributed to the production of a stand-alone device. The temperature difference was monitored for all the tests, and it resulted consistently with the results of Russell-Weibull, despite the modifications.

Methods

The thermoacoustic refrigerator discussed here was designed on the basic scheme of a standing-wave T-A refrigerator [9]. A sketch of the experimental apparatus is displayed in Figure 1, where in the left image, the three fundamental parts of the apparatus are sketched: (A) is the loudspeaker that produces a sound (pressure wave) at the resonance frequency of the (B) resonator tube, in which a standing wave is generated (one end is closed, the other end is connected to the loudspeaker), and (C) is the stack, i.e., an array of parallel channels. Figure 1b shows the refrigerator discussed in this paper. It is an improvement of the basic scheme of Figure 1a. The additional parts are (D) an aluminum cork used to seal the tube and disperse heat, (E) a cover sheet set on the woofer in which the resonator is plugged, (F) the woofer box used to contain the sound emission, and (G) the oscillator circuit used to generate a sinusoidal signal amplified by (H) the power amplifier. With these improvements, the device is able to work as a stand-alone apparatus.

For all experiments, a 160-mm, 60-W root mean square (RMS)-power woofer was used. The woofer was powered by a simple 100-W RMS Darlington amplifier implemented with a power-tension converter in order to control the electric power used by the woofer. The woofer box has internal dimensions of 180 × 180 × 210 mm; the walls and the cover sheet are made of 19-mm medium-density fiberboards (MDFs). For the design of standing-wave devices as well as traveling waves, the first parameter that needs to be set is the length of the resonator, here chosen to be 225 mm. Two different resonators were used. Because both resonators are made of PVC, modification in the resonator length could be obtained by acting on the aluminum cork. The cork displacement range needs to be small because the resonator length fixes the resonance frequencies f according to

$$f = n\frac{c}{4L} \tag{1}$$

where L is the longitudinal length of the cylinder, n is an odd number, and c is the speed of sound. Cylinders with a single open end produce only odd harmonics. When $n = 1$, the cylinder is working and produces a fundamental harmonic.

The harmonic frequency, f, sets also the thermal penetration depth δ_k:

$$\delta_k = \sqrt{\frac{k}{\pi f \rho c_p}}. \tag{2}$$

One penetration depth corresponds to the distance traveled by heat through the working fluid in 1 s. It depends on the fluid properties and working conditions. The spacing between the walls in the stack channels is vital to the good function of the refrigerator. If the spacing is too small, the stack obstructs the standing-wave setup; if the stack channels are too wide, many parcels in every channel are in weak thermal contact with the stack and do not effectively transfer heat to and from it. The ideal solution, halfway between these two extremes, is described by G.W. Swift and consists in a channel width of four thermal penetration depths [17].

The first step for a correct stack design is to choose a working fluid (in this case air) and fix the frequency. Fixing a frequency in standard working conditions consequently means fixing the tube length and the stack channels' width. A good compromise between low frequencies (that involve long tubes) and small dimensions is to use a range of lengths from 20 up to 25 cm or, in terms of frequencies, from 340 up to 430 Hz.

Circuit

In most of the available articles in thermoacoustic literature, thermoacoustic phenomena are analyzed without discussing how sound should be generated. It is common knowledge that the easiest way to generate sound is to use a multifunctional signal generator setup at the proper frequency. Alternatively, signal can be generated by a computer with an electroacoustic software and amplified through the auxiliary socket in a common home high-fidelity (Hi-Fi) sound system. These solutions are easy to set up in a laboratory but are not effective in real applications. Multifunctional signal generation, PC, and Hi-Fi are designed for other purposes; furthermore, the thermoacoustic refrigerators just need a single-frequency signal amplified for the woofer. For this reason, in this work, a harmonic oscillator and a signal amplifier were used. A simple circuit that creates a sinusoidal output is the Wien bridge (Figure 2). It is based on a network made of three resistors, two capacitors, an operational amplifier, and a bulb, which acts as a thermosensible resistor used to limit oscillation distortions through amplitude stabilization, as suggested by Hewlett [18]. The frequency is expressed as a function of R and C:

$$f = \frac{1}{2\pi RC}. \tag{3}$$

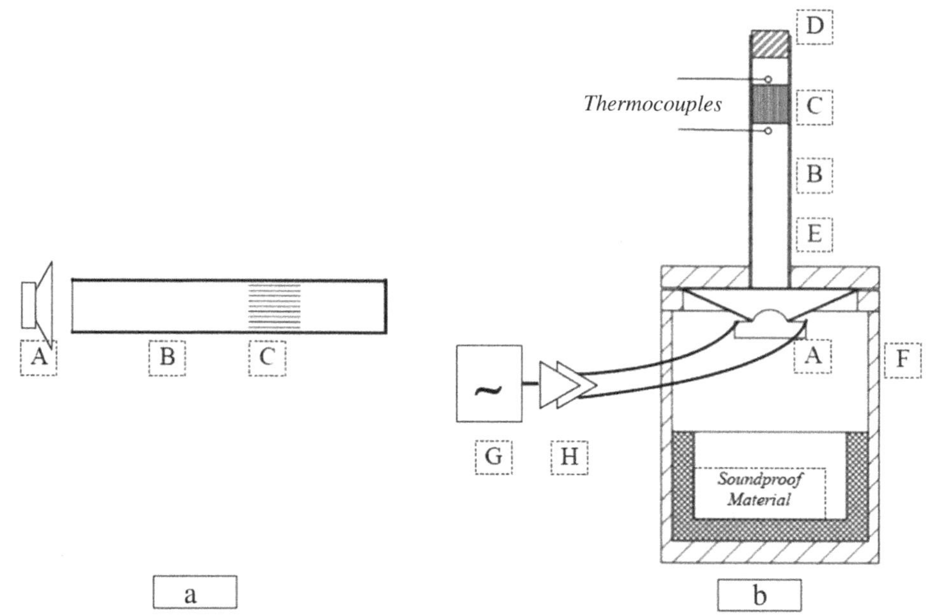

Figure 1 Device scheme. (a) Basic device scheme. **(b)** The refrigerator discussed in this paper.

The operational amplifier needs to be well chosen as a function of the operating frequency. For frequencies below 1 kHz, suitable op-amp types are 107, 108, 741, 1,556, and 8,007 [19].

A variable resistor is used to tune the oscillator. The other two resistors in the circuit are fixed and affect the stability and frequency range of the signal. The best use for a Wien bridge is to operate at a fixed frequency. For the Wien bridge oscillator, the frequency-determining elements also affect the critical gain of the positive feedback loop. Thus, any change in frequency made by altering these components must be followed by a precise adjustment of gain. If the signal is excessively noisy, or the frequency range is too narrow, or there are instabilities during long tests, a different oscillator should be employed. Fox example, Figure 3 shows the oscillator used

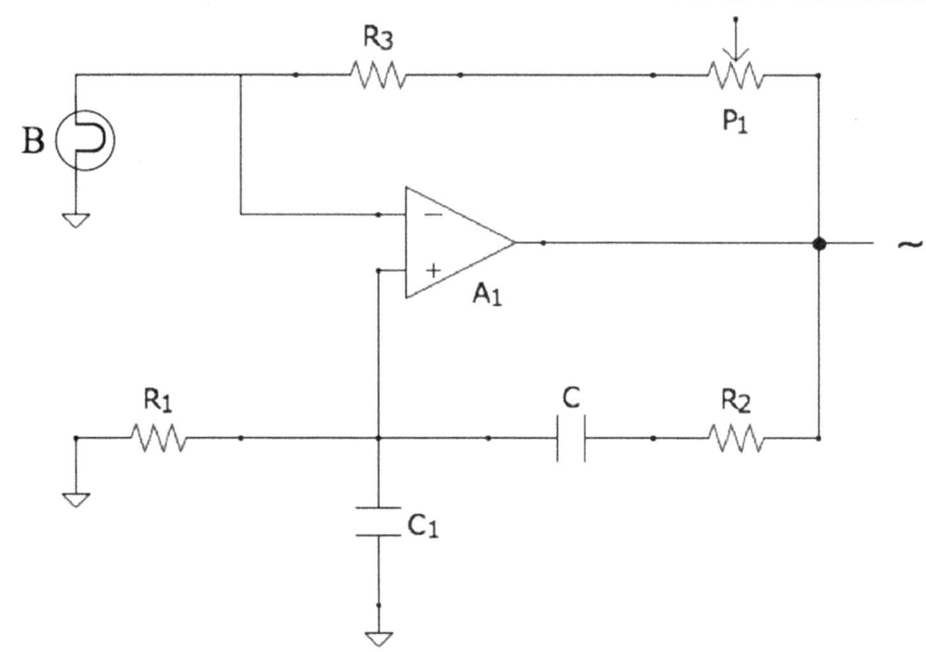

Figure 2 Wien bridge oscillator.

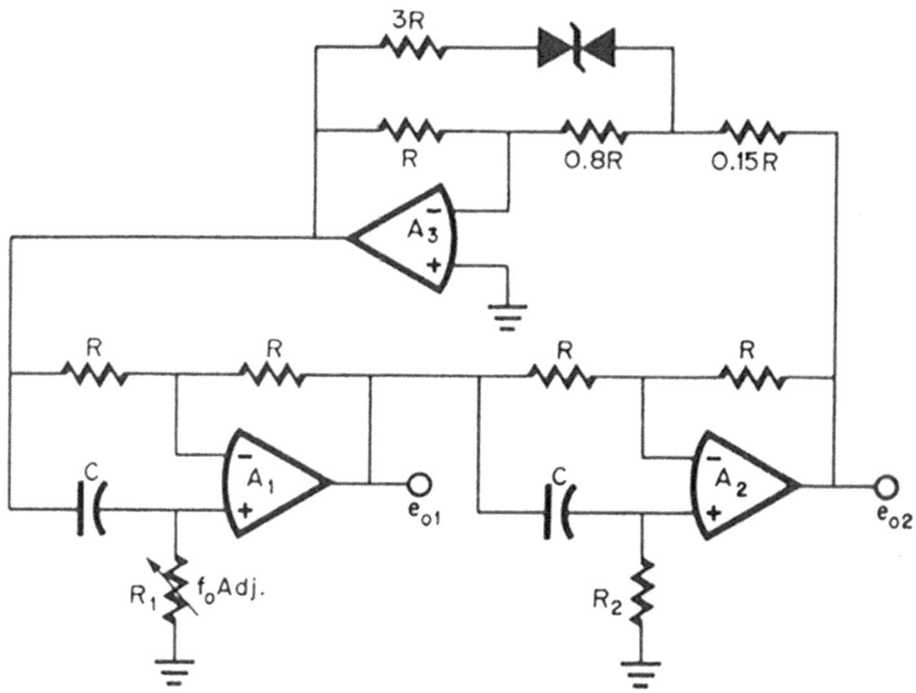

Figure 3 Op-amp oscillator [20].

in this work, where amplifiers A1 and A2 form a phase-shift circuit. This architecture allows to change the oscillation frequency without disturbing the gain magnitude of the feedback loop. The frequency equation associated with the circuit is similar to Equation 3, but R is replaced by the geometric mean value between R_1 and R_2 [20]:

$$f = \frac{1}{2\pi C\sqrt{R_1 R_2}}. \quad (4)$$

The generated signal has to be amplified before reaching the woofer. Audio amplifiers can be constructed in a number of ways; for instance, a low-cost choice consists in using a transistor amplifier. Other amplifiers, like vacuum tube or operational amplifiers, can have larger costs, and their use is appreciably advantageous only for high-fidelity sound reproduction. The purpose of this amplifier is uniquely to increase the amplitude of a perfectly sinusoidal signal; hence, there is no need to choose a high-performance or multiway amplifier. For our purpose, the few parameters that need to be controlled are maximum power and, possibly, saturation.

Woofer box

In thermoacoustic literature, boxed loudspeakers or boxed woofers are mentioned, but it is necessary to

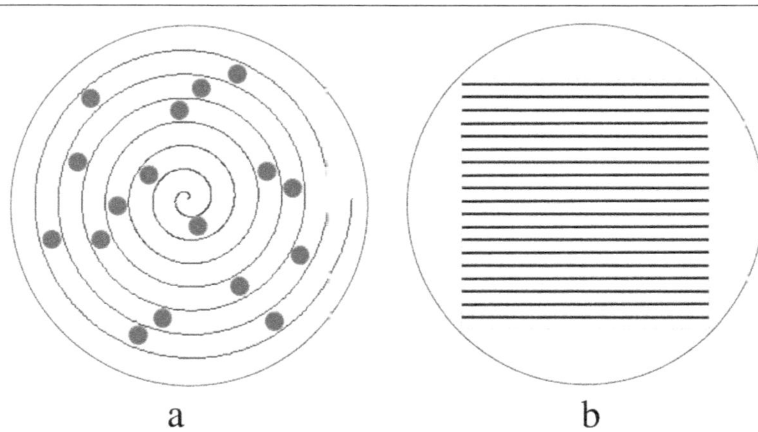

Figure 4 Rolled film stack (a) and parallel plate stack (b).

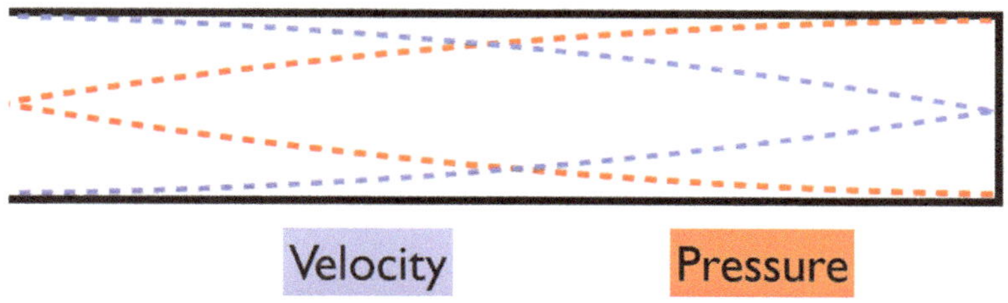

Figure 5 Standing wave in closed-end air column.

specify the dimensions, shape, and materials of the box. In fact, a good design can improve the woofer response at the operating frequency, reduce the sound emission of the device, and decrease or eliminate the rear emission that is in phase contrast with the frontal one. For low-frequency sounds, an acoustic-suspension woofer can be used. Acoustic suspension or air suspension uses the air volume sealed in the box as an elastic cushion to reinforce the oscillation of the woofer diaphragm. Clearly, the volume of the box needs to be properly calculated. In order to do this, freeware programs for air-suspension boxes designed like BassPc can be used. If the frequency is higher than 200 Hz, then the software can recommend unrealistically small volumes.

Besides, if the woofer is a two-way speaker (woofer and tweeter in the same device), there is a duct that connects the bottom of the woofer (in the box) with the top so that the air suspension cannot be used. A similar solution used for pressurized devices working with different gases is the 'gas spring system' [21]. In this case, the different properties of the working fluid allow to choose this solution even with high resonance frequencies. If the rear gas volume, type, and pressure can be adjusted, the gas spring can be tuned at the working frequency [22].

If air suspension cannot be implemented, it is impossible to exploit the rear emission to enforce the front one. Moreover, if the box is not correctly designed, the rear sound may be reflected, thus interfering with the front emission. In these cases, it is better to design the box in order to eliminate rear emission. One possibility is to remove the box and keep the woofer in mid air (e.g., attaching it to a sort of frame); in this way, the wave coming from the bottom of the membrane is dispersed in the environment. This solution involves loud sounds coming from the device and low comfort for anybody close to the woofer. Another possibility is to try to apply the infinite-baffle model and mount the loudspeaker in a very large sealed enclosure. This minimizes the change in the driver's resonant frequency caused by the enclosure. If it is not possible or convenient to use large volumes, one should choose a box large enough to contain the woofer, much deeper than wider, and with the bottom covered by soundproof material. This solution is just an approximation of the infinite-baffle model, but for the monotone wave of a thermoacoustic device, the results are satisfactory. In both air-suspension and infinite-baffle solutions, the choice of box material and thickness is necessary. An excellent material that avoids the possibility of wall resonance is the 19-mm-thick MDF, a composite wood product similar to a particleboard.

Stack and tube

With air as working fluid (c_p = 1,005 J/kg K, ρ = 1.205 kg/m^3, $k = 0.025$ W/m K) and a frequency range of 340 to 430 Hz, δ_k ranges from 1.2 to 1.4×10^{-4} m. The spacing between the stack layers has to be approximately 4 to 5×10^{-4} m. Here, two different methods were used to create the stacks; in both of them, a 4×10^{-4}-m nylon line is used for spacing.

The first method consists in spooling the nylon line on a 35-mm teflon planking in such a way that it would create a coil every 4 mm. Then, a film of spray glue is laid on the spooled line, and a 35-mm photographic film is pressed

Table 1 Cumulative function comparison

Resonator and stack	ΔT max(°C)	Error$_{\Delta T}$ (°C)	Cumulative value	Error$_{cumulative}$ (°C)
36-mm PVC tube, parallel plate stack	4.5	0.2	107	6.4
36-mm PVC tube, rolled film stack	6.1	0.2	148.5	6.4
22-mm PVC tube rolled film stack	24	0.2	688.5	6.4

Figure 6 Closed-end tube with nozzle.

on the glued line. After that, the line in excess is cut, and the film is rolled to create a cylinder (Figure 4a).

The second method consists in superimposing many microscope glass slides separated by pieces of nylon line. The slides are rectangular and have an average thickness of 1.7×10^{-4} m. The slide stack is fixed on the two lateral faces with melted ethylene-vinyl acetate (EVA). This procedure creates a parallelepiped-shaped stack that needs to fit into a circular cylinder. After the EVA consolidation, the pieces of nylon line can be taken out. The melted acetate is also used to create the cylindric frame in which the stack is set (Figure 4b).

Two different diameter resonators have been tested. The small tube has a diameter of 22 mm; the other, 36 mm. In both cases, it is important to assure a perfect sealing of the cork: the closed end of the tube is subjected to high-pressure variations, and even a small escape of air can nullify the standing wave. For this reason, the corks are ripped with teflon tape. Another contact that needs to be sealed is the woofer frame surface touching the MDF sheet: for this, a 5-mm plastic foam gasket has been used.

The fundamental standing wave associated with resonance in air columns is visualized in Figure 5 in terms of pressure and velocity variations in the column. A node for displacement is always an antinode for pressure and vice versa. When the air is constrained to a node, the air motion will be alternately compressed toward that point and expanded away from it, causing the pressure variation to reach a maximum. The fundamental harmonic of a closed-end cylinder shows that the gauge pressure variation has a maximum at the closed end and is approximately zero at the open end near the woofer (where the displacement is maximum).

T-A refrigerator working principles

Once the resonance frequency of the tube is set, the device can start working. The sound coming from the woofer creates a standing wave in the tube. The air parcels in the tube start oscillating back and forth. The standing wave is associated with a stationary pressure and velocity fields as the particles are oscillating between two points, each at a different pressure. The alternation of compression and rarefaction of the gas causes the local temperature of the gas to oscillate with it due to the adiabatic nature of sound waves. During the first half of the oscillation, the particles move and reach a high-pressure area. Due to the pressure increase, they shrink, and consequentially, the local temperature rises. In this phase, the hot and shrunk parcels are able to transfer heat to the nearby stack wall if the latter is colder than the gas parcels. Similarly, when the parcels oscillate back to a low-pressure area, the rarefaction of the gas causes the local temperature to drop, and if the temperature of the parcels becomes lower than that of the walls, heat is transferred from the walls to the parcels.

This alternation of compression and expansion results in a thermodynamic cycle in which the stack acts as a cold sink and a hot source in the terminal points of the oscillation of each parcel. Consequently, each particle of

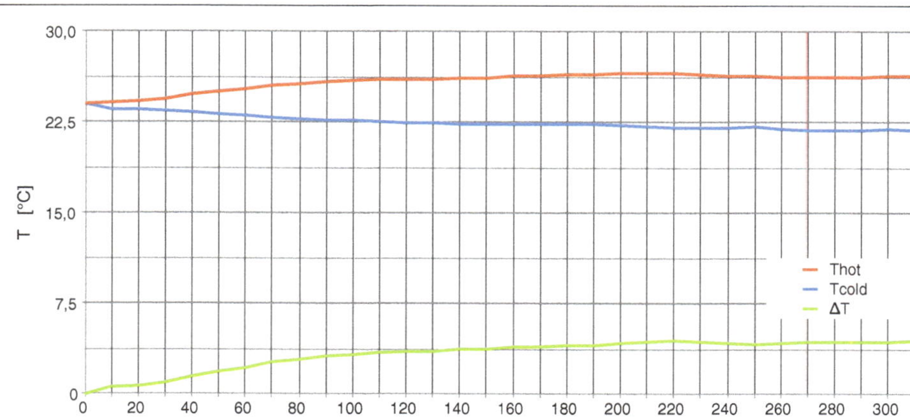

Figure 7 36-mm resonator, $L = 225$ mm, $f = 383$ Hz, parallel plate stack.

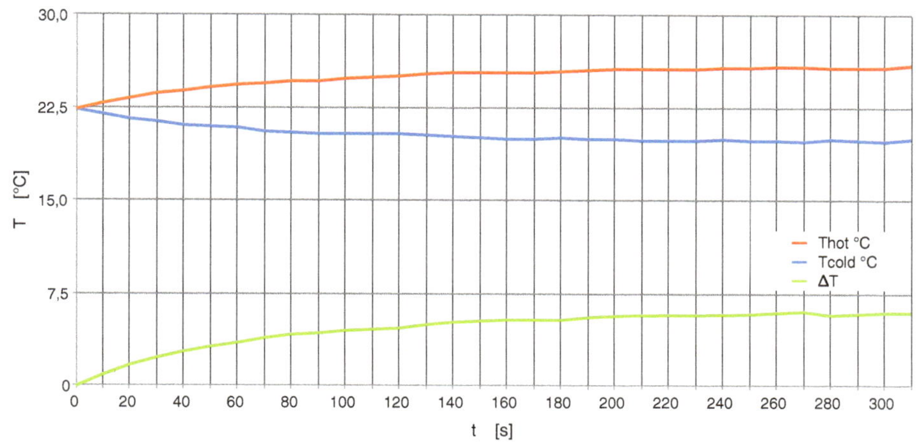

Figure 8 36-mm resonator, $L = 225$ mm, $f = 383$ Hz, rolled film stack.

gas pumps a little heat from the bottom of the stack to the top, from cold to hot, during each cycle of the sound wave. This effect is called 'bucket brigade,' to underline the contribution of every single particle cycle to the total effect. Despite the high velocities involved in the process, the thermodynamic reference cycle is the reversed Stirling cycle as reported by [23].

Results and discussion

Experimental tests

The apparatus was tested with different diameter resonators and with two different stacks. In order to compare the 22- and 35-mm stacks, they were both tested for 310 s in order to reach a steady-state operating condition. Two 1.5-mm T-type thermocouples were placed at the top and bottom faces of the stacks, acquiring temperatures every 10 s with a sensibility of 0.1°C. To detect the highest power that the woofer can run, two different phenomena must be considered. The amplification circuit is designed for amplifying music, while working with a monotone high-amplitude signal is harder. In order to ensure circuit safety, the temperature of Darlington transistors was monitored with an IR pyrometer, and the test was stopped every time the superficial transistor temperature was higher than 90°C. To keep the system working in linear conditions, as suggested by [1], it is important to increase the power until a second harmonic is barely audible. Once the highest admissible power tolerable by all the resonators was reached, the test was started. The first test compared the different stack design, then the best stack design was built in two diameters in order to compare the resonator with the influence of resonator diameter on refrigerator performance. All the tests were run, fixing the working power at an RMS of 30 W. All the tube lengths were set at 225 mm with a respective resonating frequency of 383 Hz.

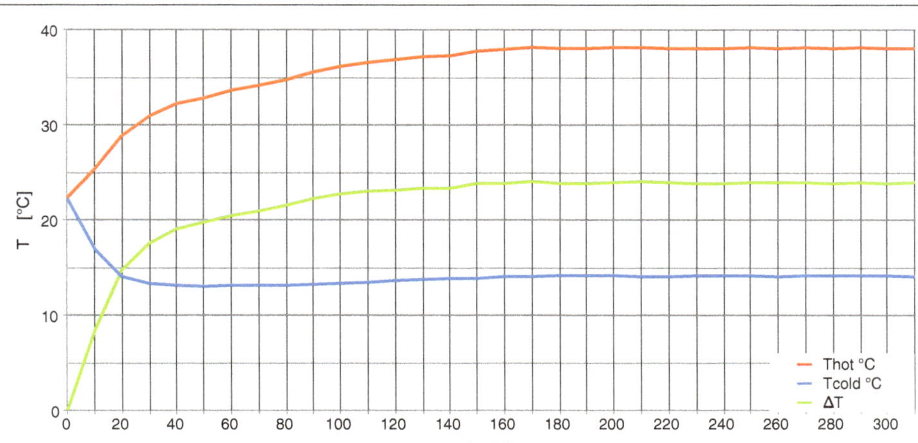

Figure 9 22-mm resonator, $L = 225$ mm, $f = 383$ Hz, rolled film stack.

Evaluation parameter

Table 1 shows the comparison between the three different tests. The comparison was made with the assistance of a cumulative function obtained as the summation of the 32 values of ΔT for each test. This value is a time-stepwise approximation of the $t - T$ function integral, and it stands as an index of the device capability to create and maintain the temperature difference between the stack faces. Because the error that affects the measurements is 0.1°C, the error on the ΔT is 0.2, while the one on cumulative function is 6.4°C.

Test results

The thermoacoustic refrigerator analyzed here resulted in just a concept device used for demonstrating the interaction between sound and heat, but it performed the duties of a real stand-alone thermoacoustic refrigerator. From a practical point of view, during the very first startup, it is necessary to tune the oscillator at the appropriate frequency; then, the volume can be increased until the maximum power is reached (see previous section), then the device can start working as standing-alone. This thermoacoustic refrigerator complexity, despite the addition of the circuits, is still less than that of a common refrigerator; moreover, it is important to highlight that the circuit complexity will not increase in the event that the resonant stack part is improved using different materials, geometries, or gases.

Table 1 shows the ΔT_{max} reached by every resonator that has been tested. Best results are obtained with the small-diameter resonator and rolled stack, for which ΔT_{max} is approximately 24°C. Such a temperature difference, compared with other simple standing-wave refrigerators in the literature [1,2,22], shows that the self-working apparatus has worked as well as the other refrigerators.

The results with different diameter resonators or different stacks showed the importance of a correct geometry and material choice. The second stack (parallel plates) was designed and built on the same fundamental parameter as the first one. The space between the plates is the same, and the wall thickness is similar. Quite surprisingly, the two stacks have led up to completely different results. In particular, the ΔT_{max} for the parallel plate stack is about 25% less than the rolled stack. These strongly different results are caused by the difference of the effective operative areas of the two stacks. The cylindrical rolled film fits perfectly in the tube, so its whole area competes to the result; the second stack is composed of two different parts: the parallelepiped-shaped stack and the plastic cylindric frame, so the part that really operates is only the central one. In this case, two different negative effects occur: the cumulative working surface of parallel slide stack is less than the rolled stack surface; moreover, the plastic frame behaves as a nozzle that can produce little variations in the resonance frequency of the tube (Figure 6).

As previously discussed, standing-wave devices are based on the resonance phenomenon; hence, little differences between the two stacks involve big variations in the thermal efficiency of the system.

More interesting is the difference between small- and large-diameter devices. Once proven that the rolled stack works better than the parallel plate stack, two devices with the same kind of rolled stack but different diameters were tested. In principle, the larger the diameter (i.e., the higher the number of channels in the stack), the more heat can be pumped out. However, the tests showed that the larger tube creates a smaller ΔT. A possible explanation for this thermal behavior is that, if the acoustic power coming from the woofer is the same, the advantage of having an increased number of channels is completely nullified by the lower specific impedance inherent to the larger duct. If the specific impedance decreases, even the Δp in every particle cycle decreases, thus creating a lower ΔT between the opposite faces of the stack. The difference between the different resonators can be summarized in Table 1; the cumulative function shows the great difference and the performance boost obtained using rolled film stacks and small diameters. These results do not suggest to use small-diameter resonators but to use small-diameter tubes in proximity of the stack, putting the stack between a convergent divergent tube.

Figures 7, 8, 9 show the temperature trends during the three different tests. All the tests show that a steady state can be reached within a time frame of 300 s, no matter if the temperature difference established between the opposite stack faces is high or low. This results guarantee short startup-time devices even for higher efficiency. Figures 7, 8, 9 also display the horizontal asymptotic trend of the temperature difference curves 'ΔT' in each of the cases analyzed. Such trends lead to two different considerations.

Firstly, the asymptotic trend is a consequence of the steady state reached by the stack: the thermoacoustic conversion allows a constant heat rate pumping from the bottom to the top of the stack. Secondly, the horizontal direction of all the temperature difference curves suggests that the whole device has reached a steady state, too. The heat coming from the bottom of the stack is correctly pumped out through the aluminum cork. If the device is not properly designed, the heat coming from the hot face of the stack cannot be discharged to the environment; consequently, the device overheats, and the temperature difference curve asymptote inclines.

Conclusions

A stand-alone standing-wave thermoacoustic refrigerator was designed, built, and tested on the basis of the

Russell-Weibull tabletop device. The oscillator, the resonator, and the power amplifier need to be tuned during the first startup, and then the device becomes self-sufficient.

Results show that the adoption of a dedicated oscillator-amplifier circuit, in place of artificial solutions (such as the use of a function generator) does not compromise the maximum ΔT achievable. Moreover, it is shown that the stack performance is influenced by the fundamental parameter δ_k as well as the extension of the stack surface. Temperature difference trends show that a steady state is reached for all the cases examined. The heat coming from the upper face of the stack is dispersed in the environment through the cork, thus avoiding the need to adopt forced ventilation systems to assist heat discharge.

In addition, results suggest that, if the cross section of the resonator is narrowed, creating a nozzle, the fundamental resonance frequency can change, driving the device out of tune. It is also seen that small tubes perform better, thanks to a higher value of the specific impedance of the resonator.

Competing interests

The authors have declared that no competing interest exists.

Author's information

Giulio Allesina holds a master's degree in Mechanical Engineer and a PhD in "High Mechanics and Automotive Design & Technology" discussing a final thesis on "Experimental and analytical evaluation of stratified downdraft gasifiers." His research and teaching focuses on renewable sources. He is teaching assistant and exams committee member for the following academic courses at the University of Modena and Reggio Emilia: Technical physics (thermodynamic and fluid dynamic), Thermodynamic and heat transfer, Energy management. He is the author of several peer-reviewed articles published on italian and international academic journals.

References

1. Russell, DA, Weibull, P: Tabletop thermoacoustic refrigerator for demonstrations. Am. J. Phys. **70**, 1231–1233 (2002)
2. Newman, J, Cariste, B, Queiruga, A, Davis, I, Plotnik, B, Gordon, M, Martin, SS: Thermoacoustic Refrigeration. GSET Research Journal, Rutgers School of Engineering, Piscataway (2006)
3. Hariharana, NM, Sivashanmugama, P, Kasthurirenganb, S: Experimental investigation of a thermoacoustic refrigerator driven by a standing wave twin thermoacoustic prime mover. Int. J. Refrigeration. **36**(8), 2420–2425 (2013)
4. Chen, M, Ju, YM: Design and experimental investigations on a small scale traveling wave thermoacoustic engine. Cryogenic. **54**, 10–15 (2013)
5. Ghorbanian, K, Karimi, M: Design and optimization of a heat driven thermoacoustic refrigerator. Appl. Thermal Eng. **62**(2), 653–661 (2014)
6. Bansal, P, Vineyard, E, Abdelaziz, O: Status of not-in-kind refrigeration technologies for household space conditioning, water heating and food refrigeration. Int. J. Sustainable Built Environ. **1**(1), 85–101 (2012)
7. Konaina, T, Yassen, N: Thermoacoustic solar cooling for domestic usage sizing software. Energy Procedia. **18**, 119–130 (2012)
8. Xu, J, Liu, B, Zhao, D: Numerical simulating for turbulent heat transfer of structural improved thermoacoustic cooler. Procedia Eng. **29**, 2177–2181 (2012)
9. Garrett, S, Hofler, T, Perkins, D: Thermoacoustic refrigeration. ASHRAE J. **34**, 28–36 (1992)
10. Hofler, T, Wheately, J, Swift, G, Migliori, A: Acoustic cooling engine. US Patent 4722201 A, 2 Feb 1988
11. Garrett, S: Resource letter: Ta-1: Thermoacoustic engines and refrigerators. Am. J. Phys. **72**, 11–17 (2004)
12. Reid, R, Swift, G: Experiments with a flow-through thermoacoustic refrigerator. J. Acous. Soc. Am. **108**, 2835–2842 (2000)
13. Landsberg, P: Thermodynamics: cool sounds. Nature (London). **394**(6694), 623–624 (1998)
14. Wu, Z, Dai, W, Man, M, Luo, E: A solar-powered traveling-wave thermoacoustic electricity generator. Solar Energy. **86**(9), 2376–2382 (2012)
15. Kirkconnell, CS, Pruitt, GR, Price, KD: Pulse tube cooler with internal MEMS flow controller. US Patent 7263838 B2, 4 Sept 2007
16. Swift, G: Thermoacoustic natural gas liquefier. In: *US Department of Energy's Natural Gas Conference Proceedings*, pp. 24–27, Houston TX, (1997)
17. Swift, G: Thermoacoustic engines and refrigerators. Phys. Today. **48**, 22–28 (1995)
18. Hewlett, W: Modified wien-bridge oscillator. US Patent 2583649 A, 29 Jan 1952
19. Berlin, H: La progettazione dei circuiti amplificatori operazionali: con esperimenti. Jackson, Milan (1979)
20. Graeme, JG: Applications of Operational Amplifiers: Third-Generation Techniques (BB electronics series). McGraw-Hill, New York (1973)
21. Tijani, MEH, Zeegers, JCH, de Waele, ATAM: A gas-spring system for optimizing loudspeakers in thermoacoustic refrigerators. J. Appl. Phys. **92**(4), 2159 (2002)
22. Tijani, M: Loudspeaker-driven thermo-acoustic refrigeration. PhD thesis, Technische Universiteit Eindhoven (2001)
23. Yazaki, T, Biwa, T, Tominaga, A: A pistonless Stirling cooler. Appl. Phys. Lett. **80**(1), 157–159 (2002)

Impacts of pH, temperature, and pretreatment method on biohydrogen production from organic wastes by sewage microflora

Jaruwan Wongthanate*, Kittibodee Chinnacotpong and Madsamon Khumpong

Abstract

Biohydrogen production could be generated from organic wastes - food and beverage processing wastewater, restaurant food waste, and raw starch waste. Fermentative hydrogen production from food and beverage processing wastewater by sewage microflora was optimized in terms of pH (4.5 to 7.0), mesophilic condition (35°C ±2°C), and thermophilic condition (50°C ±2°C). Low initial pH (6.5) and mesophilic condition favored hydrogen production (0.28 L/L) indicating that such parameters along with the wastewater characteristics were crucial to dark-fermentative hydrogen production. Pretreatment methods (methanogenic inhibitor, sterilization, sonication, and acidification) on restaurant food waste and raw starch waste to enhance biohydrogen production were also investigated in this study. Maximum hydrogen yields of 3.48 ml H_2/g COD and 2.18 ml H_2/g COD were observed in sterilization of pretreated restaurant food and raw starch wastes, respectively.

Keywords: Biohydrogen production; Organic wastes; pH; Temperature; Pretreatment method

Background

The dependence of global energy requirements on fossil fuels may eventually lead to their depletion. It is imperative to seek for an alternative energy resource as the possible successor. Hydrogen (H_2) is a clean fuel with no carbon dioxide (CO_2) emissions, in case sustainably produced, could be the ideal fuel of the future to replace the fossil fuels [1]. Biological hydrogen production (biohydrogen production) is less energy intensive and environmental friendly. Dark fermentation is a biohydrogen production process that applies anaerobic bacteria to decompose the organic materials. Moreover, organic waste such as wastewater from municipalities or industries, agricultural waste, dung, food waste, etc. is used as substrate for biohydrogen production and can reduce a quantity of waste to environment [2]. H_2 is produced by bacteria through bio-process under ambient temperature and pH regimes, and the yield can be enhanced by the manipulation of other environmental conditions. The initial pH and temperature are the important factors that influence H_2 production by bacteria. However, pretreatment methods - methanogenic inhibitor, sterilization, sonication, acidification, freezing, and thawing of organic waste - are one strategy to promote the hydrolysis and disintegration of organic compounds and augment the fermentation to enhance H_2 generation [3]. Although, the several studies demonstrated the role of those conditions on fermentative hydrogen production, there still exist different values with regard to optimum of initial pH, temperature, and pretreatment method [3,4].

Therefore, the main objectives of this study were to look for the impacts of fermentative hydrogen production from food and beverage processing wastewater in terms of initial pH (4.5 to 7.0), mesophilic (35°C ±2°C), and thermophilic (50°C ±2°C) conditions and to investigate the impacts of pretreatment method on organic wastes for enhancing the biohydrogen production.

Methods

Substrates and seed sludge

Wastewater was collected from six industrial factories in Thailand by a water sampler (grab sampling method). Food and beverage processing wastewater was used as substrate for fermentative hydrogen production. Juice processing wastewater was obtained from coconut milk

* Correspondence: jaruwan.won@mahidol.ac.th
Faculty of Environment and Resource Studies, Mahidol University, 999 Phutthamonthon 4 Road, Salaya, Phutthamonthon, Nakhon Pathom 73170, Thailand

industry (Ci) and juice industry (Ji). Food processing wastewater was obtained from the starch and rice noodle industry (Sti) and snack industry (Sni). Winery and brewery processing wastewater was from winery industry (Wi) and brewery industry (Bi), respectively. Restaurant food waste collected from the central cafeteria at Mahidol University (Salaya campus, Thailand) and raw starch waste collected from a sludge thickness tank of a noodle processing plant were use as substrates. Both waste types were grinded using a blender and were mixed in a container, as well as were sieved with a 5-mm screen. Then, they were mixed with distilled water with a volume ratio of waste to distilled water of 3:1. The physical and chemical characteristics of organic wastes - pH, Total Suspended Solid (TSS), Total Dissolved Solid (TDS), Volatile Suspended Solid (VSS), Chemical Oxygen Demand (COD), Biochemical Oxygen Demand (BOD), Total Kjeldahl Nitrogen (TKN), and Fat oil and Grease (FOG) - were analyzed [5]. Table 1 shows the physical and chemical characteristics of organic wastes.

Anaerobic sludge was taken from the Bio-fertilizer plant, Nonthaburi province, Thailand. Seed sludge was screened with sieve (2.00 mm) to eliminate the large particulate matter. Seed sludge was heated (90°C, 10 min) to eliminate the hydrogen-consuming bacteria and facilitated growth of spore-former bacteria [6].

Experimental procedures

The nutrient solution for bacterial growth contained 10 g $C_6H_{12}O_6$ (D-Glucose), 5,240 mg NH_4HCO_3, 6,720 mg $NaHCO_3$, 125 mg K_2HPO_4, 100 mg $MgCl_2 \cdot H_2O$, 25 mg $FeSO_4 \cdot 7H_2O$, 15 mg $MnSO_4 \cdot 6H_2O$, 4.37 mg $CuSO_4 \cdot 5H_2O$, 0.125 mg $CoCl_2 \cdot 5H_2O$ in 1,000 ml distilled water [4]. Regarding food and beverage processing wastewater as the substrate, a batch reactor of 500-ml serum bottle was added with 20 ml seed sludge, 50 ml nutrient solution, and 250 ml wastewater. On the other hand, batch reactor of 250-ml serum bottle consisted of 150 ml the substrate and 10 ml seed sludge as well as 40 ml of nutrient solution was set up for biohydrogen production from restaurant food or raw starch waste. The mixed liquor was purged with N_2 for 1 min to ensure anaerobic condition prior to each run and clogged with silicone rubber stoppers to avoid the gas leakage from the bottles.

The experiment was conducted to compare the biohydrogen production at uncontrolled pH at room temperature in case of food and beverage processing wastewater as the substrate. Subsequently, biohydrogen production was monitored at initial pH of 4.5, 5.0, 5.5, 6.0, 6.5, or 7.0 end under mesophilic (35°C ±2°C) and thermophilic (50°C ±2°C) conditions. In order to find out the proper pretreatment method of waste for enhancing biohydrogen production, restaurant food and raw starch wastes using as substrates were pretreated by four pretreatment methods of methanogenic inhibitor by 1 M Sodium 2-bromoethanesulfonate (BESA: $C_2H_4BrO_3SNa$, 0.2/l), sterilization by autoclave at 121°C for 20 min, sonication by an ultrasonic bath for 20 min and acidification by $HClO_4$ at initial pH 3. All bottle reactors were placed in a shaking water bath with speed 120 ± 1 (rpm) at different temperature setup. Each batch experiment was conducted in triplicate.

Analytical methods

The volume of biogas production was measured by a plunger displacement of gas-tight syringes [7]. The hydrogen concentration was determined by gas chromatography (Varian STAR 3400, Palo Alto, CA, USA) which equipped with a thermal conductivity detector (TCD) and packed stainless-steel column (Alltech Molesieve 5A 80/100 10′ × 1/8″, Nicholasville, KY, USA). Argon was used as

Table 1 Characteristics of organic wastes

Substrates	pH	TDS (mg/l)	TSS (mg/l)	VSS (mg/l)	FOG (mg/l)	TKN (mg/l)	COD (mg/l)	BOD (mg/l)
Food and beverage processing wastewater								
Juice wastewater[a]								
Ji	6.40	4,422	149	146	1,721	36	2,605	2,593
Ci	7.46	4,472	4,112	3,330	2,301	140	8,847	5,450
Food wastewater[b]								
Sti	5.30	1,602	304	327	3,010	68	4,296	1,096
Sni	5.50	5,686	4,963	4,840	175,978	14	20,000	3,353
Winery and brewery wastewater[c]								
Wi	6.10	1,163	1,684	1,559	1,910	138	2,038	1,150
Bi	6.90	2,826	348	400	1,035	56	1,185	944
Restaurant food waste	4.83	-	75,700	48,600	-	-	770	136,000
Raw starch waste	7.40	-	89,290	59,640	-	-	868	27,200

[a]Juice wastewater - juice industry (Ji) and coconut milk industry (Ci); [b]Food wastewater - starch and rice noodle industry (Sti) and snack industry (Sni); [c]Winery and brewery wastewater - winery industry (Wi) and brewery industry (Bi).

the carrier gas for analysis [8]. The temperatures of injector, detector, and column were maintained at 80°C, 90°C, and 50°C, respectively.

Hydrogen gas production was calculated from headspace measurements of gas composition and total volume of biogas produced at each time interval by using a following equation (Equation 1) [9]:

$$V_{H,i} = V_{H,i-1} + C_{H,i}(V_{G,i-1}) + V_H(C_{H,i} + C_{H,i} + C_{H,i-1}), \quad (1)$$

where $V_{H,i}$ and $V_{H,i-1}$ are cumulative hydrogen gas volumes at the current (i) and previous ($i-1$) time intervals, $V_{G,i}$ and $V_{G,i-1}$ are the total gas volumes in the current and previous time interval, $C_{H,i}$ and $C_{H,i-1}$ are the fraction of hydrogen gas in the headspace of the bottle measured using gas chromatography in the current and previous intervals, and V_H is total volume of headspace in the reactor.

A modified Gompertz equation (Equation 2) was used to calculate cumulative hydrogen data depict.

$$H = P.\exp\left\{-\exp\left[\frac{\text{Rm}.e}{P}(\lambda - t) + 1\right]\right\}, \quad (2)$$

where H (ml) is the cumulative hydrogen production, P (ml) is the hydrogen production, Rm (ml/h) is the maximum hydrogen production rate, λ (h) is the lag phase time, t (h) is the incubation time, and $e = 2.71828$.

Results and discussion

Feasibility of biohydrogen production from organic wastes

The cumulative hydrogen production from all kinds of wastewater showed that the highest production was approximately 0.53 l H_2/l from snack wastewater, 0.52 l H_2/l from coconut milk wastewater, 0.31 l H_2/l from juice wastewater, and 0.08 l H_2/1 from starch and rice noodle wastewater and the least (approximately 0.0003 l H_2/l) from winery wastewater. However, H_2 production from brewery wastewater was not observed (Figure 1). Among the six kinds of wastewater, coconut milk wastewater was best suited for continuous biohydrogen production. However, snack wastewater supported maximum cumulative hydrogen production, but it could not be used as the representative substrate because of the discontinuous biohydrogen production. As the snack wastewater had a much higher suspended solid than others, these solid remains might be non-degradable. Cumulative hydrogen production in snack wastewater and coconut milk wastewater is higher than that of other wastewaters. It is associated with the high-carbohydrate wastewater and is significantly related to the much of COD and BOD values in wastewater [10]. Presence of carbohydrate content as carbon source has positive effect on the hydrogen production in the metabolic reactions involving molecular

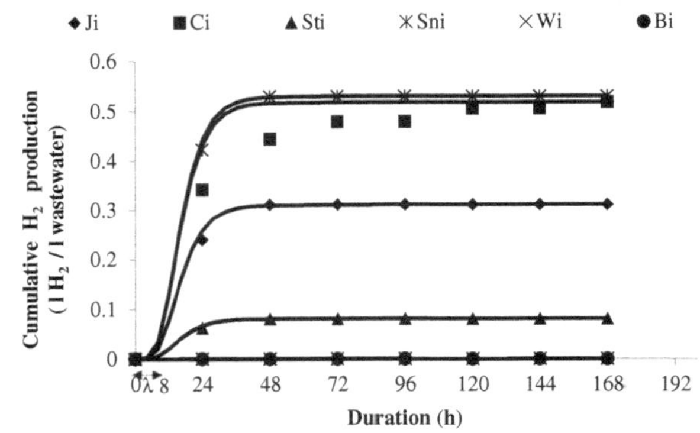

Figure 1 Cumulative hydrogen production from food and beverage processing wastewater.

hydrogen generation [11]. Thus, carbohydrate rich food and beverage processing wastewater may be further processed to convert the carbohydrate content to organic acids and then to hydrogen gas by dark fermentation [12]. The results are similar to other researches [10] as overall H_2 conversions were 0.7 to 0.9 l H_2/l for the apple wastewater, 0.1 l H_2/l for confectioner-A, 0.4 to 2.0 l H_2/l for confectioner B, and 2.1 to 2.8 l H_2/l for the potato wastewater. Moreover, cumulative hydrogen production from wastes increased in early experiments and then slowly decreased until out of biogas production in the batch reactor. Consequently, hydrogen yield of restaurant food waste (2.82 ml H_2/g COD) was higher than raw starch waste (2.02 ml H_2/g COD). These results were similar to other researches that hydrogen yield could be produced by food waste (121.6 ml H_2/g carbohydrate-COD), 250 ml H_2/g VS and 62.6 ml H_2/g VS and industrial wasted sludge (32.6 ml H_2/g carbohydrate-COD) [13,14].

Hence, it could imply that H_2 production depended on the different organic components of the wastes. Organic wastes - food and beverage processing wastewater, restaurant food, and raw starch wastes - have high COD and BOD values and are therefore suitable for fermentative hydrogen production [10].

Impacts of initial pH and temperature on biohydrogen production from food and beverage processing wastewater

Biohydrogen production from coconut milk wastewater as the substrate at the initial pH from 4.5, 5.0, 5.5, 6.0, 6.5, and 7.0 as the anaerobic dark-fermentation at room temperature revealed that lag phase was about 8 h. The lag phase was greatly affected by the initial pH. The results are different from those of others as the lag time was 74 h (at initial pH 5.0) and 41 h (at initial pH 6.0), respectively, and the lag time was longer than 19 to 28 h for initial pH in the range of 7.0 to 10.0 [15]. At the initial pH ranges of 5.5, 6.0, 6.5, 7.0, biohydrogen production was quite high only initially, and it gradually declined to the steady state of biogas production. Regarding cumulative hydrogen production, the hydrogen yield was quite low at initial pH 4.5, 5.0, 5.5, and 6.0, which produced about 0.02 l H_2/l, 0.01 l H_2/l, 0.08 l H_2/l, and 0.15 l H_2/l, respectively. However, the maximum cumulative hydrogen production was approximately 0.28 l H_2/l at the initial pH of 6.5 but was slightly decreased at initial pH 7.0 (Figure 2). Consequently, the initial pH 6.5 is the best pH value for hydrogen production because fermentative conversion of substrate to hydrogen can be increased by maintaining operating pH in and around six compared to neutral pH [1]. Moreover, the initial cultivation pHs were adjusted to 6.5 to 7.0, the pH would further decrease to a suitable value (around 6.0) for hydrogen production [16]. Nevertheless, pH control could stimulate microorganisms to achieve maximum hydrogen production ability because the activity of hydrogenase was inhibited by low or high pH in fermentation [1]. Results are consistent with those of other cites, regarding biohydrogen from starch in wastewater through anaerobic fermentation at the optimal pH of 6.5 and starch concentration at 5 g/l (37°C) with a hydrogen yield reaching 186 ml H_2/g starch [17], while achieved the conversion of organics in wastewater to H_2 in batch experiments, the highest rate of H_2 production (74.7 ml H_2/l.h) at pH 5.5 the COD of 7.5 g/l corresponding to the conversion efficiency of 38.9 ml H_2/(g COD/l) [18]. Furthermore, biohydrogen production from coconut milk wastewater under mesophilic (35°C ±2°C) and thermophilic (50°C ±2°C) conditions (initial pH 6.5) showed a lag phase of about 8 h. These results are not similar to others wherein the lag was relatively longer lasting about 11 to 13 h at initial pH 6.0 and 55°C [19], while some researchers reported biohydrogen production at mesophilic condition to start after a lag of approximately 9 h [20]. Cumulative hydrogen production from coconut milk wastewater was about 0.28 l H_2/l and 0.16 l H_2/l at mesophilic and thermophilic conditions, respectively (Figure 3). It revealed that temperature can affect the activity of hydrogen-producing bacteria by influencing the

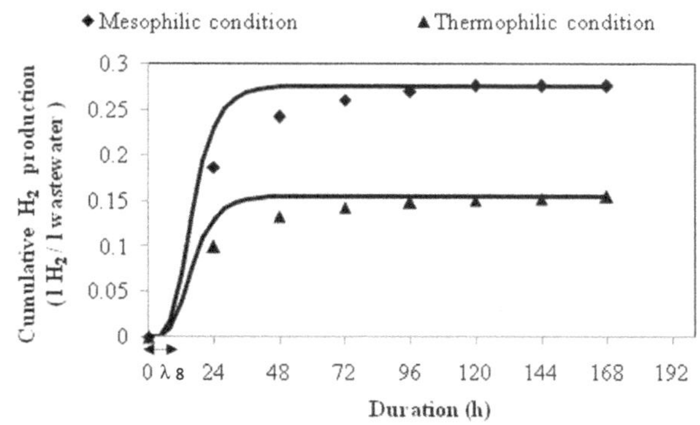

Figure 3 Cumulative hydrogen productions from coconut milk wastewater at mesophilic and thermophilic conditions.

activity of some essential enzymes such as hydrogenases for fermentative hydrogen production. In an appropriate range, increasing temperature could increase the ability of hydrogen-producing bacteria to produce hydrogen during fermentative hydrogen production, but temperature at much higher levels could decrease it with increasing levels [4]. However, operation at high temperatures is not favorable for energy recovery. Therefore, for a balance between hydrogen production and energy recovery, it seems to be reasonable to biological hydrogen-producing reactors in the mesophilic range [21]. These results are

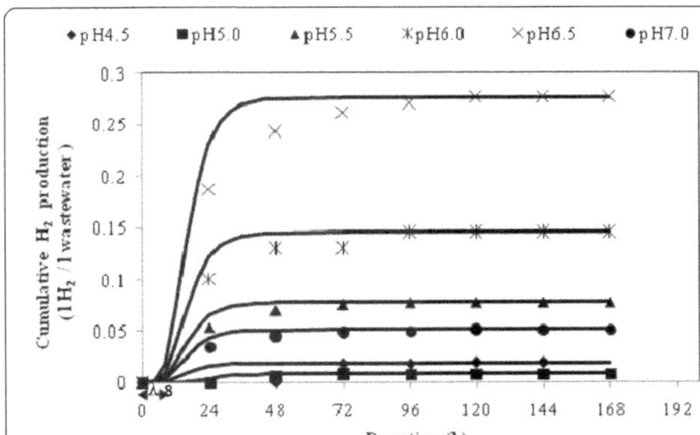

Figure 2 Impacts of initial pH on cumulative hydrogen production from coconut milk wastewater.

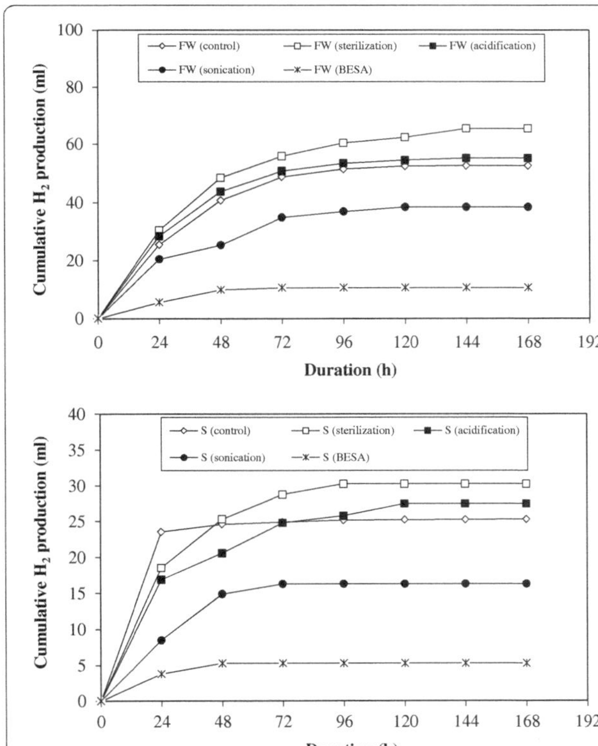

Figure 4 Impacts of pretreatment method on fermentative hydrogen production by utilizing restaurant food (FW) and raw starch (S) wastes.

different from others reported that hydrogen production for thermophilic condition at 55°C was higher (134 ml/day) compared to mesophilic condition (67 ml/day) [22]. Another report showed the effect of temperature and pH on fermentative hydrogen production from cattle wastewater by sewage sludge with the maximum hydrogen yield at pH 5.5 and at 45°C with the peak of hydrogen production (368 ml) or yield (319 ml H_2/g COD) [23]. It was also observed that the initial pH of 6.5 at temperatures of 35°C ±2°C and 50°C ±2°C dropped finally to pH 4.5 during 7 days of fermentation. The final pH of fermentation was comparable to that (about 4.7) reported by Zhang and Shen [24]. Some possible reasons for this may be that hydrogen production occurs in acidification stage of metabolic process and the hydrogen-producing bacteria has a high conversion rate of carbohydrate to hydrogen, then the high concentrations of metabolites may cause the pH to drop to such low level [4]. Moreover, acid accumulation in the system causes a sharp drop of the pH, thus inhibiting biohydrogen production. The lowering by pH suggests inactivation of hydrogen producers. The bacteria involved could not sustain its metabolic activity at pH values less than 5.0 and complete inhibition was reported in the pH range of 4.0 to 5.0 [11].

Hence, it may be concluded that the initial pH and temperature affected the biohydrogen production from coconut milk wastewater. The optimum of initial pH of 6.5 and mesophilic condition was suggested for biohydrogen production. It is also suggested that the pH control induced the system for hydrogen production more effectively than the uncontrolled sets.

Impacts of pretreatment method on restaurant food and raw starch wastes for biohydrogen production

Figure 4 illustrates the impacts of pretreatment method on fermentative hydrogen production by utilizing restaurant food and raw starch wastes. In case of methanogenic inhibitor (BESA), the cumulative hydrogen productions (hydrogen yields) were 10.50 ml H_2 (0.57 ml H_2/g COD) and 5.28 ml H_2 (0.48 ml H_2/g COD) for food and starch wastes, respectively. In sterilization, there were the cumulative hydrogen production (hydrogen yield) of 65.50 ml H_2 (3.48 ml H_2/g COD) for food waste and 30.28 ml H_2 (2.18 ml H_2/g COD) for starch waste, respectively. In sonication, the cumulative hydrogen productions (hydrogen yields) were 38.49 (2.09 ml H_2/g COD) and 16.28 ml H_2 (1.43 ml H_2/g COD) for food and starch wastes, respectively. In acidification, the cumulative hydrogen productions (hydrogen yields) of 55.10 (2.96 ml H_2/g COD) and 27.50 ml H_2 (2.20 ml H_2/g COD) were found using food and starch wastes, respectively. However, the maximum hydrogen production was found in sterilization from restaurant food and raw starch wastes that agreed with previous researches [3,25]. It might explain to this phenomenon that the substrate after sterilization is nearly complete breakdown from polymeric to monomeric molecules until they release to other nutrients. Consequently, hydrogen-producing bacteria of seed sludge could be easily produced hydrogen gas from waste [26]. On the other hand, biohydrogen production using waste pretreated by sonication and methanogenic inhibitor was much lower than other pretreatment methods. It might imply that both pretreatment methods can also suppress the activity of some hydrogen-producing bacteria in fermentation [27].

Conclusion

In this study, organic wastes - some kinds of food and beverage processing wastewater and restaurant food and raw starch wastes - supported fermentative hydrogen production. The maximum cumulative hydrogen production from coconut milk wastewater was at initial pH 6.5 under mesophilic condition. It is also concluded that characteristics of wastewater, initial pH, and temperature conditions affected the fermentative biohydrogen production. Among pretreatment methods of restaurant food and raw starch wastes, it revealed that sterilization was the best method and could enhance the maximum hydrogen yield from restaurant food waste.

Competing interests

The authors declare that they have no competing interests.

Authors' contributions

JW drafted the manuscript. All authors read and approved the final manuscript.

Acknowledgements

This research was supported by Energy Conservation Promotion Fund, Energy Policy and Planning Office, Ministry of Energy, Thailand (MU-IRB 2011/115.2009).

References

1. Mohan, SV, Babu, VL, Sarma, PN: Effect of various pretreatment methods on anaerobic mixed microflora to enhance biohydrogen production utilizing dairy wastewater as substrate. Bioresour. Technol. **99**, 59–67 (2008)
2. Wang, J, Wan, W: Factors influencing fermentative hydrogen production: a review. Int. J. Hydrogen Energy **34**, 799–811 (2009)
3. Guo, L, Li, XM, Bo, X, Yang, Q, Zeng, GM, Liao, DX, Liu, JJ: Impacts of sterilization, microwave and ultrasonication pretreatment on hydrogen producing using waste sludge. Bioresour. Technol. **99**, 3651–3658 (2008)
4. Wang, J, Wan, W: Effect of temperature on fermentative hydrogen production by mixed cultures. Int. J. Hydrogen Energy **33**(20), 5392–5397 (2008)
5. APHA: Standard methods for the examination of water and wastewater, 21st edn. APHA, Washington, DC (2005)
6. Valdez-Vazquez, I, Poggi-Varaldo, HM: Hydrogen production by fermentative consortia. Renewable and Sustainable Energy Rev. **13**, 1000–1013 (2009)
7. Owen, WF, Stuckey, DC, Healy, JB, Young, LY, Mccarty, PL: Bioassay for monitoring biochemical methane potential and anaerobic toxicity Water Res. **13**(6), 485–492 (1979)
8. Selembo, PA, Merrill, MD, Logan, BE: The use of stainless steel and nickel alloys as low-cost cathodes in microbial electrolysis cells. J. Power Sources **190**(2), 271–278 (2009)
9. Sreela-or, C, Imai, T, Plangklang, P, Reungsang, A: Optimization of key factors affecting hydrogen production from food waste by anaerobic mixed cultures. Int. J. Hydrogen Energy **36**, 14120–14133 (2011)

10. Van Ginkel, SW, Oh, SE, Logan, BE: Biohydrogen gas production from food processing and domestic wastewaters. Int. J. Hydrogen Energy **30**(15), 1535–1542 (2005)
11. Mohan, SV, Bhaskar, YV, Krishna, PM, Rao, NC, Babu, VL, Sarma, PN: Biohydrogen production from chemical wastewater as substrate by selectively enriched anaerobic mixed consortia: influence of fermentation pH and substrate composition. Int. J. Hydrogen Energy **30**, 2286–2295 (2007)
12. Kapdan, IK, Kargi, F: Bio-hydrogen production from waste materials. Enzyme Microb. Tech. **38**, 569–582 (2006)
13. Kim, SH, Han, SK, Shin, HS: Feasibility of biohydrogen production by anaerobic co-digestion of food waste and sewage sludge. Int. J Hydrogen Energy. **24**, 1607–1616 (2004)
14. Zhu, H, Parker, W, Basnar, R, Proracki, A, Falletta, P, Béland, M, Seto, P: Biohydrogen production by anaerobic co-digestion of municipal food waste and sewage sludges. Int. J. Hydrogen Energy **33**, 3651–3659 (2008)
15. Fang, HHP, Liu, H, Zhang, T: Phototropic hydrogen production from acetate and butyrate in wastewater. Int. J. Hydrogen Energy **30**(7), 785–793 (2005)
16. Kawano, T, Wada, K, Li, YY, Noike, T: Effects of substrate concentration and pH on hydrogen fermentation of mixed substrate by microflora. J. Soc. Water. Environ. **27**(7), 473–479 (2004)
17. Wei, J, Liu, ZT, Zhang, X: Biohydrogen production from starch wastewater and application in fuel cell. Int. J. Hydrogen Energy **35**(7), 2949–2952 (2010)
18. Van Ginkel, SW, Sung, S, Lay, JJ: Biohydrogen production as a function of pH and substrate concentration. Environ. Sci. Technol. **35**(24), 4726–4730 (2001)
19. Zhang, T, Liu, H, Fang, HHP: Biohydrogen production from starch in wastewater under thermophilic conditions. J. Envi. Manage. **69**(2), 149–156 (2003)
20. Baghchehsaraee, B, Nakhla, G, Karamanev, D, Margaritis, A: Fermentative hydrogen production by diverse microflora. Int. J. Hydrogen Energy **35**(10), 5021–5027 (2009)
21. Mu, Y, Zheng, XJ, Yu, HQ, Zhu, RF: Biological hydrogen production by anaerobic sludge at various temperatures. Int. J. Hydrogen Energy **31**, 780–785 (2006)
22. Cheong, DY, Hansen, CL: Feasibility of hydrogen production in thermophilic mixed fermentation by natural anaerobes. Biores. Technol. **98**, 2229–2239 (2007)
23. Tang, GL, Huang, J, Zhen, JS, Tang, Q, Yan, C, Liu, G: Biohydrogen production from cattle wastewater by enriched anaerobic mixed consortia: influence of fermentation temperature and pH. J. Biosci. Bioeng. **106**(1), 80–87 (2008)
24. Zhang, Y, Shen, J: Effect of temperature and iron concentration on the growth and hydrogen production of mixed bacteria. Int. J. Hydrogen Energy **31**, 441–446 (2006)
25. Wang, CC, Chang, CW, Chu, CP, Lee, DJ, Chang, BV, Liao, CS: Producing hydrogen from wastewater sludge by *Clostridium bifermentans*. J. Biotechnol. **102**, 83–92 (2003)
26. Cakir, A, Ozmihci, S, Kargi, F: Comparison of bio-hydrogen production from hydrolyzed wheat starch by mesophilic and thermophilic dark fermentation. Int. J. Hydrogen Energy **35**, 13214–13218 (2010)
27. Kim, SH, Shin, HS: Effects of base-pretreatment on continuous enriched culture for hydrogen production from food waste. Int. J. Hydrogen Energy **33**, 5266–5274 (2008)

Briquettes from maize cobs and *Ceiba pentandra* at room temperature and low compacting pressure without a binder

Stephen J Mitchual[1*], Kwasi Frimpong-Mensah[2], Nicholas A Darkwa[2] and Joseph O Akowuah[3]

Abstract

This paper reports the findings of densifying ground maize cobs and *Ceiba pentandra* sawdust at room temperature using low compacting pressure without a binder. The maize cobs were crushed using a hammer mill. Particle sizes of maize cobs and *C. pentandra* used for the study were ≤1 mm. The two materials were combined at mixing percentages of 90:10, 70:30 and 50:50 (*C. pentandra*/maize cobs). Briquettes were produced using a laboratory hydraulic press. Compacting pressure was varied from 20 to 50 MPa at an interval of 10 MPa. The results indicated that the relaxed density of briquettes produced from particles of maize cobs only ranged from 541 to 659 kg/m^3 whilst that made from a mixture of maize cobs and *C. pentandra* ranged from 565 to 774 kg/m^3. Compressive strength in cleft of briquettes produced from maize cobs only ranged from 0.12 to 0.54 N/mm whilst that produced from a combination of maize cobs and *C. pentandra* ranged from 7.72 to 59.22 N/mm. Additionally, at all compacting pressure levels, briquettes made from maize cobs only had an impact resistance index of 0%. Whilst those made from a combination of maize cobs and *C. pentandra* ranged from 115% to 500%. Thus, briquettes with adequate physical and mechanical characteristics could be produced from maize cobs at room temperature using low compacting pressure when maize cobs are combined with sawdust of *C. pentandra*. These findings could enhance the existing technology for densifying agricultural residues, for example, maize cobs, especially in rural communities.

Keywords: Briquettes; *Ceiba pentandra*; Compressive strength; Impact resistance index; Maize cobs; Relaxed density

Background

The solution to the current and future energy needs of Ghana requires diversifying the energy sources. Previous studies suggested that energy production methods should be best matched with the available natural resources in the region [1]. Ghana is an agricultural country and has an estimated workforce of over 60% employed by the agricultural sector [2]. This culminates into the cultivation of various crops such as millet, rice, maize, sugar cane and cocoa by farmers. This leads to the generation of large volumes of agricultural crop residues such as maize cobs, rice husk, millet stalks, baggasse and cocoa bean shells as presented in Table 1. These agricultural residues normally obtained from field and processing sites are often left to rot or burnt inefficiently in their loose form causing air pollution [3].

As shown in Table 1, by the end of the year 2010, about 1,872,000 tonnes of maize residue was generated in Ghana. This constitutes about 41% of the total agricultural crop residues generated. Maize cob, a residue of the maize crop, is a lignocellulosic biomass material which contains high amounts of organic constituents and energy. Therefore, it is recognised as a potential source of renewable energy [5]. However, its use as a domestic fuel presents its own challenges. Its structure is porous and has a low bulk density. The bulk density of a crushed maize cob is 227 kg/m^3 and it is more than double the bulk density of uncrushed cobs which is 104 kg/m^3 [6]. This low bulk density makes it difficult to handle, store and transport and also results in low energy density. Densification which employs high pressure to compress biomass raw materials in order to increase

* Correspondence: Stephen.mitchual@yahoo.com
[1]Department of Design and Technology Education, University of Education, Winneba, Kumasi Campus, P. O. Box 1277, Kumasi, Ghana
Full list of author information is available at the end of the article

Table 1 Production of selected agricultural crops and estimated potential residues in Ghana in the year 2010

Crop	Residue	Residue-to-product ratio (tonnes/tonnes of crop)	Total crop production (tonnes)	Residue production (tonnes)
Maize	Cob	1.00	1,872,000	1,872,000
Millet	Stalk	3.00	219,000	657,000
Rice	Straw	1.50	429,000	643,500
Sugar cane	Bagasse	0.30	145,000	43,500
Coconut	Shell	0.60	298,000	178,800
Oil palm fruits	EFB	0.25	2,004,000	501,000
Cocoa	Pods, husk	1.00	632,000	632,000
Coffee	Husk	2.10	1200	2,520
Total			5,600,200	4,530,320

Source: [4,5].

their density could be used to mitigate these challenges. Densification of maize cobs will result in improved energy density, lower volume (higher bulk density), and better mechanical handling [7,8]. However, maize cobs, is noted to have low lignin content (5.6%), low water soluble carbohydrates (1.1%) and low protein (2.5%). These chemicals are largely responsible for forming solid bridge bonds during densification or briquetting [9,10]. Therefore, densification of maize cobs would require a high compacting pressure and/or an external binder. A considerable amount of research on briquetting technology on maize cobs has been conducted. Previous studies showed that pressing maize cob particles at a temperature of 25°C, using a compacting pressure of 150 MPa, resulted in the production of briquettes having a durability of 0% and a relaxed density of 940 kg/m^3 [9]. Further studies indicated that when maize cob particles were preheated to 85°C before pressing at 150 MPa, the durability and density were raised to 90% and above 1,100 kg/m^3, respectively [9]. Even though improved binding of maize cob particles due to preheating to 85°C is feasible, this could invariably increase production cost as a result of increased energy input therefore limiting its utilisation. Thus, it is imperative that one selects the most economic technology for densification or briquetting of maize cob particles to reduce the cost of the finished product. *Ceiba pentandra* is a low-density species (409.22 kg/m^3) with acid-insoluble lignin and alpha-cellulose content of 24.34% and 41.24%, respectively [11]. Briquettes produced from sawdust of *C. pentandra* at room temperature (28°C) and low compacting pressure (20 to 50 MPa) without a binder had adequate relaxed density, high compressive strength in cleft and high impact resistance index (IRI) [11,12]. Studies on densified fuels derived from blends of two biomass materials indicate that the mechanical strength of briquettes produced from only one type of biomass can be improved by blending that biomass with another biomass material [13]. Therefore, this study seeks to investigate the effect of combining maize cob particles and sawdust of *C. pentandra* on the relaxed density, compressive strength in cleft and impact resistance index of briquettes produced at room temperature using low compacting pressure.

Methods

Materials and material preparation

Sawdust of *C. pentandra* and maize cobs were used for the study. Both the maize cobs and sawdust were sun-dried at an average relative humidity and temperature of 75% and 28°C, respectively, for 5 to 7 days. The maize cobs were crushed using a hammer mill. Particle sizes of maize cobs and sawdust used for the study were 1 mm or less. The two materials were combined at mixing percentages of 90:10, 70:30 and 50:50 (*C. pentandra*/maize cobs).

Moisture content

The moisture content, on oven-dry basis, of the crushed maize cobs and sawdust was determined in accordance with [14]. Five samples of sawdust, maize cobs particles and their combination, each weighing 2 g were weighed and placed in a laboratory oven at a temperature of (103°C ± 2°C). Each sample was dried until the difference in mass between two successive weighings separated by an interval of 2 h was 0.01 g or less. The moisture content of the specimen was then computed as follows:

$$\text{Moisture content (\%)}\ db = \frac{M_1 - M_o}{M_o} \times 100, \qquad (1)$$

where M_1 and M_o were masses (g) of test samples before drying and after oven drying, respectively. On the average, the moisture content of the maize cob particles, *C. pentandra* and the mixed samples were 9.00%, 13.27% and 9.21%, respectively.

Briquetting process

A 55.3-mm ID × 52.5-cm height cylindrical mould was used to produce the briquettes. Ninety grammes of each biomass material was weighed and filled into the mould. A manual hydraulic press with a gauge and piston was used to compress the biomass raw material without a binder to form the briquettes. A clearance of about 0.1 mm was provided between the piston and the inner wall of the mould to allow for air escape. The samples were then pressed using the following predetermined compacting pressure levels: 20, 30, 40 and 50 MPa. The dwelling time for each press was maintained at 10 s. This process was repeated for all the biomass materials.

Physical and mechanical properties of briquettes

The relaxed density, compressive strength in cleft and impact resistance index of the briquettes produced were investigated using standard testing methods.

Relaxed density

Relaxed density (RD) of the briquettes was determined 30 days after removal from the press in accordance with [15]. The mass of briquettes were determined using a laboratory electronic balance with an accuracy of 0.01 g. The diameter and length of a briquette were measured at three points with a digital vernier calliper. Relaxed density was then computed as follows:

$$\mathrm{RD}\left({}^{\mathrm{g}}/_{\mathrm{cm}^3}\right) = \frac{108000 \times M(\mathrm{g})}{\pi[d_1(\mathrm{mm}) + d_2(\mathrm{mm}) + d_3(\mathrm{mm})]^2 \times [l_1(\mathrm{mm}) + l_2(\mathrm{mm}) + l_3(\mathrm{mm})]}, \quad (2)$$

where d_1, d_2 and d_3 were diameters (mm) measured at three different points on the briquettes. L_1, L_2 and L_3 were lengths (mm) measured at three different points on the briquettes. M (g) is the mass of briquette.

Compressive strength

Compressive strength in cleft of briquettes was determined in accordance with [16] using an Instron universal strength testing machine (Norwood, MA, USA) with a load cell capacity of 100 kN. The crosshead speed was 0.305 mm/min. A sample of briquette to be tested was placed horizontally in the compression test fixture and a load was applied at a constant rate of 0.305 mm/min until the briquette failed by cracking. The compressive strength in cleft was then computed as follows:

$$\text{Compressive strength in cleft } \left({}^{\mathrm{N}}/_{\mathrm{mm}}\right) = \frac{3 \times \text{The load at fracture point (N)}}{[l_1(\mathrm{mm}) + l_2(\mathrm{mm}) + l_3(\mathrm{mm})]} \quad (3)$$

Impact resistance index

IRI of the briquettes produced was determined in accordance with [17], using the drop shatter test for coal. Five drops were set as the standard. Briquettes were released from a vertical height of 2 m and allowed to freely fall onto a concrete floor. After five drops, the broken pieces of briquettes as a result of the impact were collected and weighed using an electronic balance with an accuracy of 0.01 g. Only the number of pieces which weighed 5% or more of the initial weight was recorded for the purpose of calculating the impact resistance index. The impact resistance index was computed as follows:

$$\mathrm{IRI}(\%) = \frac{N}{n} \times 100, \quad (4)$$

where N is the number of drops and n is the number of pieces that weighed 5% or more of the initial weight of briquette after N drops.

Results and discussions

Relaxed density of briquettes produced from maize cob particles, *Ceiba pentandra* and their mixture

Relaxed density of briquettes produced from maize cobs, *C. pentandra* and their combined materials are presented in Table 2. Relaxed density of briquettes produced from only *C. pentandra* ranged from 523 to 716 kg/m^3 while that produced from maize cob particles only ranged from 541 to 659 kg/m^3.

Additionally, the relaxed density of briquettes obtained from a mixture of maize cob particles and *C. pentandra* ranged from 565 to 774 kg/m^3. The relaxed density of briquettes obtained from all the biomass materials used for the study could be considered adequate since they fall within the recommended values of relaxed density for briquette made using the hydraulic press. It has been previously reported that briquettes made from hydraulic piston press are usually less than 1,000 kg/m^3 and fall between 300 to 600 kg/m^3 [18,19]. It could be further inferred from this study that the relaxed density of all the briquettes produced from the combination of maize cobs and *C. pentandra* particles were higher than that of their corresponding values for the pure biomass materials. This may be due to the variations in the composition and structure of these two materials (*C. pentandra* and maize cob) which resulted in better rearrangement and densification when mixed together and pressed.

Correlation analysis between proportion of maize cob particles in the mixture and relaxed density of briquettes produced indicated that there was no significant correlation between the two parameters (Pearson's $r = 0.166$, p value = 0.102; $N = 60$, $\alpha = 0.05$; one-tailed). However, the compacting pressure was found to have a high significant positive correlation with the relaxed density of the briquettes produced from the mixture of *C. pentandra*

Table 2 Relaxed density (kg/m^3) of briquettes produced from maize cobs, *C. pentandra* and their combination

Biomass material	Mixing percentages	Compacting pressure			
	(weight basis)	20 MPa	30 MPa	40 MPa	50 MPa
Maize cobs only	Pure (100%)	541	611	636	659
C. pentandra only	Pure (100%)	523	622	666	716
C. pentandra/Maize cobs	90:10	565	641	695	742
C. pentandra/Maize cobs	70:30	584	645	708	749
C. pentandra/Maize cobs	50:50	588	661	730	774

Source of data for *C. pentandra* only: [11,12].

and maize cob particles (Pearson's $r = 0.969$, p value = 0.000; $N = 60$, $\alpha = 0.05$; one-tailed). Increased compacting pressure level resulted in a reduction in intermolecular distance as well as increased crushed cell walls, therefore resulting in the formation of denser briquettes.

Analysis of variance (Table 3) on the effect of biomass raw material (maize cobs, *C. pentandra*, and their combination) and compacting pressure on relaxed density of briquettes produced indicated that at 5% level of significance, the biomass raw material, compacting pressure and their interactions have significant effect on the relaxed density of the briquettes produced (p value < 0.05). The multiple coefficient of determination (R^2) and root means square error of the ANOVA model were 0.9730 and 12.93, respectively. The R^2 value of 0.9730 indicates that the biomass raw material and compacting pressure could explain about 97.30% of the variance in the relaxed density of briquettes produced.

Compressive strength in cleft of briquettes produced from maize cobs, *Ceiba pentandra* and their combination

Table 4 shows the results of compressive strength in cleft of briquettes produced from sawdust of maize cob particles, *C. pentandra* and their combination. The result indicates that at all compacting pressure levels, the compressive strength in cleft of briquettes produced from maize cob particles only was very low, ranging from 0.12 to 0.54 N/mm. Compressive strength in cleft of briquettes produced from *C. pentandra* only, was very high and ranged from 29.23 to 44.58 N/mm. Furthermore, the compressive strength in cleft of briquettes produced from combination of maize cob particles and *C. pentandra* ranged from 7.72 to 59.22 N/mm.

This result suggest that briquettes produced only from maize cob particles at low compacting pressure and room temperature will not have adequate compressive strength in cleft for handling (compressive strength in cleft <19.6 N/mm). However, the compressive strength in cleft of briquettes produced from maize cob particles could be improved significantly when it is combined with sawdust from *C. pentandra*. According to Rahman et al. [20], the briquettes surface compressive strength (i.e. compressive strength in cleft) of 19.6 N/mm is reasonably adequate for handling or can be used as fuel for domestic purposes. With the exception of briquettes produced from the 70:30 mixing ratio and pressed at a compacting pressure of 20 MPa, all the briquettes produced from 90:10 and 70:30 mixing ratios had adequate compressive strength in cleft (compressive strength in cleft >19.6 N/mm). Additionally, briquettes produced from the 90:10 (*C. pentandra*/maize cobs) mixing ratio and pressed at 40 and 50 MPa had compressive strength in cleft higher than their corresponding values for *C. pentandra* only. Thus, the addition of 10% maize cob particles to sawdust of *C. pentandra* and pressed at compacting pressures of 40 and 50 MPa will significantly improve the compressive strength in cleft of briquettes produced from *C. pentandra*. This result confirms the research finding of other researchers who undertook a study on densified fuels derived from a blend of two biomass materials. Previous studies [13,21,22] indicated that the durability and mechanical strength of briquettes produced from only one type of biomass materials could be improved by blending that biomass with another biomass material. For instance, the durability of wheat straw briquette was enhanced by blending the straw with wood

Table 3 ANOVA of effect of biomass raw material and compacting pressure on the relaxed density of briquettes

Source	*df*	ANOVA SS	Mean square	*F* ratio	*p* value
Biomass raw material	4	76,341.54	19,085.3850	114.13	<0.0001*
CP	3	392,450.76	130,816.92	782.28	<0.0001*
Biomass material × CP	12	12,562.94	1,046.9117	6.26	<0.0001*
Error	80	13,378.00	167.225		

Briquettes are produced from maize cobs, *C. pentandra* and their combination. *Statistically significant at 0.05 level of significance. Legend: *df*, degree of freedom; CP, compacting pressure.

Table 4 Compressive strength in cleft (N/mm) of briquettes produced using compacting pressure levels from 20 to 50 MPa

Biomass material	Mixing percentages (weight basis)	Compacting pressure			
		20 MPa	30 MPa	40 MPa	50 MPa
Maize cobs only	Pure (100%)	0.12	0.12	0.41	0.54
C. pentandra only	Pure (100%)	29.23	39.26	40.40	44.58
C. pentandra/Maize cobs	90:10	27.29	37.33	44.98	59.22
C. pentandra/Maize cobs	70:30	16.66	22.82	30.00	33.47
C. pentandra/Maize cobs	50:50	7.72	13.02	19.46	24.04

Briquettes are produced from maize cobs, *C. pentandra* and their combination. Source of data for *C. pentandra* only: [11,12].

waste [13]. Correlation analysis between proportions of maize cob particles in the mixing ratio and compressive strength in cleft of briquettes produced, indicated a strong negative correlation between the two parameters (Pearson's $r = -0.770$, p value = 0.000; $N = 60$, $\alpha = 0.05$; one-tailed). This implies that for the mixed biomass briquettes, an increase in proportion of maize cob particles in the mixture resulted in a decrease in compressive strength in cleft of briquettes produced. Compacting pressure was also found to have a high significant positive correlation with the compressive strength in cleft of briquettes produced from the combination of maize cob and *C. pentandra* particles (Pearson's $r = 0.582$, p value = 0.000; $N = 60$, $\alpha = 0.05$; one-tailed).

Two-way analysis of variance (Table 5) on the effect of biomass raw material and compacting pressure on the compressive strength in cleft of briquettes produced indicated that at 5% level of significance the biomass raw material, compacting pressure and their interaction had significant effect on compressive strength in cleft of briquettes produced (p value < 0.05).

The multiple coefficient of determination (R^2) and root mean square error of the ANOVA model were 0.9813 and 2.5939, respectively. Thus, about 98.13% of the variability in compressive strength in cleft of briquettes produced could be explained by the biomass raw material and compacting pressure.

Impact resistance index of briquettes produced from maize cobs, *Ceiba pentandra* and their combination

The impact resistance index of briquettes produced from maize cobs, *C. pentandra* and their combination is presented in Table 6. The results show that at all compacting pressure levels, briquettes produced from only maize cob particles had a very weak or no impact resistance index (%). This may be due to the low lignin content, low water soluble carbohydrates and low protein in the maize cobs. The impact resistance index of briquettes produced from the combination of maize cob and *C. pentandra* particles ranged from 115% to 500%. All the briquettes produced from this combination of biomass materials had adequate impact resistance index, that is, the impact resistance index greater than 100%, the minimum value set for this study.

According to the Italian standard for briquettes/pellets (CTI-R04/5) as cited in [23], briquette durability ≥97.7% is adequate. Additionally, the durability characteristics of briquettes is classified as high when impact resistance index is greater than 80% [24,25]. From the results obtained in this study, it was observed that when maize cob particles is combined with the sawdust of *C. pentandra*, the impact resistance index of the briquettes produced was significantly improved. Briquettes produced from the combination of *C. pentandra* and maize cob particles at 90:10 mixing ratio showed a very high impact resistance index than that produced from just the *C. pentandra* alone. The impact resistance index ranged from 200% to 500%. Furthermore, at the same mixing ratio of 90:10 and compacting pressure of 50 MPa, the briquettes produced did not show any disintegration after being dropped five times from a height of 2 m onto the ground. Correlation analysis between proportions of maize cobs in the mixing ratio and impact resistance index of briquettes produced indicated a strong negative correlation between the two

Table 5 ANOVA of effect of biomass raw material and compacting pressure on compressive strength in cleft of briquettes produced

Source	*df*	ANOVA SS	Mean square	*F* ratio	*p* value
Biomass raw material	4	23,287.04	5,821.76	865.24	<0.0001*
CP	3	3,530.87	1,176.96	174.92	<0.0001*
Biomass material × CP	12	1,437.39	119.78	17.80	<0.0001*
Error	80	538.28	6.73		

Briquettes are produced from maize cobs, *C. pentandra* and their combination. *Statistically significant at 0.05 level of significance. Legend: *df*, degree of freedom; CP, compacting pressure.

Table 6 Impact resistance index (%) of briquettes pressed using compacting pressure levels from 20–50 MPa

Biomass material	Mixing percentages	Compacting pressure			
	(weight basis)	20 MPa	30 MPa	40 MPa	50 MPa
Maize cobs only	Pure (100%)	0	0	0	0
C. pentandra only	Pure (100%)	200	283	316	350
C. pentandra/Maize cobs	90:10	200	300	400	500
C. pentandra/Maize cobs	70:30	142	192	233	450
C. pentandra/Maize cobs	50:50	115	133	150	217

Briquettes are made from maize cobs, *C. pentandra* and their combination. Source of data for *C. pentandra* only: [11,12].

parameters (Pearson's $r = -0.577$, p value = 0.000; $N = 60$, $\alpha = 0.05$; one-tailed). Thus, for a mixture of *C. pentandra* and maize cob particles, increases in the proportions of maize cob particles in the mixture could result in a decrease in the impact resistance index of the briquettes produced. Additionally, compacting pressure was found to have a high significant positive correlation with the impact resistance index of briquettes produced from the combination of maize cob and *C. pentandra* particles (Pearson's $r = 0.614$, p value = 0.000; $N = 60$, $\alpha = 0.05$; one-tailed).

A two-way analysis of variance (Table 7) to determine the effect of biomass raw material and compacting pressure on impact resistance index of briquettes produced indicated that at 5% level of significance, the biomass raw material, compacting pressure and their interactions had significant effect on the impact resistance index of briquettes produced (p value < 0.05). The multiple coefficient of determination and root mean square error of the ANOVA model were 0.8152 and 77.42, respectively. Thus, about 81.52% of the variance in the impact resistance index of briquettes produced could be explained by the biomass raw material and compacting pressure. This result confirms that characteristics of the raw material used significantly have an effect on the quality of briquettes produced.

Conclusion

This study examined the characteristics of briquettes produced from a combination of maize cob particles and *C. pentandra* sawdust at room temperature and low compacting pressure without a binder. From the study, it can be concluded that

1. Briquettes produced from maize cob particles at room temperature using low compacting pressure does not have adequate compressive strength in cleft (compressive strength in cleft <19.6 N/mm) and impact resistance index (0%) for handling, storage and transporting. However, it has adequate relaxed density.
2. Combining *C. pentandra* and maize cob particles at mixing proportions 90:10 and 70:30 and pressed with compacting pressure 30 MPa or more at room temperature will produce briquettes with adequate compressive strength in cleft (compressive strength in cleft ≥19.6 N/mm).
3. Combining *C. pentandra* and maize cob particles at mixing proportions 90:10, 70:30 and 50:50 and pressed with compacting pressure 30 MPa or more at room temperature will produce briquettes with adequate impact resistance index (impact resistance index >100%).
4. The proportion of maize cob particles in a mixture of maize cobs and *C. pentandra* sawdust significantly affects the compressive strength in cleft and impact resistance index of briquettes produced.
5. The type of biomass material and compacting pressure has a significant effect on the relaxed density, compressive strength in cleft and impact resistance index of briquettes produced from maize cobs and sawdust of *C. pentandra*.

This study therefore reveals that it is possible to produce briquettes with adequate physical and mechanical properties from maize cobs when it is combined with sawdust of *C. pentandra*.

Table 7 ANOVA of effect of biomass raw material and compacting pressure on impact resistance index of briquettes

Source	*df*	ANOVA SS	Mean square	*F* ratio	*p* value
Biomass raw material	4	1,496,860.70	374,215.175	62.44	<0.0001*
CP	3	394,655.95	131,551.983	21.95	<0.0001*
Biomass material × CP	12	223,064.90	18,588.742	3.10	0.0012*
Error	80	479,451.20	5,993.140		

Briquettes are produced from maize cobs, *C. pentandra* and their combination. *Statistically significant at 0.05 level of significance. Legend: *df*, degree of freedom; *CP*, compacting pressure.

Competing interests
The authors declare that they have no competing interests.

Authors' contributions
SJM conceived of the study, participated in the design of the study, drafted the manuscript and participated in the sequence alignment. KFM and JOA participated in the design of the study and sequence alignment. NAD participated in the design of the study, performed the statistical analysis and participated in the sequence alignment. All authors read and approved the final manuscript.

Acknowledgements
The authors are grateful to the Civil Engineering Department of Kwame Nkrumah University of Science and Technology, Kumasi, Ghana and Forest Research Institute of Ghana for providing laboratory support for this study. Special thanks also go to Miss Agnes Ankomah of Crop Research Institute of Ghana for providing statistical software for this research work.

Author details
[1]Department of Design and Technology Education, University of Education, Winneba, Kumasi Campus, P. O. Box 1277, Kumasi, Ghana. [2]Department of Wood Science and Technology, Kwame Nkrumah University of Science and Technology, Kumasi, Ghana. [3]Department of Agricultural Engineering Kwame Nkrumah University of Science and Technology, Kumasi, Ghana.

References
1. Zych, D: The viability of corn cobs as a bioenergy feedstock. Literature review, University of Minnesota, Minnesota, USA (2008)
2. Amesimeku, EK: Mechanising Ghana's agricultural sector - a sure way to ending poverty. Ghana oil and Gas summit. La Palm Royal Beach Hotel Accra, Ghana (2012)
3. Maninder, Kathuria, RS, Grover, S: Using agricultural residues as a biomass briquetting: an alternative source of energy. IOSR J. Elec. Electronics. Engin. **1**(5), 11–15 (2012)
4. FAOSTAT: Crop production in Ghana in 2010. Food and Agriculture Organisation of the UN, Rome, Italy (2010). Web site http://faostat.fao.org/site/339/default.aspx
5. OECD/IEA: Sustainable production of second-generation biofuels, potential and perspectives in major economies and developing countries. Information paper, Paris (2010). Web site http://www.iea.org/publications/freepublications/publication/biofuels_exec_summary.pdf
6. Martinov, ML, Veselinov, BV, Bojić, SJ, Djatkov, DM: Investigation of maize cobs crushing - preparation for use as a fuel. Therm. Sci. **15**(1), 235–243 (2011)
7. Bhattacharya, SC, Sett, S, Shrestha, RM: State of the art for biomass densification. Energy Sources **11**, 161–182 (1989)
8. O'Grady, M, Farley, R, Arnold, R: Palletized wood. In: Levy, MP, O'Gracy, MJ (eds.) Decision Maker's Guide to Wood Fuel for Small Industrial Energy Users. Solar Energy Research Institute, Golden, CO (1980)
9. Kaliyan, N, Morey, RV: Densification characteristics of corn cobs. Fuel Process Technol. **91**(5), 559–565 (2010)
10. Mullen, CA, Boateng, AA, Goldberg, NM, Lima, IM, Laird, DA, Hicks, DA: Bio-oil and bio-char production from corn cobs and stover by fast pyrolysis. Biomass Bioenergy **34**, 67–74 (2010)
11. Mitchual, SJ: Densification of sawdust of tropical hardwoods and maize cobs at room temperature using low compacting pressure without a binder. PhD Theses, School of Graduate Studies, Kwame Nkrumah University of Science and Technology, Kumasi, Ghana (2013)
12. Mitchual, SJ, Frimpong-Mensah, K, Darkwa, NA: Effect of species, particle size and compacting pressure on relaxed density and compressive strength of fuel briquettes. Int. J. Ener. Environ. Engin. **4**(30), 6 (2013)
13. Wamukonya, L, Jenkins, B: Durability and relaxation of sawdust and wheat-straw briquettes as possible fuels for Kenya. Biomass Bioenergy **8**(3), 175–179 (1995)
14. European standard EN 13183-1: Moisture content of a piece of sawn timber. Determination by oven-dry method, BSI, London (2002)
15. ISO: International standard ISO 3131: Standard test method for density of regular solids. ISO, Geneva (1975)
16. ASTM International: Standard test method of compressive strength of wood. ASTM standard D2166-85, West Conshohocken (2008)
17. ASTM International: Standard test method of drop shatter test for coal. ASTM standard D440, West Conshohocken (2007)
18. Tumuluru, JS, Wright, CT, Kenny, KL, Hess, R: A review on biomass densification technologies for energy application. US Department of Energy, Washington, USA (2011).
19. Saeidy, EE: Technological fundamentals of briquetting cotton stalks as a biofuel. DissertationHumboldt-Universität, zu Berlin, Berlin (2004)
20. Rahman, ANE, Aziz, MM, Prasad, CSN, Venkatesham, M: Influence of size and shape on the strength of briquettes. Fuel Process Technol. **23**, 185–195 (1989)
21. Yaman, S, Sahan, M, Haykiri-Acma, H, Sesen, K, Kucukbayrak, S: Fuel briquettes from biomass-lignite blends. Fuel Process Technol. **72**, 1–8 (2001)
22. Yaman, S, Sahan, M, Haykiri-Acma, H, Sesen, K, Kucukbayrak, S: Production of fuel briquettes from olive refuse and paper mill waste. Fuel Process Technol. **68**, 23–31 (2000)
23. Hahn, B: Existing guidelines and quality assurance for fuel pellets - pellets for Europe project. UMBERA, Umweltorientierte Betriebsberatungs-, Forschungs- und Entsorgungs-Gesellschaft m.b.H, St. Polten, Austria (2004)
24. Tabil, L, Sokhansanj, S: Process conditions affecting the physical quality of alfalfa pellets. Appl. Eng. Agric. **12**(3), 345–350 (1996)
25. Adapa, PK, Schoenau, GJ, Tabil, LG, Sokhansanj, S, Crerar, B: Pelleting of fractionated alfalfa products. ASAE Paper 036069, ASABE, St. Joseph, Mich, USA (2003)

Selection of organisms for systems biology study of microbial electricity generation: a review

Longfei Mao* and Wynand S Verwoerd

Abstract

A microbial fuel cell (MFC) is a device that uses microorganisms as biocatalysts to transform chemical energy or light energy into electricity. However, the commercial applications of MFCs are limited by their performance. This review presents the perspective that *in silico* metabolic modelling based on genome-scale metabolic networks can be used for understanding the metabolisms of the anodic microorganisms and optimizes the performance of their metabolic networks for MFCs. This is in contrast to conventional research that focuses on engineering designs and study of biological aspects of MFCs to improve interactions of anode and microorganisms. Four categories of biocatalysts - microalgae, cyanobacteria, geobacteria and yeast - are nominated for future *in silico* constraint-based modelling of MFCs after taking into account the cell type, operation mode, electron source and the availability of metabolic network specifications. In addition, the advantages and disadvantages of each organism for MFCs are discussed and compared.

Keywords: Microbial fuel cell (MFC), Flux balance analysis, *Synechocystis*, *Chlamydomonas*, *Saccharomyces*, *Geobacter*, Cyanobacteria, Algae, Genome-scale metabolic network, Constraint-based modelling

Review

Introduction

The technology for extracting an electrical current for use in external circuits from the metabolic processes of living microbes has been in development for more than a century [1]. The resulting devices, termed microbial fuel cells (MFCs), have several potential advantages over more prominent sustainable energy technologies such as solar or wind power. For example, they can directly convert organic waste into electricity [2] without pollution or inefficient intermediate steps that involve mechanical generators. This feature, energy recovery from solid wastes, has been exploited in proposed national strategies for many Asian countries [3]. It may be possible to achieve the same goal by inorganic catalysts or enzymes, but using living cells makes it possible to exploit their adaptability to environmental conditions and avoids the high capital cost of installation for other waste-to-energy systems reviewed by Eddine and Salah [4]. The whole organisms used in MFCs contain various enzymes and therefore allow different substrates (or mixed substrates) to be used. The organisms in the fuel cell system can be considered as micro-reactors and provide optimal conditions for different enzymes. Because the organisms are self-replicating, the organic matter oxidation implemented by these bio-catalysts are self-sustaining [5] and not subject to catalytic poisoning like metallic catalysts or degradation of enzyme catalysts. By selecting photosynthetic microbes, solar energy could be converted at the same time. One can envisage portable electronics powered by MFCs that are 'charged' by feeding them nutrients rather than electric current, or medical implants that derive their power directly from nutrients circulating in the bloodstream. Perhaps the process can be reversed, and external electrical power supplied to an MFC converted into biomass, as a temporary storage, to overcome the intermittent nature of many other sustainable energy sources - a possibility currently under serious consideration [6-8].

However, these future possibilities are still severely hampered by the low energy yields per mass or volume that are currently achieved. Generally MFC energy output is reported in milliwatts per square metre of electrode area or per cubic metre of electrolyte volume [9].

* Correspondence: longfei.mao@lincolnuni.ac.nz
Centre for Advanced Computational Solutions, Wine, Food and Molecular Bioscience Department, Lincoln University, Ellesmere Junction Road, Lincoln 7647, New Zealand

Scientific research has increased the densities of MFCs to over 1 kW m^{-3} (reactor volume) and to 6.9 W m^{-2} (anode area) under optimal conditions in the laboratory [10]. However, these values still cannot meet the needs of many applications, which require a power output larger than 100 kW m^{-2} [11]. For this reason large-scale waste water treatment is the application closest to industrial realisation and is the domain of much current MFC research.

A variety of designs are under development to improve the efficiency and potential application of MFCs to industry. Areas under investigation include the selection of electrode materials for optimal electrochemical performance and maximising electrode surface to volume ratios; improving charge transfer between microbes and electrodes either chemically or by mechanical design; and finding and maintaining optimal living conditions for microbe colonies, efficient supply of nutrients and removal of effluent. Different configurations are being investigated for extracting current, sometimes in combination with production of hydrogen or other metabolites of further use in energy generation and with or without exploitation of photosynthesis. The choice of process configuration and engineering design is also closely linked with the selection of the most suitable organism for a particular design or for whether overall priority is given to energy generation, waste disposal or some other objective.

A schematic representation of MFC research activity is shown in Figure 1. While there is a large volume of biochemical research literature on, for example, electron transfer chains and redox processes in cell metabolism that is relevant to MFC, relatively few studies focus specifically on MFC. This is exacerbated by the fact that ongoing research continues to identify new mechanisms for electron exchange between microbes and electrodes, new design strategies to exploit these and consequently new candidate organisms. Such organisms have not necessarily been well studied experimentally before.

In silico modelling is well suited to bridge this gap and extend knowledge in the biochemical interface between MFC biology and engineering design. Externally, electron flow (in the external circuit) and the counterflow of protons in the electrolyte make up the current that carries useful electrical power. Internally, both of these are comprehensively woven into the fabric of metabolism: electrons being transported by redox carrier molecules such as reduced nicotinamide adenine dinucleotide (NADH) that participate in a large fraction of all biochemical reactions, and protons that, for example, drive adenosine triphosphate (ATP) synthesis needed for energy transport are also ubiquitously involved in a great many reactions. This clearly calls for a systems-level approach rather than the reductionist strategy of pathway-oriented, conventional biochemistry.

This is the domain of systems biology [12]. Systems biology provides *in silico* models that incorporate biological data, metabolic flux data and different physicochemical constraints such as the conservation of mass and energy, thermodynamics, redox balance, etc. [12] and thus provides an opportunity to identify the bottlenecks hidden in a complex network of interactions and cellular compartmentation [13].

The kinetic behaviour of a metabolic network at a whole-genome level can be constructed and analysed through a mathematical model [14,15]. However, the characterization of metabolic networks is still far from comprehensive in databases [16] and even in the best-understood organisms, the majority of kinetic parameters are undetermined.

The development of new computational methods allows for the whole-network modelling of metabolism and conduction of compelling and testable predictions even without many parameters. The key idea is to incorporate stoichiometry and other fundamental principles as mathematical constraints, which separate feasible and infeasible metabolic behaviours. Compared with kinetic parameters, these constraints are much easier to

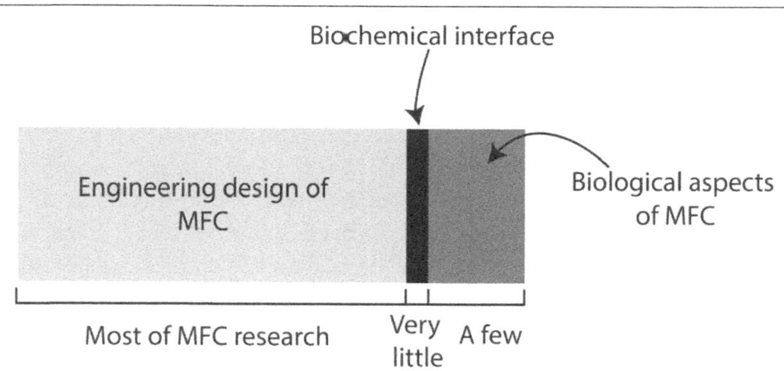

Figure 1 Areas covered in current microbial fuel cell research.

identify and make it possible to build a large-scale model [17]. These constraint-based modelling approaches allow integration of high-throughput post-genomic data but describe steady states and generally offer no information about metabolite concentrations or the temporal dynamics of the system [18-20].

Genome-scale metabolic modelling requires high-quality metabolic network reconstructions [14]. The reconstructions are based on sequenced genomes and are generally built manually using information from metabolic databases (e.g. KEGG and BRENDA) and primary literature. The metabolic network reconstruction process is described in detail elsewhere [21]. Recently, due to the development of the high-throughput technologies, the reconstructions have now been built for various organisms [22].

Flux analysis can be combined with cell biology and sub-cellular biochemistry to reveal the functionality and efficiency of the enzymes associated with cell biological components or structures [23]. Metabolic regulation can be deeply understood only when multiple system components are examined simultaneously. This kind of analysis has been conducted in microbial and medical research in recent years.

Flux determinations can produce results that are hardly predictable from observed changes in transcript or protein levels because most of metabolic control takes place at post-translational levels and enzyme activities are often not correlated with changes in transcript or protein levels [23]. The incorporation of data from enzyme platforms should make the functionality of genomics strategies more clear. System-wide metabolic flux characterization is an important part of metabolic engineering [23].

Nevertheless, there is no published literature that uses genome-scale flux models to study the metabolic behaviour of biocatalysts in MFCs. The only attempt to use a genome-scale model to study biocatalyst behaviour in an MFC was presented as a conference abstract paper in the 17th European Symposium on Computer Aided Process Engineering (2007) [24]. However, this paper is an immature work that did not provide the source of the central metabolic network, describe the gene knockout methodology or discuss the results in relation to other MFC experimental work. Therefore, future research activities are urgently needed to fill the research gap indicated in Figure 1, with recently advanced constraint-based modelling approaches.

An essential first step in applying constraint-based analysis to microbial fuel cells is to choose appropriate organisms for further study. Due to the varied strategies and designs alluded to above, no single organism can serve as a suitable model, and a major advantage of *in silico* modelling is that different organisms can be studied in the same framework to facilitate mutual comparisons. This paper reviews the background against which such choices can be made and proposes a set of four organisms for the purpose.

The 'Microbial fuel cells' section explains the construction, operation and classification of current MFC designs, and the 'Current directions of MFC research' section reviews the issues being addressed in current MFC research. Based on this, the 'Microorganisms for *in silico* study of MFC functioning' section discusses a selection of organisms that are representative of various combinations of biological aspects that can be exploited in MFCs, while also featuring well-established metabolic network reconstructions, suitable for the computational analysis.

Microbial fuel cells

MFCs are unique devices that can use microorganisms as catalysts for transforming chemical energy directly into electricity. The biggest advantage of an MFC is that it can generate combustion-less, pollution-free bioelectricity directly from the organic matter in biomass [2]. In an MFC the energy stored in chemical bonds in organic compounds is converted to electrical energy through enzymatic reactions by microorganisms. Thus, the electricity production by MFC is associated with the normal living processes of bacteria capturing and processing energy.

In a typical MFC configuration (Figure 2), microorganisms are situated in the anodic compartment and use the biomass for growth while forming electrons and protons [25]. The electrons are transported out of cells to an electrode using redox mediators or directly expelled by some microorganisms for reducing the substrate. The protons or H^+ ions are diffused through the electrolyte to the cathode where it is oxidised to water. The cathode can be in a separate chamber (i.e. double-chambered MFCs) or in the same chamber (i.e. single-chambered MFCs). A single-chambered MFC eliminates the need for the cathodic chamber by exposing the cathode directly to the air. The only by-product released by MFCs is carbon dioxide, which can be fixed by plants for photosynthesis.

MFCs require running under conditions predefined by the optimum growth and living conditions of the used microorganisms. Thus, factors affecting the MFC's efficiency include electrode material, pH buffer and electrolyte, proton exchange system and operating conditions in both the anodic chamber and the cathodic chamber. MFCs are usually operated at ambient temperature, at atmospheric pressure and at pH conditions that are neutral or only slightly acidic [26].

MFCs harness the electrons from these systems in three main operation modes: mediated electron transfer

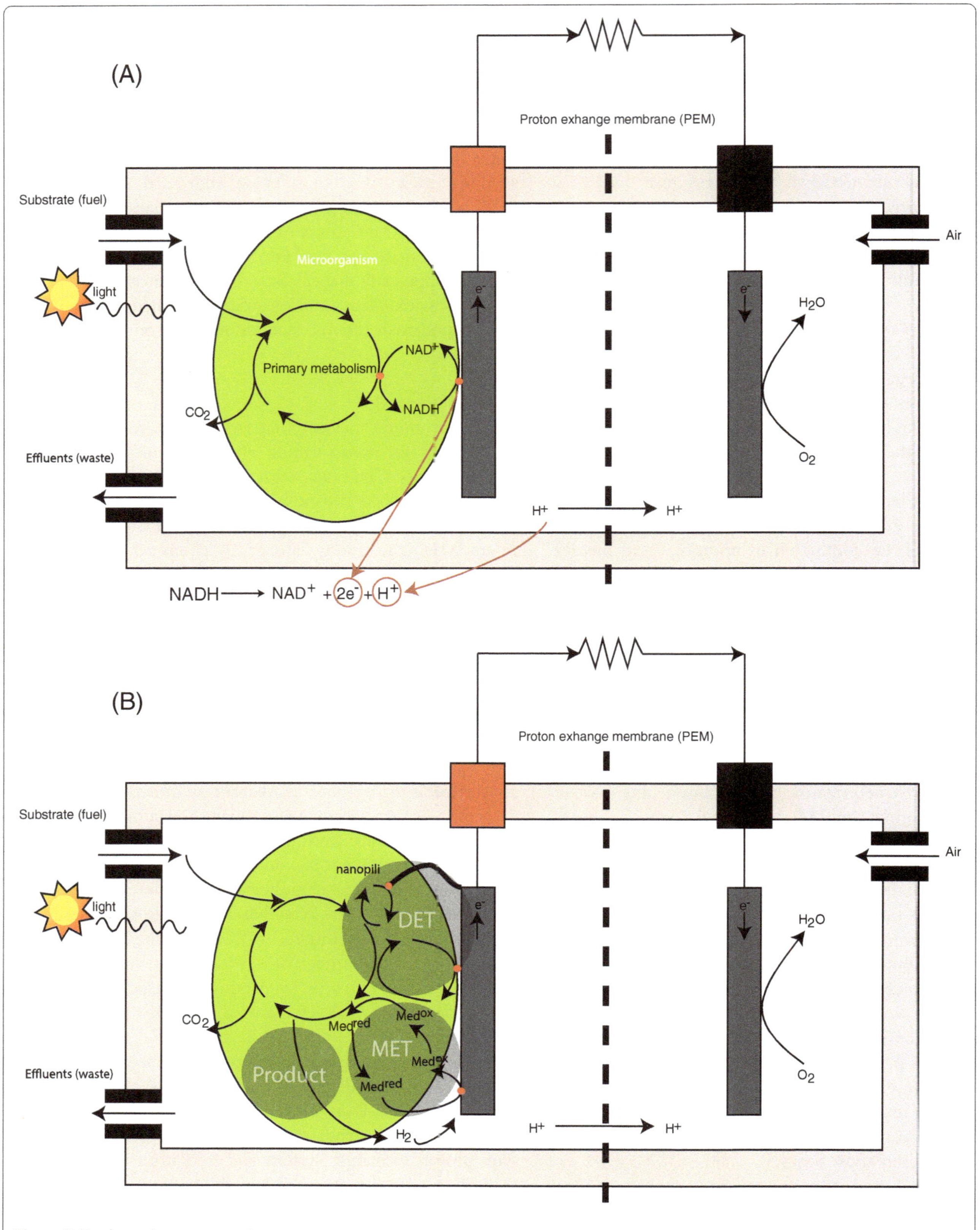

Figure 2 (See legend on next page.)

(See figure on previous page.)

Figure 2 The working principle of a microbial fuel cell. (**A**) A bacterium in the anode compartment transfers electrons obtained from an electron donor (glucose or light in the case of photosynthetic organisms) to the anode electrode. Protons are also produced in excess during electron production. These protons flow through the proton exchange membrane (PEM) into the cathode chamber. The electrons flow from the anode through an external resistance (or load) to the cathode where they react with the final electron acceptor (oxygen) and protons. (**B**) Three electron transfer modes: (1) directly via membrane-associated components (DET), (2) mediated by soluble electron shuttles (MET) or (3) primary product (Product), e.g. H_2 can act as a fuel to be oxidised to provide electrons for the electricity circuit. Med, redox mediator; Red oval, terminal electron shuttle in or on the bacterium.

(MET), direct electron transfer (DET) and product mode. Photosynthetic MFCs use photosynthesis as the electron source and can also be operated in the same modes.

Mediated electron transfer

MET is defined as where a mediator molecule acts an electron relay that repeatedly cycles between the reaction sites and the electrode [11]. MET is the most common electron transfer mode used in MFCs and can be classified into two sub-types [11]:

- Indirect transfer systems that involve freely diffusing mediator molecules (i.e. diffusive MET)
- Indirect transfer systems in which the mediator is integrated into the electrode or the cell membrane (i.e. non-diffusive MET)

In diffusive MET, the mediators enter the cell membrane and exchange electrons between cellular metabolism inside the cell and the electrode outside it. In the non-diffusive MET, the mediator can collect the electrons from the cell membrane without penetrating the cell.

Based on the type of mediators, diffusive MET can also be classified into three sub-categories:

- MET via exogenous (artificial) redox mediators
- MET via secondary metabolites
- MET via primary metabolites

The detailed mechanisms in those three classes are discussed in [27]. Because the terminal electron transfer to or from the electrode determines the overall cell potential, potential (voltage) losses can be minimised by using a mediator that has a reaction potential near that of the biological component.

Direct electron transfer

DET is defined as the case where electrons cycle directly between a microorganism and an electrode. DET can be achieved through two naturally occurring mechanisms:

- Membrane-bound c-type cytochromes, which exist in the cell membrane in some organisms [28,29] to provide electron transfer capacity. For example, multi-heme proteins have especially evolved in sediment-inhabiting metal-reducing microorganisms such as *Geobacter* [30], *Rhodoferax* [31] and *Shewanella* [32]. In their natural environment, iron (III) oxides act as the solid terminal electron acceptors, but in the case of MFC, the anode is used as the solid electron acceptor.
- Electronically conducting nanowires. The DET via outer membrane cytochromes requires the cytochrome (the bacterial cell) to be physically adhered to the fuel cell anode. When a biofilm is formed, only bacteria in the first monolayer at the anode surface are electrochemically active [30]. Thus, the maximum cell density in this bacterial monolayer usually influences the MFC performance. However, it has been shown that some *Geobacter* and the *Shewanella* strains can evolve electronically conducting molecular pili (nanowires of 2 to 3 μm long, made of fibrous protein structures [33]) that make the microorganism able to reach and use more distant solid electron acceptors [34,35]. The pili are connected to the membrane-bound cytochromes and allow transference of the electrons to the distant electron acceptors without cellular contact (Figure 2B). Thus, thicker electroactive biofilms can be formed to increase anode performance. It was shown that fuel cell performance can be increased up to tenfold upon nanowire formation of *Geobacter sulfurreducens* [35].

Product type

In product-type MFCs, microbes metabolize the substrate, releasing a secondary fuel product such as hydrogen that then diffuses to the electrode and is oxidised or reduced (as appropriate) to form a final waste product, which is discharged [11]. The product operation is similar to conventional fermentation processes, in that products of the microbial metabolism are used as the fuel at the electrode.

The product system is made up of two independent stages: one is storage of the microbial reaction product, and the other one is the product being fed to a conventional fuel cell process driven by non-biological catalysis, such as in the case of a proton exchange membrane fuel cell, where H_2 is converted into electricity [36]. These stages may also be physically separated in different containment vessels. However, a product system only truly belongs to a biofuel cell system when the microbes and the electrode are together in the same anode compartment [11]. Nowadays, the fermentation (mostly to hydrogen gas) usually takes place in the fuel cell itself [37,38].

Product systems have two main drawbacks, one is that the efficiency of the conversion of the biological substrate to hydrogen is quite low, and the other is that hydrogen oxidation requires high fuel cell temperatures. Also, the produced biofuel gas is always contaminated with other by-products such as CO, H_2S and (poly)siloxanes making it not sufficiently pure for direct use in a fuel cell [39].

Photosynthetic MFCs

Photosynthetic MFCs are MFCs that generate electricity from a light source rather than a fuel substrate and require the mediator involved to be light stable [40]. Conventionally two operating modes exist for photosynthetic MFCs:

- Energy is produced and stored by the microorganism during illumination and then released and processed in the same way as in a non-photosynthetic biofuel cell.
- The energy produced during illumination may be directly extracted in the form of electrons for creating an external electrical circuit.

A single photosynthetic MFC may possess both of these two modes of action. However, it is recently thought better to classify photosynthetic MFCs into categories based on seven approaches that integrate photosynthesis with MFCs - photosynthetic MFCs [40]:

1. Photosynthetic bacteria at the anode with artificial mediators
2. Hydrogen-generating photosynthetic bacteria with an electrocatalytic anode
3. A mixed culture, with photosynthetic bacteria supplying organic matter to heterotrophic electroactive bacteria at the anode
4. Photosynthesis in plants, supplying organic matter via rhizodeposits to heterotrophic electroactive bacteria at the anode
5. An external photosynthetic bioreactor, where only biomass or metabolic products are transferred to the anode compartment to feed heterotrophic electroactive bacteria
6. Direct electron transfer between photosynthetic bacteria and electrodes
7. Photosynthesis at the cathode to provide oxygen

These sub-types have been discussed in detail by Rosenbaum et al. [40].

Microorganisms suitable for MFCs

Most microorganisms are unable to donate sufficient electrons outside of cells to produce usable currents, because the outer layers of most microbial species are made up of non-conductive lipid membrane, peptididoglycans and lipopolysaccharides which restrain electron transfer to the anode [41]. Since the 1980s, it has been found that artificial water-soluble electron shuttles (i.e. methylene blue, thionine, neutral red and 2-hydroxy-1,4-naphthoquinone) can be used as mediators that transport the electrons from electron carrier molecules inside the cell (e.g. NADH, NADPH or reduced cytochromes) to the anode surface [41]. For example, an MFC based on *Proteus vulgaris* used thionine as a mediator to generate electricity from sucrose [42].

Since the 1990s, some bacterial species such as *Pseudomonas aeruginosa* [43] and *Clostridium butyricumcan* [44] have been found to be able to self-mediate extracellular electron transfer using their own metabolic products. Meanwhile, direct transfer of electrons (DET) that involves use of electrochemically active redox enzymes (i.e. cytochromes) has been discovered in a number of bacterial species such as *Shewanella putrefaciens* [28,29,45], *Shewanella oneidensis* [46], *Geobacter sulfurreducens*[a] [30,47], *Rhodoferax ferrireducens* [31], and the oxygenic phototrophic cyanobacterium *Synechocystis* sp. PCC 6803[a] [34]. These microorganisms are termed as exoelectrogens, and among them *S. oneidensis* and *G. sulfurreducens* have evolved electronically conducting molecular pili (nanowires) to further facilitate the DET [34,35]. Besides DET mode, *S. oneidensis* can conduct MET using a self-produced mediator [48]. The exoelectrogens in MFCs are thought to actively use electrodes to conserve electrochemical energy required for their growth and thus ensure high rates of fuel oxidation and electron transfer for the production of electrical energy [5].

In most of the previous MFC studies, bacteria have been used for electricity generation. On the other hand, since 2000, eukaryotes such as microalgae (e.g. *Chlamydomonas reinhardtii*[a]) and yeast (e.g. *Saccharomyces cerevisiae*[a]) have also emerged as good choices for MFC use, because they have been studied as model microorganisms in the lab and have been widely used in the industry for a long time.

Current directions of MFC research

Engineering design and biological aspects

Most previous studies tended to improve power densities of MFCs by optimizing the reactor configuration and operation parameters [49,50], such as modifications of the electrode materials to incorporate metals that contain current collectors [51,52], use of metals highly optimized for bacterial adhesion and metals possessing high electrical conductivity to minimise ohmic losses

[10], and application of a biocathode that can increase MFC performance by improved oxidation of hydrogen at the cathode [53]. Applications of chemical treatments and precious metals to electrodes in order to increase power production in the laboratory have also been investigated [54,55]. However, these modifications, like the use of larger laboratory reactors, may increase the cost and lead to compromises on performance based on material costs. Many bottlenecks also exist for improving those physical and chemical properties.

Since the fundamental source of electrons is the cellular metabolism, it is particularly important to focus on biological processes that take place in the microbial cells. In this regard, further clarification is needed of the factors relevant to the anodic catalysis process, such as the diversity of the electrochemically active microorganisms [56] and especially the electricity generation mechanisms in relation to normal metabolic states.

Mediator-less, mediator-self-producing and artificial mediator-based MFCs

Mediator-less MFCs are a more recent development relying on evolved ability of exoelectrogens for disposal of electrons originating from substrate oxidation. This type of MFCs has been increasingly preferred, because use of mediators complicates the cell design and these mediators are usually toxic, costly and unsustainable, limiting MFC development [27,57-60]. In addition, mediator-involved MFCs usually produce low current densities (0.1 to 1 A m^{-2}) [27]. Unfortunately, mediator-less MFCs may not yet find a wide range of applications since the discovered exoelectrogens are still few in number and it is non-exoelectrogens that are largely used in the agricultural and industrial areas [61]. Thus, an important direction of MFC research is development of MFCs using non-exoelectrogens without exogenous mediators [61].

However, problems arise from the fact that redox molecules used in electron transfer reactions are not situated on the outer membrane, but in the cytoplasmic membrane. One way is to develop direct electron transfer using carbon nanoparticles that can contact the redox centres that are incorporated in the interior cell membrane [61]. Another way is to identify and develop self-produced mediators (e.g. in the case of *Shewanella* species mentioned above) through engineering methods [56].

Conventional photosynthetic MFCs

It is appealing to study whether phototrophic microbes can be used in an MFC generating electricity from sunlight, because sunlight is an unlimited energy resource and more solar energy reaches the Earth in 1 h (4.1×10^{20} J) than the energy consumed on the planet in a year [56,62,63]. In addition, the development of a self-sustainable photosynthetic MFC is important to meeting energy requirements at remote locations, where routine addition of fuel would be technically difficult and expensive [56].

Photosynthetic MFCs can generate electricity indirectly or directly. For example, in the indirect way, *Rhodobacter sphaeroides* in the MFC can produce H_2 that is oxidised at a platinum coated anode to generate electricity [64,65], whereas in the direct way, *Rhodopseudomonas palustris*, a photosynthetic purple non-sulphur bacterium, can generate electricity in a biofilm anode MFC by direct electron transfer [66].

There are also more traditional photosynthetic MFC configurations where photosynthetic organisms live with other microbes and supply products to heterotrophs [67]. Photosynthetic microorganisms (e.g. cyanobacteria or microalgae) and heterotrophic bacteria exhibit synergistic interactions [68] that can be used in self-sustained phototrophic MFCs [62]. An indirect synergistic relationship between photosynthetic organisms and electricigens has been exemplified in a recent study, in which algal photobioreactors were used to supply organic matter produced via photosynthesis to an MFC for electricity generation [69]. The operation of this type of photosynthetic MFCs is CO_2 neutral and does not need buffers or exogenous electron transfer mediators [67]. However, the photosynthetic MFC power densities obtained are quite low when compared with those that are currently reported for conventional MFCs, e.g. 0.95 mW m^{-2} for polyaniline-coated and 1.3 mW m^{-2} for polypyrrole-coated anodes [70] versus values in the watt per square metre range for conventional cells.

MFCs based on the photosynthetic electron transfer chain

Recently, the photosynthetic electron transfer chain is considered as a source of the electrons harvested on the anode surface, which is different from the previously designed anaerobic MFCs, sediment MFCs or anaerobic photosynthetic MFCs [70]. A single-chamber photosynthetic MFC based on two photosynthetic cultures, planktonic cyanobacteria *Synechocystis* sp. PCC 6803 and a natural freshwater biofilm, has shown a positive light response (i.e. immediate increase in current upon illumination) [70]. This phenomenon proves that it is possible to extract electrons directly from the photosynthetic electron transfer chain, and not only from the respiratory transfer chain or through oxidation of hydrogen [71].

Microorganisms for *in silico* study of MFC functioning

Categories and representatives

Because the electricity generation in MFCs is based on the metabolic activity of living microorganisms, experimental screening of different microorganisms for better anodic activity has long been recognised as a fundamental way to improve MFC performance. It is also possible to improve the performance by culturing microorganisms under selective pressure for enhanced power production [30,72].

Compared to experimental studies, *in silico* modelling is less constrained to a particular MFC design and operating mode. Clearly no single organism is likely to be optimal for all of the varied designs discussed before. To date, every microorganism used in previous MFC studies has advantages and disadvantages. Selection of the microorganism depends on a variety of factors such as types of application, the capability of power generation, the availability of types of energy source for bacterial survival and the ability of extracellular electron transfer, in that electrodes are not natural electron acceptors.

From the modelling perspective, a broader view is possible. Categories of microbes can be identified to cover the range of operating modes and, within these, individual organisms selected that will allow different modes to be individually studied and also compared quantitatively.

MFC microbial communities can be divided into three groups: heterotrophic cells, photoheterotrophic cells and sedimentary cells [9]. The distinction between phototrophic and heterotrophic metabolism is fundamentally important in determining the operating mode. Another key distinction is between prokaryotes and eukaryotes. Compared to prokaryotic species and mixed cultures that have been mostly studied for different MFC applications, fewer studies involve eukaryotes as biocatalysts in MFC operations [73]. This is because the metabolic processes of eukaryotic cells take place in the membrane surrounded cell organelles (e.g. chloroplasts) and is thus putative to be difficult for some commonly used redox mediators such as 2-hydroxy-1,4-nepthoquinone to get access to [57,74]. While prokaryotes have the advantage that their simpler cell membranes and internal structure are more amenable for physical electron extraction, the more complex metabolism of eukaryotes may be more efficient and be able to support a larger diversion of redox carrier flux without undue harmful effects on the organism.

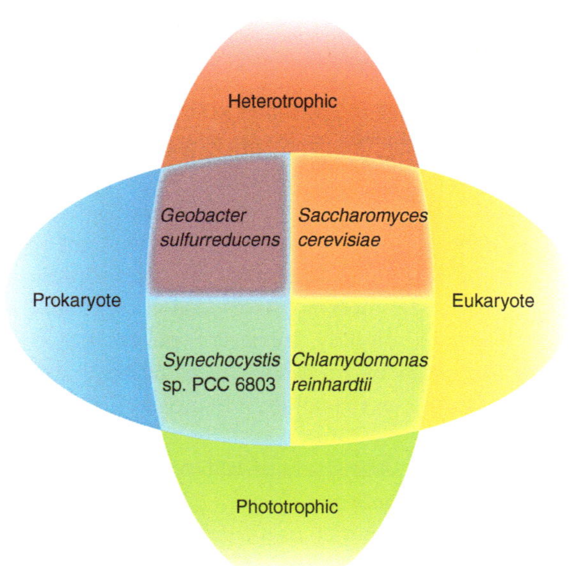

Figure 3 Classification of proposed microbes.

The four anodic microorganisms: *C. reinhardtii*, *Synechocystis* sp. PCC 6803, *S. cerevisiae* and *G. sulfurreducens*, each combine a different pair of key features and are proposed as good candidates for MFC *in silico* characterization. As illustrated in Figure 3, *C. reinhardtii* and *S. cerevisiae* are eukaryotes, whereas *Synechocystis* sp. PCC 6803 and *G. sulfurreducens* are prokaryotes. The four organisms also cover the three groups of the MFC microbial communities mentioned above, i.e. *C. reinhardtii* and *Synechocystis* sp. PCC 6803 are photoheterotrophic cells, *S. cerevisiae* belongs to heterotrophic cells and *G. sulfurreducens* is a typical sedimentary cell.

C. reinhardtii and *Synechocystis* sp. PCC 6803 are photosynthetic organisms that are also capable of producing hydrogen. The comparison of organisms with/without photosynthesis can be used to study exploitation of photosynthetic and respiratory electron transport chains to supply MFC current.

A further consideration in a modelling study is whether the data are available and computationally manageable. All four microorganisms have been studied extensively as model organisms and used in various industries for a long time, and thus, related molecular tools and biological mechanisms are abundant. In particular, genome-scale metabolic networks have been reconstructed and are regularly updated for these four organisms. Based on a literature review, the most updated models are shown in Table 1. These models include the natural redox mediators (i.e. NADH) that are well balanced for the cellular energy metabolisms (e.g. oxidative phosphorylation, glycolysis, Calvin cycles and tricarboxylic acid (TCA) cycles) and are thus practicable for MFC modelling.

It is noted that two models were published in the same period for *C. reinhardtii* [75,76] and *Synechocystis* [77,78]. These models for the same organisms are not compared with each other by the authors, and thus, their limitations can only be revealed during the MFC modelling.

The biological features of the four microorganisms are summarized in Table 2, and the relevance of each of these microbes is reviewed in more detail in the following sections.

Chlamydomonas reinhardtii

C. reinhardtii is a unicellular green alga that belongs to the Chlorophytes division, which diverged from the

Table 1 The scope of the genome-scale models of the four selected organisms

	Chlamydomonas reinhardtii		*Synechocystis* sp. PCC 6083		*Geobacter sulfurreducens*	*Saccharomyces cerevisiae*
Gene	2,249	1,080	1,811	678	617	918
Metabolite	1,862	1,068	465	795	644	1,655
Reaction	1,725	2,190	493	863	709	2,110
Compartment	4	10	3	4	2	17
Date	December 2011	August 2011	October 2011	January 2012	2009	2012
Reference	[75]	[76]	[77]	[78]	[79]	[80]

The number indicates the counts of relative items in the network models.

Streptophytes division (including land plants) more than one billion years ago [85]. *C. reinhardtii* is an approximately 10-μm, unicellular, soil-dwelling green alga with an eyespot, a nucleus, multiple mitochondria, two anterior flagella for motility and mating, and a single cup-shaped chloroplast that accommodates the photosynthetic apparatus [86,87].

Like plants, *C. reinhardtii* has a cell wall and can grow in a medium lacking carbon and energy sources when illuminated [87]. Unlike angiosperms (flowering plants), this microorganism has functional photosynthetic apparatus even when in dark conditions and with an organic carbon source [87]. In the dark, acetate is the sole carbon source used by wild-type *C. reinhardtii in vivo* [88].

Because of the relative adaptability and quick generation time, *C. reinhardtii* has been used as a model to study eukaryotic photosynthesis, eukaryotic flagella and basal body functions and the pathological effects of their dysfunction [89,90], and investigated for water bioremediation and biofuel generation [87,91-93]. The cDNA, genomic sequence and mutant strains of *C. reinhardtii* are publicly available through the *Chlamydomonas* Center [94].

Advantages of algae *C. reinhardtii* inherits all potential advantages of algae for industrial use and scientific study, including [95-97] the following: (1) Algae biomass can potentially be produced at extremely high volumes, and this biomass can yield a much higher oil (1,000 to 4,000 gal/acre/year) than soybeans and other oil crops [97]. (2) Algae do not compete with traditional agriculture because they are a non-food-based resource which can be cultivated in large open ponds or in closed photobioreactors located on low-productive or non-arable land. (3) Algae have a good adaption to different climate and water conditions and can be grown in a wide range of water sources, such as brackish, saline, or fresh and waste water. (4) Algae can make use of resources that would otherwise be considered waste as substrate for growth [97]. (5) Algae can use and sequester CO_2 from

Table 2 Comparison of facts of four selected microorganisms

Name	*Chlamydomonas reinhardtii*	*Synechocystis* sp. PCC 6083	*Saccharomyces cerevisiae*	*Geobacter sulfurreducens*
Domain	Eukaryote	Prokaryote (Gram-negative)	Eukaryote	Prokaryote (Gram-negative)
Mitochondria	Multiple	N/A	Multiple	N/A
Chloroplast	Single chloroplast occupies two thirds of the cell	Chloroplast analogy	N/A	N/A
Hydrogen synthesis enzyme	Fe hydrogenase	NiFe hydrogenase	N/A	N/A
MFC mode	Product mode	Photosynthetic DET	MET DET (output extremely low)	DET
MFC performance	0.4 W m^{-2}	0.02 W m^{-2}	1.5 W m^{-2}(MET)	1.88 W m^{-2}
	3.3 W m^{-3}	0.007 W m^{-3}	90 W m^{-3}(MET)	43 W m^{-3}
			0.003 W m^{-2} (DET)	
Optimum growth temperature	20°C to 25°C [81]	30°C to 33°C [82]	25°C to 35°C [83]	30°C to 35°C [84]
Growth mode	Autotrophic	Autotrophic	Heterotrophic	Heterotrophic and sedimentary (soil inhabitant)
	Heterotrophic	Heterotrophic		
	Mixotrophic	Mixotrophic		

N/A, not applicable.

many sources such as flue gases of fossil fuel power plants and other waste streams. (6) Algae can be processed into a wide range of products such as biodiesel, bioethanol, methane, bio-oil and biochar and high-protein animal feed. (7) The 'simple' photosynthetic alga *C. reinhardtii* is an excellent model organism for a systems biology approach compared to a complex vascular plant [12].

Biofuel and electricity production by algae Due to the advantages listed above, algae have been examined in many studies for the generation of energy products, such as bio-oil, methane, methanol and hydrogen [98]. Nevertheless, these technologies have one disadvantage in that the fuel produced must be stored, transported and further processed to produce electricity. To circumvent these problems, MFC is used as an alternative way to directly generate electricity in only one process unit by means of hydrolysis and fermentation of algae and makes use of energy originating from sunlight.

However, algae are not exoelectrogens and the conventional mediators do not perform well in the extraction of the redox potential for algae-based MFCs because the redox species are produced through the metabolic mechanisms that take place in membrane-surrounded cell organelles in algae such as mitochondria and chloroplasts [74]. Thus, neither DET nor MET mode has been applied in MFCs using algae.

Previous studies tend to use algae in MFCs of the product mode that depend on the production of hydrogen molecules, which is then oxidised at the anode for electron transfer to the MFC circuit. Another mechanism is where algae produce organic matter that is used as a substrate for electrochemically active bacteria, which then supply electrons for MFCs from oxidization of the organic matter [69].

In one landmark study of MFC using algae, *C. reinhardtii* was used in a product-mode MFC to produce hydrogen for oxidation at the anode. A maximum hydrogen production rate of 7.6 ml/l culture h^{-1} [99] was achieved, at a current yield of 9 mA at a constant electrode potential of 0.2 V. Using the culture volume and electrode dimensions, this corresponds to power densities of 0.4 W m^{-2} and 3.3 W m^{-3}. In another study [98], *Chlorella vulgaris* microalgae were used as a biomass source to feed a mixed microbial culture, producing a maximum power density of 0.98 W m^{-2} (277 W m^{-3}).

Furthermore, algae have been inoculated into a bioelectrode to generate oxygen as the electron acceptor [100]. Under illumination, algae produced oxygen as the electron acceptor for the MFC cathodic reactions, changing the bioelectrode into biocathode mode, while in darkness, the algal oxygen production stops and the bioelectrode mainly functioned as the bioanode. The reversible bioelectrode can relieve the pH membrane gradient generated by the acidification at the anode and the alkalisation at the cathode during normal MFC operation [100,101].

Hydrogen production by C. reinhardtii Only a specific group of green microalgae and cyanobacteria, e.g. microalga *C. reinhardtii*, have evolved the additional ability to harness the huge solar energy resource to drive molecular H_2 production [102-108]. The release of hydrogen by *C. reinhardtii* under light exposure was first reported in 1942 [105]. In 2000, sustained hydrogen production was achieved using induced sulphur depletion in a culture medium containing acetate, a carbon source that is used to cause the shift from aerobic to anaerobic state [107].

C. reinhardtii is one of the best eukaryotes for H_2 production [109]. The available experimental information, including genomics, indicates that *C. reinhardtii* possesses a complex metabolic network containing aerobic respiration and molecular flexibility associated with fermentative metabolism. The molecular flexibility is accomplished with adjustments in the rates of accumulation of organic acids, ethanol, CO_2 and H_2 [86,110-115] and underlies the cell's adaptive ability for hypoxic and anoxic conditions.

Compared to other H_2-producing organisms such as chemotrophic and phototrophic bacteria, *C. reinhardtii* is more practical for H_2 production as it can be easily and efficiently grown in bioreactors using solar light, grows rapidly (doubling times of the order 6 h or less) and has a flexible metabolism [116]. The genome of this model microorganism was fully sequenced in 2007 [86], which makes it possible to increase production yields of H_2 from water by optimization of cell metabolism.

Limitation of hydrogen production by C. reinhardtii In fact, hydrogen production by *C. reinhardtii* can still not meet the commercial requirement because of several biochemical and engineering shortcomings, for example, hydrogen production demands anoxia because oxygen can suppress transcription and function of hydrogenase(s). However, the anoxia is constrained by the function of the photosystem II (PSII), which provides electron and protons from water and conducts oxygen evolution in the photosynthetic electron transport chain. Economic assessments have suggested that microalgae should achieve an efficiency of 10% in the conversion of solar energy to bioenergy to be competitive with other H_2 production methods, such as biomass gasification or photovoltaic electrolysis [117]. This is a more than a fivefold increase in efficiency from current levels. Exploiting hydrogen production directly in a product-type MFC may help to bridge this gap.

Synechocystis sp. PCC 6803

Synechocystis sp. PCC 6803 is a unicellular cyanobacterium, one of the earliest groups of microbes to evolve on earth. The first primitive bacteria on Earth are dated at 3.8 to 3.6 billion years ago [118]. It is thought that cyanobacteria flourished during the period from 3.5 to 1.8 billion years ago, consuming CO_2 and providing Earth with oxygen, making possible the development of the different forms of aerobic life. At present, cyanobacteria deliver amounts of oxygen to the atmosphere similar to those that are produced by higher plants [118]. Moreover, cyanobacteria harness 0.2% to 0.3% of the total solar energy (178,000 TW) that reaches the Earth [106] and convert the solar energy into biomass-stored chemical energy at the rate of approximately 450 TW, contributing to 20% to 30% of Earth's primary photosynthetic productivity [119].

Until 1982, the cyanobacteria were called blue-green algae because they can photosynthesize and look like chloroplasts. Since then, cyanobacteria were re-classified as prokaryotes [120]. It is suggested that cyanobacteria entered into a symbiosis with cells, which were not capable of absorbing CO_2 and releasing oxygen, and later became photosynthetic organelles of plants [121]. Nowadays many species of cyanobacteria, e.g. *Synechocystis* sp. PCC 6803, are widely distributed in nature.

Synechocystis sp. PCC 6803 and plants have similar oxygen-evolving apparatus and are thus used for studying photosynthesis in plant cells. The difference is that *Synechocystis* sp. PCC 6803 grow much faster than plants, and they are relatively easy organisms for genetic manipulations [118]. Also, plants are fixed at places where they grow and they have less adaptation abilities for their growth and propagation than cyanobacteria.

Synechocystis sp. PCC 6803 grow photoautotrophically on carbon dioxide and light, as well as heterotrophically on glucose. Like *C. reinhardtii*, *Synechocystis* sp. PCC 6803 is one of several hydrogen-yielding species of cyanobacteria [122]. After its genome was fully sequenced in the 1990s [123,124], this cyanobacteria species has become a popular model photosynthetic organism studied by many researchers.

Advantages of cyanobacteria Cyanobacteria, besides other photosynthetic microorganisms such as microalgae, can establish synergistic relationships with heterotrophic bacteria, for instance, in a microbial mat [68]. Thus, they could be potentially manipulated to establish an indirect synergistic relationship with electricigens in phototrophic MFC [69]. However, phototrophic MFCs usually have low conversion efficiency [62], and the study of phototrophic MFCs is in its nascent stages [62].

Since *Synechocystis* sp. PCC 6803 is a photoautotroph that divides rapidly, it has been enlisted as a platform for production of biofuels by using sunlight as an inexpensive energy source [125,126]. This feature makes this species suitable as a candidate for MFCs of product mode.

Electrogenic activity of cyanobacteria Unlike other exoelectrogens, such as *G. sulfurreducens*, in which the electrons are derived from biochemical oxidation of organic compounds via the respiratory electron transfer chain [127], cyanobacterial electrogenic activity does not need exogenous organic fuel and is entirely dependent on the energy of light, which drives the biophotolysis of water through the photosynthetic electron transfer chain in the cyanobacteria, releasing electrons [128,129]. The electrogenic activity of cyanobacteria may represent a form of overflow metabolism to protect cells under high-intensity light [70,129]. This light-driven electrogenic activity is conserved in diverse genera of cyanobacteria and is an important microbiological channel of solar energy into the biosphere [129].

The electrogenic activity of *Synechocystis* sp. PCC 6803 has been captured in an MFC for electricity generation. The MFC can achieve a steady power density of 6.7 mW m^{-3} (peaking at 7.5 mW m^{-3}) [130,131]. These power densities are still much lower than the values achieved by the other microbes under discussion. Despite that, it is included in the selection list because it offers a unique combination of photosynthetic activity that is plausibly accessible to direct-mode electron transfer. The quoted measurements are quite recent, and it is worth exploring if this organism has the potential to deliver competitive power densities in the future.

Hydrogen production by cyanobacteria Cyanobacteria have a similar process for hydrogen production as algae, except that they use NiFe hydrogenases rather than Fe hydrogenases in microalgae and the hydrogenase of cyanobacteria is 100 times less active than those of the green algae *C. reinhardtii* [106]. These hydrogenases contain [Ni-Fe] catalytic centres that are extremely sensitive to inactivation by O_2, one of the major barriers to hydrogen production. Natural mechanisms such as consumption by respiration, chemical reduction via PSI and reversible inactivation of PSII O_2 evolution can reduce intracellular O_2 content and thereby increase H_2 production.

Saccharomyces cerevisiae

The *Saccharomyces* genus currently contains eight species [132]. *Saccharomyces cerevisiae*, *Saccharomyces bayanus* and *Saccharomyces pastorianus* are associated with anthropic environments, whereas *Saccharomyces paradoxus*, *Saccharomyces kudriavzevii*, *Saccharomyces cariocanus*, *Saccharomyces mikatae* and the recently described *Saccharomyces arboricolus* are mostly isolated from natural environments [133,134]. These *Saccharomyces* species can play a major role in food or beverage fermentation. However, the ale yeasts involved in alcoholic fermentation mostly belong to the species *S. cerevisiae* [132]. Besides its important role in baking and brewing, this yeast species has been used as a eukaryotic model organism in molecular and cell biology, for example, the characteristic of many proteins can be discovered by studying their homologs in *S. cerevisiae*.

S. cerevisiae cells are round to ovoid, are 5 to 10 μm in diameter, reproduce by a budding process and can grow aerobically on glucose, maltose and trehalose but not on lactose or cellobiose. In the presence of oxygen, it is even able to operate in a mixed fermentation/respiration mode. The ratio of fermentation to respiration varies slightly among strains but is approximately 80:20 [135]. Furthermore, *S. cerevisiae* can be processed to produce potential advanced biofuels such as long-chain alcohols and isoprenoid- and fatty acid-based biofuels, which have physical properties that more closely resemble petroleum-derived fuels [136].

Advantages of yeast for MFC Yeast is sometimes thought to be impractical as a biocatalyst, due to difficulties with transferring electrons out of cellular organelles [137]. However, since yeasts are robust, are easily handled, are mostly non-pathogenic, have high catabolic rates and grow on substrate spectrum, they are well worth considering as promising biocatalysts for MFCs [138]. In addition, several other merits may exist for using *S. cerevisiae* in MFCs. First, *S. cerevisiae* can survive and function in an anaerobic condition that is required for the anode compartment of traditional MFC. Second, the optimal growth temperature for *S. cerevisiae* is around 30°C, which is a convenient ambient temperature. Third, the metabolism of this species is well understood, which helps locate mechanisms responsible for electricity generation in MFCs. Lastly, yeast-based fuel cells could be retrofitted into ethanol plants for *in situ* power generation [139].

Yeast for in situ power generation In an anaerobic condition, yeasts usually switch to fermentation reactions where one glucose molecule is consumed for the production of two molecules of pyruvates. Pyruvate is further transformed into alcohol or organic acid by recycling NADH to NAD^+, which is a key step to sustain the glycolysis process [140]. This glycolysis reaction takes place in the cytosol of the cell rather than in the mitochondria, so NADH could be easily accessed by the mediator molecule present in the cell membrane of the yeast [73]. The glycolysis and the oxidation of NADH to NAD^+ are not influenced by the energy extraction process in the MFCs. Based on these characteristics, MFCs using yeast can be directly applied in fermenters for *in situ* power generation [139].

Limitation of S. cerevisiae for MFC use Limitations exist for *S. cerevisiae* to be used in MFCs. First, *S. cerevisiae* has a weak ability to oxidise the substrate to supply the maximum number of electrons available for yeast-based MFC. In the mitochondrial process of *S. cerevisiae*, there is a total of only 14 ATP per glucose molecule produced, which is much less than a net of 28 to 30 ATP typically achieved by most aerobes [138]. Also, mediators are commonly required to facilitate the transfer of electrons to the anode, which makes exogenous mediators necessary to MFCs based on *S. cerevisiae* because this yeast is thought incapable of producing such mediators indigenously [139].

The output of S. cerevisiae-based MFCs In general, yeast-based MFCs perform better than cyanobacteria but still have a lower power output than bacterial fuel cells [138]. It was shown that methylene blue-mediated *S. cerevisiae* MFC can give a power density of 1.5 W m^{-2} [141], which is less than the maximum of 6.86 W m^{-2} reported by Fan et al. [135] for a mixed-culture MFC. The corresponding volumetric density, based on the specified anodic chamber volume of 10 ml, is 90 W m^{-3}.

A recent MFC that employs *S. cerevisiae* as the electron donor in the anodic half-cell and *C. vulgaris* as the electron acceptor in the cathodic half-cell can reach a maximum power at 90 mV and a load of 5,000 Ω, giving a power density of 0.95 mW m^{-2} of electrode surface area [142]. This power density is still very low.

Another study investigated the possibility of *S. cerevisiae* to transfer electrons to an extracellular electron acceptor through DET mode and found that the cells that adhered to the anode were able to sustain power generation in a mediator-less MFC. However, the power performance of this MFC was extremely low (0.003 W m^{-2}) [143].

Geobacter sulfurreducens

G. sulfurreducens are comma-shaped Gram-negative, anaerobic bacteria capable of coupling oxidization of organic compounds to reduction of metals. This organism is one of the predominant metal-reducing bacteria in soil and hence plays an important ecological role in biotechnologically exploitable bioremediation. The activity of *Geobacter* species in sub-surface can be stimulated to remove organic and metal contaminants such as aromatic hydrocarbons and uranium from groundwater [144-146].

The genome sequence of *G. sulfurreducens* is available, and a system for genetic manipulation has been developed for this organism [147]. Since it was discovered in

1994 [148], this bacterium has been extensively studied for MFC applications. It has been reported that (1) *G. sulfurreducens* can completely oxidise electron donors by using only an electrode as the electron acceptor, (2) it can quantitatively transfer electrons to electrodes in the absence of electron mediators, and (3) this electron transfer is similar to those observed for electron transport to Fe(III) citrate [47].

Advantages of G. sulfurreducens *G. sulfurreducens* is the most abundant species on anode surfaces in MFCs grown with more than one bacterial species [149-151]. It can form biofilms on the anodes, which make all the cells participate in electron transport to the anode and thus increase the current production [152]. *G. sulfurreducens* is an anaerobe but can withstand low levels of oxygen and may use oxygen as an electron acceptor to support growth under aerobic conditions [153].

This *Geobacter* species can produce large amounts of electrical energy since it possesses multiple mechanisms that involve either pili or c-type cytochromes to facilitate the electron transfer to electrode in MFCs (discussed before in the 'Microbial fuel cells' section) [149]. Also, with the electron transfer to electrodes, the *Geobacter* species can effectively oxidise acetate [47,154]. A current density of 4.56 A m^{-2}, corresponding to power densities of 1.88 W m^{-2} and 43 W m^{-3}, measured for *G. sulfurreducens* is among the highest reported for a pure culture [155]. By reducing the anode compartment volume to a fraction of a millilitre, the volumetric density was in fact increased to 2.15 kW m^{-3}. While the lower value is more realistic for comparison to other studies, this does show that very high densities are achievable in principle with this organism. In addition, *G. sulfurreducens* converts acetate to current with coulombic efficiencies of over 90% [151,155].

Previous studies have shown that when a high selective pressure for high rates of current production at high coulombic efficiencies is imposed on complex microbial communities, it is the organisms closely related to *G. sulfurreducens* that are routinely enriched on anodes of the MFCs [55,154,156-158]. Thus, *G. sulfurreducens* can also be used to study adaptation for enhanced power production.

Metabolism of Geobacter species The metabolism of *G. sulfurreducens* was investigated by constraint-based modelling [159]. In contrast to *Escherichia coli*, which primarily produces energy and biosynthetic precursors through sugar fermentation, *Geobacter* completely oxidises acetate and other electron donors via the TCA cycle [160,161], which makes it necessary to transfer electrons to terminal electron acceptors for regeneration of cytoplasmic and intramembrane electron acceptors and ATP synthesis. In *G. sulfurreducens*, this is accomplished by electron transfer to extracellular electron acceptors, i.e. Fe(III) oxides [162].

Since the rate of cytoplasmic proton consumption is lower than that of proton production during the reduction of extracellular electron acceptors such as Fe(III), the energy consumption with extracellular electron acceptors is lower compared to that associated with intracellular acceptors [159]. The use of extracellular electron shuttles makes the *Geobacter* species circumvent the metabolic cost of producing the electron shuttles and consequently more energetically competitive than shuttle-producing Fe(III) reducers in sub-surface environments [159].

In silico analysis suggested that the metabolic network of *G. sulfurreducens* contains pyruvate-ferredoxin oxidoreductase, which catalyzes synthesis of pyruvate from acetate and carbon dioxide in a single step, indicating that the synthesis of amino acids in *G. sulfurreducens* is more efficient than in *E. coli* [159].

Limits and applications MFCs powered by *G. sulfurreducens* are far away from being commercialized as a practical biofuel source [152], because up until now the current levels of these MFCs are around 14 mA which could be used to power very simple components [149] but still not big enough to drive complex mechanisms. However, the actual current densities that could be generated from MFCs based on *G. sulfurreducens* are still unclear and require further investigation [151].

Comparison of Geobacter sp. and Shewanella sp. Geobacteraceae and Shewanellaceae are classic models in MFC research as their metabolism and versatility have been studied extensively [72,163]. As mentioned before in the 'Microbial fuel cells' section, they are both capable of being used for DET mode in MFCs, because both *Shewanella* sp. and *Geobacter* sp. possess nanowires, electrically conductive bacterial appendages, to transport electrons from cells to solid electron acceptors such as graphite anodes in MFCs [34,35,162]. Despite those similarities, differences also exist when compared regarding the engineering design and performance of the MFCs.

Geobacter-based MFCs generate high coulombic efficiencies [164] but require strict anaerobic conditions which limit their applicability. In contrast, *Shewanella*-based MFCs can be operated with air-exposed cultures [27]. Unlike *Geobacter* sp. that requires direct contact to the electrode surface [72], *Shewanella* sp. can use additional mediators to facilitate electron transfer outside the cell membrane [27]. Importantly, besides utilizing nanowires to mediate the electron transport [165], they can synthesize their own redox mediators (i.e.

flavins) for extracellular electron transfer under diverse environmental conditions [163,166]. These two electron-mediated mechanisms determine the efficiency of the current generation in *Shewanella*-containing MFCs [167].

A maximum power density of 24 mW m^{-2} (in the presence of an additional mediator) was reported for *Shewanella* [168]. This value appears low in comparison to bacterial cells, but that is because it was referred to the true microscopic area of a porous electrode. When expressed, as customary, in terms of macroscopic MFC dimensions, the equivalent power densities are 3 W m^{-2} and 500 W m^{-3}. These values compare favourably with *Geobacter*. When dissolved oxygen was deliberately fed into the anode chamber, *Shewanella*-based MFC was still able to produce a power output of 6.5 mW m^{-2} and 13 mA m^{-2}. The MFC used lactate as the fuel source and relied on self-excreted mediators of *Shewanella* [169].

In fact, the previously described *Shewanella*-based system would not be directly applicable to powering electronics and is required to use aerobic water [170]. For instance, a complex pumping system is necessary to continuously recirculate the anolyte between the anode and the large anolyte reservoir, but this pumping system at the anode could consume more power than the *Shewanella*-based MFC produces. Since *Shewanella* sp. cannot use oxygen as the electron acceptor, ferricyanide needs to be added as catholyte. However, ferricyanide is a non-renewable and toxic electron acceptor and can thus not be deployed in the field in the long term. Moreover, the coulombic efficiencies were found to be low (<6%), when calculated based on the incomplete oxidation of lactate to acetate [170].

Conversely, *G. sulfurreducens* can effectively oxidise acetate with electron transfer to electrodes [47,154] and convert acetate to current with coulombic efficiencies of more than 90% [151,155]. *G. sulfurreducens* is an anaerobe that can withstand low levels of oxygen and may use oxygen as an electron acceptor [153]. It has recently been shown that with a new configuration, MFCs based on *G. sulfurreducens* can become 100% aerobic, allowing for floating and/or untethered applications. At the same time, the performance of the MFCs is similar to their anaerobic/aerobic counterparts [170]. It is expected that with this aerobic configuration, power could be produced in a *G. sulfurreducens* MFC suspended in aerobic seawater [170].

Conclusions

Electricity generation in MFCs is based on the metabolic activity of living microorganisms at the anode. The selection of microorganisms is based on many criteria, but the power output, electron transfer ability and biological functions such as photosynthesis and hydrogen production are particularly important. These important properties, in different combinations, are exemplified by the four representative microorganisms discussed above, and their referential facts for modelling are compared in Table 2. Studying these individually, and in combination, should reveal significant insights in the quest for higher power output MFCs.

Most MFC researchers have been active in engineering designs, i.e. how to create scalable and economical architectures and engineer more efficient hardware and how different microbes interact with the anodes/cathodes when transporting electrons [127]. Such research covers optimizing anodic conditions, housing constructs and component materials, learning more about microbial community ecology and isolating vigorous biocatalysts [9]. Biological aspects of MFCs have also received some attention, such as the anodic activity of different organisms. However, very little research has been done on the biochemical interface between the engineering design and the biological aspects (see Figure 1).

We conclude that future studies are required to work on that interface, i.e. how to enhance the anodic activity by means of adjusting the metabolic activity of biocatalysts, for example, utilizing metabolic network analysis. The genome-scale metabolic networks are quite new concepts and have only been produced in the last few years. The analysis of the metabolic network through modelling approaches, such as flux balance analysis, plays an important role in filling the gap between genotypes and phenotypes of microorganisms to provide a full picture of the biological system.

Endnote

[a]Representative microorganisms chosen by this article and discussed in the 'Microorganisms for *in silico* study of MFC functioning' section.

Competing interests

The authors declare that they have no competing interests.

Authors' contributions

LM and WV co-conceived the idea. LM conducted the review and drafted the manuscript. WV supervised the work and corrected the manuscript. Both authors read and approved the final manuscript.

Authors' information

LM is a PhD candidate in the Department of Wine, Food and Molecular Biosciences, Lincoln University. He holds a B.Sc. degree with First Class Honours in biochemistry from Lincoln University. WV is a senior lecturer in computational biology at the Centre for Advanced Computational Solutions at Lincoln University. He holds one PhD degree in computational systems modelling from Lincoln University and another PhD degree in theoretical physics and was previously a professor of physics at the University of South Africa.

References

1. Potter, MC: Electrical effects accompanying the decomposition of organic compounds. Proc. R. Soc. London Series B **84**(571), 260–276 (1911)

2. Rittmann, BE: Opportunities for renewable bioenergy using microorganisms. Biotechnol. Bioeng. **100**(2), 203–212 (2008)
3. Pandyaswargo, A, Onoda, H, Nagata, K: Energy recovery potential and life cycle impact assessment of municipal solid waste management technologies in Asian countries using ELP model. Int. J. Energy Environ. Eng. **3**(1), 28 (2012)
4. Eddine, B, Salah, M: Solid waste as renewable source of energy: current and future possibility in Algeria. Int. J. Energy Environ. Eng. **3**(1), 17 (2012)
5. Sharma, V, Kundu, PP: Biocatalysts in microbial fuel cells. Enzyme Microb. Technol. **47**(5), 179–188 (2010)
6. Harnisch, F, Schröder, U: From MFC to MXC: chemical and biological cathodes and their potential for microbial bioelectrochemical systems. Chem. Soc. Rev. **39**(11), 4433–4448 (2010)
7. Girguis, PR, Nielsen, ME, Figueroa, I: Harnessing energy from marine productivity using bioelectrochemical systems. Curr. Opin. Biotechnol. **21**(3), 252–258 (2010)
8. Li, H, Opgenorth, PH, Wernick, DG, Rogers, S, Wu, T-Y, Higashide, W, Malati, P, Huo, Y-X, Cho, KM, Laio, JC: Integrated electromicrobial conversion of CO2 to higher alcohols. Science **335**(6076), 1596 (2012)
9. Schwartz, K: Microbial fuel cells: design elements and application of a novel renewable energy source. MMG 445 Basic Biotechnology eJournal **3**(1), 20–27 (2007)
10. Logan, BE: Scaling up microbial fuel cells and other bioelectrochemical systems. Appl. Microbiol. Biotechnol. **85**(6), 1665–1671 (2010)
11. Bullen, RA, Arnot, TC, Lakeman, JB, Walsh, FC: Biofuel cells and their development. Biosens. Bioelectron. **21**(11), 2015–2045 (2006)
12. Rupprecht, J: From systems biology to fuel–*Chlamydomonas reinhardtii* as a model for a systems biology approach to improve biohydrogen production. J. Biotechnol. **142**(1), 10–20 (2009)
13. Mukhopadhyay, A, Redding, AM, Rutherford, BJ, Keasling, JD: Importance of systems biology in engineering microbes for biofuel production. Curr. Opin. Biotechnol. **19**(3), 228–234 (2008)
14. Price, ND, Reed, JL, Palsson, BO: Genome-scale models of microbial cells: evaluating the consequences of constraints. Nature Reviews Microbiology **2**(11), 886–897 (2004)
15. Rocha, I, Forster, J, Nielsen, J: Design and application of genome-scale reconstructed metabolic models. Methods Mol. Biol. **416**, 409–431 (2008)
16. Jaqaman, K, Danuser, G: Linking data to models: data regression. Nature Reviews Molecular Cell Biology **7**(11), 813–819 (2006)
17. Terzer, M, Maynard, ND, Covert, MW, Stelling, J: Genome-scale metabolic networks. Wiley Interdiscip. Rev. Syst. Biol. Med. **1**(3), 285–297 (2009)
18. Lee, JM, Gianchandani, EP, Eddy, JA, Papin, JA: Dynamic analysis of integrated signaling, metabolic, and regulatory networks. PLoS Comput. Biol. **4**(5), e1000086 (2008)
19. Covert, MW, Xiao, N, Chen, TJ, Karr, JR: Integrating metabolic, transcriptional regulatory and signal transduction models in *Escherichia coli*. Bioinformatics **24**(18), 2044–2050 (2008)
20. Jamshidi, N, Palsson, B: Mass action stoichiometric simulation models: incorporating kinetics and regulation into stoichiometric models. Biophys. J. **98**(2), 175–185 (2010)
21. Thiele, I, Palsson, BO: A protocol for generating a high-quality genome-scale metabolic reconstruction. Nat. Protoc. **5**(1), 93–121 (2010)
22. Orth, JD, Thiele, I, Palsson, BO: What is flux balance analysis? Nat Biotech **28**(3), 245–248 (2010)
23. Fernie, AR, Geigenberger, P, Stitt, M: Flux an important, but neglected, component of functional genomics. Curr. Opin. Plant Biol. **8**(2), 174–182 (2005)
24. Saha, R, Suresha, S, Park, W, Lee, D-Y, Karimi, IA: Strain improvement and mediator selection for microbial fuel cell by genome scale in silico model. In: Pleşu, V, Agachi, PS (eds.) 17th European Symposium on Computer Aided Process Engineering – ESCAPE17., Bucharest (2007)
25. Rabaey, K, Verstraete, W: Microbial fuel cells: novel biotechnology for energy generation. Trends Biotechnol. **23**(6), 291–298 (2005)
26. Du, Z, Li, H, Gu, T: A state of the art review on microbial fuel cells: a promising technology for wastewater treatment and bioenergy. Biotechnol. Adv. **25**(5), 464–482 (2007)
27. Schröder, U: Anodic electron transfer mechanisms in microbial fuel cells and their energy efficiency. Phys. Chem. Chem. Phys. **9**(21), 2619–2629 (2007)
28. Kim, HJ, Hyun, MS, Chang, IS, Kim, BH: A microbial fuel cell type lactate biosensor using a metal-reducing bacterium, *Shewanella putrefaciens*. J. Microbiol. Biotechnol. **9**(3), 365–367 (1999)
29. Kim, HJ, Park, HS, Hyun, MS, Chang, IS, Kim, M, Kim, BH: A mediator-less microbial fuel cell using a metal reducing bacterium, *Shewanella putrefaciens*. Enzyme Microb. Technol. **30**(2), 145–152 (2002)
30. Lovley, DR: Microbial fuel cells: novel microbial physiologies and engineering approaches. Curr. Opin. Biotechnol. **17**(3), 327–332 (2006)
31. Chaudhuri, SK: Lovley. DR: Electricity generation by direct oxidation of glucose in mediatorless microbial fuel cells. Nat. Biotechnol. **21**(10), 1229–1232 (2003)
32. Chang, IS, Moon, H, Bretschger, O, Jang, JK, Park, HI, Nealson, KH, Kim, BH: Electrochemically active bacteria (EAB) and mediator-less microbial fuel cells. J. Microbiol. Biotechnol. **16**(2), 163–177 (2006)
33. Schaetzle, O, Barriere, F, Baronian, K: Bacteria and yeasts as catalysts in microbial fuel cells: electron transfer from micro-organisms to electrodes for green electricity. Energy Environ. Sci. **1**(6), 607–620 (2008)
34. Gorby, YA, Yanina, S, McLean, JS, Rosso, KM, Moyles, D, Dohnalkova, A, Beveridge, TJ, Chang, IS, Kim, BH, Kim, KS, Culley, DE, Reed, SB, Romine, MF, Saffarini, DA, Hill, EA, Shi, L, Elias, DA, Kennedy, DW, Pinchuk, G, Watanabe, K, Ishii, S, Logan, B, Nealson, KH, Fredrickson, JK: Electrically conductive bacterial nanowires produced by *Shewanella oneidensis* strain MR-1 and other microorganisms. Proc. Natl. Acad. Sci. **103**(30), 11358–11363 (2006)
35. Reguera, G, Nevin, KP, Nicoll, JS, Covalla, SF, Woodard, TL, Lovely, DR: Biofilm and nanowire production leads to increased current in *Geobacter sulfurreducens* fuel cells. Appl. Environ. Microbiol. **72**(11), 7345–7348 (2006)
36. Li, P, Ki, J-P, Liu, H: Analysis and optimization of current collecting systems in PEM fuel cells. Int. J. Energy Environ. Eng. **3**(1), 2 (2012)
37. Cooney, MJ, Roschi, E, Marison, IW, Comninellis, C, von Stockar, U: Physiologic studies with the sulfate-reducing bacterium *Desulfovibrio desulfuricans*: evaluation for use in a biofuel cell. Enzyme Microb. Technol. **18**(5), 358–365 (1996)
38. Scholz, F, Schröder, U: Bacterial batteries. Nat. Biotechnol. **21**(9), 3–4 (2003)
39. Rabaey, K, Lissens, G, Verstraete, W: Microbial fuel cells: performances and perspectives. In: Lens, P, Westermann, P, Haberbauer, M, Moreno, A (eds.) Biofuels for Fuel Cells: Biomass Fermentation Towards Usage in Fuel Cells, pp. 377–399. IWA, London (2005)
40. Rosenbaum, M, He, Z, Angenent, LT: Light energy to bioelectricity: photosynthetic microbial fuel cells. Curr. Opin. Biotechnol. **21**(3), 259–264 (2010)
41. Davis, F, Higson, SPJ: Biofuel cells–recent advances and applications. Biosens. Bioelectron. **22**(7), 1224–1235 (2007)
42. Bennetto, HP, Delany, GM, Mason, JR, Roller, SD, Stirling, JL, Thurston, CF: The sucrose fuel cell: efficient biomass conversion using a microbial catalyst. Biotechnol. Lett. **7**(10), 699–704 (1985)
43. Rabaey, K, Boon, N, Höfte, M, Verstraete, W: Microbial phenazine production enhances electron transfer in biofuel cells. Environ. Sci. Technol. **39**(9), 3401–3408 (2005)
44. Park, HS, Kim, BH, Kim, HS, Kim, HJ, Kim, GT, Kim, M, Chang, IS, Park, YK, Chang, HI: A novel electrochemically active and Fe(III)-reducing bacterium phylogenetically related to *Clostridium butyricum* isolated from a microbial fuel cell. Anaerobe **7**(6), 297–306 (2001)
45. Kim, BH, Park, HS, Kim, HJ, Kim, GT, Chang, IS, Lee, J, Phung, NT: Direct electrode reaction of Fe(III)-reducing bacterium, *Shewanella putrefaciens*. J. Microbiol. Biotechnol. **9**(2), 127–131 (1999)
46. Logan, BE, Regan, JM: Electricity-producing bacterial communities in microbial fuel cells. Trends Microbiol. **14**(12), 512–518 (2006)
47. Bond, DR, Lovley, DR: Electricity production by Geobacter sulfurreducens attached to electrodes. Appl. Environ. Microbiol. **69**(3), 1548–1555 (2003)
48. Lanthier, M, Gregory, KB, Lovley, DR: Growth with high planktonic biomass in *Shewanella oneidensis* fuel cells. FEMS Microbiol. Lett. **278**(1), 29–35 (2008)
49. Logan, BE, Regan, JM: Microbial fuel cells, challenges and applications. Environ. Sci. Technol. **40**(17), 5172–5180 (2006)
50. Fan, Y, Hu, H, Liu, H: Enhanced Coulombic efficiency and power density of air-cathode microbial fuel cells with an improved cell configuration. J. Power. Sources **171**(2), 348–354 (2007)
51. Zhang, F, Cheng, S, Pant, D, Van Bogaert, G, Logan, BE: Power generation using an activated carbon and metal mesh cathode in a microbial fuel cell. Electrochem. Commun. **11**(11), 2177–2179 (2009)
52. Zuo, Y, Cheng, S, Logan, BE: Ion exchange membrane cathodes for scalable microbial fuel cells. Environ. Sci. Technol. **42**(18), 6967–6972 (2008)
53. Chen, G-W, Choi, SJ, Lee, TH, Lee, GY, Cha, JH, Kim, CW: Application of biocathode in microbial fuel cells: cell performance and microbial community. Appl. Microbiol. Biotechnol. **79**(3), 379–388 (2008)

54. Cheng, S, Logan, BE: Ammonia treatment of carbon cloth anodes to enhance power generation of microbial fuel cells. Electrochem. Commun. **9**, 492–496 (2006)
55. Liu, JL, Lowy, DA, Baumann, RG, Tender, LM: Influence of anode pretreatment on its microbial colonization. J. Appl. Microbiol. **102**(1), 177–183 (2007)
56. Cao, X, Huang, X, Boon, N, Liang, P, Fan, M: Electricity generation by an enriched phototrophic consortium in a microbial fuel cell. Electrochem. Commun. **10**(9), 1392–1395 (2008)
57. Yang, Y, Sun, G, Xu, M: Microbial fuel cells come of age. J. Chem. Technol. Biotechnol. **86**(5), 625–632 (2011)
58. Lovley, DR: Extracellular electron transfer: wires, capacitors, iron lungs and more. Geobiology **6**(3), 225–231 (2008)
59. Shi, L, Squier, TC, Zachara, JM, Fredrickson, JK: Respiration of metal (hydr) oxides by *Shewanella* and *Geobacter*: a key role for multihaem c-type cytochromes. Mol. Microbiol. **65**(1), 12–20 (2007)
60. Oh, ST, Kim, JR, Premier, GC, Lee, TH, Kim, C, Sloan, WT: Sustainable wastewater treatment: how might microbial fuel cells contribute. Biotechnol. Adv. **28**(6), 871–881 (2010)
61. Yuan, Y, Ahmed, J, Zhou, L, Zhao, B, Kim, S: Carbon nanoparticles-assisted mediator-less microbial fuel cells using *Proteus vulgaris*. Biosens. Bioelectron. **27**(1), 106–112 (2011)
62. He, Z, Kan, J, Mansfeld, F, Angenent, LT, Nealson, KH: Self-sustained phototrophic microbial fuel cells based on the synergistic cooperation between photosynthetic microorganisms and heterotrophic bacteria. Environ. Sci. Technol. **43**(5), 1648–1654 (2009)
63. Malik, S, Drott, E, Grisdela, P, Lee, J, Lee, C, Lowy, DA, Gray, S, Tender, LM: A self-assembling self-repairing microbial photoelectrochemical solar cell. Energy Environ. Sci. **2**(3), 292–298 (2009)
64. Cho, YK, Donohue, TJ, Tejedor, I, Anderson, MA, McMahon, KD, Noguera, DR: Development of a solar-powered microbial fuel cell. J. Appl. Microbiol. **104**(3), 640–650 (2008)
65. Rosenbaum, M, Schröder, U, Scholz, F: In situ electrooxidation of photobiological hydrogen in a photobioelectrochemical fuel cell based on *Rhodobacter sphaeroides*. Environ. Sci. Technol. **39**(16), 6328–6333 (2005)
66. Xing, D, Zuo, Y, Cheng, S, Regan, JM, Logan, BE: Electricity generation by *Rhodopseudomonas palustris* DX-1. Environ. Sci. Technol. **42**(11), 4146–4151 (2008)
67. Larsen, K, Ibrom, A, Beier, C, Jonasson, S, Michelsen, A: Ecosystem respiration depends strongly on photosynthesis in a temperate heath. Biogeochemistry **85**(2), 201–213 (2007)
68. Stal, LJ, van Gemerden, H, Krumbein, WE: Structure and development of a benthic marine microbial mat. FEMS Microbiol. Lett. **31**(2), 111–125 (1985)
69. Strik, D, Terlouw, H, Hamelers, HV, Buisman, CJ: Renewable sustainable biocatalyzed electricity production in a photosynthetic algal microbial fuel cell (PAMFC). Appl. Microbiol. Biotechnol. **81**(4), 659–668 (2008)
70. Zou, Y, Pisciotta, J, Billmyre, RB, Baskakov, IV: Photosynthetic microbial fuel cells with positive light response. Biotechnol. Bioeng. **104**(5), 939–946 (2009)
71. Yagishita, T, Horigome, T, Tanaka, K: Effects of light, CO_2 and inhibitors on the current output of biofuel cells containing the photosynthetic organism *Synechococcus* sp. J. Chem. Technol. Biotechnol. **56**(4), 393–399 (1993)
72. Lovley, DR: Bug juice: harvesting electricity with microorganisms. Nature Reviews. Microbiology **4**(7), 497–508 (2006)
73. Raghavulu, SV, Goud, RK, Sarma, PN, Mohan, SV: *Saccharomyces cerevisiae* as anodic biocatalyst for power generation in biofuel cell: influence of redox condition and substrate load. Bioresour. Technol. **102**(3), 2751–2757 (2011)
74. Rosenbaum, M, Schröder, U: Photomicrobial solar and fuel cells. Electroanalysis **22**(7–8), 844–855 (2010)
75. Dal'Molin, CG, Quek, LE, Palfreyman, RW, Nielsen, LK: AlgaGEM - a genome-scale metabolic reconstruction of algae based on the *Chlamydomonas reinhardtii* genome. BMC Genomics **12**(Suppl 4), S5 (2011)
76. Chang, RL, Ghamsari, L, Manichaikul, A, Hom, EF, Balaji, S, Fu, W, Shen, Y, Hao, T, Palsson, BØ, Salehi-Ashtiani, K, Papin, JA: Metabolic network reconstruction of *Chlamydomonas* offers insight into light-driven algal metabolism. Mol. Syst. Biol. **7**, 518 (2011)
77. Yoshikawa, K, Kojima, Y, Nakajima, T, Furusawa, C, Hirasawa, T, Shimizu, H: Reconstruction and verification of a genome-scale metabolic model for *Synechocystis* sp. PCC6803. Appl. Microbiol. Biotechnol. **92**(2), 347–358 (2011)
78. Nogales, J, Gudmundsson, S, Knight, EM, Palsson, BO, Thiele, I: Detailing the optimality of photosynthesis in cyanobacteria through systems biology analysis. Proc. Natl. Acad. Sci. **109**(7), 2678–2683 (2012)
79. Sun, J, Sayyar, B, Butler, JE, Pharkya, P, Fahland, TR, Famili, I, Schilling, CH, Lovley, DR, Mahadevan, R: Genome-scale constraint-based modeling of *Geobacter metallireducens*. BMC Syst. Biol. **3**, 15 (2009)
80. Heavner, BD, Smallbone, K, Barker, B, Mendes, P, Walker, LP: Yeast 5 - an expanded reconstruction of the *Saccharomyces cerevisiae* metabolic network. BMC Syst. Biol. **6**, 55 (2012)
81. Adams, M: General growth and mating. http://nutmeg.easternct.edu/~adams/ChlamyTeach/growingchlamy.html (2012). Accessed 12 Feb 2013
82. Sheng, J, Kim, HW, Badalamenti, JP, Zhou, C, Sridharakrishnan, S, Krajmalnik-Brown, R, Rittmann, BE, Vannela, R: Effects of temperature shifts on growth rate and lipid characteristics of *Synechocystis* sp. PCC6803 in a bench-top photobioreactor. Bioresour. Technol. **102**(24), 11218–11225 (2011)
83. Watson, K: Temperature relations. In: Rose, AH, Harrison, JS (eds.) The Yeasts, pp. 41–71. Academic, London (1987)
84. Trinh, N, Park, J, Kim, B-W: Increased generation of electricity in a microbial fuel cell using *Geobacter sulfurreducens*. Korean J. Chem. Eng. **26**(3), 748–753 (2009)
85. Yoon, HS, Hackett, JD, Ciniglia, C, Pinto, G, Bhattacharya, D: A molecular timeline for the origin of photosynthetic eukaryotes. Mol. Biol. Evol. **21**(5), 809–818 (2004)
86. Merchant, SS, Prochnik, SE, Vallon, O, Harris, EH, Karpowicz, SJ, Witman, GB, Terry, A, Salamov, A, Fritz-Laylin, LK, Maréchal-Drouard, L, Marshall, WF, Qu, LH, Nelson, DR, Sanderfoot, AA, Spalding, MH, Kapitonov, VV, Ren, Q, Ferris, P, Lindquist, E, Shapiro, H, Lucas, SM, Grimwood, J, Schmutz, J, Cardol, P, Cerutti, H, Chanfreau, G, Chen, CL, Cognat, V, Croft, MT, Dent, R, et al.: The *Chlamydomonas* genome reveals the evolution of key animal and plant functions. Science **318**(5848), 245–250 (2007)
87. Harris, EH: *Chlamydomonas* as a model organism. Annu. Rev. Plant Physiol. Plant Mol. Biol. **52**(1), 363–406 (2001)
88. Manichaikul, A, Ghamsari, L, Hom, EF, Lin, C, Murray, RR, Chang, RL, Balaji, S, Hao, T, Shen, Y, Chavali, AK, Thiele, I, Yang, X, Fan, C, Mello, E, Hill, DE, Vidal, M, Salehi-Ashtiani, K, Papin, JA: Metabolic network analysis integrated with transcript verification for sequenced genomes. Nat. Methods **6**(8), 589–592 (2009)
89. Keller, LC, Romijn, EP, Zamora, I, Yates 3rd, JR, Marshall, WF: Proteomic analysis of isolated chlamydomonas centrioles reveals orthologs of ciliary-disease genes. Current Biol.: CB **15**(12), 1090–1098 (2005)
90. Pazour, GJ, Agrin, N, Walker, BL, Witman, GB: Identification of predicted human outer dynein arm genes: candidates for primary ciliary dyskinesia genes. J. Med. Genet. **43**(1), 62–73 (2006)
91. Vilchez, C, Garbayo, I, Markvicheva, E, Galván, F, León, R: Studies on the suitability of alginate-entrapped *Chlamydomonas reinhardtii* cells for sustaining nitrate consumption processes. Bioresour. Technol. **78**(1), 55–61 (2001)
92. Ghirardi, ML, Posewitz, MC, Maness, PC, Dubini, A, Yu, J, Seibert, M Hydrogenases and hydrogen photoproduction in oxygenic photosynthetic organisms. Annu. Rev. Plant Biol. **58**, 71–91 (2007)
93. Kosourov, SN, Seibert, M: Hydrogen photoproduction by nutrient-deprived Chlamydomonas reinhardtii cells immobilized within thin alginate films under aerobic and anaerobic conditions. Biotechnol. Bioeng. **102**(1), 50–58 (2009)
94. The Chlamydomonas Center: *Chlamydomonas* connection. http://www.chlamy.org (2013). Accessed 12 Feb 2013
95. Chen, P, Min, M, Chen, Y, Wang, L, Li, Y, Chen, Q, Wang, C, Wan, Y, Wang, X, Cheng, Y, Deng, S, Hennessy, K, Lin, X, Liu, Y, Wang, Y, Martinez, B, Ruan, R: Review of biological and engineering aspects of algae to fuels approach. Int. J. Agric. Biol. Eng. **2**(4), 1–30 (2009)
96. Pienkos, PT, Darzins, A: The promise and challenges of microalgal-derived biofuels. Biofpr. **3**, 431–440 (2009)
97. Campbell, MN: Biodiesel: algae as a renewable source for liquid fuel. Guelph Eng. J. **1**(1916–1107), 207 (2008)
98. Velasquez-Orta, SB, Curtis, TP, Logan, BE: Energy from algae using microbial fuel cells. Biotechnol. Bioeng. **103**(6), 1068–1076 (2009)
99. Rosenbaum, M, Schröder, U, Scholz, F: Utilizing the green alga *Chlamydomonas reinhardtii* for microbial electricity generation: a living solar cell. Appl. Microbiol. Biotechnol. **68**(6), 753–756 (2005)
100. Strik, DPBTB, Hamelers, HVM, Buisman, CJN: Solar energy powered microbial fuel cell with a reversible bioelectrode. Environ. Sci. Technol. **44**(1), 532–537 (2009)
101. Harnisch, F, Schröder, U: Selectivity versus mobility: separation of anode and cathode in microbial bioelectrochemical systems. ChemSusChem **2**(10), 921–926 (2009)

102. Melis, A, Happe, T: Hydrogen production. Green algae as a source of energy. Plant Physiol. **127**(3), 740–748 (2001)
103. Rupprecht, J, Hankamer, B, Mussgnug, JH, Ananyev, G, Dismukes, C, Kruse, O: Perspectives and advances of biological H_2 production in microorganisms. Appl. Microbiol. Biotechnol. **72**(3), 442–449 (2006)
104. Hankamer, B, Lehr, F, Rupprecht, J, Mussgnug, JH, Posten, C, Kruse, O: Photosynthetic biomass and H2 production by green algae: from bioengineering to bioreactor scale-up. Physiol. Plant. **131**(1), 10–21 (2007)
105. Gaffron, H, Rubin, J: Fermentative and photochemical production of hydrogen in algae. J. Gen. Physiol. **26**(2), 219–240 (1942)
106. Kruse, O, Rupprecht, J, Mussgnug, JH, Dismukes, GC, Hankamer, B: Photosynthesis: a blueprint for solar energy capture and biohydrogen production technologies. Photochem. Photobiol. Sci. **4**(12), 957–970 (2005)
107. Melis, A, Zhang, L, Forestier, M, Ghirardi, ML, Seibert, M: Sustained photobiological hydrogen gas production upon reversible inactivation of oxygen evolution in the green alga *Chlamydomonas reinhardtii*. Plant Physiol. **122**(1), 127–136 (2000)
108. Ghirardi, ML, Zhang, L, Lee, JW, Flynn, T, Seibert, M, Greenbaum, E, Melis, A: Microalgae: a green source of renewable H2. Trends Biotechnol. **18**(12), 506–511 (2000)
109. Esquìvel, MG, Amaro, HM, Pinto, TS, Fevereiro, PS, Malcata, FX: Efficient H2 production via *Chlamydomonas reinhardtii*. Trends Biotechnol. **29**(12), 595–600 (2011)
110. Mus, F, Dubini, A, Seibert, M, Posewitz, MC, Grossman, AR: Anaerobic acclimation in *Chlamydomonas reinhardtii*. J. Biol. Chem. **282**(35), 25475–25486 (2007)
111. Dubini, A, Mus, F, Seibert, M, Grossmman, AR, Posewitz, MC: Flexibility in anaerobic metabolism as revealed in a, mutant of Chlamydomonas reinhardtii lacking hydrogenase activity. J. Biol. Chem. **284**(11), 7201–7213 (2009)
112. Posewitz, MC, Dubini, A, Meuser, JE, Seibert, M, Ghirardi, ML: Hydrogenases, hydrogen production, and anoxia. In: Elizabeth, HH, Stern, DB, Witman, GB (eds.) The Chlamydomonas Sourcebook, 2nd edn, pp. 217–255. Academic, London (2009)
113. Timmins, M, Thomas-Hall, SR, Darling, A, Zhang, E, Hankamer, B, Marx, UC, Schenk, PM: Phylogenetic and molecular analysis of hydrogen-producing green algae. J. Exp. Bot. **60**(6), 1691–1702 (2009)
114. Doebbe, A, Keck, M, La Russa, M, Mussgnug, JH, Hankamer, B, Tekçe, E, Niehaus, K, Kruse, O: The interplay of proton, electron, and metabolite supply for photosynthetic H2 production in *Chlamydomonas reinhardtii*. J. Biol. Chem. **285**(39), 30247–30260 (2010)
115. Grossman, AR, Catalanotti, C, Yang, W, Dubini, A, Magneschi, L, Subramanian, V, Posewitz, MC, Seibert, M: Multiple facets of anoxic metabolism and hydrogen production in the unicellular green alga *Chlamydomonas reinhardtii*. New Phytol. **190**(2), 279–288 (2011)
116. Markov, SA, Eivazova, ER, Greenwood, J: Photostimulation of H2 production in the green alga *Chlamydomonas reinhardtii* upon photoinhibition of its O2-evolving system. Int. J. Hydrogen Energy **31**(10), 1314–1317 (2006)
117. Seibert, M, King, PW, Posewitz, MC, Melis, A, Ghirardi, ML: Photosynthetic water-splitting for hydrogen production. In: Wall, JD, Harwood, CS, Demain, A (eds.) Bioenergy, pp. 273–291. ASM, Washington D.C (2008)
118. Zorina, A, Mironov, K, Stepanchenko, N, Sinetova, M, Koroban, N, Zinchenko, V, Kupriyanova, E, Allakhverdiev, S, Los, D: Regulation systems for stress responses in cyanobacteria. Russian J. Plant Physiol. **58**(5), 749–767 (2011)
119. Waterbury, JB, Watson, SW, Guillard, RL, Brand, LE: Widespread occurrence of a unicellular, marine, planktonic, cyanobacterium. Nature **277**(5694), 293–294 (1979)
120. Carr, NG, Whitton, BA: The Biology of Cyanobacteria. University of California Press, Berkeley (1982)
121. Martin, W, Rujan, T, Richly, E, Hansen, A, Cornelsen, S, Lins, T, Leister, D, Stoebe, B, Hasegawa, M, Penny, D: Evolutionary analysis of *Arabidopsis*, cyanobacterial, and chloroplast genomes reveals plastid phylogeny and thousands of cyanobacterial genes in the nucleus. Proc. Natl. Acad. Sci. U.S. A. **99**(19), 12246–12251 (2002)
122. Dutta, D, De, D, Chaudhuri, S, Bhattacharya, SK: Hydrogen production by Cyanobacteria. Microb. Cell Fact. **4**, 36 (2005)
123. Kaneko, T, Sato, S, Kotani, H, Tanaka, A, Asamizu, E, Nakamura, Y, Miyajima, N, Hirosawa, M, Sugiura, M, Sasamoto, S, Kimura, T, Hosouchi, T, Matsuno, A, Muraki, A, Nakazaki, N, Naruo, K, Okumura, S, Shimpo, S, Takeuchi, C, Wada, T, Watanabe, A, Yamada, M, Yasuda, M, Tabata, S: Sequence analysis of the genome of the unicellular cyanobacterium *Synechocystis* sp. strain PCC6803. II. Sequence determination of the entire genome and assignment of potential protein-coding regions. DNA Res. **3**(3), 109–136 (1996)
124. Nakamura, Y, Kaneko, T, Hirosawa, M, Miyajima, N, Tabata, S: CyanoBase, a www database containing the complete nucleotide sequence of the genome of *Synechocystis* sp. strain PCC6803. Nucleic Acids Res. **26**(1), 63–67 (1998)
125. Atsumi, S, Higashide, W, Liao, JC: Direct photosynthetic recycling of carbon dioxide to isobutyraldehyde. Nat. Biotechnol. **27**(12), 1177–1180 (2009)
126. Johnson, CH, Stewart, PL, Egli, M: The cyanobacterial circadian system: from biophysics to bioevolution. Annu. Rev. Biophys. **40**, 143–167 (2011)
127. Lovley, DR: The microbe electric: conversion of organic matter to electricity. Curr. Opin. Biotechnol. **19**(6), 564–571 (2008)
128. Pisciotta, JM, Zou, Y, Baskakov, IV: Role of the photosynthetic electron transfer chain in electrogenic activity of cyanobacteria. Appl. Microbiol. Biotechnol. **91**(2), 377–385 (2011)
129. Pisciotta, JM, Zou, Y, Baskakov, IV: Light-dependent electrogenic activity of cyanobacteria. PLoS One **5**(5), e10821 (2010)
130. Madiraju, KS, Lyew, D, Kok, R, Raghavan, V: Carbon neutral electricity production by *Synechocystis* sp. PCC6803 in a microbial fuel cell. Bioresour. Technol. **110**, 214–218 (2012)
131. McCormick, AJ, Bombelli, P, Scott, AM, Philips, AJ, Smith, AG, Fisher, AC, Howe, CJ: Photosynthetic biofilms in pure culture harness solar energy in a mediatorless bio-photovoltaic cell (BPV) system. Energy Environ. Sci. **4**(11), 4699–4709 (2011)
132. Dequin, S, Casaregola, S: The genomes of fermentative *Saccharomyces*. Comptes Rendus Biologies **334**(8–9), 687–693 (2011)
133. Kurtzman, CP: Phylogenetic circumscription of *Saccharomyces*, *Kluyveromyces* and other members of the Saccharomycetaceae, and the proposal of the new genera *Lachancea*, *Nakaseomyces*, *Naumovia*. Vanderwaltozyma and Zygotorulaspora. FEMS Yeast Res. **4**(3), 233–245 (2003)
134. Vaughan-Martini, A, Martini, A: Facts, myths and legends on the prime industrial microorganism. J. Ind. Microbiol. **14**(6), 514–522 (1995)
135. Fan, Y, Sharbrough, E, Liu, H: Quantification of the internal resistance distribution of microbial fuel cells. Environ. Sci. Technol. **42**(21), 8101–8107 (2008)
136. Peralta-Yahya, PP, Keasling, JD: Advanced biofuel production in microbes. Biotechnol. J. **5**(2), 147–162 (2010)
137. Wilkinson, S: "Gastrobots" - benefits and challenges of microbial fuel cells in food powered robot applications. Auton. Robots **9**(2), 99–111 (2000)
138. Haslett, ND, Rawson, FJ, Barriëre, F, Kunze, G, Pasco, N, Gooneratne, R, Baronian, KH, Haslett, ND, Rawson, FJ, Barriëre, F, Kunze, G, Pasco, N, Gooneratne, R, Baronian, KH: Characterisation of yeast microbial fuel cell with the yeast *Arxula adeninivorans* as the biocatalyst. Biosens. Bioelectron. **26**(9), 3742–3747 (2011)
139. Gunawardena, A, Fernando, S, To, F: Performance of a yeast-mediated biological fuel cell. Int. J. Mol. Sci. **9**(10), 1893–1907 (2008)
140. Feldmann, H: Yeast Molecular Biology: A Short Compendium on Basic Features and Novel Aspects. Adolf-Butenandt-Institut, University of Munich, Munich (2005)
141. Ganguli, R, Dunn, BS: Kinetics of anode reactions for a yeast-catalysed microbial fuel cell. Fuel Cells **9**(1), 44–52 (2009)
142. Powell, EE, Evitts, RW, Hill, GA, Bolster, JC: A microbial fuel cell with a photosynthetic microalgae cathodic half cell coupled to a yeast anodic half cell. Energy Sources Part a-Recovery Utilization Environ. Effects **33**(5), 440–448 (2011)
143. Sayed, ET, Tsujiguchi, T, Nakagawa, N: Catalytic activity of baker's yeast in a mediatorless microbial fuel cell. Bioelectrochemistry **86**, 97–101 (2012)
144. Anderson, RT, Vrionis, HA, Ortiz-Bernad, I, Resch, CT, Long, PE, Dayvault, R, Karp, K, Marutzky, S, Metzler, DR, Peacock, A, White, DC, Lowe, M, Lovley, DR: Stimulating the in situ activity of *Geobacter* species to remove uranium from the groundwater of a uranium-contaminated aquifer. Appl. Environ. Microbiol. **69**(10), 5884–5891 (2003)
145. Holmes, DE, Finneran, KT, O'Neil, RA, Lovley, DR: Enrichment of members of the family Geobacteraceae associated with stimulation of dissimilatory metal reduction in uranium-contaminated aquifer sediments. Appl. Environ. Microbiol. **68**(5), 2300–2306 (2002)
146. Lovley, DR, Anderson, RT: Influence of dissimilatory metal reduction on fate of organic and metal contaminants in the subsurface. Hydrogeol. J. **8**, 77–88 (2000)
147. Coppi, MV, Leang, C, Sandler, SJ, Lovley, DR: Development of a genetic system for *Geobacter sulfurreducens*. Appl. Environ. Microbiol. **67**(7), 3180–3187 (2001)

148. Caccavo Jr, F, Lonergan, DJ, Lovley, DR, Stolz, JF, McInerney, MJ: *Geobacter sulfurreducens* sp. nov., a hydrogen- and acetate-oxidizing dissimilatory metal-reducing microorganism. Appl. Environ. Microbiol. **60**(10), 3752–3759 (1994)
149. Nevin, KP, Kim, BC, Glaven, RH, Johnson, JP, Woodard, TL, Methé, BA, Didonato, RJ, Covalla, SF, Franks, AE, Liu, A, Lovley, DR: Anode biofilm transcriptomics reveals outer surface components essential for high density current production in *Geobacter sulfurreducens* fuel cells. PLoS One **4**(5), e5628 (2009)
150. Chae, KJ, Choi, MJ, Lee, JW, Kim, KY, Kim, IS: Effect of different substrates on the performance, bacterial diversity, and bacterial viability in microbial fuel cells. Bioresour. Technol. **100**(14), 3518–3525 (2009)
151. Yi, H, Nevin, KP, Kim, BC, Franks, AE, Klimes, A, Tender, LM, Lovley, DR: Selection of a variant of *Geobacter sulfurreducens* with enhanced capacity for current production in microbial fuel cells. Biosens. Bioelectron. **24**(12), 3498–3503 (2009)
152. Salgado, CA: Microbial fuel cells powered by *Geobacter sulfurreducens*. MMG 445 Basic Biotechnology eJournal **5**, 96–101 (2009)
153. Lin, WC, Coppi, MV, Lovley, DR: *Geobacter sulfurreducens* can grow with oxygen as a terminal electron acceptor. Appl. Environ. Microbiol. **70**(4), 2525–2528 (2004)
154. Bond, DR, Holmes, DE, Tender, LM, Lovley, DR: Electrode-reducing microorganisms that harvest energy from marine sediments. Science **295**(5554), 483–485 (2002)
155. Nevin, KP, Richter, H, Covalla, SF, Johnson, JP, Woodard, TL, Orloff, AL, Jia, H, Zhang, M, Lovley, DR: Power output and columbic efficiencies from biofilms of *Geobacter sulfurreducens* comparable to mixed community microbial fuel cells. Environ. Microbiol. **10**(10), 2505–2514 (2008)
156. Gregory, KB, Bond, DR, Lovley, DR: Graphite electrodes as electron donors for anaerobic respiration. Environ. Microbiol. **6**(6), 596–604 (2004)
157. Holmes, DE, Bond, DR, O'Neil, RA, Reimers, CE, Tender, LR, Lovley, DR: Microbial communities associated with electrodes harvesting electricity from a variety of aquatic sediments. Microb. Ecol. **48**(2), 178–190 (2004)
158. Tender, LM, Reimers, CE, Stecher, HA, Holmes, DE, Bond, DR, Lowy, DA, Pilobello, K, Fertig, SJ, Lovley, DR: Harnessing microbially generated power on the seafloor. Nat. Biotechnol. **20**(8), 821–825 (2002)
159. Mahadevan, R, Bond, DR, Butler, JE, Esteve-Nuñez, A, Coppi, MV, Palsson, BO, Schilling, CH, Lovley, DR: Characterization of metabolism in the Fe(III)-reducing organism *Geobacter sulfurreducens* by constraint-based modeling. Appl. Environ. Microbiol. **72**(2), 1558–1568 (2006)
160. Champine, JE, Underhill, B, Johnston, JM, Lilly, WW, Goodwin, S: Electron transfer in the dissimilatory iron-reducing bacterium *Geobacter metallireducens*. Anaerobe **6**(3), 187–196 (2000)
161. Galushko, AS, Schink, B: Oxidation of acetate through reactions of the citric acid cycle by Geobacter sulfurreducens in pure culture and in syntrophic coculture. Arch. Microbiol. **174**(5), 314–321 (2000)
162. Reguera, G, McCarthy, KD, Mehta, T, Nicoll, JS, Tuominen, MT, Lovley, DR: Extracellular electron transfer via microbial nanowires. Nature **435**(7045), 1098–1101 (2005)
163. Fredrickson, JK, Romine, MF, Beliaev, AS, Auchtung, JM, Driscoll, ME, Gardner, TS, Nealson, KH, Osterman, AL, Pinchuk, G, Reed, JL, Rodionov, DA, Rodrigues, JLM, Saffarini, DA, Serres, MH, Spormann, AF, Zhulin, IB, Tiedje, JM: Towards environmental systems biology of *Shewanella*. Nature Reviews Microbiology **6**(8), 592–603 (2008)
164. Call, DF, Wagner, RC, Logan, BE: Hydrogen production by *Geobacter* species and a mixed consortium in a microbial electrolysis cell. Appl. Environ. Microbiol. **75**(24), 7579–7587 (2009)
165. El-Naggar, MY, Gorby, YA, Xia, W, Nealson, KH: The molecular density of states in bacterial nanowires. Biophys. J. **95**(1), L10–L12 (2008)
166. Marsili, E, Baron, DB, Shikhare, ID, Coursolle, D, Gralnick, JA, Bond, DR: *Shewanella* secretes flavins that mediate extracellular electron transfer. Proc. Natl. Acad. Sci. **105**(10), 3968–3973 (2008)
167. Newton, GJ, Mori, S, Nakamura, R, Hashimoto, K, Watanabe, K: Analyses of current-generating mechanisms of *Shewanella loihica* PV-4 and *Shewanella oneidensis* MR-1 in microbial fuel cells. Appl. Environ. Microbiol. **75**(24), 7674–7681 (2009)
168. Ringeisen, BR, Henderson, E, Wu, PK, Pietron, J, Ray, R, Little, B, Biffinger, JC, Jones-Meehan, JM: High power density from a miniature microbial fuel cell using *Shewanella oneidensis* DSP10. Environ. Sci. Technol. **40**(8), 2629–2634 (2006)
169. Ringeisen, BR, Ray, R, Little, B: A miniature microbial fuel cell operating with an aerobic anode chamber. J. Power. Sources **165**(2), 591–597 (2007)
170. Nevin, KP, Zhang, P, Franks, AE, Woodard, TL, Lovley, DR: Anaerobes unleashed: aerobic fuel cells of *Geobacter sulfurreducens*. J. Power. Sources **196**(18), 7514–7518 (2011)

22

Empirical model assessment of commercial aircraft emissions according to flight phases

Enis T Turgut[1*], Oznur Usanmaz[2] and Marc A Rosen[3]

Abstract

The quantities of common emissions are investigated for a specific type of commercial aircraft. Actual flight data and International Civil Aviation Organization emission data are used. All flight phases are considered, including landing and takeoff phases. The investigation is carried out for the domestic flights only and considers relevant parameters, such as engine type, flight phase, and ground or air operation of the flight. The findings suggest that the quantities of emissions of unburned hydrocarbon (HC) and carbon monoxide (CO) during the descent phase can exceed those for the taxi phases and the idle operation of the engines, depending on the approach procedure. The main source of nitrogen oxide (NO_x) is usually the climb phase, while the mean total flight emissions are calculated as 6 to 8 kg of HC, 60 to 75 kg of CO, and 28 to 31 kg of NO_x. The effect of the duration of taxi phase on the production of HC and CO emissions is also discussed.

Keywords: Aircraft emission, Environment, Turbofan engine, Flight data records

Background

Since the advent of turbofan engines in 1960s, there has been an increase in the commercial flights and, correspondingly, air traffic movement in the vicinity of airports. The increase in air traffic has led to some problems, e.g., noise and environmental impact through emissions. Noise is perceived easily and can be monitored and regulated. Regulations related to aircraft emissions began to be put in place by the Environmental Protection Agency (EPA) in 1974, starting with smoke monitoring and followed by regulations for hydrocarbon emissions (1984), nitrogen oxide (NO_x) and carbon monoxide (CO) (1997), and further controls for NO_x (2005).

In the last 20 years, much effort has been expended to improve understanding and reduction of the environmental impact of aircraft by such organizations as the Federal Aviation Administration, National Aeronautics and Space Administration, European Organisation for the Safety of Air Navigation, International Civil Aviation Organization (ICAO), International Air Transport Association, and Civil Air Navigation Services Organisation. Databases involving fuel consumption and emissions of NO_x, CO, and hydrocarbon (HC) for the scheduled civil aviation were developed in 1992 [1] and 1996 [2]. Studies have included investigations of aircraft NO_x emissions within the ANCAT/EC project [3], analyses of emissions resulting from civil and military aviation within the project of Global Aviation Emissions Inventories for 2002 and 2025 (AERO2K) [4], and theoretical aircraft fuel consumption and emission modeling within the project of System for Assessing Aviation's Global Emissions [5].

Moreover, aircraft emissions have become an important issue not only for aviation authorities but also environmental scientists and the public. According to the EPA, aircraft engines were responsible for 2% of the total US mobile emission sources in 1997. However, it is reported that aircraft can contribute up to 4% of emissions in the vicinity of airports [6]. In addition, difficulties in measuring the altitude of flight emissions have motivated numerous studies focusing on the generation and identification of emissions in the vicinity of the airports such as Atlanta, Heathrow, New York, Los Angeles, Zurich, and Copenhagen [7-13].

Nevertheless, improved information is needed on aircraft emissions according to flight phases. The objective of this study is to improve understanding on common emissions (HC, CO, and NO_x) resulting from aircraft, based on the ICAO databank [14]. This databank includes exhaust

* Correspondence: etturgut@anadolu.edu.tr
[1]Department of Airframe and Powerplant Maintenance, Faculty of Aeronautics and Space Sciences, Anadolu University, Eskisehir 26470, Turkey
Full list of author information is available at the end of the article

emissions and fuel flow rates for currently used turbofan engines during landing and takeoff (LTO) phases. The investigation is carried out in such a way that the actual flight data are considered. In order to perform this task, a novel method is developed, in which interpolation and extrapolation of the relationship between the fuel flow and the emission indices data of specific types of engine are provided. Data are obtained from flights of ten randomly selected B737-800 (hereafter B738) commercial aircraft. Two types of turbofan engines are used in these aircraft: CFM56-7B26 and CFM56-7B26/3.

Analysis setup

The most frequently used domestic routes, aircraft types, and the engine types are considered for the data selection. For this purpose, ten randomly selected B738 commercial aircraft are used. Five of them are for flights between the Antalya International (AYT) and Sabiha Gokcen International (SAW) airports, and five are for flights between the Izmir Adnan Menderes International (ADB) and the SAW airports. The SAW airport is the arrival airport for each group. The airports are defined in Table 1. All flights occurred in 2009.

The route selections relate to the frequency of flights to SAW. In 2009, for domestic flights, the two most frequent arrivals to SAW are from ADB and AYT, with 2,956 and 2,684 total arrivals, respectively. For B738 aircraft only, the numbers of arrivals reduce to 1,540 for AYT and 1,134 for ADB. Of the total arrivals to SAW with B738 aircraft, flights from AYT and ADB account for 15% and 11%, respectively. Other arrival percentages are shown in Figure 1.

According to the traffic statistics of the SAW airport in 2009, the most used aircraft type is the B738. Taking into account all aircraft traffic to or from the SAW airport (including international flights), those utilizing the B738 aircraft account for the largest proportion at 47%. Therefore, ten randomly selected B738 flights are evaluated here, all from 2009. Actual flight data of these flights are obtained from Pegasus Airlines, an international airline based in Turkey. Considered in the analysis are aircraft flight parameters such as aircraft altitude, ground speed and weight, as well as engine parameters such as fuel flow rate, rotation speeds of the low pressure system (N1) and the high pressure system (N2) on percentage bases, and exhaust gas temperature (EGT).

With respect to engines, six of the assessed flights were powered by CFM56-7B26 (hereafter 7B26) engines, while the remaining four were powered by CFM56-7B26/3 (hereafter 7B26/3) turbofan engines. Moreover, for the flights of each route, three 7B26 and two 7B26/3 engines are utilized.

Methods

In the ICAO emission databank, there are 17 different models of the CFM56-7B series engine, which can be tabulated according to bypass ratio, overall pressure ratio, and thrust parameters (see Table 2). The engine family can be classified into three groups: CFM56-7B*X* (single annular combustion), CFM56-7B*X*/2 (double annular combustion), and CFM56-7B*X*/3 (improved emission single annular combustion), where *X* denotes model numbers such as 18, 20, 22, 24, 26, and 27 [15]. The classification is mainly based on the combustion chamber design.

As can be seen from Table 2, there is an inverse relationship between the model number (which increases moving down the list in Table 2) and the bypass ratio, and a direct relationship between the model number and the pressure ratio and the thrust in each group. This pattern is valid for all three engine groups.

The ICAO emission database has a disadvantage in that its emission indices are obtained only for limited fuel flow rates. For instance, a single fuel flow rate is accepted for the entire flight phase. This prevents a precise identification of the emissions generated from commercial aircraft since the fuel flow rate is not constant. To address this problem here, the approach used in this study is developed based on linear and polynomial extrapolation and interpolation for three types of emissions: HC, CO, and NO_x. Nonetheless, it is noted that emission production mechanisms generally are not simple. In this regard, equivalence ratio, combustion temperature, and compressor pressure ratio, all of which vary with engine power, or the ambient conditions might significantly influence the emissions, and numerous studies of these influences have been reported [7,8,16-18].

Table 1 Airport information

Aerodrome location indicator and name	ARP geographical coordinates	Distance to SAW (km)	Aircraft movement[a]
SAW - Istanbul Sabiha Gokcen International	40°53′54″N, 29°18′33″E	-	63,749
ADB - Izmir Adnan Menderes International	38°17′21″N, 27°09′18″E	344	54,197
AYT - Antalya International	36°54′01″N, 30°47′34″E	462	127,236

ARP aerodrome reference point. [a]Aircraft movement is defined at http://icaodata.com/Terms.aspx# AircraftMovement%28airports%29 as 'an aircraft take-off or landing at an airport. For airport traffic purposes one arrival and one departure is counted as two movements.' The airport movement statistics for 2009 are retrieved from http://dhmi.gov.tr/istatistik.aspx.

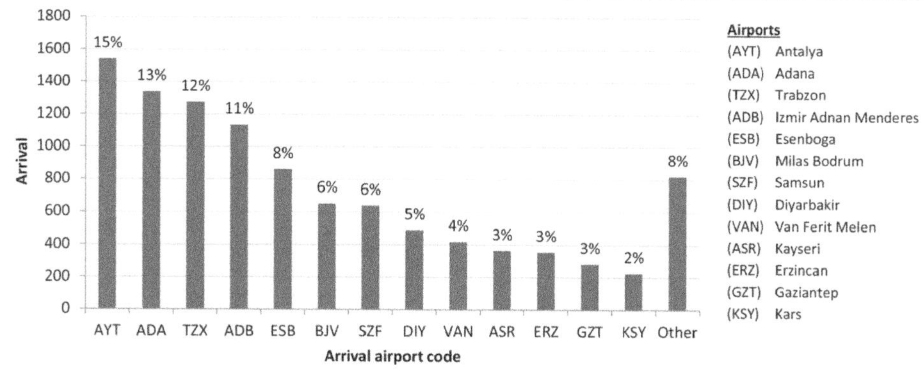

Figure 1 Breakdown of domestic arrivals to Sabiha Gokcen International (SAW) airport of B738 aircraft for 2009. The 'other' category denotes flights from the other 12 domestic airports, which each have an arrival number of less than 200.

In the ICAO emission database, fuel flow rates and emission indices are given by engine model for the following flight phases: takeoff, climb, approach, and idle. Excluding the 7B*X*/2 series, it is noted that there are 24 (6 × 4) fuel flow rate and emission indices observations for either 7B*X* or 7B*X*/3 series engines. These 24 data groups generate the observation domain for each engine group, which is used for establishing the relationships between the fuel flow and the specific emission indices. Since six of the flights are performed with the aircraft having a 7B*X* series engine and four of the flights are performed with the aircraft having a 7B*X*/3 series engine, two engine models are considered here. Therefore, the results are different from those in which a single engine type is considered.

Table 2 Specific performance parameters of CFM56-7B series engines (modified from [14])

Engine model	Bypass ratio	Pressure ratio	Thrust (kN)
CFM56-7B18	5.5	21.6	86.7
CFM56-7B20	5.4	22.6	91.6
CFM56-7B22	5.3	24.4	101.0
CFM56-7B24	5.2	25.8	107.7
CFM56-7B26	*5.1*	*27.6*	*117.0*
CFM56-7B27	5.0	28.6	121.4
CFM56-7B20/2	5.4	22.8	91.6
CFM56-7B22/2	5.3	24.6	101.0
CFM56-7B24/2	5.2	25.9	107.7
CFM56-7B26/2	5.1	27.8	117.0
CFM56-7B27/2	5.0	28.8	121.4
CFM56-7B18/3	5.5	21.4	86.7
CFM56-7B20/3	5.5	22.4	91.6
CFM56-7B22/3	5.3	24.2	101.0
CFM56-7B24/3	5.3	25.6	107.6
CFM56-7B26/3	*5.1*	*27.7*	*117.0*
CFM56-7B27/3	5.1	29.0	121.4

Italic font indicates the sample engines.

The variation of common emission indices with fuel flow rate indicates that linear relationships are present for certain fuel flow fragments, where fuel flow fragments are the fuel flow rate regions for a given flight phase for all of the engines. For instance, the given fuel flow rates for the climb phases of all the engine models constitute a fuel flow region, while there are other specific fuel flow ranges for the other three flight phases: takeoff, approach, and landing. In other words, there is specific fuel flow region for each flight phase for the engines of each group. For these regions, the relationship between the emission indices and the fuel flow rates can be straightforwardly identified. However, for fuel flow rates outside the specified regions, particularly between the two regions, the approach based on the aforementioned relationship can lead to incorrect results. Therefore, one needs additional methods which can be obtained utilizing the sequential regions. For instance, if the overall fuel flow ranges are divided into three parts (highest, medium, and lowest regions indicating the amount of fuel flow), then the region between the highest and medium fuel flow rates and also between the medium and the lowest fuel flow rates can be described by additional relationships. The required model descriptions are presented in the following section.

Results and discussion

The relationships are developed for two sections, representing both engine series, and given in Tables 3 and 4. The fuel flow range is also given in these tables. The coefficients of determination (R^2) for the regression models are given in the last columns of Tables 3 and 4. All of the regression models exhibit high coefficients of determination, mostly over 0.960 (with the exceptions of 0.904 and 0.910, which relate to the fuel flow range of CO emissions). This means that these models explain more than 96.0% of the variation in emission index when compared to the total variation.

The relationships between the emission index (EI) for common emissions and the fuel flow rate (ff) which are

Table 3 Model emission indices for CFM56-7B*X* (flights 1, 3, 4, and 6 to 8)

Emission index (EI) (g)	ff range (kg s^{-1})	Model[a]	R^2
EI (HC)	ff ≥ 0.260	0.1	0.978
	0.260 > ff > 0.116	$0.0029 \times (ff)^{-2.988}$	0.979
	0.09 ≤ ff ≤ 0.116	−92.07 × (ff) + 12.34	NA[b]
EI (CO)	ff ≥ 0.714	$9.10 \times (ff)^3 - 30.46 \times (ff)^2 + 32.52 \times (ff) - 10.61$	0.975
	0.714 > ff > 0.349	$0.42 \times (ff)^{-1.446}$	0.904
	0.349 ≥ ff ≥ 0.260	−24.64 × (ff) + 9.95	0.995
	0.260 > ff > 0.116	$0.17 \times (ff)^{-2.194}$	0.995
	ff ≤ 0.116	−536.95 × (ff) + 79.89	0.981
EI (NO_x)	ff > 0	20.29 × (ff) + 2.83	0.989

ff fuel flow rate; *NA* not available. [a]Coefficients in the models are rounded to two decimals in most cases. [b]EI of HC for this range is given in the ICAO database as a constant value of 0.1 g HC/kg of fuel.

both obtained from the ICAO database are shown in Figures 2, 3, 4 and 5. Regarding the HC emissions for the two types of engines considered, it can be seen that the EI for HC emissions can be divided into three parts for 7B*X* engines: 0.09 to 0.116 kg/s (lower ff), 0.116 to 0.260 kg/s (moderate ff), and above 0.260 kg/s (higher ff). On the other hand, there are five parts for HC emissions for 7B*X*/3 engines. However, if we focus on the lower ff range, such as below 0.110 kg/s, it can be seen that the EI for HC emissions of 7B*X* engines is low compared to 7B*X*/3 engines, while between the top and bottom limits at this part (0.09 to 0.110 kg/s), the slope of the curve for 7B*X*/3 engines is steeper (see Figures 2a and 3a). Generally, it can be noted that the EI for HC

Table 4 Model emission indices for CFM56-7B*X*/3 (flights 2, 5, 9, and 10)

Emission index (EI) (g)	ff range (kg s^{-1})	Model[a]	R^2
EI (HC)	ff ≥ 0.702	$0.07 \times (ff)^2 - 0.15 \times (ff) + 0.11$	0.972
	0.702 > ff > 0.343	$0.03 \times (ff)^{-0.77}$	0.964
	0.343 ≥ ff ≥ 0.256	−0.39 × (ff) + 0.18	0.973
	0.256 > ff > 0.110	$0.001 \times (ff)^{-3.41}$	0.992
	ff ≤ 0.110	−161.40 × (ff) + 19.09	0.966
EI (CO)	ff ≥ 0.702	$1.56 \times (ff)^2 - 3.03 \times (ff) + 1.63$	0.972
	0.702 > ff > 0.343	$0.18 \times (ff)^{-2.43}$	0.910
	0.343 ≥ ff ≥ 0.256	−31.11 × (ff) + 13.37	0.991
	0.256 > ff > 0.110	$0.32 \times (ff)^{-2.07}$	0.999
	ff ≤ 0.110	−944.89 × (ff) + 132.56	0.985
EI (NO_x)	ff > 0	14.79 × (ff) + 3.04	0.984

[a]Coefficients in the models are rounded to two decimals in most cases.

emissions of 7B*X*/3 engines is greater than for 7B*X* engines. In order to understand the effect of these small values, note that the idle ff value of these types of engines is in the order of 0.070 kg/s (no aircraft movement) and 0.097 kg/s (during taxi) per engine (although values can differ due to other factors such as the mass of aircraft or ambient temperature).

Similar behavior is exhibited by the EI for CO emissions. The EI for CO emissions of 7B*X*/3 is usually higher than for 7B*X* engines, except those for higher ff values (above 0.714 kg/s). Also, in this ff segment, the relation between EI to ff is significantly changed (see Figure 4).

As seen in Figures 2, 3 and 4, there can be interruptions between certain data values. Reasonable ff values in the interrupted region can be estimated by considering front and rear regions such as done in Figure 2b. Therefore, since there is no ICAO information regarding the EI for corresponding ff values in these regions, the precision of the predictions for these regions is likely lower. Nonetheless, estimates for these regions are often useful.

Lastly, regarding the EI for NO_x emissions, it is seen in Figure 4 that the 7B*X*/3 engines produce less emissions per kilogram of fuel. This result is expected, since these engines produce less NO_x due to combustor upgrades (recall that it was found earlier that EIs for HC and CO emissions are higher compared to other types of engines). Furthermore, a higher ff leads to higher NO_x emissions, indicating a direct relationship.

Considering overall ff ranges, it appears from the ICAO results that the relationships between ff values and HC and CO emissions are nonlinear (which for the HC case is also stressed in [18] and elsewhere), while it is linear for the EI for NO_x emissions.

In Figures 6 and 7, the variations of ff vs. flight phase and flight time are shown. It can be seen that the critical phases for HC and CO emissions are phases 1 and 2 as well as 9 to 14. Of these, phases 2 and 9 in particular produce more HC and CO emissions due to having high durations and less engine power. Regarding phase 9, controlling the production of these types of emissions might be possible by following different descent patterns leading to increased engine power; however, this action would increase NO_x emissions and also the fuel consumption. From a NO_x perspective, on the other hand, phases 3 to 7 appear to be more critical.

To identify the emissions of each flight, a detailed analysis is carried out in this section. Accordingly, the emissions are determined for each route due to the different distances between the departure and the arrival airports, and thereby the fuel consumptions. Each route involves five flights. As stated earlier, two of the five aircraft on each route use the CFM56-7B26/3 engine, while the other three use the CFM56/7B26. As shown in Table 2,

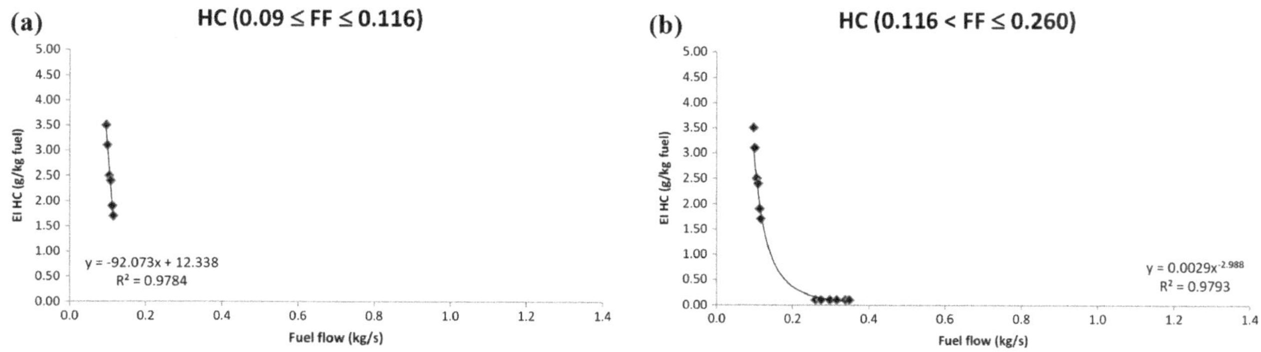

Figure 2 Effect of fuel flow rate on emission index for HC emissions for 7BX series engines. (**a**) HC ($0.09 \leq ff \leq 0.116$) and (**b**) HC ($0.116 < ff \leq 0.260$).

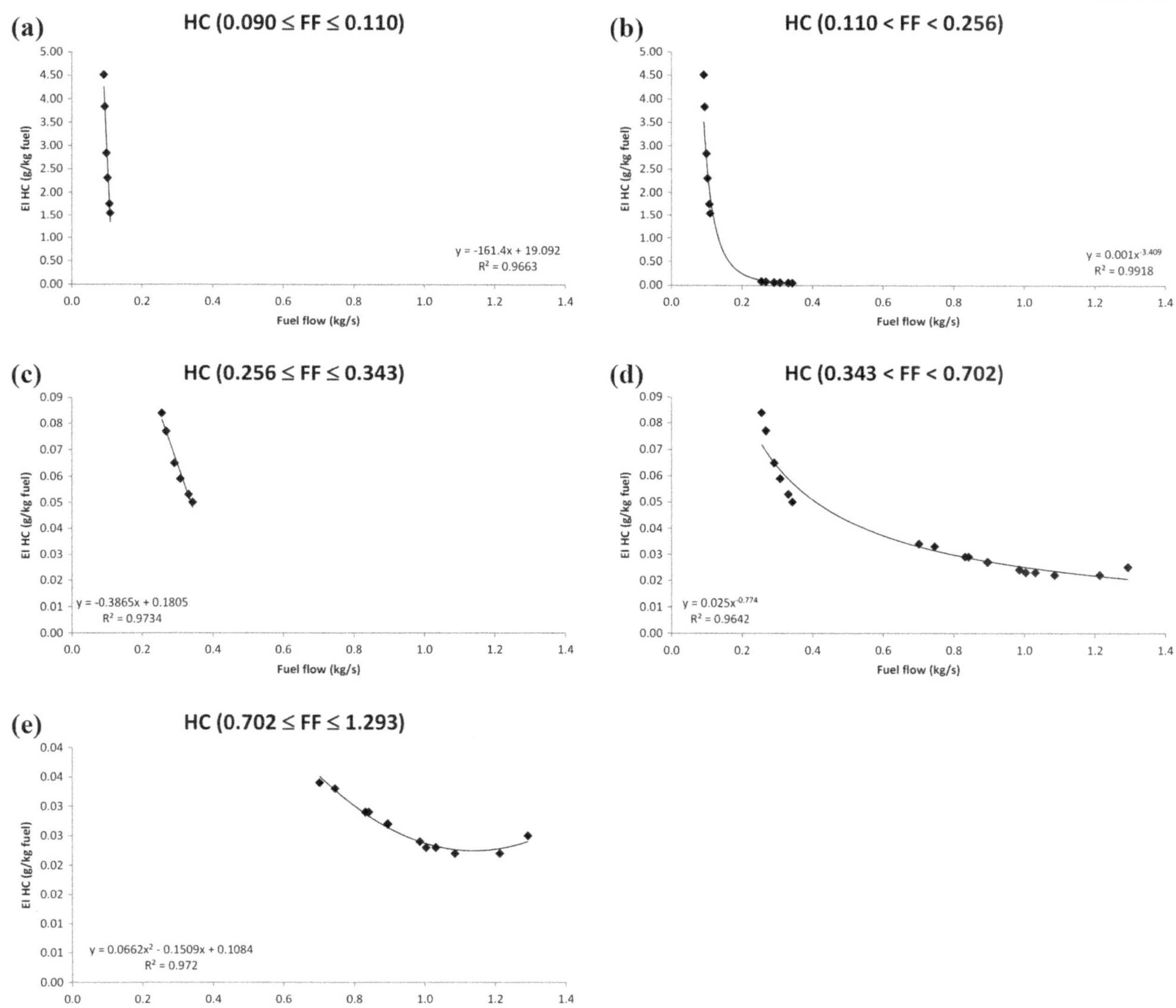

Figure 3 Effect of fuel flow rate on emission index for HC emissions for 7BX/3 series engines. (**a**) HC ($0.090 \leq ff \leq 0.110$), (**b**) HC ($0.110 < ff < 0.256$), (**c**) HC ($0.256 \leq ff \leq 0.343$), (**d**) HC ($0.343 < ff < 0.702$), and (**e**) HC ($0.702 \leq ff \leq 1.293$).

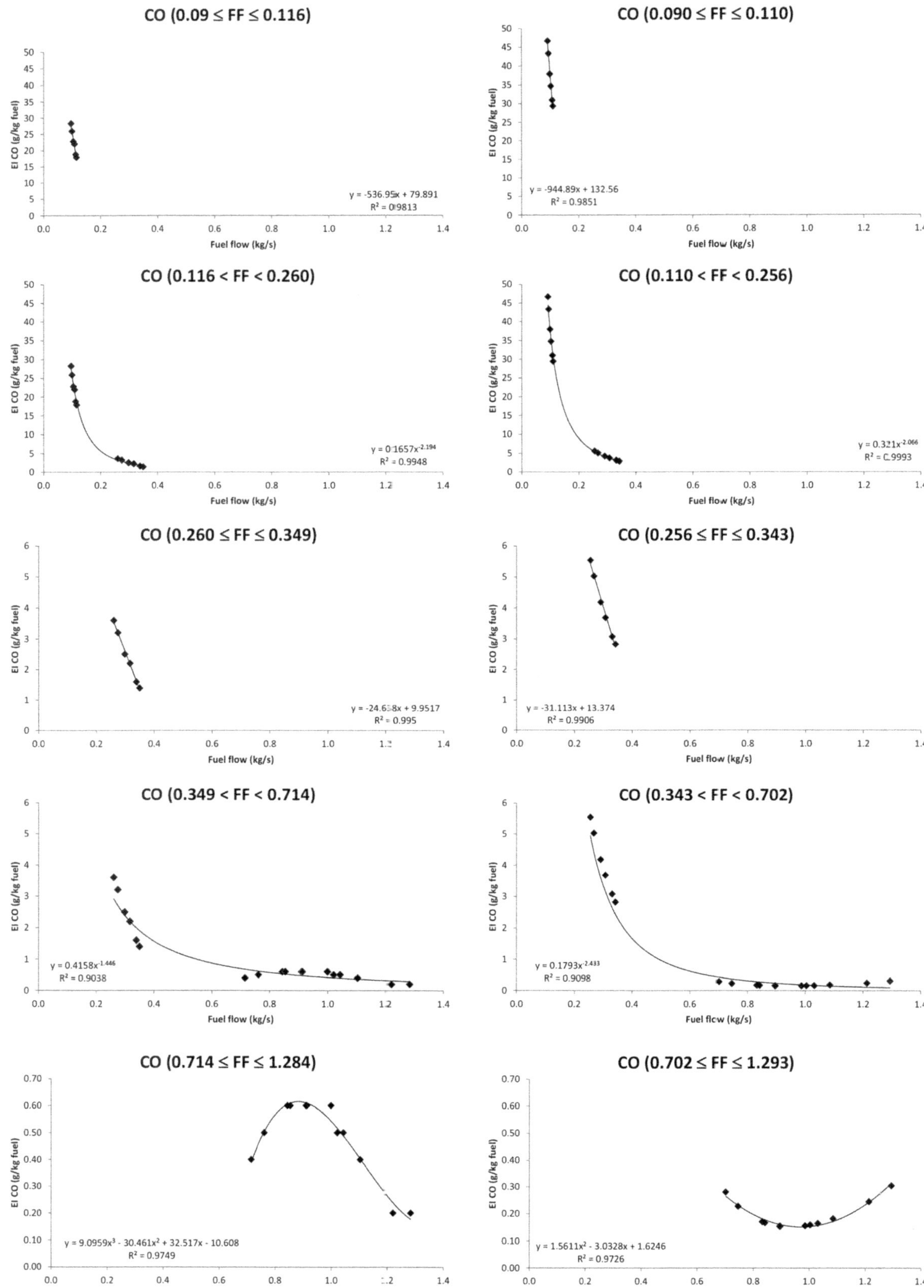

Figure 4 Effect of fuel flow rate on emission index for CO emissions. *7BX* (left) and *7BX/3* (right) series engines.

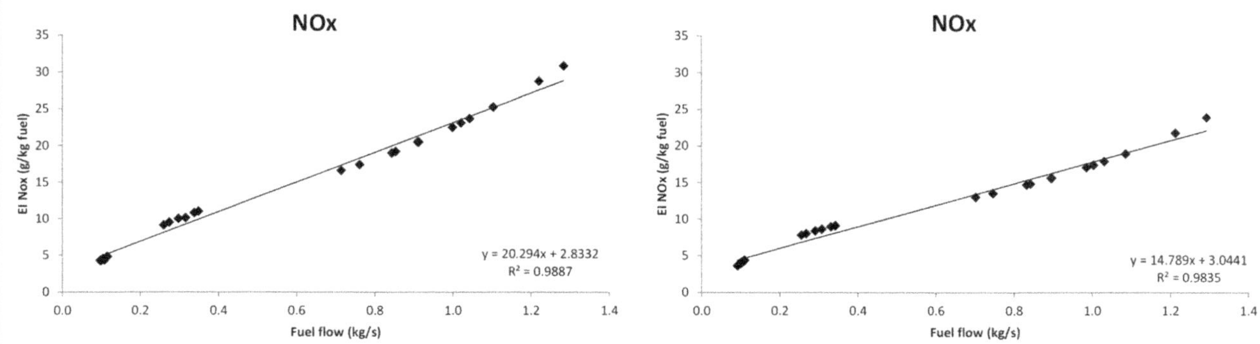

Figure 5 Effect of fuel flow rate on emission index for NO_x emissions. 7BX (left) and 7BX/3 (right) series engines.

the bypass ratio, the pressure ratio, and the thrust of these two kinds of engines are the same. Nonetheless, the combustion chamber designs differ, impacting the emission production mechanism. According to Brasseur et al., rich mixtures and the local flame extinctions in the combustion chamber are likely to induce CO and HC production, respectively [19]. This addresses the importance of the design of the primary zone of the combustion chambers, residence time of the flame, and the flame temperature [20,21]. Also, ambient temperature could affect the engine thermal efficiency and thereby certain emission types [18].

Consequently the following analyses are performed in the following subsections: 'Emission breakdown by flight phase' and 'Flight phase duration vs. amount of emissions.'

Emission breakdown by flight phase

In the ICAO database, emission indices are given for only four flight phases. Utilizing the models developed in this study, one can obtain information related to other flight phases, such as cruise and taxi.

The duration of the flight phases are depicted in Figure 8. Since the flights are domestic and short range, the LTO phases constitute the larger part of the total flight duration. The duration of the phases can change significantly for international flights.

From the perspective of the flight phase where emission production takes place, it can be useful to know the breakdown of aircraft utilization on the ground and in the air. In this context, all flights are examined in terms of the utilization duration. The results indicate that the percentage of ground utilization of the aircraft ranges between 22% to 29% for AYT and 18% to 27% for ADB departures, respectively. The averages of ground utilization of the aircraft by route are around 25% and 22%, respectively. That is, almost one fourth of the operation duration takes place on the ground.

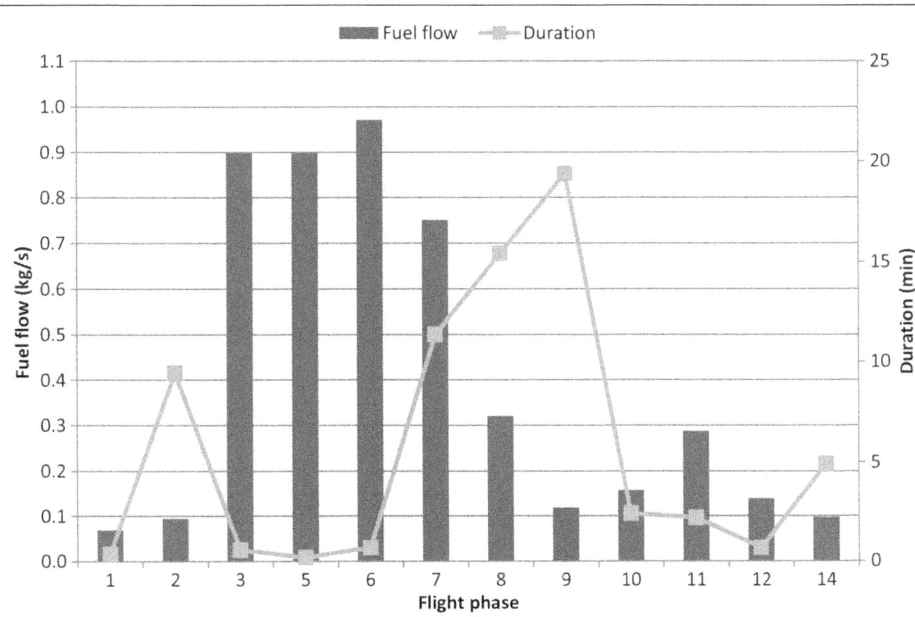

Figure 6 Mean ff vs. flight phase for single engine (flight 1, 7BX engine).

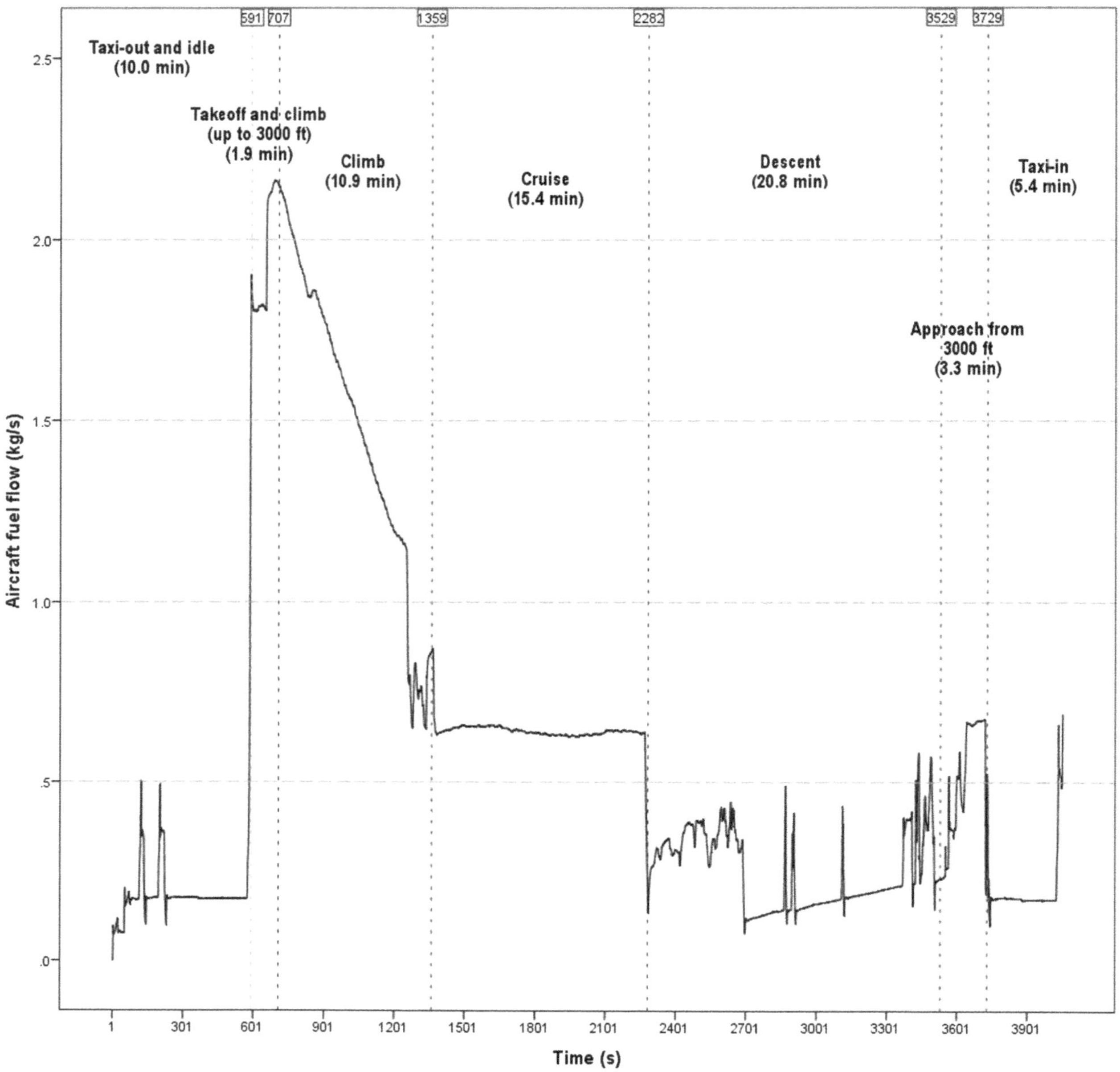

Figure 7 Aircraft fuel flow vs. elapsed time for flight 1 (7BX engine).

The breakdowns of the HC, CO, and NO_x emissions for the AYT-SAW route are shown in Figure 9a. It is seen that certain emission types produced at certain flight phases can vary greatly from those for the other phases. On the other hand, the same flight phase can exhibit a relatively lower amount of some kinds of emission types and a relatively higher amount of other types of emissions. Here, it is noted that the engine power setting and phase duration significantly affect emission quantities.

Due to incomplete combustion, the emissions of HC and CO for lower engine power settings are observed to be much greater than those for higher power settings [2,12,22,23]. Therefore, such emissions obtained for flight phases such as taxi and descent are found to be higher than the phases such as takeoff and climb. This pattern is observed from the graphs in Figure 9. Accordingly, HC and CO emissions resulting from aircraft operation during taxi (P2 for taxi out and P14 for taxi in) and descent (P12), where the engine power setting is relatively low, are observed to be at higher levels. For instance, the HC emissions are calculated as 3.7 and 3.8 kg for the sum of P2 and P14, and P9, respectively. Since the phase duration can have a great effect on the emissions, the duration should be considered along with the amount of the emissions.

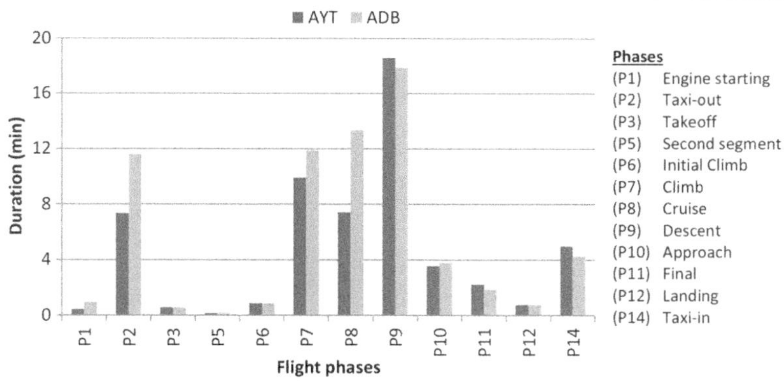

Figure 8 Durations of flight phases. The average total flight times are 67.9 and 56.8 min for AYT and ADB, respectively.

As can be seen from the graphs in Figure 9, for relatively lower power settings as for the idle and taxi (P2 and P14) phases, higher quantities of HC and CO emissions are observed, depending on the phase duration. For instance, the highest levels of HC emissions are observed for the phases of descent, taxi out, and taxi in, for which the corresponding durations are 18, 12, and 4 min. Considering the air quality in the vicinity of the airport (related to flight phases P1 to P3 and P12 to P14), it is found that 2.9 kg of HC is produced in AYT and 0.8 kg in SAW. Although the emission production mechanism during taxi in is not fundamentally different from that for taxi out, the difference in HC emissions results from the longer taxi duration at AYT. HC emissions are much lower for the other flight phases.

In the above discussions, the flight phases P3 (takeoff) and P12 (landing) are assumed to occur in the vicinity of the airport since the flight phases occur at a low height over the runway. For instance, the average heights for AYT-SAW routes are determined to be 19 m for P3 and 12 m for P12. As a result, the durations of the idle and taxi phases have significant effects on the quantities of HC emissions. For the above example, the longer ground operation duration (P1 to P3) at the departure airport AYT yields a higher HC emission (2.9 kg), while the shorter ground operation (P12 to P14) in the arrival airport SAW yields a lower HC emission (0.8 kg). Thus, for the high air traffic in busy airports, where aircraft often waits in taxi sequence or holding point for takeoff in long queues, or for airports with long taxi ways, there may be significant HC emissions in the vicinity of the airport.

The descent flight phase (P9) is the longest phase. In this phase, depending on the descent procedure, the power settings of the engine can be idle or at a value slightly higher than the idle. For instance, the N1 revolutions per minute (RPM) of the engine for the 9th flight is calculated at around 30% to 35% of the actual RPM at full speed, after 89% of the descent time has elapsed. During the same phase, the EGT values are concentrated at 426°C to 429°C. By comparison, the distributions of the N1 RPM and the EGT values for the taxi phase are concentrated at around 18% to 24% and 470°C to 530°C, respectively. Although the lower power settings may lead to higher HC emissions, the net benefit can be positive due to reduced fuel consumption and amounts of other emissions, such as NO_x and carbon dioxide (CO_2).

The emission breakdown of CO by flight phase is similar to those for HC emissions. However, the mass of CO emissions can be an order of magnitude higher than that of HC emissions. As seen in Figure 9a, the highest CO emissions are observed for the flight phases of descent (34.0 kg), taxi in (25.3 kg), and taxi out (7.4 kg), which are similar to the case for HC emissions. CO emissions are found to be quite low for the remaining flight phases, for which power settings are relatively high.

The total HC and CO emissions for the AYT-SAW route are calculated as 8 and 76 kg, respectively. According to [2], approximately 70% of CO and HC releases are emitted below 9-km altitude. In this study, this range is further refined. Dividing the entire flight into the two categories as air and ground, it can be observed that almost half of the total emissions of both types are produced during ground operations, with flight phases of takeoff (P3) and landing (P12) included in ground operations. Considering two categories, the percentage of the average flight phase durations and fuel consumptions of the ground category for these five flights are determined to be 27% and 11%, respectively.

The highest levels of NO_x emissions are observed during the climb, cruise, and descent phases. The average NO_x emissions for the five flights is 31.3 kg, of which 16.0% is produced in ground operations, 34.6% in climb, 20.6% in cruise, and 15.9% in descent. Taxi in exhibits higher NO_x emissions than taxi out due to the longer phase duration.

The emissions and the fuel consumptions for the second route (ADB-SAW) are shown in Figure 9b. Although the duration of the flight phases differs from flights of

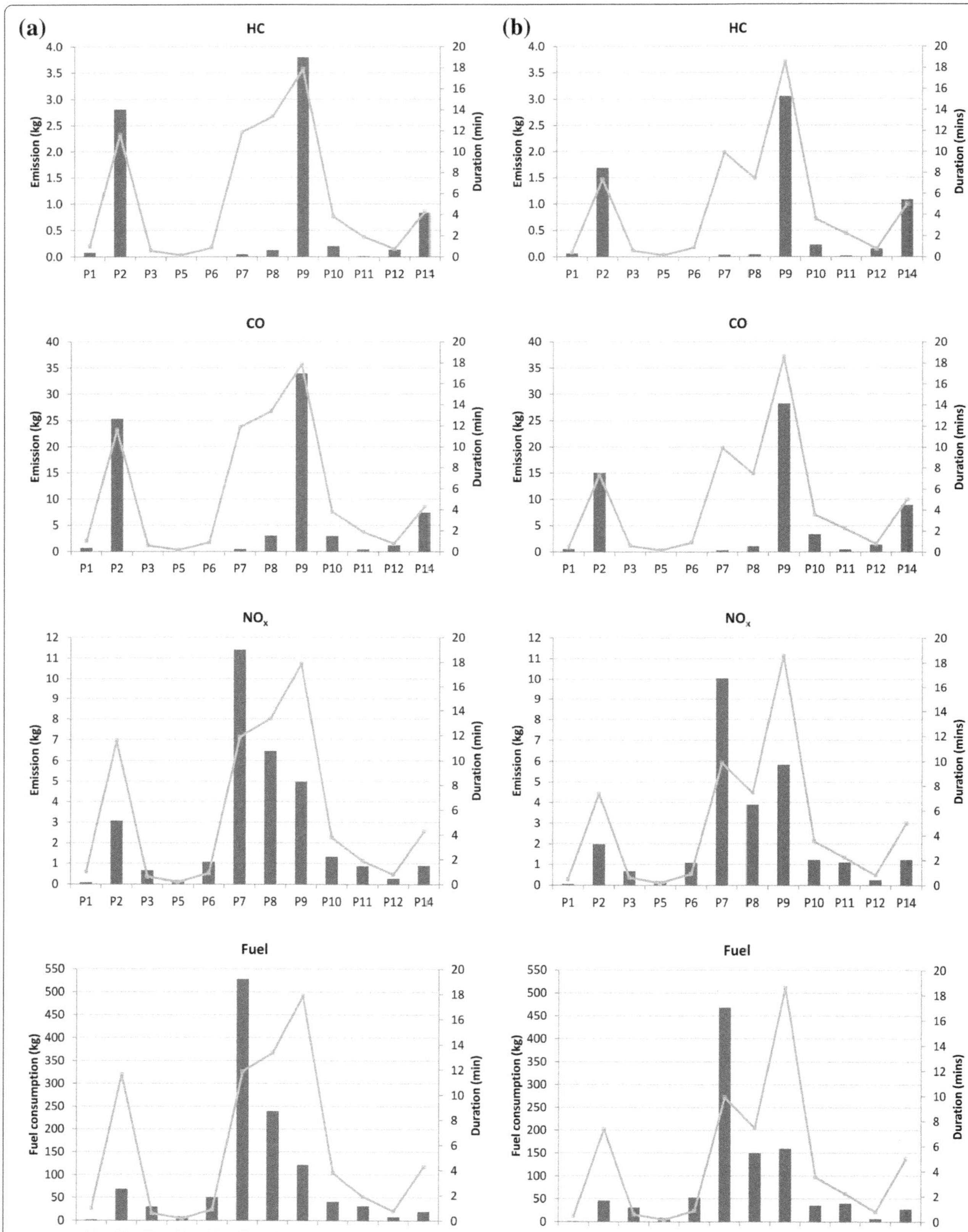

Figure 9 Emission breakdown by flight phase for (a) AYT and (b) ADB departures. The gray line in each graphic indicates the duration of the related phase.

the first route, the emission trends are similar to those for the first route, particularly for HC and CO. That is, the descent and the taxi phases exhibit the highest HC and CO emissions. Similarly, the highest HC and CO emissions are observed for the descent phase. Due to different route and fuel consumption values, the emissions are found to be lower than those for the previous route, as expected. However, the decrease in the emissions is not proportional to the decrease in the fuel consumptions. Accordingly, the average of the fuel consumption of ADB-SAW flights is 11% lower than the fuel consumption of AYT-SAW flights, while the HC and CO emissions are found to be 21% lower than those for the second route. However, the NO_x emissions are found to be 12% lower than the ADB-SAW flights. That is, there is a similar relationship between fuel flow and NO_x emissions, but the effect of fuel consumption on HC and CO emissions appears to be more than significant than those for the NO_x emissions.

Flight phase duration vs. amount of emissions

The impact is investigated for the flight phase duration on the large emissions seen in certain flight phases. For instance, the flight phase P9 exhibits the highest HC and CO emissions and the highest flight phase duration for the flights in both routes. However, such a relation is not observed between NO_x emissions and fuel consumption. Accordingly, close correlation coefficients for the relationship between the flight phase duration and HC and CO emissions are observed for flights on both routes (the correlation coefficients of HC and CO emissions, respectively, are found to be 0.709 and 0.731 for AYT-SAW and 0.803 and 0.816 for ADB-SAW). The correlation between NO_x emissions and flight phase duration is found to be 0.771 for AYT-SAW and 0.761 for ADB-SAW. As expected, there is no significant relationship between fuel consumption and flight phase.

No significant relationship is found between fuel consumption per flight phase and phase duration. This is an expected result since the time in mode of ground operation of aircraft is greatly affected by the traffic volume of the airport. For instance, the engine can run for a long time in the idle power setting, and due to the low fuel flow rate, the total fuel consumption for this phase would be much lower compared to those for other phases. On the other hand, although short in duration, the total fuel flow rates during takeoff could be 8 to 10 times higher than those for idle, leading to a higher total fuel consumption. The corresponding correlation coefficients are found to be low, at 0.607 and 0.586 for AYT and ADB, respectively.

It is noted that two common methodologies are used to estimate the fuel flow rate of commercial aircraft, namely, the Base of Aircraft Data (BADA) and Boeing Fuel Flow methods. In the BADA method, a total energy model is used for a wide range of aircraft types which involves all the forces acting on the aircraft and the variations of potential and kinetic energies. In the Boeing method, fuel flow data from the ICAO emission databank, which are calibrated to temperature and pressure ratios to obtain corrected fuel flows for the related cruise altitude and engine power, are used. However, since each method is based on fuel flow estimates considering relatively limited operating conditions, there exists the potential for high levels of error. This study, on the other hand, is believed to generate better estimates since it is based on actual flight data.

Conclusion

For the common emission species CO, HC, and NO_x, linear and nonlinear models are developed in this study for all flight phases, for two routes, and two engine models. The models are based on the ICAO emission measurements. Actual flight data are obtained from flight data records. The models permit evaluation of emissions for all of flight phases for various fuel flow rates.

The models can be used for other aircraft types using the same types of the engines. That is, for the same engine type and known fuel flow rates, emission quantities can be calculated using the developed models. A breakdown of emissions by flight phase is obtained, and the findings agree well for flights on both routes. Accordingly, the highest CO and HC emissions are found in the descent phase followed by the taxi phases (in and out). The highest NO_x emissions are found in the climb phase followed by the cruise and descent phases. There are less but not negligible NO_x emissions in the taxi phase due to the high taxi duration. The mean total flight emissions are calculated as 8 kg of HC, 75 kg of HC, and 31 kg of NO_x for the AYT route and 6 kg of CO, 60 kg of HC, and 28 kg of NO_x for the ADB route. Lastly, the average ground time of the flights are calculated as 22% to 25% of the total flight time for both routes. Since the HC and CO emissions are mostly produced at the lower power settings of the engine, decreasing the taxiing time provides significant abatement of those two emissions.

The authors believe that in order to characterize emissions in a broader sense, future research is merited to perform emission tests in a widespread manner, while the ICAO time in mode assumptions could be updated and better estimations introduced. In the same manner, the ICAO fuel flow assumptions could be revised, and instead of absolute values, certain intervals could be used.

Abbreviations

ADB: Izmir Adnan Menderes International; AYT: Antalya International; CO: Carbon monoxide; EGT: Exhaust gas temperature; EI: Emission index; EPA: Environmental Protection Agency; ff: Fuel flow rate (kg/h, kg/s); HC: Hydrocarbon; ICAO: International Civil Aviation Organization; LTO: Landing and takeoff; NO_x: Nitrogen oxide; RPM: Revolutions per minute; R^2: Coefficient of determination; SAW: Istanbul Sabiha Gokcen International..

Competing interests

The authors declare that they have no competing interests.

Authors' contributions

ETT and OU carried out the analysis and drafted the manuscript. MAR checked the analysis and reviewed the revised manuscript. All authors read and approved the final manuscript.

Authors' information

ETT is a doctor who is an assistant professor in the Faculty of Aeronautics and Space Sciences at Anadolu University. He received his BSc and MSc in the School of Civil Aviation and Graduate School of Sciences at Anadolu University in 1999 and 2003, respectively. He received his PhD in 2007. OU received her BSc degree in Electrical and Electronics Engineering from Anadolu University in 1990 and her MSc degree in Exploitation Aeronautique from Ecole Nationale de l'Aviation Civile. She received her PhD degree from the Graduate School of Science, Department of Civil Aviation at Anadolu University in 1999 and has been working at the Department of Air Traffic Control, Faculty of Aeronautics and Space Sciences, Anadolu University as an assistant professor. MAR is a professor in the Faculty of Engineering and Applied Science at the University of Ontario Institute of Technology, where he was a founding dean from 2002 to 2008. He was the former president of the Engineering Institute of Canada. He served as president of the Canadian Society for Mechanical Engineering and is a fellow of that society as well as the Engineering Institute of Canada, the Canadian Academy of Engineering, the American Society of Mechanical Engineers, and the International Energy Foundation. Previously, he was the chair of the Department of Mechanical Engineering at Ryerson University.

Acknowledgments

The authors thank Captain Cahit Tasbas, Captain Mustafa Kemal Helvacioglu, and FDM Specialist Ms. Sercin Ozen from Pegasus Airlines for their co-operation, and the Sabiha Gokcen Airport Authority for its kind assistance in data acquisition. The authors also thank Anadolu University for the financial support.

Author details

[1]Department of Airframe and Powerplant Maintenance, Faculty of Aeronautics and Space Sciences, Anadolu University, Eskisehir 26470, Turkey. [2]Department of Air Traffic Control, Faculty of Aeronautics and Space Sciences, Anadolu University, Eskisehir 26470, Turkey. [3]Faculty of Engineering and Applied Science, University of Ontario Institute of Technology, 2000 Simcoe Street North, Oshawa, Ontario L1H 7K4, Canada.

References

1. Baughcum, SL, Tritz, TG, Henderson, SC, Pickett, DC: Scheduled Aircraft Emission Inventories for 1992: Database Development and Analysis. NASA contract report no. 4700. NASA Langley Research Centre, Hampton, VA (1996)
2. Sutkus, DJ, Baughcum, SL, DuBois, DP: Scheduled Civil Aircraft Emission Inventories for 1999: Database Development and Analysis. Report NASA/CR-2001-211216. NASA Center for Aerospace Information, Hanover, MD (2001)
3. Gardner, RM, Adams, K, Cook, T, Deidewig, F, Ernedal, S, Falk, R, Fleuti, E, Herms, E, Johnson, CE, Lecht, M, Lee, DS, Leech, M, Lister, D, Masse, B, Metcalfe, M, Newton, P, Schmitt, A, Vandenbergh, C, Drimmelen, R: The ANCAT/EC global inventory of NO_x emissions from aircraft. Atmos. Environ. **31**(12), 1751–1766 (1997)
4. Eyers, CJ, Addleton, D, Atkinson, K, Broomhead, MJ, Christou, R, Elliff, T, Falk, R, Gee, I, Lee, DS, Marizy, C, Michot, S, Middel, J, Newton, P, Norman, P, Plohr, M, Raper, D, Stanciou, N: AERO2k Global Aviation Emissions Inventories for 2002 and 2025. European Commission, Contract No. G4RD-CT-2000-00382, QinetiQ Ltd, Farnborough, Hampshire (2004)
5. Kim, BY, Fleming, GG, Balasubramanian, S, Malwitz, A, Lee, J, Ruggiero, J, Klima, K, Lock, M, Holsclaw, CA, Morales, A, McQueen, E, Gillette, W: System for assessing aviation's global emissions (SAGE), version 1.5, global aviation emissions inventories for 2000 through 2004. Report FAA-EE-2005-02, Federal Aviation Administration, Office of Environment and Energy, Washington, DC (2005)
6. U.S. Environmental Protection Agency: Environmental Fact Sheet, Adopted Aircraft Engine Emission Standards, EPA 420-F-97-010. http://www.epa.gov/oms/regs/nonroad/aviation/aircr-fr.pdf (1997). Accessed 28 April 2011
7. Schafer, K, Jahn, C, Sturm, P, Lechner, B, Bacher, M: Aircraft emission measurements by remote sensing methodologies at airports. Atmos. Environ. **37**(37), 5261–5271 (2003)
8. Herndon, SC, Shorter, JH, Zahniser, MS, Nelson, DD, Jayne, J, Brown, RC, Miake-Lye, RC, Waitz, I, Silva, P, Lanni, T, Demerjian, K, Kolb, CE: NO and NO_2 emission ratios measured from in-use commercial aircraft during taxi and takeoff. Environ. Sci. Technol. **38**(22), 6078–6084 (2004)
9. Unal, A, Hu, Y, Chang, ME, Odman, MT, Russell, AG: Airport related emissions and impacts on air quality: application to the Atlanta International Airport. Atmos. Environ. **39**(32), 5787–5798 (2005)
10. Herndon, SC, Rogers, T, Dunlea, EJ, Jayne, JT, Miake-Lye, R, Knighton, B: Hydrocarbon emissions from in-use commercial aircraft during airport operations. Environ. Sci. Technol. **40**(14), 4406–4413 (2006)
11. Winther, M, Kousgaard, U, Oxbel, A: Calculation of odour emissions from aircraft engines at Copenhagen Airport. Sci. Total Environ. **366**(1), 218–232 (2006)
12. Schurmann, G, Schafer, K, Jahn, C, Hoffmann, H, Bauerfeind, M, Fleuti, E, Rappenglück, B: The impact of NO_x, CO and VOC emissions on the air quality of Zurich airport. Atmos. Environ. **41**, 103–118 (2007)
13. Westerdahl, D, Fruin, SA, Fine, PL, Sioutas, C: The Los Angeles International Airport as a source of ultrafine particles and other pollutants to nearby communities. Atmos. Environ. **42**(13), 3143–3155 (2008)
14. Civil Aviation Authority (CAA): ICAO Emission Databank. http://easa.europa.eu/environment/edb/docs/edb-emissions-databank.xls (2013). Accessed 10 April 2013
15. U.S. Department of Transportation, Federal Aviation Administration: CFM56-7B type certification, Type Certificate Data Sheet E00055EN. http://rgl.faa.gov/Regulatory_and_Guidance_library/rgMakeModel.nsf/0/a25bbd09030a2d0d862574fa00744f01/$FILE/E00056EN.pdf (2008). Accessed 28 April 2011
16. Herndon, S: Measurement of Gaseous HAP Emissions from Idling Aircraft as a Function of Engine and Ambient Conditions. Transportation Research Board, Report 63, Washington, USA (2012)
17. Timko, MT, Herndon, SC, Wood, EC, Onasch, TB, Northway, MJ, Jayne, JT, Canagaratna, MR, Miake-Lye, RC, Knighton, WB: Gas turbine engine emissions—part I: volatile organic compounds and nitrogen oxides. Journal of Engineering for Gas Turbines and Power **132**, 061504 (2010)
18. Yelvington, PE, Herndon, SC, Wormhoudt, JC, Jayne, JT, Miake-Lye, RC, Knighton, WB, Wey, C: Chemical speciation of hydrocarbon emissions from a commercial aircraft engine. Journal of Propulsion and Power **23**(5), 912–918 (2007)
19. Brasseur, GP, Cox, RA, Hauglustaine, D, Isaksen, I, Lelieveld, J, Lister, DH, Sausen, R, Schumann, U, Wahner, A, Wiesen, P: European scientific assessment of the atmospheric effects of aircraft emissions. Atmos. Environ. **32**(13), 2329–2418 (1998)
20. Giampaolo, T: The Gas Turbine Handbook: Principles and Practices, 2nd edn. Fairmont Press, Lilburn (2003)
21. Hunecke, K: Jet Engines: Fundamentals of Theory, Design and Operation, 6th edn. Motorbooks International Publishers, Osceola (2003)
22. Mazaheri, M, Johnson, GR, Morawska, L: Particle and gaseous emissions from commercial aircraft at each stage of the landing and takeoff cycle. Environ. Sci. Technol. **43**, 441–446 (2009)
23. Anderson, BE, Chen, G, Blake, DR: Hydrocarbon emissions from a modern commercial airliner. Atmos. Environ. **40**, 3601–3612 (2006)

Investigation of the effect of organic working fluids on thermodynamic performance of combined cycle Stirling-ORC

Mohammad Bahrami[1*], Ali A Hamidi[2] and Soheil Porkhial[3]

Abstract

This paper presents thermodynamic investigation and environmental consideration of combined Stirling-organic Rankine cycle (ORC) power cycle. Combined cycle can be assisted by solar energy and an ORC used as an annular cold-side heat rejector for a free piston Stirling cycle. ORC can increase the power output efficiency by 4% to 8% compared to that of a Stirling standard cycle. Operating temperatures of ORC are between 80°C and 140°C. The main objective of this work is to model the combined cycle for performance optimization in respect to the use of several different working fluids with relevant temperature ranges. Total power efficiency in the range of 34% to 42% was observed for different cases. Several working fluids in the ORC were investigated from a thermal, operational, and environmental point of view. Working fluids considered were FC72, FC87, HFE7100, HFE7000, Novec649, *n*-pentane, *n*-decane, R245fa, and toluene. Practical issues like thermodynamic cycle efficiency, latent heat, density, toxicity, flammability, ozone depletion potential, global warming potential, and atmospheric lifetime are considered. Considering the cycle efficiency, *n*-decane shows the best performance at both levels of temperature supposed. However, this fluid has the highest saturated vapor specific volume (resulting in a larger condenser) and the lowest condenser saturation pressure (higher infiltration of non-condensable gases). The best candidates for the cycle regarding all the considered aspects were found to be toluene, HFE7100, and *n*-pentane. Comparing these three fluids, toluene presents the highest efficiency, the highest impact on the environment, the biggest vapor specific volume, and the minimum mass flow rate in Rankine cycle, therefore decreasing the pump power consumption. *N*-pentane exhibits the lowest cycle efficiency and vapor specific volume, but this fluid has super-atmospheric saturation pressure advantage. HFE7100 is a good working fluid from environmental and safety point of view.

Keywords: Thermodynamic investigation, Organic working fluid, Combined cycle Stirling-ORC, Environmental consideration

Background

The quest to reduce environmental impacts of conventional energy resources and, more importantly, to meet the growing energy demand of the global population had motivated considerable research attention in a wide range of environmental and engineering application of renewable form of energy 1]. The need to preserve fossil fuels and the use of renewable energies have led to the use of Stirling engines which have excellent theoretical efficiency, equivalent to the related Carnot cycle. They can consume any source of thermal energy (combustion energy, solar energy, etc.) and result in less pollution than the traditional engines [2]. High heat conversion efficiency, reliability, low noise operation, and ability of Stirling engines to use many fuels meet the demand of the effective use of energy and environmental preservation [3,4]. In most instances, the engines operate with hotter and cooler temperatures of 923 and 338 K, respectively. Engine efficiency ranges from about 30% to 40%, resulting from a typical hotter temperature range of 923 to 1073 K and normal operating speed range of 2,000 to 4,000 rpm [5].

* Correspondence: mohamad.bahramee@gmail.com
[1]Department of Energy Engineering, Graduate School of the Environment and Energy, Science and Research Branch, Islamic Azad University, Tehran 1341766394, Iran
Full list of author information is available at the end of the article

Generally, Stirling cycle development has focused on two areas of improvement: reducing losses and increasing power output. The proposed method of Ingram and Peterson [6] had an improvement on Stirling system performance. The total power output may be improved with the proper implementation of a combined power cycle with a bottoming organic Rankine cycle (ORC).

There is a sufficient amount of heat and availability to run a bottoming cycle from the rejected heat of a Stirling engine. Since the Stirling cycle under consideration in this study already makes use of a cooling loop as shown in Figure 1, a Rankine cycle may be readily integrated into the design in place of the cooling loop as shown in Figure 2. The proposed system [6] investigates the theoretical conditions that create maximum power output from a combined Stirling-Rankine cycle by increasing the Stirling heat rejection temperature. This reduces the power produced from the Stirling cycle itself but allows for more total power with an efficient low-temperature Rankine cycle. Rankine cycle expanders have traditionally had to be large for any attractive efficiency level. The measured isentropic efficiency of the scroll expander reaches 84% when the pressure ratio is close to the scroll intrinsic expansion ratio [7]. However, in experimental testing, the generator and scroll expanders produced 2.07 and 2.96 kW

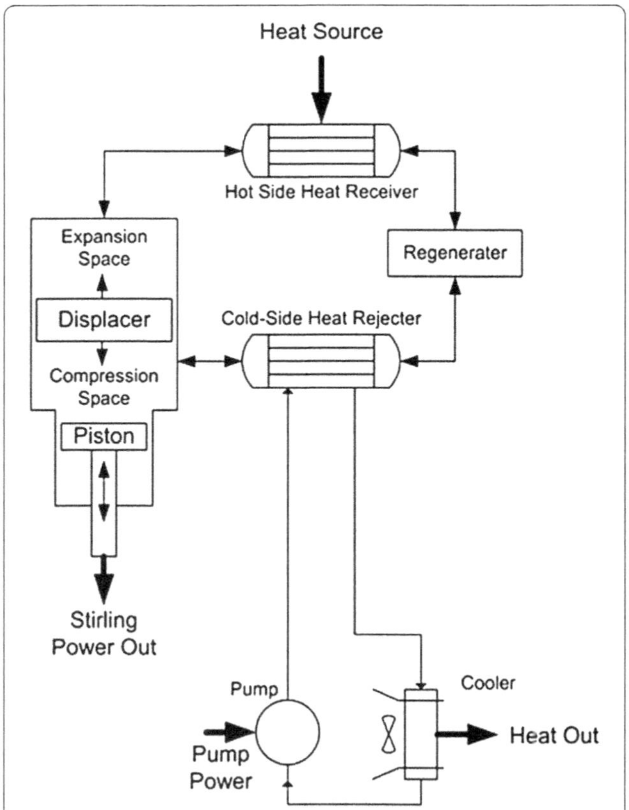

Figure 1 Schematic of a free piston Stirling engine and loop cooling exhausting wasted heat [6].

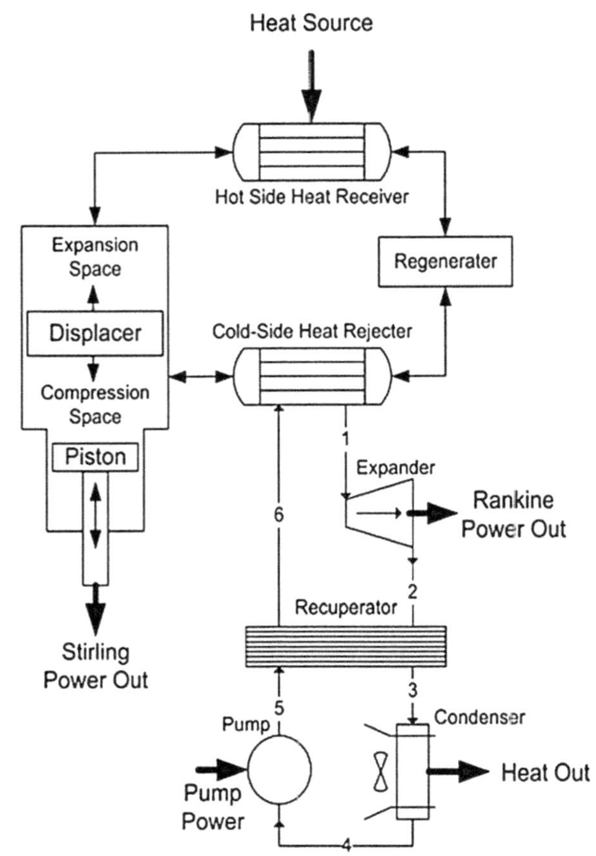

Figure 2 Schematic of a free piston Stirling engine with a Rankine bottoming cycle [6].

and had isentropic efficiencies of 0.85 and 0.83, respectively [8]. Turbine expanders sized for 1.5 kW operated with 85% isentropic efficiency [9]. The Rankine recuperator recovers heat that would otherwise be rejected from the cycle in the condenser. It is possible to use a microchannel recuperator with actual heat transfer effectiveness significantly higher than 0.85 [10]. For the sake of simplicity, the pressure drop of the recuperator is ignored.

ORC generates power from low-grade heat sources by replacing water with organic working fluids such as refrigerants or hydrocarbons [11]. To investigate any improvement when using these fluids, the effects of the properties of nine different working fluids on the overall combined cycle efficiency have been considered. In this respect, the economic, environmental, and operating performances of an ORC depend on the properties of the working fluids as well as the design and operating characteristics of the cycle. A suitable fluid for an ORC must exhibit favorable physical, chemical, environmental, safety, and economic properties such as low specific volume, liquid specific heat, viscosity, toxicity, flammability, ozone depletion potential (ODP), global warming potential (GWP), and cost, as well as favorable process characteristics such as high efficiency, latent heat, density, molecular weight, Mollier diagram close to isentropic, suitable thermal stability limits, compatible with turbine materials and lubricating oil, non-corrosive, non-inflammable, and moderate pressure in the heat exchangers [11-13]. The saturated vapor specific volume gives an indication of condenser size, which is related to the system's initial cost. A super-atmospheric (>100 kPa) saturation pressure eliminates infiltration of gases, which is important for operational reasons, since gas infiltration reduces system efficiency. The working fluid having high latent heat and high density is preferred. To achieve the maximum output power of the combined cycle, Rankine cycle efficiency must be increased. For this reason, the effects of the properties of several working fluids on the Rankine cycle efficiency have been studied.

Among all possible alternative energy options, for example, wave energy, geothermal energy, solar energy, wind energy, and hydro energy, solar energy is becoming more popular in the world. This is mainly due to the availability of plenty of sunlight in many countries [14]. Therefore, in this study, the combined cycle is assisted by solar energy, and nine working fluids have been chosen as candidates for the bottoming ORC. These are FC72, FC87, HFE7100, HFE7000, Novec649, *n*-pentane, *n*-decane, R245fa, and toluene. Two levels of operating temperatures of ORC were considered, namely 80°C and 140°C. The solar dish used with 700°C receiver temperature was considered as the heat source. Results are presented per kilowatt of heat input to the system. Therefore, a 40% total system efficiency means 0.4-kW output power.

Methods

Working fluid classification

A working fluid can be classified as dry, isentropic, or wet fluid depending on the slope of the saturation vapor curve on a T-s diagram. Vapor saturation curve of wet fluid (e.g., water) has a negative slope, resulting in a two-phase mixture upon isentropic expansion. A dry fluid (e.g., *n*-pentane) has a positive slope, while an isentropic (e.g., R245fa) has an infinitely large slope, the fact that the vapor saturation line on a T-s diagram is vertical for these fluids [12].

Water is a suitable working fluid for the larger scale fossil fuel-fired Rankine cycle plants. Water is appropriate for those high temperature applications, but it has its limitations that become more significant during lower temperature operations. Organic fluids are generally extracted from petroleum. The main difference between organic fluids and water is the lower evaporation energy of the former, and therefore less heat is needed to evaporate the organic fluid. The thermodynamic and chemical characteristics of these fluids no longer require superheating. For most of organic fluids, an isentropic expansion from saturated vapor results in superheated vapor through a turbine at low and moderate temperatures, rather than a two-phase mixture as with water, hence avoiding complications to the turbine and cycle design [13,15]. These characteristic curves are shown in Figure 3 for the three types of fluids. The degree to which organic fluids are dry is generally related to their molecular weight or molecular complexity. For organic fluids, the larger the molecular weight, the greater is the slope of the T-s curve.

Practical issues relating to the operation and maintenance of the cycle should also be taken into account. Maizza and Maizza [16] suggested that high latent heat, high density, and low liquid specific heat are preferable. A fluid with a high latent heat and density absorbs more energy from the source in the evaporator, thus reduces the required fluid flow rate, resulting in reduced size of the facility and the pump power consumption [12]. The saturated vapor specific volume gives an indication of condenser size, which is related to system initial cost. Figures 4 and 5 respectively present the saturation pressure and vapor specific volume as a function of saturation temperature of the organic fluids considered in the analysis. It should be noted that organic fluid vapor volume varies four orders of magnitude between FC87 and *n*-decane. Organic fluids with low saturation volumes require smaller condensing equipment. Therefore, fluids with high saturation volumes are not suitable for application due to the large size of the required condensing equipment [13].

Regarding saturation pressures, it should be noted that R245fa, FC87, HFE7000, and *n*-pentane operate at super-atmospheric pressure at 40°C (condenser temperature). Super-atmospheric operation is profitable for small-scale applications by eliminating infiltration of non-condensable gases. The two main devices that are used to eliminate the non-condensable gases with a cycle operating with

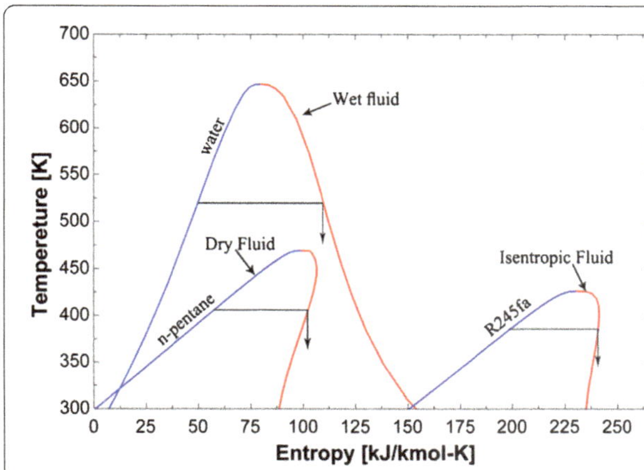

Figure 3 Three types of working fluids: dry, isentropic, and wet.

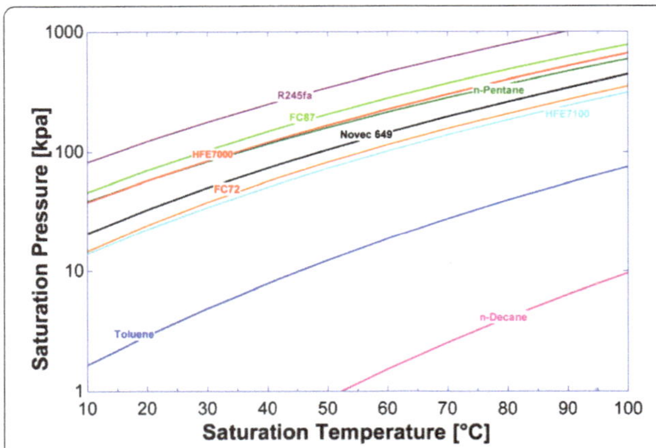

Figure 4 Saturation pressure as a function of saturation temperature.

sub-atmospheric fluids are steam jet air ejectors (SJAE) and liquid ring vacuum pumps. SJAE use high pressure motive steam to evacuate the non-condensables from the condenser (jet pump). Liquid ring vacuum pumps use a liquid compressant to compress the evacuated non-condensables and then discharge them to the atmosphere [17].

The substances that can serve as working fluids in an ORC are shown in Table 1. The table presents a classification of working fluids according to the slope of the saturation vapor line. The critical point suggests the possible operating temperature and pressure range. The molecular weight and latent heat have influence on the required mass flow rate.

Working fluid environmental impact and risk statement

The three environmental factors applicable for fluids comparison are ODP, GWP, and atmospheric lifetime (ALT). The ODP is the ratio of the impact on ozone of a chemical compared to the impact of a similar mass of

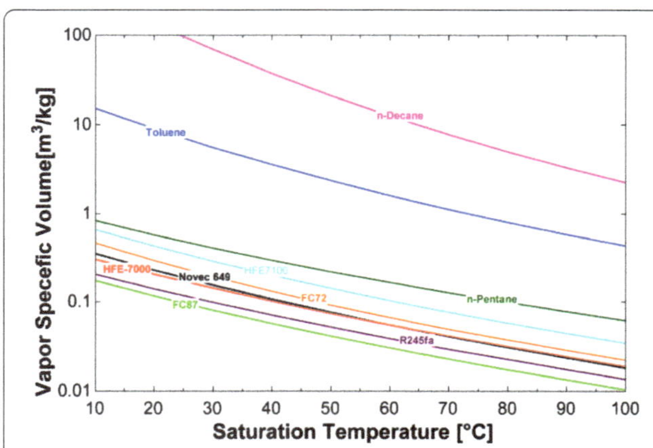

Figure 5 Saturated vapor specific volume as a function of saturation temperature.

R-11 or CFC-11. Thus, the ODP of R-11 is defined to be 1.0. Normally, fluids have ODPs between 0.1 and 1 [18]. GWP represents how much a given mass of a chemical contributes to global warming over a given time period as compared to the same mass of carbon dioxide. Carbon dioxide's GWP is defined as 1.0. A GWP is calculated over a specific time interval, commonly 100 years [19]. ALT is the length of time that a gas will remain in the atmosphere based on its decay rate and its tendency and likeliness to react with other gases. To complement these indices, the maximum time-average concentration for a normal 8-h work day (TOX) has also been presented as toxicological index [13,20,21].

Table 2 illustrates the environmental and safety characteristics of the working fluids under study. HFE7100, HFE7000, and Novec649 are good fluids from environmental and safety point of view. However, FC72 and FC87 perfluoro compounds (PFCs) are photo-chemically stable and are expected to persist in the atmosphere for more than 1,000 years. PFCs have high GWP, exceeding 5,000 100-year ITH. ODP for these compounds is zero [21].

Stirling-ORC combined cycle thermodynamic model description

In order to evaluate the performance of a combined cycle Stirling-ORC system, a thermodynamic model has been created. The model has been used to determine the state conditions of the working fluid in the Rankine cycle, from which the outputs from the system were analyzed.

The combined system was modeled in an idealized condition. Isentropic efficiencies were used for all components, and ideal heat exchanger effectiveness was assumed. For simplicity, the power of the fan used in the air-cooled condenser was not considered. The calculations were done at two distinct temperatures of 80°C and 140°C. The Stirling cycle was modeled as a fraction of Carnot (f_{Carnot}) thermal efficiency through Equation 1; f_{Carnot} was considered to be 0.5.

$$\eta_{Stirling} = f_{Carnot}\eta_{Carnot} \tag{1}$$

$$\eta_{Carnot,St} = 1 - \frac{T_C}{T_H}, \tag{2}$$

where, η means the efficiency, T_C is the colder temperature, and T_H is the hotter temperature. The Stirling cycle power output, W_{St}, is simply the product of the Stirling efficiency and the power into the Stirling cycle, $Q_{in,St}$. The remaining heat, $Q_{out,St}$, is rejected from the Stirling cycle and is available for the Rankine cycle.

$$W_{St} = Q_{in,St}\,\eta_{St} \tag{3}$$

$$Q_{out,St} = Q_{in,St} - W_{St}. \tag{4}$$

Table 1 Classification of working fluids according to the slope of saturation vapor line

Fluid	P_crit (kPa)	T_crit (K)	Molecular weight (g/mol)	Specific heat (J/kgK)	Latent heat (kJ/kg)	Type
FC72	1,867	448.8	338	1.100	88	Dry
FC87	2,039	421.1	228	1.100	103	Dry
HFE7000	2,478	437.7	200	1.300	142	Dry
HFE7100	2,229	468.5	250	1.183	111.6	Dry
n-decane	2,103	617.7	142.29	2.210	251.1	Dry
Novec649	1,864	441.8	316	1.103	88	Dry
n-pentane	3,364	469.7	72.15	2.363	349.00	Dry
R245fa	3,651	427.2	134.05	980.90	177.08	Isentropic
Toluene	318.6	4126	92.14	1720	399.52	Isentropic

*P*_crit, critical pressure; *T*_crit, critical temperature.

The Rankine cycle is modeled by evaluating fluid properties at state points between components. These states correspond to states 1 to 6 in Figure 2. The two cycles are connected in the model with the Stirling cycle, cold-side heat rejection being the source of heat input to the boiler of the Rankine cycle. As the ideal model is used in the study, it is assumed that all of the heat rejected from the Stirling cycle goes into the Rankine cycle. The highest Rankine cycle temperature, T_1, is 5°C below the lowest Stirling cycle temperature, and isobaric conditions are used.

$$\dot{m}_{\mathrm{ORC}}(h_1 - h_6) = Q_{\mathrm{out,St}}, \tag{5}$$

where $\dot{m}$ is the mass flow rate (g/s), and h is the specific enthalpy (J/kg). For the Rankine expander, an isentropic efficiency ($\eta_{s,\mathrm{exp}}$) of 0.9 was considered.

$$h_2 = h_1 - \eta_{s,\mathrm{exp}}\left(h_1 - h_{2,s}\right). \tag{6}$$

The recuperator is considered if the pump outlet temperature (point 5) is less than or equal to the outlet turbine temperature (point 2). For the present study, recuperator effectiveness (ε_{rec}) was assumed to be 0.9.

$$\varepsilon_{rec} = \frac{h_2 - h3}{h_2 - h(P2, T5)}, \tag{7}$$

where P is means pressure, and T is temperature. The pump is modeled with a constant isentropic efficiency of 50%, which accounts for motor efficiency and the actual isentropic efficiency of the pump [22].

$$h_5 = h_4 + \frac{1}{\eta_{s,\mathrm{pump}}}\left(h_{5,s} - h_4\right). \tag{8}$$

The model parameters represent reasonable values for physical systems and are listed with their values in Table 3.

Analysis

To evaluate the performance of the combined cycle, a numerical model was developed using thermodynamics analysis, and parametric studies were performed by

Table 2 Safety and environmental data of the working fluids

Fluid	ALT (yr)	ODP	GWP	TOX (ppm)	Risk statements
FC72	n.a.	0	n.a.	n.a.	NR
FC87	1,000	0	5,000	n.a	NR
HFE7000	4.9	0	370	75	NR
HFE7100	4.1	0	320	750	NR
n-decane	n.a.	n.a.	n.a.	1,369	HF and T
Novec649	0.014	0	1	150	NR
n-pentane	n.a.	0	20	700	EF and T
R245fa	n.a.	0	950	300	NR
Toluene	n.a.	n.a.	n.a.	100	HF and T

EF extremely flammable; *HF* highly flammable; *T* toxic; *NR* no risk statements; *n.a.* not applicable.

Table 3 Parameters for thermodynamic systems model combined Stirling-ORC power cycle

Parameter	Parameter description	Value	Units
T_H	Hot side temperature of Stirling cycle	700	°C
Tc	Cold side temperature of Stirling cycle	85 to 145	°C
To	Ambient temperature	25	°C
f_{Carnot}	Fraction of Carnot	0.50	
$Q_{in,St}$	Heat into Stirling cycle	1	kW
$\eta_{s,exp}$	Expander isentropic efficiency	90%	
ε_{rec}	Recuperator effectiveness	0.90	
$\eta_{s,pump}$	Pump isentropic efficiency	50%	
T_{cond}	Condenser temperature	40	°C
$\Delta T_c - T_1$	Boiler temperature difference	5	°C

varying different parameters in the system. The net power output from the combined cycle is considered to be the Stirling cycle power output, W_{St}, plus the Rankine cycle expander power output, W_{ORC}, less the Rankine cycle pumping power, W_{pump} (Equation 9). The total efficiency and the second law efficiency of combined cycle could be calculated through Equations 10 and 11. The Carnot efficiency is the maximum limit combined cycle efficiency and is a function of high temperature source and low temperature sink, Equation 12. The Rankine cycle and Stirling cycle efficiency could be calculated through Equations 13 and 14. The maximum power output is derived by specifying and varying particular parameters and allowing the system model to calculate all the rest. The irreversibility rate for a cycle in steady state, steady flow condition can be defined by Equation 17, where I, T_0, S_{gen}, EX_{in}, and EX_{out} are the irreversibility, ambient temperature, the entropy generation rate in Equation 16, and the incoming and outgoing exergy flows, respectively. q_j is the heat transfer per unit mass, and Tj is the temperature of the j_{th} component of the cycle. The exergy efficiency of the combined cycle is defined by Equation 15 [23]. In fact, second law and exergy efficiency are often considered the same; in other words, the distinction between exergy and second law efficiencies is not clearly recognized by many other studies [24]. Exergy based on both useful work and useful heat are the outcomes of available energy of a thermodynamic system with respect to a reservoir. Furthermore, some studies propose a definition of exergy based on the Carnot cycle [25].

$$W_{net} = W_{St} + W_{ORC} - W_{pump} \tag{9}$$

$$\eta_{total} = \frac{W_{net}}{1kW} \tag{10}$$

$$\eta_{2law} = \frac{\eta_{total}}{\eta_{Carnot,total}} \tag{11}$$

$$\eta_{carnot,total} = 1 - \frac{T_{cond}}{T_H} \tag{12}$$

$$\eta_{ORC} = \frac{W_{ORC} - W_{pump}}{Q_{out,St}} \tag{13}$$

$$\eta_{St} = \frac{W_{St}}{Q_{in,St}} \tag{14}$$

$$\eta_{ex} = \frac{\text{Exergy recovered}}{\text{Exergy supplied}} = \frac{W_{net}}{W_{rev}} = \frac{W_{net}}{W_{net} + T_0\, S_{gen}} \tag{15}$$

$$S_{gen} = \frac{Q_{cond}}{T_{cond}} - \frac{Q_{in,St}}{T_H} \tag{16}$$

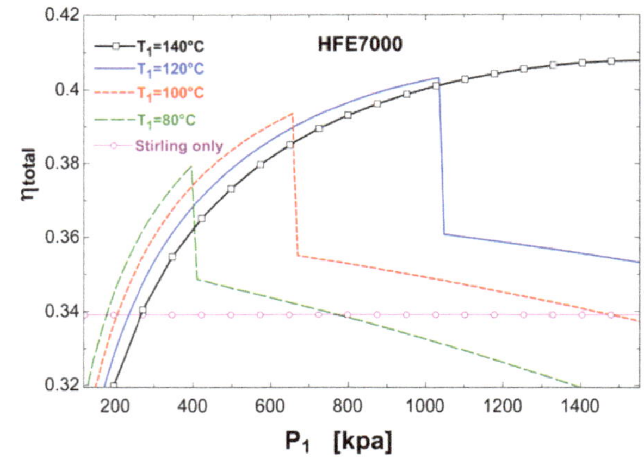

Figure 6 **The total efficiency line is shown where the optimal pressure is found at each Rankine cycle operating temperature.**

$$I = T_0\, S_{gen} = T_0 \sum_j \left(\frac{q_j}{T_j}\right) = EX_{in} - EX_{out} \tag{17}$$

Results and discussion

T_c is the temperature at which the Stirling cycle is rejecting heat from the cold-side heat exchanger. As this temperature rises, the maximum possible efficiency, the Carnot efficiency, from the Stirling cycle decreases. If the fraction of Carnot efficiency remains constant for the Stirling cycle, then the Stirling efficiency drops. In Rankine cycle, the turbine inlet pressure was optimized to maximize the thermal efficiency. This optimization makes it possible to decide whether it is better to have saturated vapor or superheated vapor at turbine inlet. The optimization was done with EES software [26]. The selection of the fluid for the cycle was complicated. For this cycle, where the operating temperatures are 80°C to 140°C, it is necessary to consider that all the fluids have

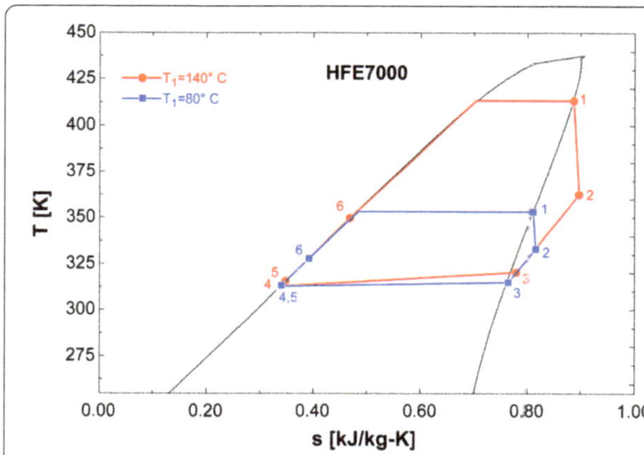

Figure 7 T-s process diagram for a Rankine cycle. Operated at maximum power conditions for the combined cycle operating temperatures of 80°C and 140°C.

Table 4 Thermal performance of several fluids in combined cycle for the ORC operating at 80°C

Fluid	P_1 (kpa)	P_2 (kpa)	V_3 (m^3/kg)	SHΔT	$Q_{out,St}$ (W)	Q_{cond} (W)	η_{ST} (%)	η_{ORC} (%)	η_{total} (%)	η_{2law} (%)	η_{ex} (%)	W_{pump} (W)	$\dot{m}$ (g/s)
FC72	208.4	56.36	0.13	0.06	684	621.3	31.6	9.16	37.87	55.83	57.04	1.42	6.78
FC87	484.8	147.7	0.06	0.07	684	623.1	31.6	8.9	37.69	55.57	56.78	3.46	6.73
HFE7000	400	119.7	0.10	0.03	684	620.3	31.6	9.31	37.97	56	57.19	2.18	4.66
HFE7100	184.6	50.48	0.20	0.05	684	619.7	31.6	9.40	38.03	56.08	57.28	1.09	5.16
n-decane	4.097	0.49	37.7	0.01	684	617.9	31.6	9.67	38.21	56.35	57.55	0.02	1.73
Novec649	261.6	72.96	0.11	0.06	684	620.9	31.6	9.22	37.91	55.89	57.1	1.87	6.72
n-pentane	366.6	116.3	0.29	0.06	684	620.3	31.6	9.3	37.97	55.98	57.18	1.43	1.73
R245fa	789.1	249.6	0.07	0.08	684	622.3	31.6	9.02	37.77	55.7	56.9	2.84	3.40
Toluene	38.41	7.907	3.55	0.53	684	618.5	31.6	9.57	38.15	56.25	57.45	0.11	1.53

P_1, inlet turbine pressure; P_2, outlet turbine pressures; V_3, specific volume at condenser inlet; $\dot{m}$, mass flow rate.

a critical temperature above 140°C. In Figure 6, for example, the total efficiency of combined cycle for HFE7000 is plotted as a function of pressure. The T_c is held constant, while pressure is varied; the system total efficiency levels are compared. Each constant T_1 line in the Figure 6 exhibits the same behavior. In other words, total efficiency increases with rising pressure until a maximum is reached, then total efficiency drops off sharply, and continues to decline as the pressure increases. The total efficiency of combined cycle eventually drops below the horizontal line representing the efficiency from the original standard Stirling system. The maximum possible total efficiency always increases with increasing pressure for the full range shown. The peaks occur when saturation conditions are reached at the Rankine boiler. The lines then drop down in power level immediately after saturation because the pressure has risen until the limited temperature is no longer sufficient to boil the fluid. Up to the point of reaching the saturation conditions, there is a decrease in the amount of superheat going into the expander with increasing pressure.

As shown in Figure 6, the combined cycle total efficiency lines do not drop below the original standard Stirling cycle efficiency until long after the saturation pressure was passed. Moving from saturated liquid to saturated vapor by boiling the Rankine fluid does not change the Stirling cycle efficiency, or Stirling power output, since it is an isothermal process. Because there is more energy in the saturated vapor, a smaller mass flow rate is required to absorb all the rejected heat from the Stirling cycle. In this way, there is less pumping power input to the system. With this fluid, there is also a much greater change in entropy expanding with a higher fraction of the fluid vaporized (higher quality). T-s process diagram in two levels of temperature for the different system state points is shown for HFE7000 in Figure 7. The states in the processes illustrated correspond to the system schematic in Figure 2.

Tables 4 and 5 present a comparison of different fluids for the Rankine cycle operating at 80°C and 140°C, respectively. The inlet and outlet turbine pressures, ΔT of superheating at point 1, specific volume at condenser inlet, rejected heat into Rankine cycle, condenser heat, Stirling cycle efficiency, Rankine cycle efficiency, total efficiency of the combined cycle, second law efficiency of the combined cycle, exergy efficiency of the combined cycle, Rankine cycle pumping power, and the mass flow rate of working fluid in the Rankine cycle have all been presented.

When operating at 80°C, *n*-decane presents the best thermal performance compared to all other fluids. This

Table 5 Thermal performance of several fluids in combined cycle for the ORC operating 140°C

Fluid	P_1 (kpa)	P_2 (kpa)	V_3 (m^3/kg)	SHΔT	$Q_{out,St}$ (W)	Q_{cond} (W)	η_{ST} (%)	η_{ORC} (%)	η_{total} (%)	η_{2law} (%)	η_{ex} (%)	W_{pump} (W)	$\dot{m}$ (g/s)
FC72	884.7	56.36	0.13	0.03	714.8	590.1	28.52	17.45	40.99	60.44	61.61	8.39	6.15
FC87	1747	147.7	0.06	0.02	714.8	606.6	28.52	15.14	39.34	58	59.19	20.86	6.18
HFE7000	1545	119.7	0.11	0.92	714.8	592.2	28.52	17.15	40.78	60.12	61.3	12.73	4.29
HFE7100	775.5	50.48	0.20	0.02	714.8	583.9	28.52	18.32	41.61	61.35	62.51	6.2	4.69
n-decane	38.82	0.5	38.18	0.02	714.8	574.8	28.52	19.59	42.52	62.7	63.84	0.17	1.58
Novec649	1072	72.96	0.11	0.02	714.8	592.6	28.52	17.11	40.74	60.08	61.25	11.07	6.09
n-pentane	1329	116.3	0.30	0.02	714.8	585.4	28.52	18.11	41.46	61.13	62.29	6.42	1.61
R245fa	2337	249.6	0.07	10.05	714.8	598.9	28.52	16.21	40.11	59.13	60.32	10.31	3.21
Toluene	218.9	7.907	3.58	0	714.8	577.4	28.52	19.22	42.26	62.31	63.45	0.70	1.42

P_1, inlet turbine pressures; P_2, outlet turbine pressures; V_3, specific volume at condenser inlet; $\dot{m}$, mass flow rate.

can be seen in Table 4. Again, when operating at 140°C, *n*-decane, with 0.02°C of superheating, presents the best combined cycle efficiency as it is illustrated in Table 5. Toluene, HFE7100, *n*-pentane, Novec649, and HFE7000 present acceptable efficiencies, but Toluene and *n*-pentane are toxic and flammable. *N*-decane requires minimum pumping power in Rankine cycle, and HFE7000 and *n*-pentane have super-atmospheric saturation pressure.

Conclusion

The main conclusion of this work is maximization of combined cycle efficiency with the choosing of suitable working fluid for ORC as bottoming cycle operating at 80°C to 140°C. It is difficult to find the best fluid, which has simultaneously high cycle efficiency, low vapor specific volume at turbine outlet, super-atmospheric saturation pressure, high latent heat, high density and low liquid specific heat, low environmental impact, low toxicity, and has non-combustion characteristics.

For the ORC operating at the two levels of temperature, *n*-decane presented the highest cycle efficiency between the dry and isentropic fluids. But this fluid presents the highest vapor specific volume, leading to a very large condenser and high initial cost. It also presents the lowest saturation pressures, leading to infiltration of non-condensable gases.

There are only four fluids with super-atmospheric saturation pressure: R245fa, FC87, HFE7000, and *n*-pentane. The best candidates for the cycle regarding all the aspects are toluene, HFE7100, and *n*-pentane. *N*-pentane does not require special sealing system because the cycle operates at super-atmospheric saturation. Toluene presents the highest vapor specific volume, the lowest saturation pressure, and the least mass flow rate, leading to the smallest size of utilities, equipments, and the pump consumption among the three fluids. From the environmental and safety point of view, HFE7100 has a small impact compared with toluene and *n*-pentane. For the system operating at 80°C, toluene presents total combined cycle efficiency of 38.15% against 38.03% of HFE7100 and 37.97% of *n*-pentane. These values of cycle efficiency are similar. For the Rankine cycle operating at 140°C, the difference in the total combined cycle efficiency is bigger. Total efficiencies for different fluids investigated are the following: toluene 42.26%, HFE7100 41.61%, and *n*-pentane 41.46%. A small efficiency leads to a high heat input in the combined cycle; therefore, a big capital investment in solar dish area is required.

Abbreviations

0: Ambient; 2law: Second law; ΔT: Temperature difference (°C); ε_{rec}: Recuperator effectiveness; η: Efficiency (%); ALT: Atmospheric lifetime (years); C: Cold side of Stirling cycle; Carnot: Carnot; cond: Condenser; crit: Critical; Ex: Exergy (W); Exp: Expander; *f*: Fraction constant; GWP Global warming potential (100-year period); *h*: Specific enthalpy (J/kg); H: Hot side of Stirling cycle; *I*: Irreversibility (W); in: Input; *j*: Cycle component index; $\dot{m}$: Mass flow rate (g/s); net: Met output; ODP: Ozone depletion potential (CFC as a reference); ORC: Organic Rankine cycle; out: Output; *P*: Pressure (Pa); *Q*: Heat transfer rate (W); S_{gen}: Entropy generation rate (W/K). S: Isentropic; st: Stirling cycle; *T*: Temperature (°C); total: Total; W: Power (W).

Competing interest

The authors declare that they have no competing interests.

Authors' contributions

MB carried out all the computation, performed the modeling, modeling analysis, software simulation, and drafted the manuscript. AAH checked the equations and analysis, reviewed the revised manuscript, and helped in correcting the manuscript. All authors read and approved the final manuscript.

Authors' information

MB graduated the MSc in Energy Engineering in 2012 from the Sciences and Research Branch Islamic Azad University, Tehran, Iran. His current research focuses on thermodynamic cycles and refrigeration. AAH is a professor at the Chemical Engineering Department of Tehran University, Iran. He received his Ph.D. from the Department of Chemical Engineering and Fuel Technology of Sheffield University, England in 1986. Ever since returning to Iran, he has been working at Tehran University teaching heat transfer and combustion and fuel-related courses. His main research interests are heat transfer phenomena, heat exchangers design and performance, and recently, the application of nanofluids in cooling systems. He has co-authored over 70 papers published in national and international journals and conferences. He is also very active in translating and writing books (in Persian) on Chemical Engineering subjects specially on heat transfer. SP is an assistant professor at the Mechanical Engineering Department of Islamic Azad University, Karaj Branch, Iran. He received his Ph.D. from the Department of Mechanical Engineering at the Amir Kabir University of Technology, Tehran, Iran with GPA in 2002. His areas of interest include heat transfer, cooling, and renewable energy. He also teaches courses on dynamics, heat transfer, thermodynamics, and refrigeration at the Islamic Azad University, Karaj Branch. He is the head of the Iranian Geothermal Energy Agency (IGEA) and the Technology Transfer in Renewable Energy Organization (SUNA), Tehran, Iran. He has published 12 conference papers, 3 ISI papers, and 3 books.

Acknowledgments

The authors would like to express their sincere gratitude to the editor (Professor Farivar Fazelpour) and anonymous reviewers for their invaluable comments and suggestions, which substantially improved the overall quality of the manuscript. The support from the Department of Energy Engineering, Science and Research Branch, Islamic Azad University, Tehran, Iran, is gratefully acknowledged.

Author details

[1]Department of Energy Engineering, Graduate School of the Environment and Energy, Science and Research Branch, Islamic Azad University, Tehran 1341766394, Iran. [2]Department of Chemical Engineering, Tehran University, Tehran 1417613131, Iran. [3]Department of Mechanical Engineering, Islamic Azad University (Karaj Branch), Alborz, Karaj 1998833871, Iran.

References

1. Sunday, O, Oyedepo, SO, Adaramola, MS, Paul, S: Analysis of wind speed data and wind energy potential in three selected locations in south-east Nigeria. IJEEE **3**, 7 (2012)
2. Timoumi, Y, Tlili, I, Ben Nasrallah, S: Performance optimization of Stirling engines. Renew. Energy **33**, 2134–2144 (2008)
3. Thombarea, DG, Verma, SK: Technological development in the Stirling cycle engines. Renew. Sustain. Energy Rev. **12**, 1–38 (2008)
4. Wei, F, Wang, R, Diguet, G, Lin, G: Performance optimization of irreversible ferromagnetic Stirling heat pumps. IJEEE **2**(2), 77–83 (2011)
5. Kongtragool, B, Wongwises, S: Thermodynamic analysis of a Stirling engine including dead volumes of hot space, cold space and regenerator. J. Renew. Energy **31**, 345–359 (2006)

6. Ingram-Goble, R, Peterson, R: Modeling and optimization of a combined cycle Stirling-ORC system and design of an integrated microchannel Stirling heat rejector. Mechanical Engineering Department, Oregon State University, MSc Thesis (2010)
7. Wang, H, Peterson, R, Harada, K, Miller, E, Ingram-Goble, R, Fisher, L, Yih, J, Ward, C: Performance of a combined organic Rankine cycle and vapor compression cycle for heat activated cooling. Energy **36**(1), 447–458 (2011)
8. Mathias, JA, Johnston, J, Cao, J, Priedeman, DK, Christensen, RN: Experimental testing of gerotor and scroll expanders used in, and energetic and exergetic modeling of, an organic Rankine cycle. J. Energ. Resour-ASME **131**, 012201–012209 (2009)
9. Yagoub, W, Doherty, P, Riffat, S: Solar energy-gas driven micro-chip system for an office building. Appl. Therm. Eng. **26**, 1604–1610 (2006)
10. Wang, H, Peterson, RB: Performance Enhancement of a Thermally Activated Cooling System Using Microchannel Heat Exchangers. Oregon State University Corvallis, Oregon (2011)
11. Papadopoulos, A, Stijepovic, M, Linke, P: On the systematic design and selection of optimal working fluids for Organic Rankine Cycles. Appl. Therm. Eng. **30**, 760–769 (2010)
12. Chen, H, Goswami, D, Stefanakos, EK: A review of thermodynamic cycles and working fluids for the conversion of low-grade heat. Renew. Sustain. Energy Rev. **14**, 3059–3067 (2010)
13. Facão, J, Oliveira, AC: Analysis of energetic, design and operational criteria when choosing an adequate working fluid for small ORC systems. In: ASME 2009 International Mechanical Engineering Congress & Exposition, pp. 13–19. Lake Buena Vista, Florida (2009)
14. Sharma, P, Harinarayana, T: Enhancement of energy generation from two layer solar panels. IJEEE **3**, 12 (2012)
15. ORCycle: From waste heat to electricity. (2012). http://www.orcycle.be/index.php/en/orctheorie. Accessed 09 March 2012
16. Maizza, V, Maizza, A: Working fluids in non-steady flows for waste energy recovery systems. Appl. Therm. Eng. **16**, 579–90 (1996)
17. Vosough, A, Falahat, A, Vosough, S, Nasr Esfehani, H, Behjat, A, Naseri Rad, R: Improvement power plant efficiency with condenser pressure. IJMSE **2**, 38–43 (2011)
18. US Environmental Protection Agency: Ozone Layer Protection Glossary. USEPA, Washington DC (2010)
19. US Environmental Protection Agency: Global Warming Potentials of ODS Substitutes. USEPA, Washington DC (2011)
20. Facão, J, Palmero-Marrero, A, Oliveira, AC: Analysis of a solar assisted micro-cogeneration ORC system. Int. J. Low. Carbon. Tech. **3**(4), 254–264 (2008)
21. 3M: Innovative technology for a changing world. (2012). http://www.3m.com. Accessed 18 March 2012
22. Brasz, JJ: Assessment of C6F as working fluid for organic Rankine cycle applications. Paper presented at the international refrigeration and air conditioning conference, pp. 10–12. Purdue University, Columbia City (2008)
23. Rayegan, R, Tao, YX: A procedure to select working fluids for solar organic Rankine cycles (ORCs). Renew. Energy. **36**, 659–670 (2011)
24. Lior, N, Zhang, N: Energy, exergy, and second law performance criteria. Energy **32**, 281–296 (2007)
25. Palazzo, P: Thermal and mechanical aspect of entropy-exergy relationship. IJEEE **3**, 4 (2012)
26. Klein, SA: Engineering equation solver. F-chart software, McGraw Hill, New Jersey (2010)

Permissions

All chapters in this book were first published in IJEEE, by Springer; hereby published with permission under the Creative Commons Attribution License or equivalent. Every chapter published in this book has been scrutinized by our experts. Their significance has been extensively debated. The topics covered herein carry significant findings which will fuel the growth of the discipline. They may even be implemented as practical applications or may be referred to as a beginning point for another development.

The contributors of this book come from diverse backgrounds, making this book a truly international effort. This book will bring forth new frontiers with its revolutionizing research information and detailed analysis of the nascent developments around the world.

We would like to thank all the contributing authors for lending their expertise to make the book truly unique. They have played a crucial role in the development of this book. Without their invaluable contributions this book wouldn't have been possible. They have made vital efforts to compile up to date information on the varied aspects of this subject to make this book a valuable addition to the collection of many professionals and students.

This book was conceptualized with the vision of imparting up-to-date information and advanced data in this field. To ensure the same, a matchless editorial board was set up. Every individual on the board went through rigorous rounds of assessment to prove their worth. After which they invested a large part of their time researching and compiling the most relevant data for our readers.

The editorial board has been involved in producing this book since its inception. They have spent rigorous hours researching and exploring the diverse topics which have resulted in the successful publishing of this book. They have passed on their knowledge of decades through this book. To expedite this challenging task, the publisher supported the team at every step. A small team of assistant editors was also appointed to further simplify the editing procedure and attain best results for the readers.

Apart from the editorial board, the designing team has also invested a significant amount of their time in understanding the subject and creating the most relevant covers. They scrutinized every image to scout for the most suitable representation of the subject and create an appropriate cover for the book.

The publishing team has been an ardent support to the editorial, designing and production team. Their endless efforts to recruit the best for this project, has resulted in the accomplishment of this book. They are a veteran in the field of academics and their pool of knowledge is as vast as their experience in printing. Their expertise and guidance has proved useful at every step. Their uncompromising quality standards have made this book an exceptional effort. Their encouragement from time to time has been an inspiration for everyone.

The publisher and the editorial board hope that this book will prove to be a valuable piece of knowledge for researchers, students, practitioners and scholars across the globe.

List of Contributors

Abhishek Joshi
Petroleum Engineering Program, Department of Ocean Engineering, Indian Institute of Technology Madras, Chennai 600 036, India

Jitendra S Sangwai
Petroleum Engineering Program, Department of Ocean Engineering, Indian Institute of Technology Madras, Chennai 600 036, India

Kousik Das
Petroleum Engineering Program, Department of Ocean Engineering, Indian Institute of Technology Madras, Chennai 600 036, India

Nagham Amer Sami
Department of Petroleum Engineering, College of Engineering, University of Baghdad, Baghdad, Iraq
Department of Chemical Engineering, AC Tech Campus, Anna University, Chennai 600 025, India

Sundaramurthy Suresh
Department of Chemical Engineering, Maulana Azad National Instituteof Technology Bhopal, Bhopal 462 051, India

Vimal Chandra Srivastava
Department of Chemical Engineering, Indian Institute of Technology Roorkee, Roorkee 247 667, India

Indra Mani Mishra
Department of Chemical Engineering, Indian Institute of Technology Roorkee, Roorkee 247 667, India

Saritha Vara
Department of Environmental Studies, GITAM Institute of Science, GITAM University, Visakhapatnam, Andhra Pradesh 530045, India

Ok-Youn Yu
Appalachian State University, Boone, NC 28608, USA

Zenon Medina-Cetina
Texas A&M University, College Station, TX 77843, USA

Seth D Guikema
Johns Hopkins University, Baltimore, MD 21218, USA

Jean-Louis Briaud
Texas A&M University, College Station, TX 77843, USA

David Burnett
Texas A&M University, College Station, TX 77843, USA

Hamdi Mohamed
Laboratoire d'Etudes des Systèmes Thermiques et Energétique, EcoleNationale d'Ingénieurs de Monastir, University of Monastir, Monastir 5000, Tunisia

Bentîcha Hmaeid
Laboratoire d'Etudes des Systèmes Thermiques et Energétique, EcoleNationale d'Ingénieurs de Monastir, University of Monastir, Monastir 5000, Tunisia

Punnose P Kovoor
Institute of Product Design & Manufacturing, IPROM, Universiti Kuala Lumpur, Kuala Lumpur 56100, Malaysia.

Mohd Razif Idris
Institute of Product Design & Manufacturing, IPROM, Universiti Kuala Lumpur, Kuala Lumpur 56100, Malaysia.

Masjuki Haji Hassan
Department of Mechanical Engineering, University of Malaya, Kuala Lumpur 50603, Malaysia

Tengku Fazli Tengku Yahya
Malaysian Institute of Chemical & Bio Engineering Technology, MICET, Universiti Kuala Lumpur, Alor Gajah, Melaka 78000, Malaysia

Stuart J Self
Faculty of Engineering and Applied Science, University of Ontario Institute of Technology, 2000 Simcoe Street North, Oshawa, ON L1H 7K4, Canada

Bale V Reddy
Faculty of Engineering and Applied Science, University of Ontario Institute of Technology, 2000 Simcoe Street North, Oshawa, ON L1H 7K4, Canada

Marc A Rosen
Faculty of Engineering and Applied Science, University of Ontario Institute of Technology, 2000 Simcoe Street North, Oshawa, ON L1H 7K4, Canada

Sandeep Lal
School of Engineering and Physics, The University of the South Pacific, Suva, Fiji Islands

Atul Raturi
School of Engineering and Physics, The University of the South Pacific, Suva, Fiji Islands

Sunday O Oyedepo
Mechanical Engineering Department, Covenant University, Ota, Ogun State 112101, Nigeria

Muyiwa S Adaramola
Department of Energy and Process Engineering, Norwegian University of Science and Technology, Trondheim, NO-7491, Norway

Samuel S Paul
Department of Mechanical and Manufacturing Engineering, University of Manitoba, Winnipeg, Manitoba R3T 5V6, Canada

Pierfrancesco Palazzo
Technip, 68 Viale Castello della Magliana, Rome 00148, Italy

Kumarasamy Sudhakar
Energy Centre, Maulana Azad National Institute of Technology Bhopal,Madhya Pradesh 462 051, India

Tulika Srivastava
Energy Centre, Maulana Azad National Institute of Technology Bhopal,Madhya Pradesh 462 051, India

Guddy Satpathy
Energy Centre, Maulana Azad National Institute of Technology Bhopal,Madhya Pradesh 462 051, India

Manicam Premalatha
CEESAT, National Institute of Technology Tiruchirappalli, Tamil Nadu 642051, India

Masashi Kotobuki
Department of Material and Environmental Engineering, Hakodate National College of Technology, 14-1 Tokura-cho, Hakodate, Hokkaido 042-8501, Japan

Oloketuyi S Idowu
Department of Mechanical Engineering, University of Ibadan, Ibadan, Nigeria

Oyewola M Olarenwaju
Department of Mechanical Engineering, University of Ibadan, Ibadan, Nigeria

Odesola I Ifedayo
Department of Mechanical Engineering, University of Ibadan, Ibadan, Nigeria

Julio F San José Alonso
Departamento Ingeniería Energética y Fluidomecánica, E. Ingenierías Industriales, Universidad de Valladolid, Paseo del Cauce 59, 47014 Valladolid, Spain

I Gobernado Arribas
Departamento Ingeniería Energética y Fluidomecánica, E. Ingenierías Industriales, Universidad de Valladolid, Paseo del Cauce 59, 47014 Valladolid, Spain

S Alonso Miñambre
Departamento Ingeniería Energética y Fluidomecánica, E. Ingenierías Industriales, Universidad de Valladolid, Paseo del Cauce 59, 47014 Valladolid, Spain

Nicholas B DeFelice
Department of Environmental Sciences and Engineering, Gillings School of Global Public Health, University of North Carolina Chapel Hill, Campus Box 7431, Chapel Hill 27599-7431, NC, USA

Jacqueline MacDonald Gibson
Department of Environmental Sciences and Engineering, Gillings School of Global Public Health, University of North Carolina Chapel Hill, Campus Box 7431, Chapel Hill 27599-7431, NC, USA

Orlando A Aguilar
Department of Mechanical Engineering, Technological University of Panama, Panama, Panama

Ronaldo Maghirang
Department of Biological and Agricultural Engineering, Kansas State University, Manhattan, KS 66506, USA

Steven L Trabue
USDA Agricultural Research Service, National Laboratory of Agriculture and the Environment, Ames, IA 50011, USA

Charles W Rice
Department of Agronomy, Kansas State University, Manhattan, KS 66506, USA

Larry E Erickson
Department of Chemical Engineering, Kansas State University, Manhattan, KS 66506, USA

Ok-Youn Yu
Appalachian State University, Boone, NC 28608, USA

Brian Raichle
Appalachian State University, Boone, NC 28608, USA

Sam Sink
Strata Solar, Chapel Hill, NC 27517, USA

Giulio Allesina
BEELab, Department of Engineering 'Enzo Ferrari', University of Modena and Reggio Emilia, Via Vignolese 905/b, Modena 41125, Italy

Jaruwan Wongthanate
Faculty of Environment and Resource Studies, Mahidol University, 999 Phutthamonthon 4 Road, Salaya, Phutthamonthon, Nakhon Pathom 73170, Thailand

Kittibodee Chinnacotpong
Faculty of Environment and Resource Studies, Mahidol University, 999 Phutthamonthon 4 Road, Salaya, Phutthamonthon, Nakhon Pathom 73170, Thailand

Madsamon Khumpong
Faculty of Environment and Resource Studies, Mahidol University, 999 Phutthamonthon 4 Road, Salaya, Phutthamonthon, Nakhon Pathom 73170, Thailand

Stephen J Mitchual
Department of Design and Technology Education, University of Education, Winneba, Kumasi Campus, P. O. Box 1277, Kumasi, Ghana

Kwasi Frimpong-Mensah
Department of Wood Science and Technology, Kwame Nkrumah University of Science and Technology, Kumasi, Ghana

Nicholas A Darkwa
Department of Wood Science and Technology, Kwame Nkrumah University of Science and Technology, Kumasi, Ghana

Joseph O Akowuah
Department of Agricultural Engineering, Kwame Nkrumah University of Science and Technology, Kumasi, Ghana

Longfei Mao
Centre for Advanced Computational Solutions, Wine, Food and Molecular Bioscience Department, Lincoln University, Ellesmere Junction Road, Lincoln 7647, New Zealand

Wynand S Verwoerd
Centre for Advanced Computational Solutions, Wine, Food and Molecular Bioscience Department, Lincoln University, Ellesmere Junction Road, Lincoln 7647, New Zealand

Enis T Turgut
Department of Airframe and Powerplant Maintenance, Faculty of Aeronautics and Space Sciences, Anadolu University, Eskisehir 26470, Turkey

Oznur Usanmaz
Department of Air Traffic Control, Faculty of Aeronautics and Space Sciences, Anadolu University, Eskisehir 26470, Turkey

Marc A Rosen
Faculty of Engineering and Applied Science, University of Ontario Institute of Technology, 2000Simcoe Street North, Oshawa, Ontario L1H 7K4, Canada

Mohammad Bahrami
Department of Energy Engineering, Graduate School of the Environment and Energy, Science and Research Branch, Islamic Azad University, Tehran 1341766394, Iran

Ali A Hamidi
Department of Chemical Engineering, Tehran University, Tehran 1417613131, Iran

Soheil Porkhial
Department of Mechanical Engineering, Islamic Azad University (Karaj Branch), Alborz, Karaj 1998833871, Iran